Laser Safety

Laser Safety

Roy Henderson

Bioptica, Cambridge, UK

and

Karl Schulmeister

ARC Seibersdorf Research, Seibersdorf, Austria

Taylor & Francis
Taylor & Francis Group
New York London

Published in 2004 by
Taylor & Francis Group
270 Madison Avenue
New York, NY 10016

Published in Great Britain by
Taylor & Francis Group
2 Park Square
Milton Park, Abingdon
Oxon OX14 4RN

Printed in the United States of America on acid-free paper
10 9 8 7 6 5 4

International Standard Book Number-10: 0-7503-0859-1 (Hardcover)
International Standard Book Number-13: 978-0-7503-0859-5 (Hardcover)

The image of the laser eye-protection reproduced on the cover was kindly provided by the NoIR Laser Company (www.noirlaser.com) and used with permission.

Library of Congress Cataloging-in-Publication Data

Catalog record is available from the Library of Congress

informa
Taylor & Francis Group
is the Academic Division of Informa plc.

Visit the Taylor & Francis Web site at
http://www.taylorandfrancis.com

To Ruth (RH) and Gabi (KS)

Contents

Preface

Laser safety, for some, can be a tedious business that gets in the way of the 'real' work of using lasers. While in many cases it is straightforward, at other times the need to evaluate laser hazards and determine the necessary precautions can seem to involve quite difficult concepts that the safety standards do not really manage to explain very clearly. For others, laser safety can be a fascinating subject, a challenging combination of optical physics, biological phenomena, engineering design and human behaviour. New issues constantly occur, as technology advances and applications spread.

Our aim in writing this book has been to provide the reader, at whatever level of involvement in the manufacture or use of laser equipment, with a comprehensive handbook that explains in detail both the background to laser safety and its practical implementation. It discusses the safety of laser equipment for manufacturers and the establishment of safe working practices for laser users. It explains the biophysical basis for emission and exposure limits, and describes in detail the revised system of laser product classification.

The book is heavily based on the international standard for laser safety, IEC 60825-1 (Edition 1.2), adopted in Europe as EN 60825-1 and increasingly relevant to laser safety in the United States as well. We also discuss the application of the IEC standard to LEDs, which are included within its scope but can lead to the necessity for quite complex evaluations. In addition, we include discussion of other relevant standards, such as the European laser eye protection standards EN 207 and EN 208. Where requirements in the US differ, under ANSI user standards or the CDRH product standard, we explain what these differences are. Our intention throughout is to give guidance to the reader on the application of the various safety standards, and this book should therefore be seen as supplementary to those standards, not a replacement for them.

We discuss terminology and the misuse of terminology, and point out common pitfalls and misunderstandings. A large section of the book is devoted to the evaluation of laser emission and laser exposure. It is our experience, as practitioners in laser safety, that misunderstandings can be widespread and sometimes lead to significant underestimation or overestimation of the level of hazard, resulting in the adoption of safety measures that are either inadequate (and thus potentially hazardous) or overprotective (and therefore unduly restrictive).

While we have limited the book primarily to the safety of lasers and LEDs, and discuss topics related to broadband incoherent sources only briefly, much of the book also has relevance to the safety of broadband sources, especially the chapter on units and radiometry, and that on the interaction of laser radiation with human tissue.

Readers will find that our discussions extend from detailed theoretical considerations to practical issues or workplace safety; our aim of making this book as comprehensive as possible inevitably means that our coverage varies considerably in depth and content, and not every reader will find all of the book of direct relevance to their needs. Nevertheless, it is our hope that we have structured the material in such a way as to enable people to readily find the information they seek, at the level which they require. Moreover, we trust that we have given enough detail to enable people to recognize when their own particular problem may not be as straightforward as they may have originally thought.

Both authors are heavily involved in the work of the international laser safety committee, and our own understanding of laser safety has grown over the years through discussions and debate (and sometimes argument!) with numerous professional colleagues. We are indebted to them all, but in particular we would like to thank Jack Lund, David Sliney, Bruce Stuck, Steve Walker and Joe Zuclich for many helpful discussions. We would also like to thank our wives for their continued understanding and support.

The development of this book, and in particular many meetings between the authors, was in part funded by ARC Seibersdorf research on behalf of the Austrian Ministry of Transport, Innovation and Technology, which we gratefully acknowledge. We would also like to thank many of the staff at ARC Seibersdorf research, especially Thomas Auzinger for skilful artwork, Sandra Althaus for beam propagation modelling, Ulfried Grabner for work on LEDs and line lasers, Marko Weber for data plots and Georg Vees for his comments on chapter 2.

Finally, to show that the potential dangers of optical radiation have long been recognized—

> He saw; but blasted with excess of light, closed his eyes in endless night.
>
> Thomas Gray, 1716–1771 (*on Milton, written in Cambridge*)

Roy Henderson, Cambridge
Karl Schulmeister, Seibersdorf

Chapter 1

Lasers, light and safety

1.1 Lasers: stimulating light

1.1.1 Creating light

Lasers are devices that can produce intense beams of light. First developed during the 1960s, they were originally regarded as something of a technical curiosity; a new and fascinating light source but with an unknown future. While investigations began into a number of potential uses and whole new areas of research opened up, lasers were initially dubbed 'a solution looking for a problem'. Although they captured the imagination of science-fiction writers and film makers, many early aspirations went unfulfilled, mainly due to the limited types of laser then available and the poor understanding of how such intense light beams interact with matter. Since then, however, the technology has greatly matured. New words, such as 'optronics' and 'photonics', have been coined to describe the new science of light, and lasers have found extensive application in a wide range of very different fields, ranging from manufacturing industry to medicine and from communication to creative arts.

Safety is, or should be, an integral part of using laser technology. Laser hazards can result in serious injury, even death. These hazards arise mainly, although not entirely, from the ability of lasers to produce harmful effects at a distance from the laser itself, through the intense beams of light which they generate.

The name 'laser' is an acronym, and is taken from a phrase that describes what lasers are and how they work. It stands for—Light Amplification by the Stimulated Emission of Radiation.

Lasers emit light, but can generate visible *or* invisible emission, and so the term 'light' can be misleading. It is often applied in everyday use in a more restricted sense to refer only to 'visible light', that is, to the light that we can see with our eyes. This kind of light—the light of which we are aware through our sense of sight—forms only part of the spectrum of what is known as *optical radiation*. Optical radiation encompasses both the ultraviolet and infrared regions

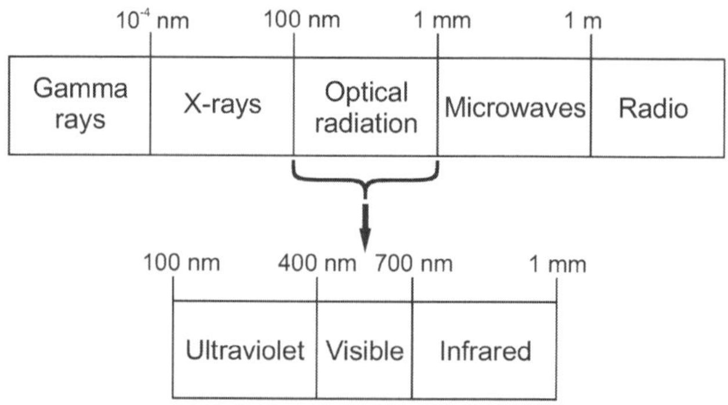

Figure 1.1. The electromagnetic (EM) radiation spectrum, indicating the wavelength boundaries of the principle wavebands.

in addition to the band of visible light. It would be more accurate, therefore, to say that lasers produce intense beams of optical radiation. This radiation may be visible (that is, visible light), but it can also be invisible ultraviolet radiation or invisible infrared radiation.

Optical radiation itself is just part of a more general kind of radiation known as *electromagnetic* (EM) radiation. EM radiation is a form of wave energy that can propagate through empty space, as well as through many material substances (in the way that visible light, for example, can pass through water or glass). Being a wave motion (consisting of oscillating electric and magnetic fields), EM radiation can be characterized by its wavelength. The EM radiation spectrum, extending from gamma rays at very short wavelengths to radio waves at very long wavelengths, is illustrated in figure 1.1. Radiation in different parts of the EM spectrum can have very different properties. Not all of this radiation can pass through the atmosphere, however, and we are therefore protected from a large part of the more harmful short-wavelength EM radiation that is emitted quite naturally by the Sun.

The wavelength of EM radiation within the optical band is usually specified in units of nanometres. One nanometre (abbreviated to nm) is one thousand-millionth, or 10^{-9}, of a metre. In the infrared region, however, micrometres (also known as microns) are also commonly used. One micrometre (abbreviated to μm) is one millionth (10^{-6}) of a metre, i.e. it is equal to one thousand nanometres.

The band of visible radiation, visible light, extends from a wavelength of 380 nm at the blue end of the visible spectrum to 780 nm at the red end. This defines the limits over which the human eye can see, and is the definition of visible light that is used by the Commission Internationale de I'Eclairage (CIE—the International Commission on Illumination). The eye's visual sensitivity to light is illustrated in figure 1.2. As can be seen, it is very non-uniform, and reaches its maximum sensitivity in the middle of the visible spectrum, in the green

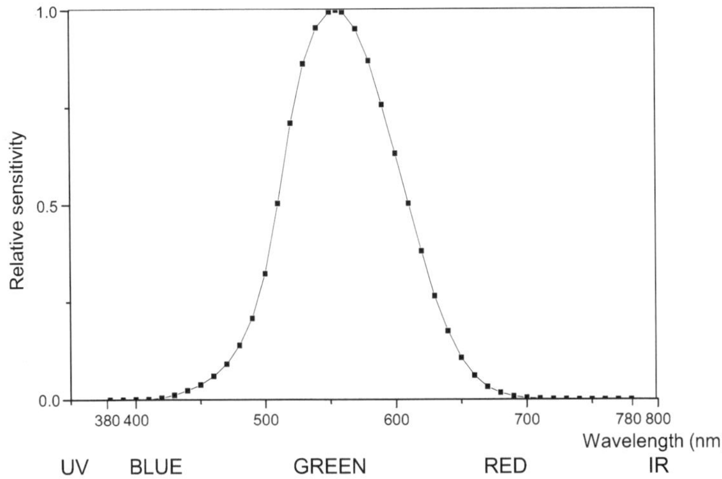

Figure 1.2. The visual sensitivity curve of the human eye.

region at a wavelength of around 555 nm. (This corresponds quite closely to the peak emission of the Sun.) At the extreme ends of the visible spectrum the eye's sensitivity is very low. For this reason, in laser safety, where a distinction often has to be made between visible and invisible laser beams, the visible band is defined as the more limited region between 400 and 700 nm, as shown in figure 1.1. Under the definition used in the majority of laser safety standards, therefore, the ultraviolet region lies below 400 nm, extending down to 100 nm (although the lowest wavelength for which safety limits are currently specified is 180 nm, the start of which is termed the vacuum ultraviolet where absorption in air is very high), while the infrared region lies above 700 nm, extending out as far as 10^6 nm, or 1 mm.

The ability to produce both visible and invisible emission is common to many ordinary light sources. A filament lamp, for example, not only generates broadband visible radiation (that is, emission right across the visible spectrum at all wavelengths between 400 and 700 nm which combines to produce the effect of white light), but in addition generates considerable quantities of infrared emission and a very small amount of ultraviolet emission. Indeed, most of the output of a conventional filament lamp is in the infrared region which, for purposes of illumination, represents wasted energy.

However, while most lamps produce broadband emission, lasers concentrate their output over an extremely narrow portion of the spectrum that may, for most practical purposes, be considered as a single wavelength. Lasers, therefore, are often referred to simply by the wavelength of their emission. Some lasers do have the ability to generate outputs at more than one wavelength, but these remain discrete, separate wavelengths that do not merge into a continuum.

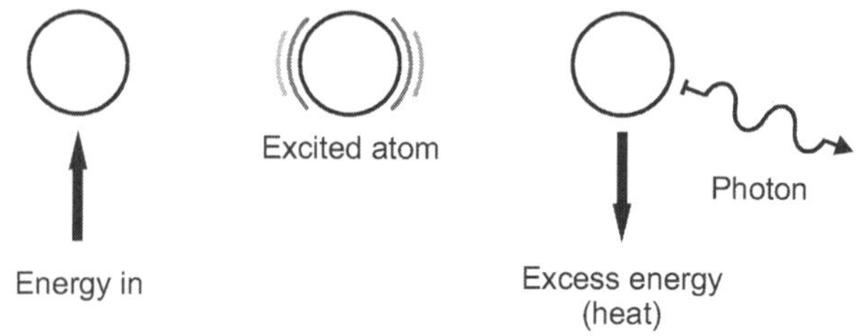

Figure 1.3. In the process of spontaneous emission an atom is first excited (energized) and then releases some of this absorbed energy in the form of a single photon. The excess energy (the difference between the absorbed energy and the photon energy) is dissipated as heat. Once the atom has returned to its initial or 'ground' state, the process can be repeated.

What distinguishes lasers from lamps, however, is not simply the spectral characteristics of the output but the fundamentally different process by which the radiation is generated. This process in turn gives rise to very special properties that make lasers unique.

Optical radiation is generated by energy transitions that occur within individual atoms or molecules. Any material that emits light must first absorb energy, and the energy that it absorbs is then contained in the atoms or molecules that make up the material. Some of this energy can then be released from these atoms or molecules in the form of photons. A photon is the smallest possible 'packet' of light energy, and can be considered as a short burst of waves or a 'light particle' having no mass (and is therefore 'light' in both senses of the word). Considering packages of light in this way can be a useful though far from perfect analogy. The energy of an individual photon is inversely proportional to the associated wavelength. Thus, ultraviolet photons (normally generated by processes involving the inner electrons of atoms), are far more energetic than infrared photons (which largely arise through changes in the energy levels that bind atoms together in molecules). At intermediate wavelengths, and consequently at intermediate photon-energy levels (that is, within or close to the visible band), the process is one involving energy exchanges of the outer electrons of the individual atoms.

In conventional light sources the photons are emitted 'spontaneously' in a process known as spontaneous emission. This means that the atom or molecule, having first absorbed energy and therefore being in an energized or 'excited' state, releases this energy quite spontaneously after a random (but very small) interval of time. Any single photon that is produced during this release of energy is emitted in a random direction (figure 1.3). In consequence, therefore, as this process is repeated, radiation is emitted in all directions away from the source, and the individual photons or 'wave packets' of which this radiation is comprised have a quite random or 'incoherent' relationship to each other.

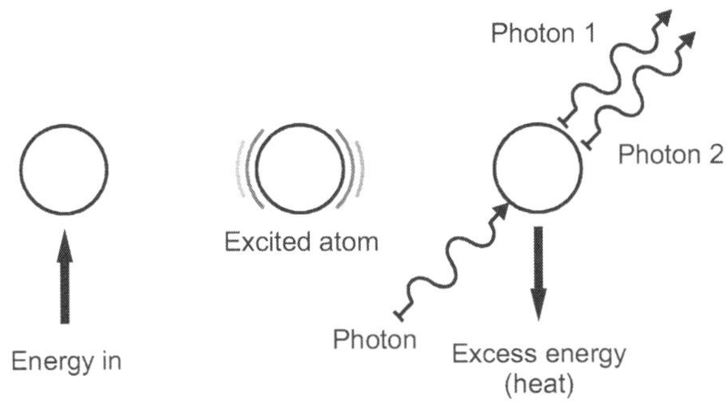

Figure 1.4. In stimulated emission, an excited atom is stimulated to emit a photon (before it would have done so by spontaneous emission) by a photon colliding with the atom. Two photons are then emitted in the same direction (the one that has caused the stimulation and one generated by the atom), and are both in phase with each other. These photons can then collide with other excited atoms to cause further stimulated emission.

In a laser, on the other hand, having first been energized, the individual atoms or molecules can be 'stimulated' to release their energy before they would have done so spontaneously. This is achieved by arranging for a photon, having the same energy as the photon that would have been released spontaneously, to collide with the atom or molecule. The result of this process of 'stimulated emission' is that two photons now exist; the original one that caused the stimulation and a second one, due to the release of energy arising from this process of stimulation. Furthermore, both these photons now travel in exactly the same direction, and the waves of which they are comprised are exactly in phase, or in step. This process of stimulated emission is illustrated in figure 1.4.

Stimulated emission by itself would be of little practical value unless it were possible to exploit this process to create gain, that is to 'amplify' the coherent emission that is generated. This is done in two ways. First, by ensuring that an efficient energizing process is utilized, such that there is a high probability of the individual atoms or molecules being in an excited state. This will allow stimulated emission to occur on a significant scale, and is known as 'population inversion', since it is the reverse of the normal or stable state in which, at any given time, most of the atoms or molecules will be in the ground or 'unexcited' state. Secondly, in order to ensure that a large number of photons pass through the material to cause stimulated emission, some form of 'feedback' is required. Feedback is needed so that a high proportion of the photons that are generated are fed back into the material. This is achieved by creating a 'resonator', formed by a pair of mirrors at each end of the material within which stimulated emission, or 'laser action', can occur (figure 1.5). One of these two mirrors is designed to have a very high reflectivity (at the laser wavelength), so directing back into the resonator the majority of photons produced along the resonator axis. The other

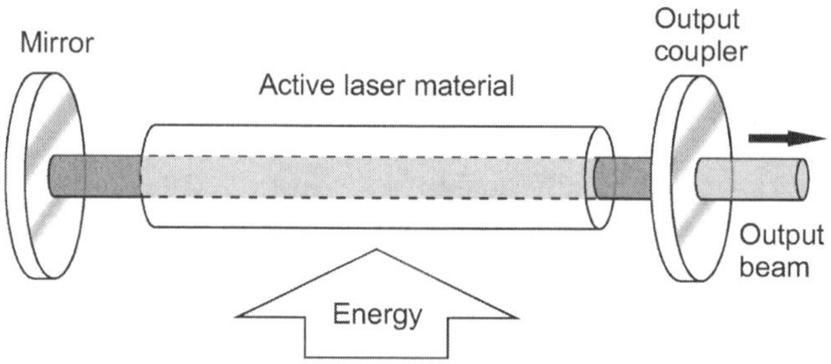

Figure 1.5. The principal components of a laser resonator. The laser material or 'medium' (which may be a solid, a gas or a liquid) is often in a cylindrical form and located between two mirrors to create a 'resonant cavity'. An energy source couples energy into the laser medium, in which the build-up of stimulated emission between the mirrors generates the laser beam.

mirror, known as the 'output coupler', also has a reasonably-high reflectivity but this is combined with some transmission, such that while the majority of incident photons are reflected by this mirror back into the laser resonator, a small fraction of them are allowed to pass through the mirror so forming the output beam of the laser.

Initially, of course, at the start of this process (when the laser is switched on and the laser material is first energized), only spontaneous emission can occur, and this emission will be in all directions. A sufficient number of photons will, however, be emitted by chance parallel to the axis of the laser resonator to initiate the process of stimulated emission. Through a cascading effect, as more and more photons are produced along the laser axis, stimulated emission rapidly grows to become the dominant mechanism of photon generation.

Many lasers generate, through this process, well-collimated and essentially parallel beams. Others, such as laser diodes, produce divergent emission. This latter effect arises because of the very small cross-sectional area of the resonator in such lasers. Optical radiation, being a wave motion, is subject to diffraction. This is the unavoidable bending of light caused by structures that are small on an optical scale (that is, with respect to the emission wavelength). Because of the very small size of laser diodes, diffraction effects produce this characteristic divergent emission. This is usually not a problem, however, because it is possible, if desired, to employ a lens system to collimate this output and thereby form a beam identical to that of other kinds of lasers. Indeed, both types of output are equivalent. A laser diode produces divergent emission from a very small, effectively point-source of emission, which can be readily formed into a collimated beam. A collimated (parallel) beam, on the other hand, appears to originate from a point-source located an infinite distance away. Both types of laser are, therefore, often and quite justifiably called *point sources*. This is very different from the majority of conventional lamps, which are *extended sources*, by virtue of the extended nature of the emitting surface.

The fact that a point source of light (whether it is a laser or not) can produce a parallel beam having a finite diameter may not be immediately obvious. Consider, however, the three illustrations shown in figure 1.6. These show a point source, radiating in all directions, where the emission passes through a circular aperture to form a beam. In figure 1.6(*a*), the source is close to the aperture and so the divergence or angular spread of the beam beyond the aperture is quite large. As the distance between the source and the aperture increases, as shown in figure 1.6(*b*), the divergence of the beam formed by the aperture decreases. At very great distances, the divergence of the beam becomes very small; in the limit, at an infinite distance, the beam that is produced by the aperture is essentially parallel, as shown in figure 1.6(*c*). This is a simplified description of a collimated laser beam; the *apparent* source (from which the radiation *appears* to originate) is a single point located, effectively, at infinity. The apparent source of a well-collimated laser beam is *not* the emission aperture of the laser, or even the inside of the laser resonator. It lies a long way *behind* the laser.

A common example of a point source producing parallel emission is that of a star. Although we know that stars are very large, they appear from Earth, even through the most powerful telescopes, to be no more than points of light because of their vast distances from us. A star approximates very well to a point source at an infinite distance. Because of this, even though stars radiate in all directions, the rays of light that reach us from a star are all parallel. (Indeed, it is only with the closest stars that there is any detectable difference in the direction from which the rays originate when observed from one side of the Earth's orbit around the Sun compared to those arriving at the opposite side; a baseline—corresponding to the aperture diameter in figure 1.6—of over 260 million kilometres!)

The size of a source can, of course, be defined in terms of its actual dimensions. The globe of a frosted filament lamp (the *apparent* source from which the light produced by such a lamp appears to originate) might be 60 mm in diameter. The diameter of the Sun, on the other hand, is almost 700 000 km. What is often more useful, however, is to express source-size in terms of the angular diameter of the source, the *angular subtense*, measured at the position from which the source is being viewed. Thus, when seen from a distance of one metre, the 60 mm lamp subtends an angle of 3.5 degrees, while the Sun, observed from the Earth, subtends an angle of 0.5 degrees. The lamp, one metre away, therefore *appears* to be seven times larger than the Sun.

The angular size of a source determines the size of its optical image, that is the image created by a focusing system (such as in a camera or by the eye). In the previous example, the image of the lamp (produced at the film in a camera or on the retina at the back of the eye) is seven times larger than the image of the Sun. A star, on the other hand, produces only a tiny spot as its image. It is, in fact, unresolvable (meaning that its angular size is less than the smallest image that can be created by the optics); the size of the focused spot is in this case governed by

a)

b)

c)

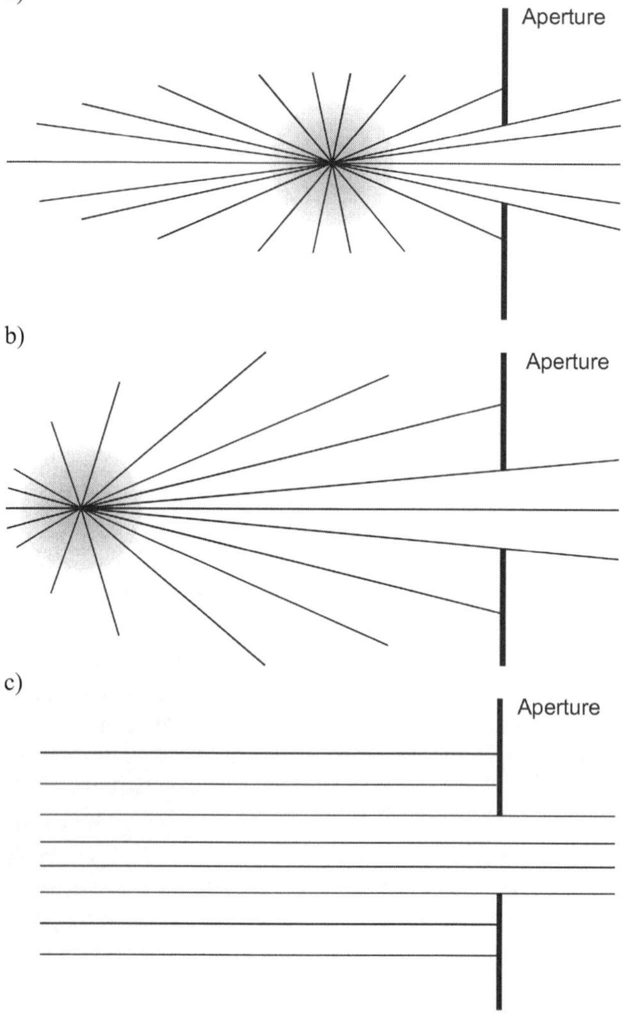

Figure 1.6. A point source (such as a distant star) produces divergent emission in all directions, and a circular aperture can be positioned some distance from the source to produce a beam of light beyond the aperture. (*a*) When the aperture is close to the source the beam is very divergent. (*b*) As the distance from the source to the aperture increases, the beam becomes less divergent. (*c*) Where the distance is very large (in comparison to the size of the aperture), the beam is effectively parallel or 'collimated'.

the fundamental limitations of the optical system, not by the angular size of the source.

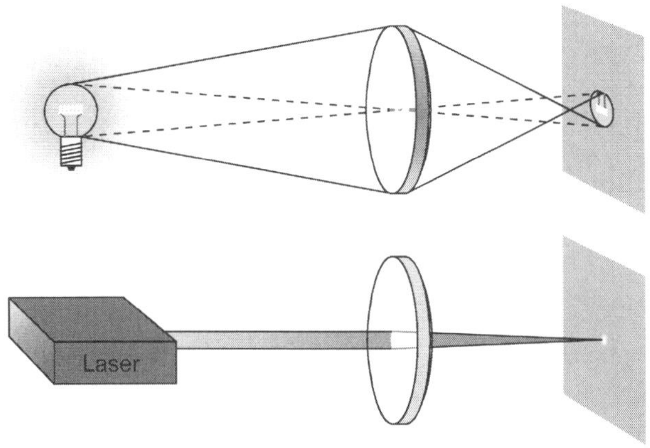

Figure 1.7. Comparison between an extended source and a point source. A conventional lamp (upper picture) is an extended source (because it has a finite emitting area), and the smallest patch of light that can be created by focusing the output of the lamp with a lens is the geometric image of the source. A laser, however, (lower picture) is effectively a point source, and so its output can be focused by a lens to create a point image. Furthermore, while only a small proportion of the output of the lamp can be collected and focused by the lens (because the lamp emits in all directions), the entire output of the laser, which is contained in a narrow collimated beam, can be focused to a small spot. Even if the total light output of the lamp and the laser were the same, the concentration of power produced by the lens would be very much higher in the case of the laser.

Lasers, like stars, are point sources because they produce point images (focused points of light), even though the laser beam, as it emerges from the laser, may have a diameter of several millimetres or more. In contrast, the light from a conventional lamp, when imaged by an optical system, cannot be focused down to produce anything smaller than the geometrical image of its emitting area. (This is the filament itself in the case of a clear-glass filament lamp, or the glass globe in the case of the frosted or 'pearl' lamp previously discussed.) This difference in the source size between a laser (a point source) and a conventional lamp (an extended source) that is apparent from their corresponding images produced by a lens is shown in figure 1.7.

If, instead of using a focusing lens to produce an image on a screen, the laser beam were to be viewed directly by the eye, then the image produced on the retina would be as shown in figure 1.8. The laser beam, if it were visible, would appear to the eye as a small spot inside the emission aperture, even though the emerging beam might have almost the same diameter as the aperture. (This obviously ignores the serious injury that could be caused to the eye through the direct viewing of a laser beam!) In fact, the image of the emission aperture and of the focused beam may not be simultaneously in focus on the retina, since the

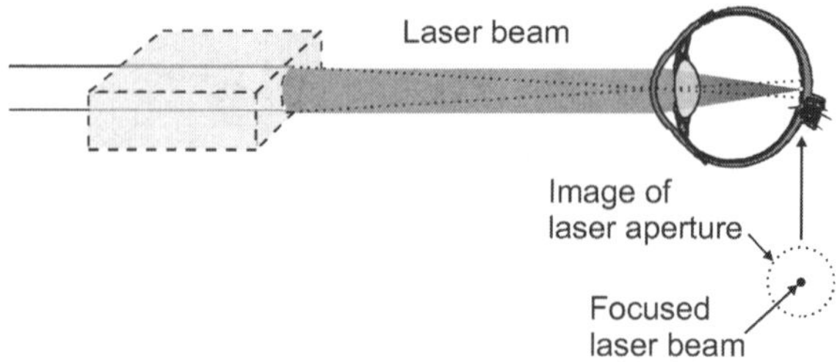

Figure 1.8. Direct viewing of a laser beam. Just as in the case of the laser in figure 1.7, a laser beam entering the eye can be focused to produce a very small spot on the retina. The image of the laser's emission aperture, however, can be very much larger. The fact that a laser beam may have an appreciable diameter as it leaves the laser does not limit its ability to form a point image.

position of the apparent source (for a collimated beam) is not at the emission aperture but at infinity, off the left-hand side of the diagram. The eye can focus on the emission aperture, in which case there will be an out-of-focus image of the laser source (appearing as a larger blurred spot), or it can focus at infinity, to produce a sharp focused spot from the laser beam surrounded by an out-of-focus image of the emission aperture.

In the case of a highly-divergent laser source, such as a bare laser diode, the apparent source position and the exit aperture can be co-located. Nevertheless, the laser can still be a point source because of the very small emitting area. As with a collimated beam, the emission can be focused by the eye to produce a very small spot on the retina, unlike a conventional lamp.

1.1.2 Quantifying light

When light is used for illumination purposes (which, after all, remains the principal application of light), light quantities are normally expressed in photometric units. These units (having names such as lumen, lux and candela) are spectrally-weighted quantities that are based on the visual response of the human eye (figure 1.2). Under the photometric system, measurements are made by using a combination of optical detector and filter (a light meter or 'photometer') that has the same spectral sensitivity as a normal human eye. This gives more 'weight' to green light, for example, than it does to blue or red light where the visual response of the eye is much lower. Ultraviolet and infrared radiation, of course, have a zero value in photometric units, regardless of the actual quantity of radiation that may be present. This is because it is invisible to the eye, and any photometric light-meter should be insensitive to it.

The usefulness of light for purposes other than illumination, and the ability of light to cause damage, are both unrelated to the process of vision: what matters is the total magnitude of the optical radiation that is present. For the majority of laser applications, therefore, and in all light-safety assessments, absolute, radiometric units are used. These are fundamental quantities of power and energy. While radiometric measurements are discussed in more detail in the next chapter, we give here an overview of the principle quantities and units that are relevant in laser safety.

The watt (W) is used as the unit of radiant power and the joule (J) as the unit of radiant energy. Power is defined as the rate of flow of energy. An emitted power of one watt is equivalent to an energy rate of one joule per second. Quantities expressed in terms of energy (in J) can thus be readily converted to power (in W) by dividing the energy by the emission duration in seconds. Similarly, quantities expressed in terms of power may be converted into corresponding energy units by multiplying the power by the emission duration (provided that the level of power remains constant throughout the emission duration). A laser beam of three watts that is emitted for ten seconds will therefore generate a total energy during this time of thirty joules.

Quantities of power or energy that are very much smaller or larger than the base units of joules and watts can be used as shown below:

- 1 milliwatt (mW) and 1 millijoule (mJ) are equal to 1/1000 W and 1/1000 J, respectively;
- 1 kilowatt (kW) and 1 kilojoule (kJ) are equal to 1000 W and 1000 J, respectively;
- 1 megawatt (MW) and 1 megajoule (MJ) are equal to 1000 000 W and 1000 000 J, respectively.

It is usual to express the emission of continuous wave (cw) or 'steady state' lasers in terms of power (P), but to measure the output of pulsed lasers in terms of the energy (Q) of each pulse. A pulsed laser will thus have a pulse energy of Q joules. If the pulse duration is t seconds (where t is normally much less than one second) and the pulse repetition rate is f hertz (f pulses per second), then the peak power of the laser, for each pulse, will be Q/t watts. The average power of the laser, however, (the average rate of energy emission) will be Qf watts, since this is the average rate of energy emission per second. The values for *peak* power and *average* power of a pulsed laser can be very different.

Typical laser output powers may vary from below one milliwatt to several kilowatts. The peak power of some pulsed lasers, given the extremely short pulse durations that are possible, can reach several megawatts.

It is interesting to examine the way in which we can view a laser beam, and to relate the emission power of the laser with the optical power levels necessary for vision. Consider a one milliwatt laser pointer. This produces a narrow, (typically red) almost parallel laser beam of about two millimetres in diameter. If it is directed at a projection screen, a small bright red spot is seen. Although we refer to the laser beam as being red, it only becomes visible when some of the beam enters our eye and is focused on the retina. This will occur when the beam strikes the reflecting matt surface of the screen. We cannot actually see the beam as it passes through the air between the laser and the screen. (Note, however, that more powerful beams can be visible. This is because of the small amount of scattering that arises from the dust and other particles floating in the air. Scattering, the redirection of light out of the beam caused by such particles, is always present, but with higher power beams the scattering can be sufficient to become visible.)

However, how do we actually see the beam where it strikes the screen? It becomes visible because not only is a white screen highly reflective (reflecting most of the light that is incident upon it) but, unlike a mirror, it reflects *diffusely*, redirecting the reflected radiation, not in a narrow beam as a flat mirror would, but in all directions away from the surface. The spot on the screen thus forms the apparent source, radiating in all directions. Wherever we sit around the screen, therefore, our eyes can intercept some of this reflected radiation, allowing us to see the red spot.

If the power striking the surface of the screen is one milliwatt, and it is all reflected (neglecting the very small absorption loss that will inevitably occur at the screen), then the re-radiated beam also has a power of one milliwatt. But this is radiated into a hemisphere centred on the laser spot on the screen. What we pick up with our eyes will only be a very small fraction of this. In fact, at a distance of two metres from the screen, the power entering the pupil of each eye will be about one nanowatt (10^{-9} W). Yet this is sufficient for the laser spot to be readily seen. Were the laser beam to be directed not at the screen but straight into our eye, the entire beam would pass through the pupil and could be focused to a small spot on the retina. The power entering the pupil would then be *one million times greater* (one milliwatt rather than one nanowatt) than in the case of indirect viewing. Even with a laser pointer, therefore, which is usually considered to be reasonably harmless, the effect of accidental direct exposure of the eye to the beam can be, literally, dazzling!

It is often necessary in laser safety to define the *concentration* of radiant power or energy that is incident at a surface, as shown in figure 1.9. This is expressed in terms of either the irradiance E, which is the power per unit area (normally specified in units of watts per square metre), or the radiant exposure H, the energy per unit area (specified in units of joules per square metre). These parameters are sometimes referred to as power density and energy density respectively. Strictly, however, this terminology is incorrect, since *density* properly relates to volume, not area.

Radiant power, radiant energy, irradiance and radiant exposure are all very important parameters in quantitative laser safety assessments. They are discussed in more detail in chapter 2.

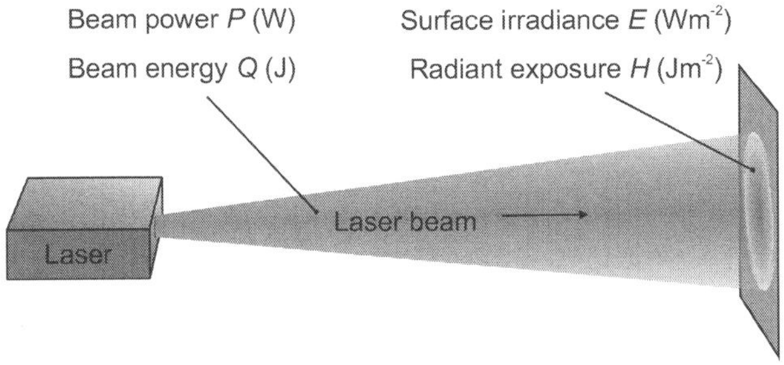

Figure 1.9. The concentration of laser power or energy at a surface. For many assessments in laser safety we need to quantify the power or energy per unit area that is incident at a surface. For radiant power, the surface concentration is called *irradiance* and is measured in units of watts per square metre; for pulse energy it is called *radiant exposure* and is measured in units of joules per square metre.

1.2 The properties of laser radiation

The term 'laser radiation' refers to the optical radiation, or 'light', that is emitted by a laser. But if lasers produce optical radiation, what are the distinctive features of this radiation that differentiate it from that produced by conventional light sources? Why is laser safety such a concern when ordinary lamp safety is much less so? Interestingly, it is not necessarily the most obvious characteristics of lasers that are the most hazardous.

It has already been noted that laser emission is monochromatic, that is, it is effectively concentrated at a single wavelength (or, sometimes, at several discrete, individual wavelengths). Most lamps, in contrast, emit broadband radiation. While this is an obvious distinction it is not, from the safety perspective, the most significant one.

Lasers are often considered to be powerful emitters of optical radiation. They can produce effects not possible using ordinary light sources. This is mainly because of the high *concentration* of the emitted power. The total emission generated by the majority of lasers (certainly in terms of average power), though hazardous, is less than that emitted by an ordinary household lamp.

Rather than monochromaticity and power, the two most important properties of lasers insofar as their hazard potential is concerned are those of *directionality* and *source-size*.

Directionality is the property that enables lasers to produce high levels of concentrated power at considerable distances from the source. Even though the output from a one-milliwatt laser pointer is much less than that produced by a pocket torch (flashlight), the concentration of this output into a narrow beam that is only a millimetre or so in diameter will produce a level of irradiance (power per unit area) much higher than that produced by the torch. Even lasers of moderate power, therefore, are capable of causing high levels of surface exposure (at the eyes or the skin) that may exceed safe limits.

This property of directionality can be characterized in terms of the angular *divergence* of the emitted beam; the angle at which the beam spreads out from the source. Though many laser beams can appear to be very collimated and therefore parallel, diffraction effects mean that no beam can be perfectly parallel and must diverge to a certain extent. (In other words, the diameter of the beam gets larger at increasing distances from the source.) For many lasers, the divergence angle is very low, only a fraction of a degree, but for other lasers (e.g. bare laser diodes) the beam divergence can be large. Special optics can be used to produce other beam geometries, such as fan-shaped beams which have a large divergence in, say, the vertical plane but a very low divergence in the horizontal plane.

Beam divergence angles can be expressed in degrees, but for small divergences are more usually quoted in milliradians (one thousandth of a radian), where one radian is the angle subtended at the centre of a circle by a an arc around the circumference equal in length to the radius of the circle. There are thus 2π radians in a complete circle, and one radian is therefore equal to about 57 degrees.

The ability of an optical source to produce given levels of exposure at a distant surface can be expressed in terms of the source *intensity*. Intensity is a measure of the emitted power per unit solid-angle, and can be expressed in units of watts per steradian (W sr^{-1}). One steradian is the solid-angle subtended at the centre of a sphere by an area on the surface of the sphere equal to the square of the sphere's radius. There are therefore 4π steradians in a complete sphere. A 65 degree cone has a solid angle at its apex of about one steradian. (Solid angular measure is defined more fully in chapter 2.) Because of their generally high levels of directionality (low levels of divergence), lasers have high levels of emitted *radiant intensity*.

The very small **source size** of most lasers enables their emission to be focused down to concentrate this power over an even smaller area, and so create much higher levels of surface exposure. This is how lasers are often used, of course, by focusing the beam to produce the required effect, whether it is an industrial laser being used for welding, or a semiconductor laser-diode whose output is being focused down into an optical fibre. Unfortunately, this can also happen, at certain wavelengths, inside the eye, creating extremely high and seriously damaging levels of exposure on the retina at the back of the eye.

For these reasons lasers can produce harmful effects, even though their output power may be well below that of considerably less harmful conventional optical sources. One useful way in which these properties can be quantified is that of brightness or *radiance*. Radiance (discussed further in chapter 2) is a measure of the intensity per unit area. It represents the power emitted into a unit solid angle from a unit area of emitting surface, and is measured in units of watts per square metre per steradian (W m^{-2} sr^{-1}). Because of their very small effective emitting areas (apparent source size) combined with their high levels of radiant intensity, lasers have extremely high values of radiance, greater than that of all other artificial sources and even exceeding the radiance of the surface of the Sun.

The importance of radiance is that, where a given optical system (of given f-number or focal ratio) is used to image or focus the light emitted by a source, it is the *radiance* of the source that determines the maximum value of *irradiance* that can be produced at the image plane of the system. In the case of the eye, therefore, the huge values of radiance that are possible with lasers of even low output power mean that lasers can produce higher levels of exposure (irradiance) at the retina of the eye than is possible from any conventional source, including the Sun.

Directionality and source size, which are related although separate aspects of the spatial distribution of the emitted radiation, therefore represent the most important safety-related properties of a laser beam. They govern the maximum level of exposure (the degree of light concentration at the surface of the body or inside the eye) that can be produced from an optical source of given power. Much is often made of the emission power of lasers, but it is actually their high levels of radiant intensity (arising from their low divergence) and radiance (arising from their small apparent source size) that in reality make them both extremely useful and potentially hazardous.

The uniqueness of lasers is often related to their *coherence*. Coherence is a measure of the degree to which the emitted waves remain in phase. Conventional light sources are extremely incoherent; the process of spontaneous emission results in the phase relationship between the photons that are generated being totally random. Lasers, on the other hand, can produce highly coherent emission. High levels of coherence are a very useful property for certain applications involving interference between separate paths of light that have travelled different distances, such as in holography. But while laser emission needs to be reasonably coherent in order that it has the important spatial properties that it possesses, coherence itself has no direct bearing on the resultant hazard. All that matters insofar as most laser injuries are concerned is the level of the incident exposure—the irradiance or radiant exposure at the surface of the particular body tissue. The incident exposure is primarily a function of the source intensity or, for retinal exposure, the source radiance. Since lasers have higher values of both of these parameters than other sources, lasers are capable of inflicting more serious harm than is possible from other sources of optical radiation.

A more detailed discussion of these exposure conditions and of the effects that lasers can cause is given in chapter 3.

1.3 The safety of laser technology

The concerns that arise over laser hazards and the need for having a formal and systematic approach to risk analysis and safety control really stem from three unique aspects of laser technology. First, laser hazards are not at all obvious. The appearance of the laser equipment or even a knowledge of its output power may give little indication to an untrained person of its ability to cause injury. Second, a person who is accidentally exposed to laser radiation may be unaware of this until a serious injury has been caused. Third, lasers can cause harm at a distance, sometimes at a considerable distance, from the laser equipment itself. There need be no direct physical contact with the laser itself.

Laser safety, as a discipline, is the task of controlling the risk of laser technology through the appropriate design and use of laser equipment. It therefore impacts on both manufacturers and users, and requires an understanding of legal requirements, laser safety standards and established principles of best practice. While the main focus of laser safety is, inevitably, on the harm that could arise from accidental human exposure to hazardous levels of laser radiation, there are other safety issues that may also need to be considered. These are often termed ancillary or associated hazards, and result from aspects of laser operation that include the interaction of the laser beam with materials, especially of concern with high-power lasers (which can ignite inflammable materials or generate fume by vaporization), or other hazards associated with the laser (such as electrical hazards or toxic materials). These additional hazards are discussed further in chapter 6.

Laser safety requires that all potential hazards are evaluated, that the impact of these hazards is assessed, and that appropriate safety precautions are adopted. Safety precautions are more usually referred to as control measures or protective measures, or sometimes simply as 'controls'. They include such aspects as physical enclosures to limit access to the hazard, written procedures that have to be followed and protective equipment (such as safety eyewear) that has to be worn.

The level of detail that safety evaluation requires, and the depth of knowledge needed to complete it, can vary widely, depending on the type of laser in use, the purpose for which it is being used, and the circumstances under which it is operated. There are, however, two broad categories into which most laser safety activities can be divided. These cover *qualitative* aspects and *quantitative* aspects of laser safety.

Qualitative aspects include overall management and policy issues, the identification of possible hazards, the use of beam enclosures and procedural methods of hazard control. In other words they require a recognition that hazards exist, but not necessarily a detailed evaluation of the magnitude of those hazards. For many of those involved with laser safety it is these aspects with which they are mainly concerned.

Quantitative aspects, however, involve numerical assessments of the levels of laser radiation and the application of the various emission and exposure limits specified in laser safety standards. Such assessments may be necessary, for example, whenever classifying a laser product, or when evaluating the exposure conditions that might exist in order to determine the distance over which the hazard extends or to specify the level of eye-protection that is needed. These assessments can require a reasonable familiarity with optical principles and radiometric parameters, a confidence in undertaking arithmetic calculations, and an understanding of the detailed measurement specifications defined in the safety standards.

Laser safety should not be seen in isolation, however, but considered as part of an overall approach to health and safety, both in the workplace and amongst the public at large. It may at times require specialist knowledge and appear to be highly technical in nature. Nevertheless, the aim is simply stated; to ensure that laser equipment is designed to be safe and that it is used in a safe manner.

The process of identifying what needs to be done in order to ensure the safe use of laser equipment is accomplished, in essence, by finding answers to the following questions.

- What can go wrong? (The hazards that might exist and the conditions under which they can arise.)
- How likely is this to happen? (The likelihood that harm will occur.)
- What are the consequences? (The severity of the injury that could be caused.)
- How can this injury be prevented? (The control measures that need to be set in place.)

This assessment process should be undertaken within the framework of general health and safety requirements using the detailed standards on laser safety that have been developed. It is the aim of this book to help with this process, and to provide much of the background understanding that is necessary in order that these questions can be successfully answered.

1.4 Safety standards

Maximum limits of safe exposure to laser radiation for both eyes and skin are issued by the International Commission for Non-ionizing Radiation (ICNIRP) [1]. These limits, called exposure limits (ELs), are incorporated into international laser safety standards and also form the basis for product classification.

The main international standard for laser safety is IEC 60825-1, published in Geneva by the International Electrotechnical Commission [2]. This standard defines the accessible emission limit (AEL) for each of several laser product classes and specifies requirements for laser products, including labelling, according to the product class. It also provides guidance to users on the safe operation of laser equipment, and defines safe limits of laser exposure, given in terms of the maximum permissible exposure (MPE), which is based on ICNIRP's ELs. The IEC standard is adopted in Europe as EN 60825-1, and is mandated to be applied to laser equipment under a number of European Product Directives, including the Low Voltage Directive, the Machinery Directive, and the Medical Devices Directive.

While the AELs define the emission limits of the various laser product classes, the MPEs are used to assess whether a given level of exposure to laser

Table 1.1. International laser safety standards published by the International Electrotechnical Commission (IEC).

Reference	Title
IEC 60825-1	Equipment classification, requirements and user's guide
IEC 60825-2	Safety of optical fibre communication systems
IEC 60825-3	TR Guidance for laser displays and shows
IEC 60825-4	Laser guards
IEC 60825-5	TR Manufacturer's checklist for IEC 60825-1
IEC 60825-6	TS Safety of products with optical sources, exclusively used for visible information transmission to the human eye
IEC 60825-7	TS Safety of products emitting 'infrared' optical radiation, exclusively used for wireless 'free air' transmission and surveillance (NOHD < 2.5 m)
IEC 60825-8	TR Guidelines for the safe use of medical laser equipment
IEC 60825-9	TR Compilation of maximum permissible exposure to incoherent optical radiation
IEC 60825-10	Laser safety application guidelines and explanatory notes

TR signifies a Technical Report and TS a Technical Specification, otherwise the document is a full standard. Users of these documents should ensure that the most recently published version or amendment is used.

radiation is safe. They can also be used to determine the hazard distance, i.e. the distance from the laser within which an exposure hazard can exist. This can be a very important factor in evaluating the risk.

IEC 60825-1 is one of a series of related laser safety standards and guidance documents. It is, however, the generic standard that defines the basic manufacturing requirements that laser products have to satisfy, and it establishes an overall framework under which laser products should be used. Other documents in the 60825 series either define additional requirements (as normative standards) or provide more detailed guidance (in the form of technical reports or specifications) on the use of lasers in specific applications, for example in optical telecommunication, in industrial processing, and in medicine. A full listing of these documents is given in table 1.1. Further documents in the IEC 60825 series are under development, and existing ones do undergo revision from time to time, and so users of these standards should always ensure that they remain up to date with the latest requirements and recommendations given in these documents. (Readers may refer to the IEC website, www.iec.ch, for up-to-date information on IEC standards.)

In the United States, all laser products sold or offered for sale must satisfy the requirements of the Federal Performance Standard for Laser Products

Table 1.2. US laser safety standards published by the Laser Institute of America (LIA) on behalf of the American National Standards Institute (ANSI).

Reference	Title
ANSI Z136.1	American National Standard for the Safe Use of Lasers
ANSI Z136.2	American National Standard for the Safe Use of Optical Fiber Communication Systems Utilizing Laser Diode and LED Sources
ANSI Z136.3	American National Standard for the Safe Use of Lasers in Health Care Facilities
ANSI Z136.5	American National Standard for the Safe Use of Lasers in Educational Institutions
ANSI Z136.6	American National Standard for the Safe Use of Lasers Outdoors

Users of these standards should ensure that the most recently published version is used. The Laser Institute of America also publish a number of practical guides on various aspects of laser safety.

(21 CFR 1040) [3]. Such products have to be registered with CDRH (the Center for Devices and Radiological Health, a division of the Food & Drug Administration), and a report submitted confirming compliance of the product with the Federal Performance Standard. The classification procedures and manufacturing requirements defined in the US laser product standard differ, in certain respects, from those of IEC, but CDRH is adopting the IEC classification scheme in changes to the Federal standard. In addition, for laser users, ANSI (the American National Standards Institute) issues a number of safety standards covering different laser applications (see table 1.2), and has adopted the ICNIRP MPEs in its latest standard for laser users (ANSI Z136.1) [4]. This standard also defines a classification scheme, differing from that of CDRH, which is intended for non-commercial lasers such as research equipment. A full listing of ANSI laser safety standards is given in table 1.2.

One important difference between IEC and US safety requirements is that both CDRH and ANSI laser safety standards generally *exclude* LEDs. The exception is ANSI Z136.2 (see table 1.2), which covers the use of lasers and LEDs in telecommunication applications.

References

[1] ICNIRP Guidelines 2000 *Health Phys.* **79** 431–40

[2] IEC 60825-1 2001 *Safety of Laser Products—Part 1: Equipment Classification, Requirements and User's Guide* (Geneva: IEC)

[3] 21 CFR 1040 1994 *Performance Standards for Light-Emitting Products: Section 1040.10 Laser Products* and *Section 1040.11 Specific Purpose Laser Products* (Maryland: FDA)

[4] ANSI Z136.1 2000 *American National Standard for Safe Use of Lasers* (Florida: LIA)

Chapter 2

Quantifying levels of laser radiation

In order to evaluate the potential hazard of exposure to laser radiation, the level of human exposure needs to be characterized (by measurement or calculation). Similarly, when the manufacturer has to classify his laser product, the level of radiation emitted from the product needs to be assessed. These measured or calculated values are then compared to appropriate exposure or emission limits. The basic concepts of quantifying light were introduced in chapter 1, here we explain the principles of units and optical measurements, generally referred to as radiometry, in more detail. Besides reviewing general radiometric terms we also discuss particular issues pertinent to laser safety where the *biologically effective* levels of exposure have to be determined in order to be compared to exposure limits. In some cases, these biologically effective quantities can differ significantly from the actual physical radiometric quantities. At the end of the chapter, relevant properties of equipment used to measure the level of laser radiation are discussed, and practical information for performing measurements is given.

2.1 Power and energy

The basic quantity to characterize the potential of optical radiation to affect a given material or tissue in terms of heating it up or inducing chemical reactions is *energy*. The physicist's definition of energy is the ability to perform work, where work has to be understood in a broad sense which includes affecting chemical changes or increasing the temperature in matter. For some effects on tissue, such as photochemical changes (discussed in detail in chapter 3), the *energy* delivered to tissue is the relevant quantity and the effect does not depend on the time taken to deliver that energy. For interactions where an increase of temperature is necessary, the *rate of energy delivery* to a given volume is important, as it has to compete against thermal conduction which drains thermal energy into surrounding matter. The rate of energy flow has its dedicated name, it is referred to as *power*, sometimes also as *radiant flux*. The exact term for power and energy

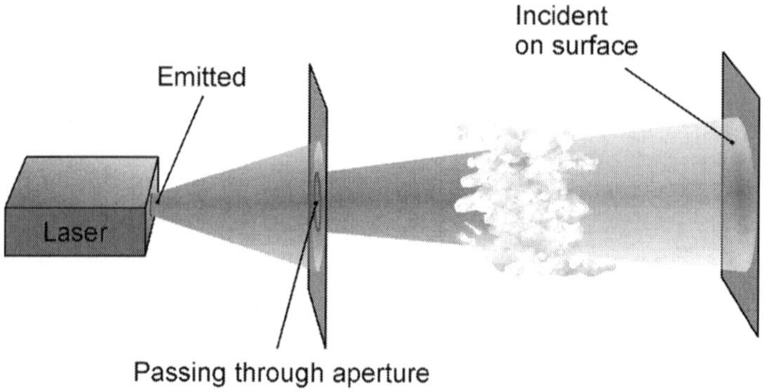

Figure 2.1. When one refers to radiant energy or power, for completeness one should also specify the geometrical reference, i.e. one might have to distinguish between levels of radiation *emitted* from a radiation source, passing through an aperture or *incident* on a surface, as there might be losses of energy or power.

when used for laser and optical radiation (and not for instance for electrical power of the equipment), is 'radiant power' and 'radiant energy'. For brevity, in this book, 'radiant' is often omitted.

Since power is the rate of energy flow, i.e. energy flow per unit of time, the two quantities *power* and *energy* are closely linked via the period of time over which they are being considered. The mathematical representation of the interdependence is, therefore,

$$\text{Power} = \frac{\text{Energy}}{\text{Period of Time}} \tag{2.1}$$

The internationally standardized units with which power and energy are measured are watts (W) and joules (J), respectively, and the relationship of the units, following equation (2.1), is

$$1\text{ watt} = \frac{1\text{ joule}}{1\text{ second}} \tag{2.2}$$

Depending on the problem at hand, one might have to differentiate between power or energy that is *emitted* by a laser and the power or energy that arrives at a target, i.e. is *incident* on a target. Some losses may have occurred between the point of emission and the point of incidence. For instance there could be an aperture which physically obstructs part of the beam, or there could be other losses (see figure 2.1).

The concept of power being equivalent to energy flow can be visualized by recalling that a laser beam, or optical radiation in general, can be seen as consisting of a stream of light particles, or 'photons'. Each photon carries a certain energy, and the more photons that are emitted in a given time, the more powerful the laser beam is. For instance, for a wavelength of 620 nm, the energy of one photon is 3.6×10^{-19} J so that a beam with a radiant power of 1mW corresponds to about 3×10^{15} emitted photons per second (3 million billion photons per second). Although it might seem that one photon carries very little energy, especially when one considers thermal interaction (i.e. heating up of material), one should also bear in mind that the energy of one visible photon and especially of an ultraviolet photon is sufficient to induce chemical changes or even to break biomolecular bonds.

The definition of *radiant power* as *flow rate of energy* (equation (2.1)) is equivalent to saying that energy equals emitted (or incident) power multiplied by time, i.e. expressed as the formula

$$\text{Energy} = \text{Power} \times \text{Time} \tag{2.3}$$

Strictly speaking, equations (2.1) and (2.3) are only correct for levels of power which do not change during the time under consideration. The general mathematical definition of power P (which may vary with time t) is

$$P = \frac{\mathrm{d}Q}{\mathrm{d}t} \tag{2.4}$$

where Q is the symbol for energy and the ratio is defined for the momentary energy flow $\mathrm{d}Q$ during an (infinitesimally) small time interval $\mathrm{d}t$, i.e. equation (2.4) is the exact definition for the momentary value of P.

The generally valid expression for equation (2.3) is an integral

$$Q = \int_{t_1}^{t_2} P(t)\,\mathrm{d}t \tag{2.5}$$

where the time t_1 is for instance the beginning of a pulse and t_2 the end of the pulse to determine the pulse energy, but there could be also several pulses between t_1 and t_2, when one considers the total energy over a longer time domain.

The relationship between energy and power can be best visualized when one plots the power as function of time, as is shown in figure 2.2. This shows laser radiation where emission of radiation commences at 1 s, and radiation is consequently emitted with a constant power of 10 mW up to 3 s, i.e. laser radiation is emitted for a duration of 2 s. Since in this example, the level of power is constant during the period of emission, the emitted energy can be calculated by multiplying the power by the emission duration to obtain the energy value of 20 mJ. Graphically, the temporal behaviour of the emission makes up a rectangle, where the power is represented by the height and the duration of the emission is represented by the width of the rectangle; the energy is therefore equivalent to the *area* of the rectangle. Pulses and emission patterns with the same graphical area have the same energy, as is also shown in figure 2.2, where the emission on the right hand side has half the power but double the duration of the emission on the left.

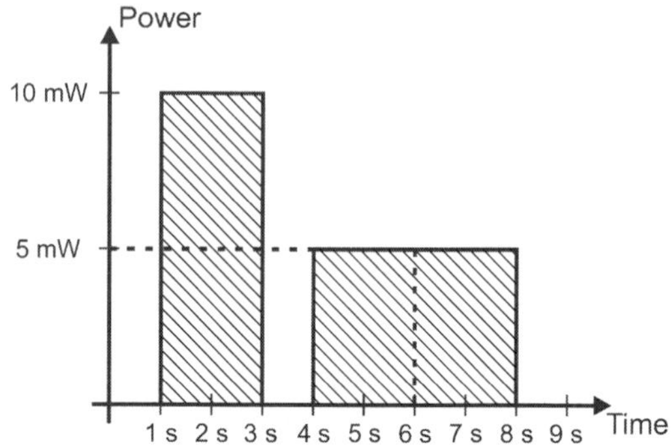

Figure 2.2. Emitted radiant power plotted as function of time. Two emissions with different peak power and duration but equal energy.

As an everyday example of the temporal relationship between power and energy, we can think of a household electricity bill, which characterizes the energy consumption in a given period of time. The energy is given in units of kilowatt-hours, kWh, where 'kW' is the unit of power, and 'h' the unit of time. If one has a 1000 W water heater on for 1 h, it has delivered an energy of 1 kWh to the water. The energy of 1 kWh could also be converted into joules: $1000 \text{ W} \times 3600 \text{ s} = 3.6 \times 10^6$ J. The 'electricity bill units' (used energy) of kilowatt-hours can be a mnemonic help for the relationship between energy and power (equation (2.3)).

When we refer to energy, it is important to specify which 'entity' contains that energy, such as the electrical energy consumption for the last month, the energy contained in a litre of milk or, when referring to laser radiation, the energy per photon, the energy per laser pulse, or the energy emitted or incident over a certain period of time. For instance, one often hears people say 'the laser emits one joule', when they actually should refer to 1 joule *per pulse*—after all, a laser pointer with a radiant power of 1 mW can also emit 1 J, but it takes 1000 s of emission. The concept of energy only makes sense when it is associated with a certain event, an entity or a period of time. In photobiology, the incident energy is often referred to as *dose*. To be precise, dose in photobiology is usually equivalent to energy per unit area, not simply energy (see section 2.2).

When laser radiation is continuously emitted for longer than about a second (with a power level which is reasonably constant), we refer to a *continuous wave* laser, abbreviated to *cw*. To characterize pulsed laser radiation, the parameters listed in table 2.1 are usually used (see also figure 2.3).

Table 2.1. Parameters usually used to characterize pulsed laser radiation.

Quantity	Symbol	Unit
Energy per pulse	Q_{pulse}	joule (J)
Peak power	P_{peak}	watt (W)
Pulse duration	t_{pulse}	second (s)
Average power	P_{aver}	watt (W)
Pulse repetition frequency, also called repetition rate	f	hertz (Hz)
Period	t_{period}	second (s)

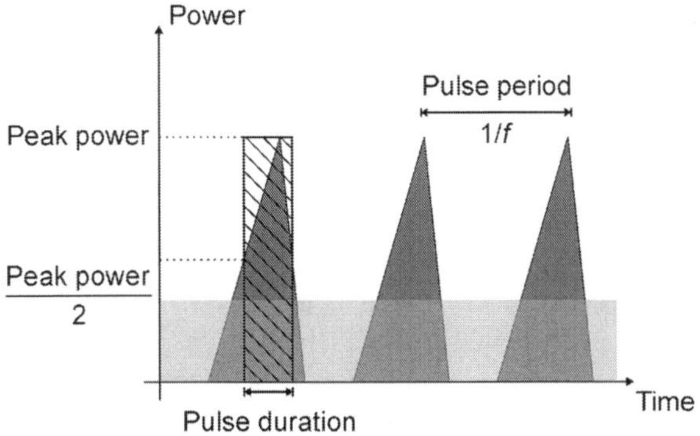

Figure 2.3. A pulse train consisting of three triangular laser pulses with a given pulse duration t_{pulse} and peak power, spaced by the period of the pulse train. Also shown is the average power level, which is the average rate of energy emission, resulting from a redistribution of the energy contained in the pulses over time.

Pulse duration and peak power

The pulse duration is usually defined as the FWHM, the Full Width of the pulse at Half the Maximum power level, which is also the applicable definition in laser safety (see figure 2.3). For rectangular and triangular pulse shapes, the peak power can easily be calculated by dividing the pulse energy with the pulse duration:

$$P_{peak} = \frac{Q}{t_{pulse}}. \tag{2.6}$$

For pulse shapes other than triangular or rectangular, there will be a varying degree of over- or underestimation of the peak power when using equation (2.6). However, since emission limits for product classification and exposure limits for eye and skin hazard evaluation of pulses are generally specified in terms of energy rather than peak power (with the exception of pulse durations shorter than 1 ns for wavelengths outside the retinal hazard area), it is usually not necessary to calculate peak power for safety purposes. In addition, the emission of pulsed lasers is normally specified by the manufacturer in terms of pulse energy, and radiometers intended for measuring pulses are also usually calibrated in terms of energy values.

Example. Excimer lasers typically have pulse durations of about 20 ns and pulse energies of about 200 mJ. The corresponding peak power for these values (assuming a triangular pulse shape) is 10 MW (megawatt). This value can seem very high, and corresponds to the electrical output of a medium sized power plant. The example shows that when a given energy, which need not be high, is emitted within such a short time, then very high peak powers result. It is these high peak powers which make such short pulses an interesting tool in medicine and technology, but also highly hazardous when they are incident on the skin or the eye.

Note: The actual pulse shape of excimer lasers is closer to a skewed Gaussian shape where the maximum value is shifted towards the beginning of the pulse. Such a shape could, for instance, be described quite well by $P(t) = t^2 \exp(-t^2/\sigma)$ where σ determines the pulse width. When this formula is used, the peak power is found to be 4.79 MW. When a Gaussian pulse shape is assumed, the peak irradiance value becomes 4.70 MW, i.e. the assumption of a triangular pulse shape slightly overestimates the peak power when calculated from the energy per pulse.

Pulse repetition frequency and period

The pulse repetition frequency (number of pulses per second) is the reciprocal value of the period of the pulse train, i.e. the time between maxima (or other characteristic points) of two consecutive pulses. It is also often called repetition rate. The concept of frequency really only applies to pulse trains that for some time exhibit a constant period.

Duty cycle

The term duty cycle (dimensionless) is sometimes used to characterize the ratio of the pulse duration to the period, i.e. it can be calculated by multiplying the pulse duration by the pulse frequency, i.e. it is a measure of how much the pulses ‘fill out’ the time; for a cw laser, the duty cycle equals 1.

Average radiant power

Radiant power can either describe a momentary power level, such as when describing the change of power as a function of time during the pulse, or it can be a value averaged over a finite period of time. The average power level is also determined by using equation (2.1), but then *energy* is the total energy emitted or incident within a given duration over which the power is averaged, and *time* is the averaging duration

$$P_{\text{aver}} = \frac{\text{Total energy within averaging duration}}{\text{averaging duration}}. \tag{2.7}$$

Equation (2.7) is the general expression for average power. For a train of pulses with constant pulse energies and with pulse repetition frequency f, the average power can be calculated by

$$P_{\text{aver}} = Q \times f. \tag{2.8}$$

This relationship can be simply inferred by considering that the number of pulses within time t is $f \times t$, so that the total energy within the time t becomes $Q \times f \times t$, which needs to be divided by time t to obtain the average power, and $Q \times f$ remains. The averaging process can be visualized as spreading out the energy contained in the pulses over the averaging time. Following this understanding, the average power does not depend on the pulse duration as long as the pulse energy remains the same. For non-constant pulse trains, i.e. when the repetition rate or the pulse energy varies, the value of the average power depends both on the duration of the averaging duration and on the section of the pulse train which is considered for averaging, i.e. over which section of the train the 'temporal frame' of averaging is laid.

Example. If the excimer laser in the previous example emits a stream of pulses at a constant pulse repetition frequency of 100 Hz, each pulse having an energy of 200 mJ, the average power equals 20 W. Obviously, when the energy of the pulses is spread over time, the average power is a lot smaller than the pulse peak power.

2.2 Irradiance and radiant exposure

The previous section has dealt with the basic quantities of energy and power, and their temporal relationship. When considering the actual laser interaction with material, however, i.e. laser radiation being incident on a surface and being absorbed at or relatively close to the surface (be it a workpiece or human tissue), then not only the power or energy contained in the incident beam is relevant, but also the surface area over which the radiation is distributed. When a given power is focused into a small spot, it is evident that the irradiated material will be affected (for instance heated) faster or more intensely than when the same power is distributed over a larger surface area. The appropriate quantity to describe this level of irradiation therefore relates the power or energy to the size of the irradiated area, and these quantities are referred to as *irradiance* and *radiant exposure*, respectively (see figure 2.4):

$$\text{Irradiance} = \frac{\text{Power incident on area}}{\text{Area}} \qquad \text{with units of W m}^{-2} \tag{2.9a}$$

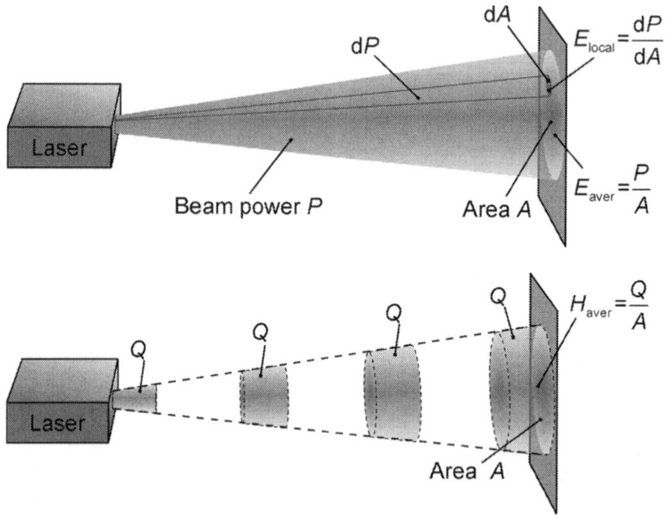

Figure 2.4. Irradiance E and radiant exposure H is derived by relating the power or energy incident on a surface to the irradiated area.

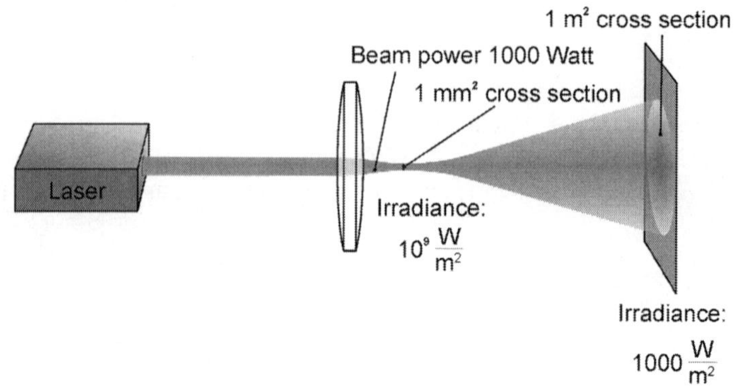

Figure 2.5. Example of a laser beam having a power of 1000 W being focused to a spot with a beam cross section of 1 mm^2 and some distance behind the focus being incident with a beam cross section of 1 m^2. The effect on a workpiece or human tissue in the two locations within the beam is quite succinct.

and

$$\text{Radiant Exposure} = \frac{\text{Energy incident on area}}{\text{Area}} \quad \text{with units of J m}^{-2}. \quad (2.9b)$$

For example, when a laser beam with 1000 W radiant power is focused to a spot having an area of 1 mm^2, the irradiance at the spot would be 1000 W per square millimetre, or 10^9 W m^{-2}, i.e. 1 gigawatt per square metre (see figure 2.5). At first glance, it might be puzzling how a one kilowatt laser could produce gigawatts per square metre, but the point is that this irradiance exists only over the irradiated area of 1 mm^2. If we wished to produce the same level of irradiance over the area of one square metre, we would need a laser beam having a power of one gigawatt. When the power of 1000 W of this laser is spread over an area of 1 m^2 (for instance at some distance behind the focus), than at this surface it would produce an irradiance of 1000 W m^{-2}. (This simple treatment assumes homogenous irradiation, i.e. a constant beam profile; the rigorous definition is given in section 5.3.) While the power of 1000 W when concentrated on 1 mm^2 produces an irradiance sufficient to melt metal and to produce deep burns in human tissue within milliseconds (lasers used for surgery have powers of about 30–50 W), the same power distributed over an area of 1 m^2 is comparable to the irradiance produced by sunlight at the Earth's surface. The power contained in the laser beam is always 1000 W, it is the cross section of the beam and hence the irradiance, which makes the difference. In practice, irradiance and radiant exposure are often related to an area of 1 cm^2, as the irradiated areas are more appropriately measured in square centimetres; 10^9 W m^{-2} would then recalculate to 100 kW cm^{-2}, and 1000 W m^{-2} would recalculate to 0.1 W cm^{-2}. However, it is recommended that parameters are always converted to the base units of m, s, W, J, W m^{-2}, J m^{-2} etc, before embarking on laser safety calculations.

2.2.1 Terminology

Irradiance and radiant exposure can be seen as quantifying the 'concentration' or 'density' of power or energy; in fact, they are often referred to as 'power density' or 'energy density', respectively. However, these terms strictly refer to power or energy per unit volume, not per unit area [1]. Following this international convention, power density would be measured in W m^{-3} and energy density would be measured in J m^{-3}. Nevertheless, the terms power density and energy density are in practice widely used (and are also easier to remember) instead of the correct terms irradiance and radiant exposure, for example in the field of low level laser therapy.

Another quantity which is often confused with radiant exposure is *fluence*, as it is also defined as energy per unit area, but it actually refers to the energy passing through a given area from both sides, which is relevant in scattering media. Radiant exposure refers to radiation that is incident on the (surface) area only from the direction of the irradiating source (see figure 2.6). For non-scattering matter, such as clear glass or metal, radiant exposure and fluence have the same value, but for scattering media, such as tissue exposed to red light, fluence can be higher than radiant exposure. In this context, fluence is used in the science of laser-tissue interaction, but in laser safety we refer only to the radiation that is incident from the 'outside' of the body, and therefore radiant exposure is used exclusively.

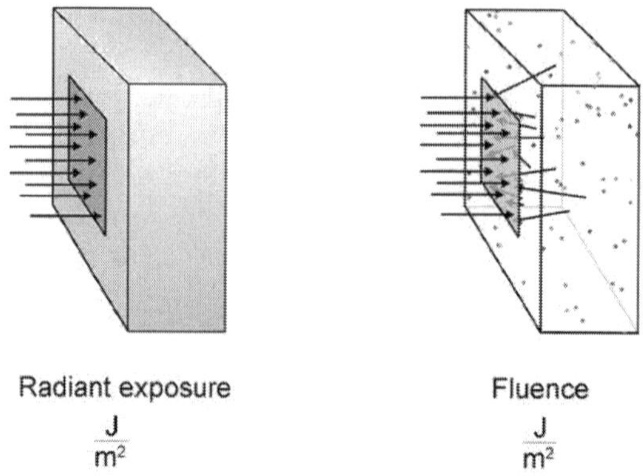

Figure 2.6. Radiant exposure is defined as energy incident or passing through an area, while fluence is defined as energy incident or passing through an area from both sides.

Intensity is another term that is frequently misused, even by optical specialists. Intensity, strictly called *radiant* intensity, is defined as the power emitted into a given solid angle of space, divided by that angle, and is measured in watts per steradian, i.e. W sr^{-1} (for a detailed definition see section 2.3.3). Often intensity is wrongly used to mean power per unit area (which properly is referred to as irradiance).

We will also introduce the quantity of *exitance* in this section, which is also defined as power per unit area, as is irradiance. Exitance, however, describes the power *emitted* per unit area from a source (using 'source' in the wider sense, for example including diffuse reflection). Exitance is a useful quantity in understanding the difference between the irradiance that is produced by a source at some distant surface and the irradiance profile that is produced in the image plane when the source is imaged. The latter is directly related to the exitance (including imaging by the eye onto the retina, or imaging by a lens for radiance measurements).

2.2.2 Averaging over area—limiting aperture

The previous example of an incident laser beam having a cross sectional area of 1 m^2 is an oversimplification, as it assumes that the irradiance profile across the reference area is uniform. For non-uniform profiles, we have to distinguish between local irradiance values and an irradiance value averaged over some finite area. The mathematically exact definition of (local) irradiance, E, and radiant exposure, H is

$$E = \frac{\mathrm{d}P}{\mathrm{d}A} \qquad \text{and} \qquad H = \frac{\mathrm{d}Q}{\mathrm{d}A} \tag{2.10}$$

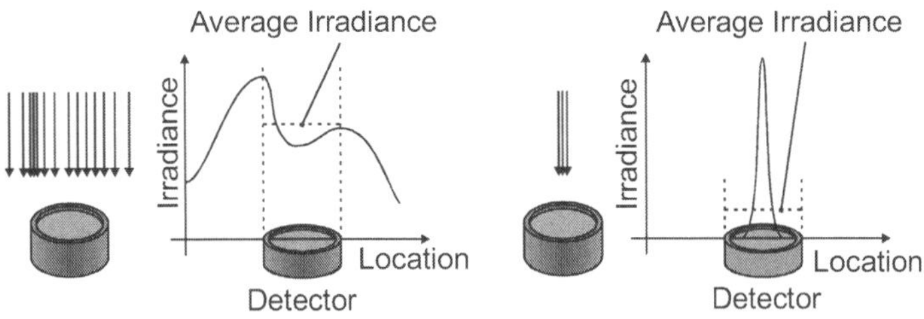

Figure 2.7. Example of irradiance profiles across a detector surface or aperture, and the corresponding averaged value. Left-hand side: inhomogeneous profile; average irradiance determined with detector, level of average irradiance depends on position of detector within beam. Right-hand side: a beam with diameter much smaller than the aperture—the averaged irradiance is much smaller than real irradiance.

which expresses that the power $\mathrm{d}P$ or the energy $\mathrm{d}Q$ (per pulse or for a certain emission or exposure duration) that is incident on the infinitesimally small area $\mathrm{d}A$ (see also figure 2.4). By relating the irradiance and radiant exposure to an infinitesimally small area $\mathrm{d}A$, the precise value of irradiance and radiant exposure at the location of $\mathrm{d}A$ is obtained. In practice, when we measure the irradiance or radiant exposure, we have to use a detector or an aperture in front of the detector with a finite area A to measure a finite power P or energy Q, and we obtain the measured irradiance or radiant exposure by dividing P or Q by the area. This practical determination of irradiance and radiant exposure represents invariably some extent of averaging over the area of the detector or the aperture, as depicted in figure 2.7. The averaging can be conceptualized as 'spreading' the total power on the detector over the detector or aperture area, as is indicated in figure 2.7.

By the nature of averaging, the averaged level of irradiance is always less than the peak irradiance within the averaging area. Only for constant irradiance profiles does the averaged value not depend on the averaging area. The aperture area, i.e. measurement area, can also be considered as the smallest spatial *resolution* with which the irradiance can be determined. Hotspots in the irradiance profile smaller than the aperture area cannot be detected.

Even a radiometer that is calibrated to measure irradiance in fact measures total radiant power incident on the sensitive area of the detector, and the division with the area of the detector is figured into the calibration factor. For such an irradiance radiometer, if we try to improve the measurement resolution for detecting hotspots in the irradiance profile by decreasing the area with an aperture, we would have to change the calibration factor correspondingly (by the ratio of the new and the original area).

It is an important principle in laser safety that an averaged value of irradiance or radiant exposure, which might be significantly smaller than the local 'true' physical irradiance, is compared to the exposure limit for the eye or the skin. In the field of laser safety and for hazard evaluation of broadband optical radiation, specific averaging apertures which are related to biological parameters such as pupil size and eye movements are defined together with the exposure limits for the eye and the skin. The specified aperture over which the irradiance or radiant exposure value needs to be averaged is referred to in laser safety guidelines and standards as the 'limiting aperture'. Because of biophysical phenomena, irradiance hot spots which are smaller than the specified apertures are not relevant for laser safety assessments. In some cases, the specified size of the aperture results in measured irradiance values that would be considered nonsense when compared to the real physical values. An example of this is when the irradiance of a laser beam having a diameter of 1 mm is averaged over an aperture of diameter 7 mm. The averaged value is about 50 times smaller than the real physical value (see example below). However, for hazard evaluations, it is this *biologically effective* value that has to be compared to the respective exposure limit for optical radiation, as will be discussed in more detail in chapter 3. When one uses an averaging area smaller than the one specified, the level of hazard to the eye or the skin would be overestimated. For practical assessments it is helpful to consider that if the beam diameter is smaller than the specified aperture diameter so that the full beam power is passing through the aperture, then there is no actual need to place an aperture in front of the detector, as the measured power will not be affected by the size of the aperture area. In this case, one just measures the power and divides the power with the area of the specified limiting aperture. It is only when the beam diameter is comparable to or larger than the specified aperture that the aperture becomes relevant, i.e. we would then need to place such an aperture in front of the detector and measure only the power (or energy) passing through the 'limiting' aperture to subsequently divide that value with the area of the aperture to obtain the irradiance value.

Example. Consider a laser pointer emitting a beam having a diameter of 1 mm (with the simplifying assumption that the beam is uniform) and having a radiant power of 1 mW. The area of the beam is therefore 7.9×10^{-7} m^2 resulting in an irradiance value of 1273 W m^{-2}. However, for ocular hazard analysis for visible wavelengths, a limiting aperture having a diameter of 7 mm is specified. With this averaging aperture, the biophysical relevant averaged irradiance equals only 26 W m^{-2}. It is this smaller value which has to be compared to the exposure limit for the eye. This example is chosen so that the averaged irradiance lies just at the exposure limit for momentary involuntary exposures. Using the actual irradiance instead of the biologically effective irradiance would overestimate the hazard by 49 times!

2.3 Angle and intensity

In laser safety, some important parameters are expressed in angular quantities, namely beam divergence, the field-of-view of a radiometer and the angular

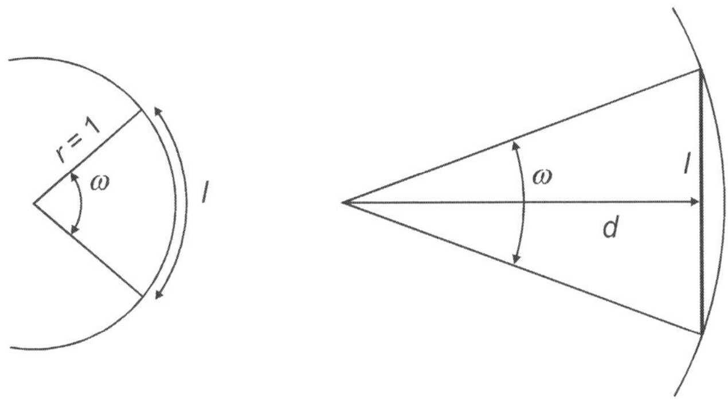

Figure 2.8. Definition of the plane angle ω (left), as well as simplified determination for small angles by dividing the extent l by the distance d.

subtense of the apparent source (which determines the irradiated area on the retina). These angles are typically measured in radians or rather milliradians (abbreviated to mrad). The field-of-view can also be measured in terms of the solid angle, in units of steradians. Beam divergence is a central parameter for beam propagation and is necessary for calculating irradiance or radiant exposure at some location in the beam, and is discussed further in chapter 6. The angular subtense of the source is one of the parameters on which the exposure limits depend and is therefore discussed in chapter 3. The measurement field-of-view is discussed in section 2.4. Here we give the basic definition of plane and solid angle, as these are also needed for the radiometric quantities radiance and intensity.

2.3.1 Plane angle

The SI unit of the plane angle is the radian, and is defined such that a full circle has an angle of 2π radians, i.e. 6.28 radians. This definition also relates the radian to the common unit of angle, the degree, as a full circle has 360°, and thus 6.28 radians = 360° or 1 radian = 57.3°. The plane angle in radians is numerically equivalent to the arc length of a circle having a radius of unity, as the full circle has a circumference of 2π, i.e. for a radius of 1 m, the circumference equals 6.28 m. For a circle having a radius other than unity, the angle is given by the arc length divided by the radius of the circle (see figure 2.8).

In laser safety, angles are typically small, so that the plane angle ω subtended by an object that is at distance d can be easily calculated in radians by dividing the height of the object, l, by the distance, d. The angle subtended by an object obviously changes with the distance of the reference point from the object. For instance, a person of height 1.8 m observed from a distance of 1000 m subtends an angle of 1.8 m/1000 m = 1.8 mrad; at a distance of 100 m the person subtends an angle of 18 mrad. An angle of 1.8 mrad is also subtended by an object which extends 0.18 mm at a distance of 10 cm from the origin or reference point. This reference point can, for instance, be the viewing position (i.e. the position of the eye) in which case the angle subtended by an object is referred to as the viewing angle, and in laser safety is also referred to as the angular subtense of the source. This quantity will be discussed in detail in chapter 3. When the angular subtense of an object which is not perpendicular to the direction of the centre of the object from the reference point is to be determined, then we have to use the projection of the object onto the perpendicular plane in order to determine the angular subtense. Although the angle ω can be approximately determined by dividing the width of the object by the distance, the exact angle, based on the definition of arc length, can be calculated by using $2 \times \arctan(\omega/2)$. The difference, however, is not significant for angles of the order of 100 mrad or less. (For 100 mrad, the difference between the exact and the simplified angle determination is about 0.1%, for 1 rad the difference is about 7%.)

The concept of angular subtense is very useful in optics because the angular subtense of the source is directly related to the size of the image. In many cases it is the angular subtense that more appropriately characterizes the 'size' of an object rather than the actual physical height or diameter. For instance, as noted in chapter 1, even without knowing the distance and the diameter of the Sun, we can easily measure the angular subtense as seen from the Earth, which is about 0.5°, that is 8.7 mrad.

2.3.2 Solid angle

The solid angle can be understood as an extension of the plane angle into three dimensional space, i.e. it can be seen as the 'angular area' subtended by a surface at a given distance from the reference point. While the plane angle in units of radians is equal to the arc length of a section of a circle divided by the radius of the circle, the solid angle, having units of steradians (sr) is defined as the area on the surface of a sphere divided by the square of the sphere's radius, as is schematically shown in figure 2.9. As the surface of a sphere of unit radius has an area of 4π (i.e. for a radius of 1 m, the area equals 12.6 m^2), the full space around an origin subtends a solid angle of 4π sr. A solid angle of 1 sr therefore represents a proportion of about 8% of the space around a given origin.

For an object with circular geometry which subtends a plane angle ω, as shown in figure 2.9, the corresponding solid angle Ω subtended by that object can be calculated using the following equation when the angles are small (i.e. of the order of 100 mrad or less)

$$\Omega = \frac{\pi \omega^2}{4}. \tag{2.11}$$

For larger angles, equation (2.11) should be replaced by

$$\Omega = 2\pi(1 - \cos(\omega/2)). \tag{2.12}$$

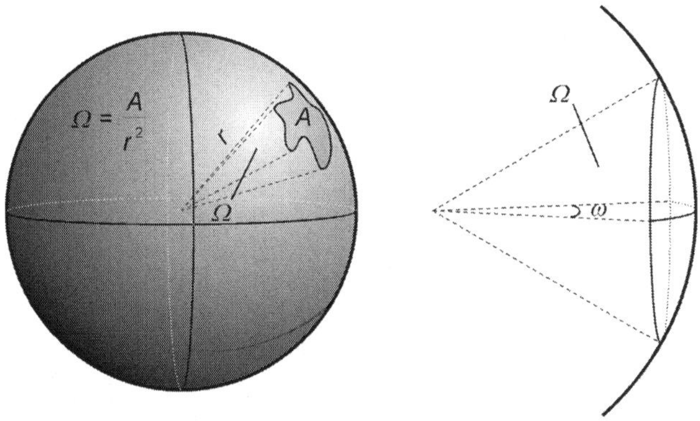

Figure 2.9. Schematic diagram for the definition of solid angle as measured in steradians (left-hand side). On the right-hand side, the drawing shows a circular object where the plane angle ω of that object can be related with the solid angle Ω.

The difference between the exact solid angle and the value obtained with the simple equation (2.11) for $\omega = 1$ rad is about 2%, for $\omega = 100$ mrad the difference is only 0.02%.

2.3.3 Radiant intensity

Radiant intensity, I, is defined as the power emitted from a point source into a given solid angle, divided by that solid angle

$$I = \frac{\mathrm{d}P}{\mathrm{d}\Omega}. \tag{2.13}$$

Intensity is measured in units of W sr^{-1}. A source which emits homogeneously (i.e. uniformly) in all directions into the surrounding space will have an intensity of 1 W sr^{-1} if it emits a total power of 4π watts, since the full sphere around the source subtends a solid angle of 4π sr. A source of high intensity emits a high level of power into a small angle of space. A high level of intensity means that the beam produces a high level of irradiance even at large distances from the source. A typical laser beam with a small divergence has a very high level of radiant intensity. Intensity is used to characterize the emission of radiation from point sources, i.e. sources of radiation which subtend a very small angle. Sources which have a finite radiating surface area, i.e. which create an image with finite size in an optical system (such as the eye), are best characterized in terms of radiance, as discussed in the following sections.

Figure 2.10. Concept of a limited field-of-view—only part of the scene is seen, corresponding to radiation being incident on the eye (or the detector) from only part of the scene. Radiation emitted from other objects is blocked (indicated in the figure by a shaded area, which in reality should obviously not be partially transparent).

2.4 Field-of-view—angle of acceptance

Besides the aperture area of the detector, its angle of acceptance (also often referred to as the field-of-view, FOV) is the second geometric detector property of relevance for the measurement of optical radiation. In simple terms, the FOV is the part of space (quantified in angular terms) which is 'seen' by the detector, or from which the detector receives radiation. If the detector were replaced by the eye, a narrow field-of-view would correspond to looking through a narrow pipe, where only a small part (small solid angle) of the surroundings area can be seen, the rest being blocked by the walls of the pipe (see figure 2.10).

For general irradiance measurements, the FOV is assumed to be 'open', i.e. the radiometer has a large FOV which at least encompasses the entire source which is to be measured, so that the full source is 'seen' by the detector. However, in laser safety, for irradiance measurements of extended sources, the FOV (angle of acceptance) may have to be limited to a certain value which may result in measuring radiation emitted from only a certain part of the source—the part which is within the field-of-view (the angle of acceptance) of the detector. For practical assessments it is helpful to note that the measurement is only affected by the FOV of the detector for the case where there is radiation emitted outside of the FOV. If the source is smaller than the FOV, the extent of the measurement FOV does not influence the measurement. Therefore, the size of the measurement FOV is only relevant for extended sources such as LED arrays, diffuse reflections or diffuse radiating surfaces.

The values of the FOV that are used in laser safety assessments are derived from biophysical parameters and are related either to thermal conduction effects in the retina or to eye movements, as is discussed in more detail in chapter 3.

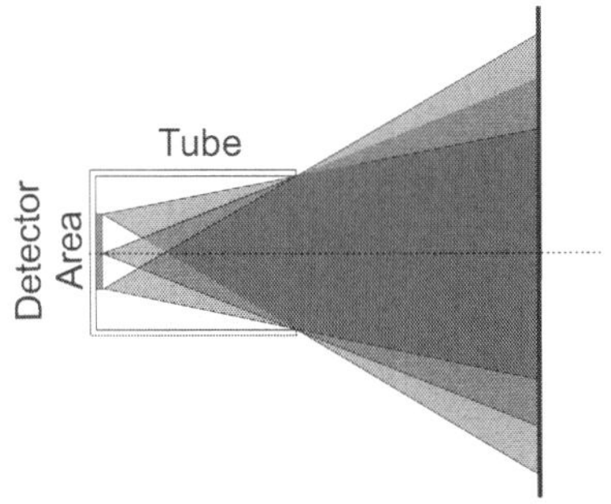

Figure 2.11. Using a tube (called a Gershun tube) with the detector to limit the field-of-view (FOV) results in a central FOV and outer areas of the detector which have a different FOV.

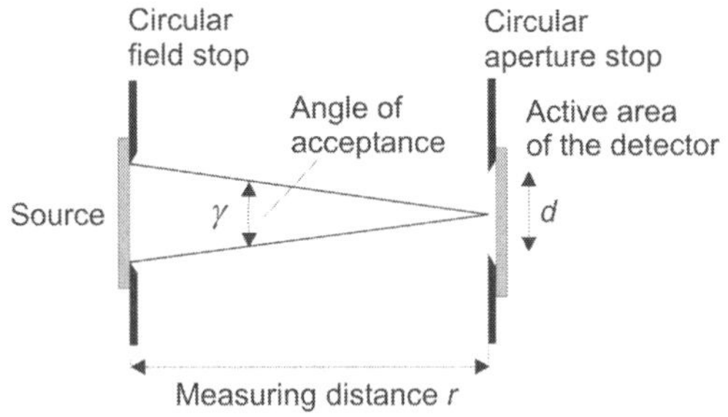

Figure 2.12. A well-defined FOV can be obtained by placing the field stop at the source.

2.4.1 Terminology and optical set-up

In laser safety and the hazard analysis of optical radiation, a circular FOV is required, which is usually specified as a plane angle, measured in units of rad (or more often mrad). The FOV may also be specified as a solid angle, in units of steradian, which can also be applied to a non-circular FOV.

The term *angle of acceptance*, expressed as a plane angle, is also often used instead of the term FOV. In ANSI Z136.1, the term *cone angle* is used.

In some optics or radiometry textbooks it is suggested that a so-called *Gershun tube* is used to limit the FOV of a detector. However, as is depicted schematically in figure 2.11, this does not result in a well-defined FOV, as different points on the surface of the detector 'see' different areas in space. Consequently, the resulting total FOV of the detector has a central part from which the whole detector surface receives radiation, and a surrounding part from which only the outer parts of the detector receive radiation. (This effect is referred to as *vignetting* in optics.)

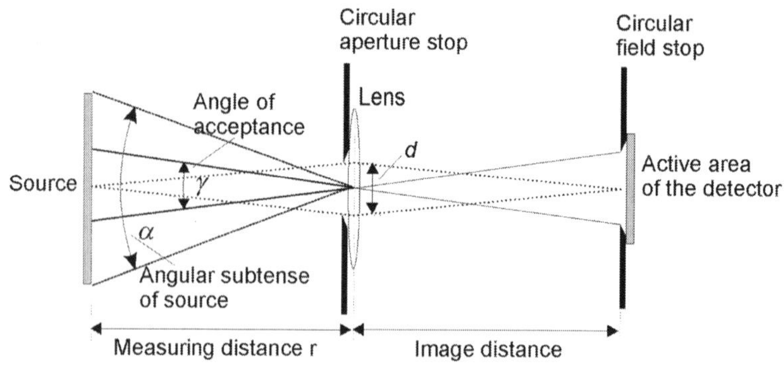

Figure 2.13. As the source is imaged onto the field stop in front of the detector, a telescopic set-up can also be used for sources which cannot be accessed or which use projecting optics.

A well-defined FOV, which is necessary for some optical radiation hazard measurements, can be obtained by either of two set-ups, which are depicted in figures 2.12 and 2.13. Assuming a circular FOV, in both set-ups the size of the FOV is determined by the size and location of the circular field stop. The FOV can be defined by placing the field stop at the source and the detector at a corresponding distance (see figure 2.12), where the plane angle (the FOV in units of radians) is given by the ratio of the diameter of the field stop to the distance between the field stop and the aperture stop. This set-up relies on the placement of the field stop at or very close to the source, which means that the source (or rather, the apparent source which is imaged onto the retina) has to be accessible. If the field stop is not placed in close proximity to the source, then the FOV is no longer well defined, and the set-up is comparable to the Gershun tube shown in figure 2.11. This set-up is therefore not applicable if the source is recessed inside a housing or, since the FOV really relates to imaging the apparent source, might even be located some distance behind the laser, and is therefore not physically accessible. The limiting aperture in these set-ups has the same function as that discussed in section 2.2.2, and the position of the limiting aperture relative to the source is what is referred to as the measurement distance. (In general optics, the limiting aperture is also referred to as the 'aperture stop'.) Quite often, projecting optics are used in front of the source or, for the case of a laser beam, the beam waist usually represents the apparent source which might be located inside the laser or even be virtual and located behind the laser, in which case not only is the source inaccessible, but the measurements must relate to the *apparent* source.

The second, more general set-up for a well defined measurement FOV is shown in figure 2.13. By imaging the source onto the field stop, the field stop does not have to be placed at the source and therefore this arrangement can also be used to measure sources that are not directly accessible or which employ projecting optics. In order to define a FOV, a lens is used to image the source onto the plane of the field stop. Since the source is imaged onto the field stop, the shape and size of the field stop directly determine the part of the source that is measured. The plane angle FOV (angle of acceptance) is determined by the ratio of the diameter of field stop to the distance of the field stop to the lens (the imaging distance). In this case, the averaging measurement aperture for the determination of the irradiance (the limiting aperture) is in front of the lens, i.e. it is the aperture stop of the input optics.

For such an imaging set-up it is important to note the difference between the irradiance level and irradiance profile that exists at the limiting aperture (i.e. at the lens), and the irradiance level and profile at the field stop (i.e. in the image plane). The imaging set-up as described is used for measurement of hazards that relate to the retina of the eye, as it is equivalent to the imaging process in the eye, where the limiting aperture is replaced by the pupil of the eye. At the limiting aperture, the irradiance profile is a superposition of the radiation emitted by those points of the source which emit radiation into the direction of the limiting aperture. The irradiance profile in the image plane, however, is not directly related to the irradiance profile at the limiting aperture, as at the field stop the irradiance profile is the optical image of the source. The irradiance profile in the image plane is, however, directly related to the exitance profile of the source (or rather, the projected exitance). Imaging the source means that rays that are emitted from one point of the source are brought together again to one point in the image plane. At the position of the limiting aperture (i.e. at the lens), the rays overlap, and any one point on the lens receives rays from many different points across the source. The size of the limiting aperture determines the amount of radiation that enters the imaging system (the detector or the eye). This can also be quantified in terms of the power that passes through the aperture, which is obtained mathematically by multiplication of the (average) irradiance by the area of the limiting aperture. It is this power (this amount of radiation) which is then 'distributed' over the image plane to form an image. While the shape of the image (the irradiance profile) is not affected by the size of the limiting aperture, a smaller aperture results in less radiation entering the imaging system and less radiation being available to form the image.

The discussion also shows that by introducing a field stop into the image plane in front of the detector (which does not need to be directly at the field stop, but needs to collect the total power that passes through the field stop), the detector only measures part of the image, or rather, only a proportion of the total power that enters the system and that forms the total image. When we divide this partial power value with the area of the limiting aperture, we calculate the proportion of the total irradiance that has its origin in the part of the source that is within the angle of acceptance (the FOV).

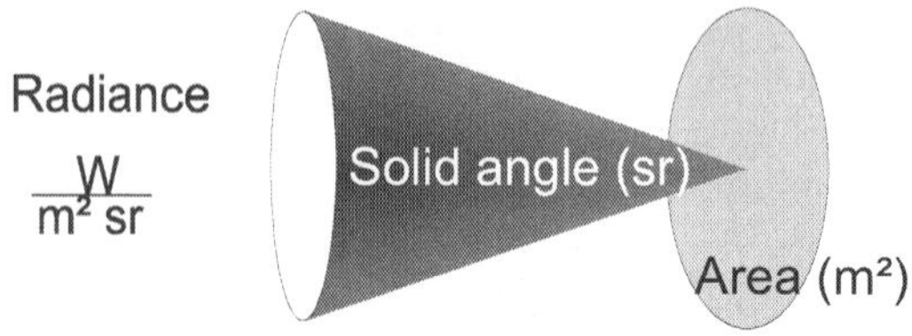

Figure 2.14. The unit of radiance is W m^{-2} sr^{-1}—for hazard measurements, radiance can be seen as the irradiance at the detector (averaged over the appropriate area) divided by the field-of-view of the detector as measured in steradians.

2.5 Radiance

Since exposure limits for the eye and skin are given in terms of radiant exposure or irradiance, and the emission limits for product classification are given in terms of energy or power, it is not really necessary to be familiar with the concept of radiance in order to perform laser hazard assessments or product classification. However, the concept of radiance can help in the understanding of the measurement requirements specified for retinal photochemical hazard evaluation and for retinal thermal hazard evaluation for sources larger than 100 mrad, as well as of the potentially increased hazard that can arise when exposure with optical viewing instruments occurs. In addition, the exposure limits for retinal hazards from broadband optical sources are given in terms of radiance values [2, 3].

Radiance can be most easily understood by considering the way in which it is measured. Radiance is the power incident on a given area of detector from a given part of space (i.e. from a certain solid angle defined by the field-of-view of the detector) divided by that area and that solid angle, as shown in figure 2.14. At this stage of the discussion, the FOV is considered small enough so that the source emission (the exitance) contained within the FOV is uniform. (For a detailed discussion on averaging, the reader should refer to section 2.5.1.) Radiance can also be regarded as the level of irradiance that arises from a certain part of the surrounding space per unit solid angle of surrounding space.

The quantities which make up radiance are therefore

$$\text{Radiance} = \frac{\text{Power}}{\text{Area and solid angle}} \qquad \text{and the units are (W m}^{-2}\text{ sr}^{-1}\text{)}. \qquad (2.14)$$

There is also an equivalent quantity based on energy, i.e. measured in units of J m^{-2} sr^{-1}, and it is referred to as the *time integrated radiance* or often just as *integrated radiance*. As the following discussion of radiance is related to geometrical aspects only, it also applies to time integrated radiance.

In contrast to irradiance, which is related to a certain position in the beam, radiance is a property of the propagating beam and has the same value wherever

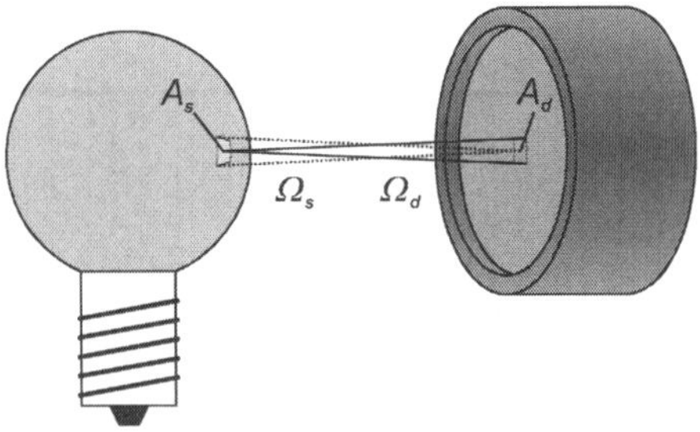

Figure 2.15. One of the advantages of radiance is that it characterizes the emission from the source L_{s} just as well as the radiation at the detector L_{d} or the human eye, i.e. $L_{\mathrm{s}} = L_{\mathrm{d}}$. (When the areas are not perpendicular to the direction of the beam, this invariance is valid for the projected areas.)

it is measured. That is, it is a property of the radiation emitted by a source and also characterizes the exposure at a detector as is depicted in figure 2.15. It can be evaluated at any point along the beam—the value of radiance is the same in every case. The generally applicable equation for the radiance L is

$$L = \frac{\mathrm{d}^2 P}{\mathrm{d}A\,\mathrm{d}\Omega} \tag{2.15}$$

(where the incremental area $\mathrm{d}A$ is measured normal to the beam axis).

When seen as a source property, radiance L is defined as the power $\mathrm{d}P$ emitted from a surface element $\mathrm{d}A$ into the (infinitesimally small) solid angle $\mathrm{d}\Omega$. The invariance of radiance with respect to the position along the beam (sometimes referred to as the *radiance theorem* or as the conservation of brightness) can be understood by considering that the extent of the area on one side, for instance at the source, is associated with a solid angle when seen from the other side, i.e. from the detector. For instance, if one had the same power emitted into the same solid angle but from a smaller area, the radiance as seen from the source will become higher. At the same time, the radiance as seen from the detector also becomes higher, as the optical radiation reaching the detector has originated from a smaller solid angle.

Radiance can vary across the emitting area of a source and can also depend on the direction away from the source. Some sources, such as a frosted lamp, have the same radiance in all directions and over the whole emitting surface (see figure 2.15), while other sources, such as an electric torch, are both highly directional and have values of radiance that vary over the area of the emitting area (see figure 2.16). An equivalent directional and positional dependence applies to the detected radiance where the role of area and direction are reversed. As shown in figure 2.16 one has to carefully position and orientate the radiance detector in order to determine the radiance over the desired area of the source and in

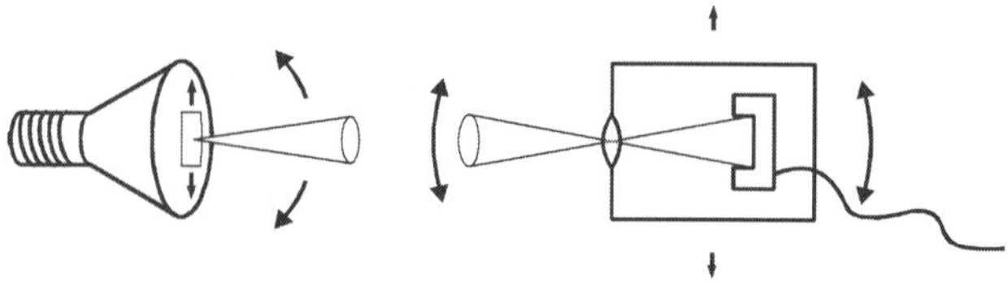

Figure 2.16. Radiance can be highly directional-dependent. The level of radiance can depend on the position of the emitting surface as well as the direction into which the radiation is emitted.

the required direction. When performing hazard assessments, it is the maximum radiance level that has to be determined, i.e. we have to 'scan' the emitter with the given measurement FOV and to move the detector in the plane perpendicular to the beam to determine the maximum level of radiance.

When the imaging set-up as shown in figure 2.13 is used to the define the field-of-view and to measure radiance, it is important that the source (or rather, the apparent source, as discussed in section 3.12.1) is imaged onto the field stop, i.e. the lens-field stop distance has to be varied so that the optical 'object' distance of the set-up is equal to the distance of the apparent source to the lens. This can be either done by first characterizing the location of the apparent source and calculating the corresponding image distance which is then chosen as the lens-field stop distance, or the lens-field stop distance is varied until the detector signal is maximized. When the source is not properly imaged onto the field stop, the image will be blurred and some radiation that would have been detected then lies outside the detector, so that the measured value is less than the actual radiance value that should have been measured.

When radiance is applied to visible light and the spectrum is weighted with the spectral sensitivity of the eye, it is generally called 'brightness'. A source appears bright when both the irradiance at the eye (at the cornea) is high (so that high levels of power enter the pupil of the eye) and the radiation originates from a small spot which results in a small image on the retina. As laser beams typically produce a very high irradiance at the eye and appear to be originating from a very small point in space (as the beam is collimated, i.e. the rays are almost parallel), lasers have extreme values of radiance (or 'brightness').

It follows from the invariance of radiance that its value cannot be changed by optical instruments (except where losses due to reflection or absorption occur, in which case the radiance can be decreased but it cannot be increased). This is an important principle when it comes to evaluating the potential hazard increase for exposure through optical viewing instruments, as is discussed in sections 4.3.5.1 and 5.6.4. However, it should be noted that the law of invariance of radiance strictly applies only to the actual, non-averaged radiance, and may not

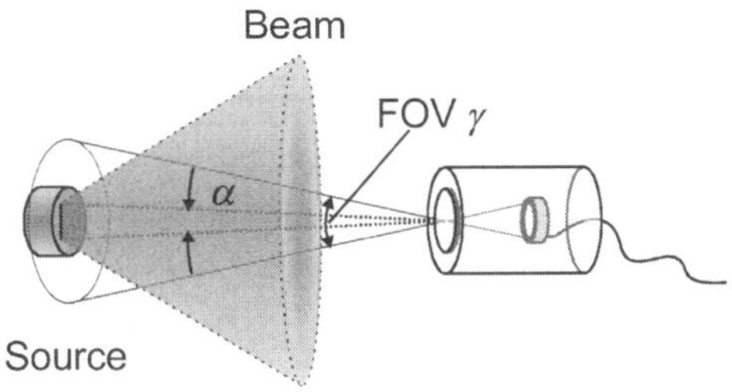

Figure 2.17. A source such as an LED subtends a certain angle α at the reference position of the detector. This source size may be smaller than the specified measurement FOV.

be applicable to the biologically effective value of radiance that is obtained when using a averaging FOV (see the following section).

2.5.1 Averaging over the FOV

Just as there are averaging apertures defined for irradiance measurements that are related to biophysical parameters, there are also averaging field-of-views defined for radiance measurements. The radiance is generally averaged over the measurement FOV when the infinitesimal small $d\Omega$ of equation (2.15) is replaced by a real, finite FOV Ω. Therefore there is a 'golden rule' of radiance measurements: the source has to overfill the detector's FOV. This is a simplified expression of the requirement that to determine an accurate physical radiance, the measurement FOV needs to be small enough to resolve the radiance profile of the source (as with the measurement of irradiance profile discussed previously). In other words, the measurement FOV needs to be small enough so that the averaging effect is not significant. If the source emission exhibits hot spots, these hotspots of radiance cannot be detected when they are smaller than the measurement FOV. The golden rule of radiance measurement (i.e. to overfill the FOV), however, does not apply in the hazard evaluation of optical radiation, where a certain FOV is specified over which the radiance is to be averaged. The size of this averaging FOV is mainly derived from eye movements which average the exposure over a certain area on the retina. The averaged radiance is directly related to an effective irradiance on the retina (see following section). It is the averaged, *biologically effective* value which needs to be compared to the exposure limit for retinal damage. In the extreme case, the angular subtense of the source α is smaller than the specified averaging FOV (see figure 2.17), and the averaged radiance is therefore much smaller than the real physical value (see example below).

Since radiance can be considered to be an irradiance measurement linked to the specific direction and solid angle from which the irradiance is received (as seen from the detector) or into which the radiation is emitted (as seen from the source), the averaging of irradiance over the limiting aperture as discussed above in section 2.2.2 also applies to radiance measurements.

Example. Consider a source which subtends an angular subtense of 1.5 mrad, equivalent to a solid angle of 1.7×10^{-6} sr (see equation (2.11)) and an averaging field-of-view of 110 mrad (one of the values specified for hazard evaluation) or 9.5×10^{-3} sr. These values result in an averaged radiance which is a factor of 5376 smaller than the actual radiance of the source.

Furthermore, if the source emits a beam which has a diameter of 1 mm at the lens of the radiance meter, and one failed to account for the averaging over the limiting aperture which might be as large as 7 mm, then the hazard would be overestimated by an additional factor of up to 50, resulting in a total overestimation of the radiance level and therefore of the hazard by a factor of about 270 000. Although this example is chosen to produce a rather extreme value, it does show that gross overestimation of the hazard can result when limiting apertures and fields of view are not considered.

2.5.2 Transforming radiance to irradiance

The basic relation between radiance, L, and irradiance, E, is

$$E = L \cdot \Omega \tag{2.16}$$

where Ω is the solid angle. (This equation is valid for small angles, which is generally applicable in laser safety.) The relationship appears simple, but due to the intricacies of optical radiation hazard measurements, care has to be taken when applying it for calculating irradiance from radiance, for instance for transforming exposure limits given in radiance units into irradiance limits and when specifying corresponding measurement requirements. The irradiance in equation (2.16) is measured at the position of the limiting aperture when the radiance is determined with the imaging set-up as shown in figure 2.13.

When the exposure limit is expressed in terms of radiance together with an averaging FOV, for example, it is that FOV which has to be used in equation (2.16) to express the exposure limit in terms of irradiance (and not the source angular subtense), but in order to make this fully equivalent, the same FOV also needs to be used for the irradiance measurement. An example of such a transformation is given in section 3.12.6.4 for the photochemical retinal laser exposure limits, which were directly derived from broadband limits specified as radiance. Carrying out an irradiance measurement with a specified FOV and comparing this value to an irradiance exposure limit (derived as just described) is identical to performing a radiance measurement and comparing the value to the radiance limit. There is one conceptual difference in the role of the FOV for the two cases, as the FOV for the radiance measurement acts as an *averaging* FOV and decreases the measured value in respect of hotspots within the FOV, particularly if the source is *smaller* than the FOV (see previous section). For an irradiance measurement, the FOV is actually a *limiting FOV*, and only affects the measured value if the source is *larger* than the FOV, as it limits the irradiance measurement to radiation which is emitted from only part of the source and excludes other radiation (that would be included for an ordinary irradiance measurement with an 'open' FOV). Therefore, as the laser safety standards are based on irradiance rather than on radiance, the term *limiting FOV* or limiting angle of acceptance is used.

As the maximum permissible exposure (MPE) values are defined at the position of the cornea of the eye, in this section we have discussed the relationship of the irradiance at the cornea (or at the limiting aperture for measurement) to the radiance. It is helpful in the understanding of radiance to see how radiance can be used to calculate the retinal irradiance level (or the irradiance at the field stop for the imaging system discussed in section 2.4.1). For the retinal irradiance, we specify the diameter of the pupil d in mm and τ refers to the transmittance of the ocular media in front of the retina. When we multiply the radiance L by the pupil area $d^2\pi/4$, we obtain the power that enters the eye per steradian of solid angle, which is also imaged onto the retina per steradian of retinal area subtended at the pupil. Thus, the retinal irradiance can be calculated by replacing the 'per steradian' by 'per retinal area that corresponds to one steradian', which is Area $= 1$ (sr) $\times$ (17 mm)^2, where 17 mm is the effective focal length of the eye, (equivalent to the distance from the pupil where the vertex of the cone of the solid angle is located, to the imaging plane). We obtain the formula for the retinal irradiance E_{retina} of

$$E_{\text{retina}} = 0.0027 d^2 \tau L.$$

2.5.3 Actual measurement FOV—simplification for small sources

The FOV actually used for the practical measurement of radiance or irradiance need not always be equal to the specified averaging (or limiting) field-of-view, i.e. we do not always have to go to the trouble of setting up the radiometer so that it has a well-defined FOV. The measurement FOV (angle of acceptance) is usually given the symbol γ when a plane angle is intended and Γ when expressed as a solid angle. In the evaluation of photochemical retinal damage, for example, the prescribed averaging FOV for radiance measurements (or the equivalent limiting FOV for irradiance measurements), is then denoted by γ_{ph} or Γ_{ph}.

The actual size of the measurement FOV does not affect the evaluation when the angular subtense α of the source which is to be characterized is smaller than the averaging or limiting FOV. As the full source size is contained within the specified FOV, any size of FOV may be used as long as it is large enough so that the whole source is 'seen' by the detector, i.e. the only requirement is that $\gamma > \alpha$ which is satisfied for most radiometers. For the case of radiance measurements, one only has to take care not to divide by the source size but by the specified averaging FOV in order to obtain the correct radiance value. It is only when the source is larger than the specified averaging or limiting FOV that the measurement

FOV should be equal to the averaging or limiting FOV, and one then needs to scan the source for maximum readings. Where the source is larger than the FOV, using a larger FOV as the specified averaging or limiting FOV would have different effects depending on whether it is a radiance measurement or an irradiance measurement. Where a radiance measurement is carried out with a FOV larger than the specified averaging FOV, the averaging effect would be too great and would result in a value which might be smaller than the appropriate value, and the hazard could therefore be underestimated. For an irradiance measurement, where the FOV acts as the limiting FOV, the use of a larger FOV than specified would mean that more of the source was included in the measurement, with the result that the irradiance value and the consequent hazard would be exaggerated.

2.6 Wavelength issues

Wavelength, usually measured in units of nanometres (nm) or micrometres (μm), has already been introduced in chapter 1. In this section we discuss quantities and concepts that relate to the wavelength of optical radiation.

2.6.1 Wavelength bands

The nomenclature used for different regions of the electromagnetic spectrum is shown in figure 1.1. The part of the electromagnetic spectrum of concern in laser safety is referred to as optical radiation, with the broad subdivision of ultraviolet, visible and infrared radiation. Following a convention developed by the international lighting commission CIE, optical radiation is further divided into wavelength bands that are characterized by different photobiological effects on the skin and the eye, mainly arising from the dependence of the absorption coefficient of different parts of the eye and the skin on wavelength, as shown in table 2.2.

2.6.2 Visible radiation

When light having wavelengths between about 400 nm and 700 nm falls onto the retina, the radiation induces a visual response and is therefore referred to as visible radiation or as light (although, as mentioned in chapter 1, the term light is also sometimes used in the wider sense of optical radiation including the ultraviolet and the infrared). While the term 'visible wavelength range' is conventionally defined by lower and upper limits such as 400 nm and 700 nm, there are actually no sharp boundaries to the visible region, i.e. to what the eye can see. It is rather that the eye's sensitivity depends strongly on the wavelength and becomes very low for wavelengths below 400 nm and above 700 nm. The variation in the sensitivity as a function of wavelength has been investigated, and a 'standard observer' sensitivity curve was defined by the CIE [4], which is shown in figure 1.2. This sensitivity function of the eye, defining the eye's ability to

Table 2.2. Wavelength bands as relevant for photobiology, following CIE notation.

CIE shorthand	Wavelength range	Photobiological effect
UV-C	100–280 nm	Absorbed in uppermost cell layers of eye and skin; highly effective in producing photokeratoconjunctivitis; germicidal. Radiation with wavelengths smaller than about 180 nm is heavily absorbed by the oxygen of the air and is also termed the 'vacuum ultraviolet' region. Vacuum UV need not usually be considered for hazard evaluation.
UV-B	280–315 nm	Intermediate absorption depth; highly effective in producing photokeratoconjunctivitis and sunburn.
UV-A	315–400 nm	Penetrates deep into eye and skin; potential damage to the lens.
vis	400–700 nm	Visible wavelength range. Following CIE standard terminology, the visible region extends from 380–780 nm. See discussion in section 2.6.2.
IR-A	700–1400 nm	Radiation focused onto the retina, but not visible; deep penetration into the skin. Following CIE standard terminology, the IR-A region extends from 780–1400 nm.
IR-B	1400–3000 nm	Decreasing penetration depth for increasing wavelength for eye and skin, from deep penetration and large volume absorption at 1400 nm to surface absorption at 3000 nm.
IR-C	3000 nm–1 mm	Radiation absorbed in uppermost cell layers of eye and skin.

perceive visible radiation, is also sometimes referred to as the 'luminosity curve' or 'spectral luminous efficiency'. The eye is most sensitive in the green part of the spectrum, at a wavelength of 555 nm, where the relative sensitivity reaches its maximum (i.e. a value of 1). The standard sensitivity curve is derived from the spectral sensitivity for colour vision (photopic vision utilizing the cones in the retina) in contrast to the spectral sensitivity for 'grey-scale' (night) vision in dim surroundings which is shifted somewhat to smaller wavelengths (scotopic vision using the rods in the retina) with a maximum sensitivity at 504 nm.

The relative sensitivity curve can be understood in terms of perceived relative brightness. For instance, when 1 nW of power from a green laser pointer at a wavelength of 532 nm enters the eye it is seen to be about three times brighter than the same power from a red laser pointer at 630 nm, and about 28 times brighter than a red laser pointer at 670 nm (with the assumption that they all produce the same retinal image size). From the curve in figure 1.2 it can be seen that there are no sharp borderlines to the visible band. However, the sensitivity decreases towards higher wavelengths and towards smaller wavelengths, so that the further away the wavelength of the radiation is from the position of maximum sensitivity (at 555 nm), the more light is needed for it to be perceived. The definition of what is 'visible' is therefore somewhat arbitrary, but in laser safety is taken as 400 nm to 700 nm. Although the sensitivity outside of this wavelength range is very small, if enough radiation is incident on the retina, the radiation will still produce a visual effect. For instance, diffuse reflections on a sheet of white paper from a 100 mW laser diode at 810 nm can be seen quite well in a dark room (one would have to make sure that the direct beam cannot enter the eye). Cynical colleagues argue that with sufficient power, even 1064 nm Nd:YAG laser radiation can be seen, 'but not for very long' (meaning that the sensitivity for visual perception at 1064 nm is so low that the retinal irradiance necessary to induce a visual response is so high that the retina is damaged).

The definition of the visible wavelength range of 400–700 nm that is used in laser safety is somewhat narrower than the visible band defined by CIE, which is 380–780 nm. The background for the narrower definition of 'visible' in laser safety is that for safety evaluation of exposure to 'visible' radiation, one generally adopts an exposure duration of 0.25 s for accidental exposure. This limited exposure duration is related to aversion responses to bright light, which protect the eye from exposure to radiation which could be hazardous for longer exposure durations. The visible range in laser safety is defined more narrowly so that radiation levels which are not hazardous for momentary exposure durations are perceived as bright enough for aversion responses to take effect and for prolonged viewing to be perceived as uncomfortable. For comparison, the sensitivity at 380 nm and 780 nm is about 0.000 03, while at 400 nm and 700 nm it is 0.004, a factor of 133 higher.

2.6.3 Spectral quantities

It is one of the characteristics of lasers that radiation is only emitted at one or more *discrete* wavelengths which are determined by the laser medium. LEDs, however, produce radiation which is not concentrated at a single wavelength, but is emitted over a finite wavelength range. An example of the spectral emission of a typical infrared LED is shown in figure 2.18.

For broadband incoherent radiation, such as from the Sun or a lamp which produces 'white' light, or from a 'white light' LED (a blue LED having a phosphorous coating to create additional longer wavelengths), the wavelength spread of the radiative output of the source is described by a *spectrum*. The appropriate radiometric quantities are consequently *spectral irradiance*, $E_\lambda(\lambda)$, in units of W m^{-2} nm^{-1}, or *spectral radiance*, $L_\lambda(\lambda)$, measured in units of W m^{-2} sr^{-1} nm^{-1}. When the spectral data is integrated over a range of wavelengths (i.e. the area underneath the curve is calculated), then we obtain the respective integrated values given in W m^{-2} or W m^{-2} sr^{-1}. The maximum solar irradiance at the Earth's surface, for example, has an integrated value of about

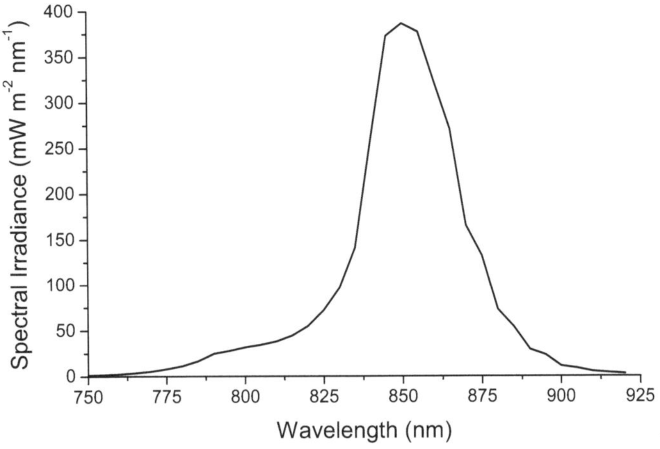

Figure 2.18. Spectral irradiance of an infrared LED at a distance of 20 cm from the LED.

1000 W m^{-2}—of that 1000 W m^{-2}, about 2% is in the UV, 45% is in the visible and 53% is in the IR wavelength range.

2.6.4 Action spectra

In terms of evaluation of the potential hazard, different wavelengths may have widely varying effects on the eye and the skin, leading to a strong wavelength dependence of exposure limits and emission limits as discussed in chapter 3. The wavelength dependence of exposure limits can be interpreted as the varying 'effectiveness' of a given level of laser radiation to produce a lesion (an observable injury). When the exposure threshold is smaller for wavelength λ_1 than for wavelength λ_2, then laser radiation at wavelength λ_1 is more 'effective' in producing an injury than at wavelength of λ_2. For laser radiation, only discrete wavelengths are usually an issue, and any wavelength dependence of the 'effectiveness' of a given level of exposure to produce a lesion can simply be accounted for in the definition of the exposure limits. The wavelength dependence of the relative effectiveness of broadband optical radiation to produce a given effect is accounted for by an *action spectrum*, $s(\lambda)$. Basically, each photobiological mechanism or effect has a distinct action spectrum, e.g. there is an action spectrum for photochemical retinal injury, another one for chlorophyll which characterizes the effectiveness of light to induce photosynthesis, etc. The concept of an action spectrum is typically used for photochemical effects, although the concept might also be applied to thermal damage where the wavelength dependence of the absorption results in wavelength dependence of the exposure limits. The numerical value of the effectiveness, i.e. the ordinate of the action spectrum, lies between 0 for wavelengths which have no effect to 1 for

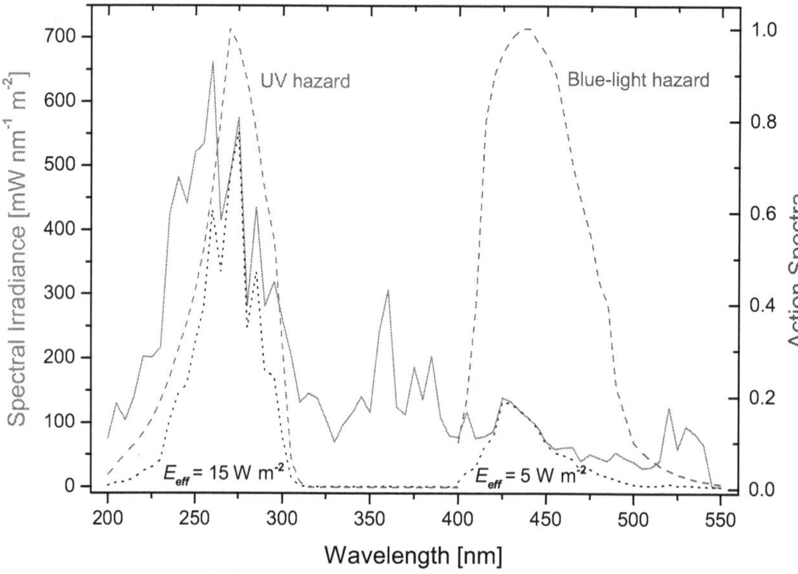

Figure 2.19. Spectral irradiance produced by the plasma plume during laser welding of steel with an 8 kW CO_2 laser beam 50 cm from the plasma. Weighting of the irradiance spectra with action spectra result in an *effective* spectral irradiance.

the wavelength having the highest relative effectiveness. Alternatively, we can consider the action spectrum as the relative *sensitivity* of the tissue to developing a lesion: where the action spectrum equals 1, the sensitivity of the tissue (for a particular effect) is highest, but where the action spectrum is 0, the tissue is not sensitive to radiation at these wavelengths. In this respect, the visual sensitivity curve presented in figure 1.2 also constitutes an action spectrum, where the effect is that of visual perception.

To perform a hazard evaluation, i.e. to characterize a given (measured) irradiance regarding its potential to produce a certain injury, the spectral irradiance $E_\lambda(\lambda)$ is *weighted* with the action spectrum in relation to the injury or effect under consideration. This weighting is simply done by multiplying the measured spectrum with the action spectrum, which produces an *effective* spectral irradiance. An example of the weighting process is shown in figure 2.19 where the measured spectral irradiance at 50 cm from a laser welding plasma is multiplied with the action spectrum for photochemical retinal hazard as well as with the action spectrum for damage of the cornea by UV radiation.

Those wavelengths that are less effective in producing the effect, i.e. are regarded as less hazardous, have a smaller *effective* irradiance. Once the effective spectral irradiance is obtained, it can be integrated over a given wavelength band to obtain the effective irradiance, E_{eff} expressed in units of W m^{-2}. The weighting process can be expressed mathematically by the equation

$$E_{\mathrm{eff}} = \int E_{\lambda}(\lambda) \cdot s(\lambda)\,\mathrm{d}\lambda. \tag{2.17}$$

This effective irradiance (or radiance) value is then compared to the exposure limit for the particular effect under consideration.

When the action spectrum is inverted, i.e. when the reciprocal values are calculated, so that the value of 1 is now the minimum, and when it is then multiplied with the respective exposure limit, then the variation of the monochromatic exposure limit with wavelength is obtained. This is the reverse of the process of how action spectra are experimentally determined: monochromatic radiation or radiation with a small bandwidth is used to determine levels of radiation which lead to the effect under study for a range of wavelengths. This results in a collection of exposure limits for specific wavelengths. The wavelength at which the exposure limit has the minimum value (where the sensitivity is greatest) is identified, and after division of all values with this minimum value and inversion (taking the reciprocal values), the action spectrum as well as the exposure limit for the effective integrated irradiance value is obtained.

Although action spectra as such are not used in laser safety, the photochemical laser exposure limits in the ultraviolet and in the visible wavelength range are directly derived from the action spectra for broadband radiation, as will be discussed in chapter 3. This origin of the laser exposure limits can be helpful in applying them to coincidental exposure to radiation with different wavelengths or to broadband LED sources, especially to white LEDs.

2.6.5 Photometric quantities and units

Because of the importance of visible light, a special system of quantities and units is defined which accounts for the wavelength dependence of the visual process as characterized by the visual sensitivity curve shown in figure 1.2. These quantities are referred to as *photometric*, in contrast to the basic (i.e. unweighted) physical quantities as defined in the previous sections, which are referred to as *radiometric*. Since the hazard potential of optical radiation depends on the power or energy and not the visual perception, photometric quantities are not relevant for laser safety, but they are reviewed here for completeness.

The main photometric quantities and their radiometric equivalence are listed in table 2.3. Luminous flux and luminous energy are the photometric equivalents of radiant power and radiant energy, respectively. Radiometric quantities are unweighted whereas photometric quantities are weighted against the visual sensitivity spectrum. The units in which these photometric quantities are measured are given particular names such as lumen and talbot to distinguish them from the unweighted radiometric units of watt and joule. As the visual sensitivity curve can be seen as an action spectrum (see previous section),

Table 2.3. List of photometric quantities and units as well as equivalent radiometric quantities and units.

Photometric quantity	Unit name (unit symbol) [relation with basic unit]	Corresponding radiometric quantity	Radiometric unit symbol
Luminous flux	lumen [lm]	Radiant power	W
Luminous energy	Talbot (no symbol) [lm s]	Radiant energy	J
Illuminance	lux (lx) [lm m^{-2}]	Irradiance	W m^{-2}
Luminous intensity	candela (cd) [lm sr^{-1}]	Radiant intensity	W sr^{-1}
Luminance ('brightness')	no unit name cd m^{-2} [lm m^{-2} sr^{-1}]	Radiance	W m^{-2} sr^{-1}

photometric quantities are actually *effective* quantities. For example, illuminance is really the visually effective irradiance.

The mathematical definition of the photometric quantity of, for instance, the illuminance E_{vis} in units of lux is based on weighting the basic radiometric quantity of (spectral) irradiance $E_\lambda(\lambda)$ in units of W m^{-2} nm^{-1} with the standard observer visual sensitivity $v(\lambda)$:

$$E_{vis} = 683 \int_{380\text{ nm}}^{780\text{ nm}} E_\lambda(\lambda) \cdot v(\lambda)\,d\lambda. \tag{2.18}$$

The factor 683 in equation (2.18) has the units of lumens per watt and is the spectral luminous efficacy of monochromatic radiation at a wavelength of 555 nm, which is the wavelength where $v(\lambda)$ has its maximum. Equation (2.18) also applies for the other 'pairs' of photometric and radiometric quantities. Following this definition, 1 W of optical radiation at a wavelength of 555 nm corresponds to a luminous flux of 683 lumen. One watt of radiation at wavelengths other than 555 nm has a smaller luminous flux.

When the spectral irradiance is measured with monochromator radiometers, then equation (2.18) is used numerically to calculate the illuminance. Often, however, an integrating radiometer is used where the spectral sensitivity of the radiometer is designed with filters to mimic the standard observer visual sensitivity, so that the weighting and integration of equation (2.18) is built into the radiometer, which directly measures lux.

2.7 Absorption, reflection and scattering

When light is incident on matter, it can be reflected, transmitted or absorbed (see figure 2.20). When the fraction of light which is reflected, transmitted or absorbed is quantified, the respective quantities are called reflectance R, transmittance τ

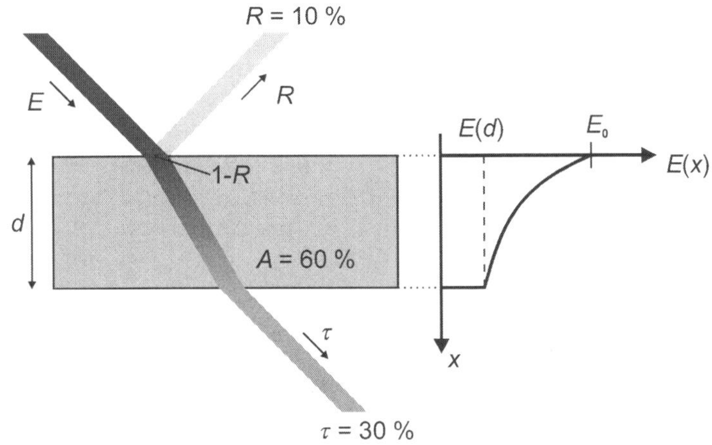

Figure 2.20. Left: radiation can be either reflected, absorbed or transmitted (the reflection from the lower surface is neglected for simplicity). Right: the irradiance profile within an optically homogeneous material (i.e. no scattering) is exponential (see also section 2.7.1).

and absorptance A and they must add up to unity (i.e. the percentage reflected, transmitted and absorbed must add up to 100%):

$$\tau + A + R = 1. \tag{2.19}$$

These parameters, for a given type of material are generally wavelength dependent, but may also be temperature dependent and can be different for different directions of polarization of light. As an example of the wavelength dependence of optical properties, window glass absorbs very little in the visible part of the spectrum (especially if it is thin), but it absorbs heavily in the far ultraviolet and in the far infrared. For most wavelengths, a glass–air interface has a reflectivity of 4%, resulting in a reflectance of about 8% for two surfaces, which is the reason why a window reflects the Sun. (When referring to a single interface, the term *reflectivity* should be used instead of reflectance.) Metals, on the other hand, have a high absorptivity (*absorptivity* is used as the term for the general bulk property rather than absorptance) at all wavelengths of optical radiation, and so practically no radiation is transmitted for anything but very thin foils. Polished metals also have a high reflectance. Almost all materials are highly absorbing (opaque) in the far infrared wavelength range, including materials that are transparent in the visible, such as window glass or Perspex. Materials that have high transmission in the infrared region and also, at least to some extent, in the visible range are rare. One example, zinc selenide (ZnSe), is used commonly as a lens material for carbon dioxide (CO_2) lasers even though it is toxic, since it transmits red wavelengths quite well, allowing an aiming beam to be used in conjunction with the infrared beam from the CO_2 laser.

2.7.1 Absorption law

We see in the previous examples that thickness plays a role in absorption, and so very thin metal foils can have some transmittance while thick window glass, which is generally considered transparent in the visible range, absorbs in the blue and red, making the glass appear green. There is obviously a thickness dependence governing the transmitted (i.e. non-absorbed) radiation. This is generally exponential and is referred to as the Beer–Lambert law

$$E(x) = E_0(1 - R)\exp(-\xi \cdot z) \tag{2.20}$$

where E_0 is the level of irradiance incident on the surface, z is the distance from the surface into the material, and ξ is the absorption coefficient. (In physics, the symbol α is often used for the absorption coefficient, however this symbol is used in laser safety for the angular subtense of the apparent source.) The absorption coefficient is usually given in units of cm^{-1} or m^{-1} (although in chapter 5 for the atmospheric attenuation coefficient the units km^{-1} are used), so that the inverse of this value, referred to as *absorption depth*, corresponds to the distance into the material at which the level of radiation has dropped to the fraction 1/e (37%) of the surface level.

Equation (2.19) can be rearranged for the absorptance A to characterize the fraction of the incident radiation that is absorbed in the material between the surface and the depth d

$$A = \frac{E_0(1 - R) - E(d)}{E_0} = 1 - R - \frac{E(d)}{E_0} = 1 - R - \tau. \tag{2.21}$$

2.7.2 Volume scattering

So far in this discussion it has been assumed that the material is optically homogenous, i.e. 'clear' such as clear glass or clean water, and the directionality (except for refraction at the surface) of a beam or of the light rays is conserved when passing through the material. This is not the case for optically inhomogeneous matter, such as material having small localized centres of differing indices of refraction, or where different kinds of material are mixed, such as small fat droplets in water (milk) or water droplets in air (steam). The inhomogeneous nature of such material causes the individual rays of light to be redirected, and the incident beam is totally broken up, and so the radiation is scattered over a larger volume. Some of the scattered radiation may be re-emitted (see figure 2.21) and, depending on the absorption coefficient and absorption depth, some may be transmitted. Scattering itself is a separate process from absorption, and some materials can scatter the incident light with little or no absorption, although generally some absorption will occur. When the absorption is very strong, so that the photon or the 'light ray' is absorbed before it can be scattered, then obviously scattering is not relevant.

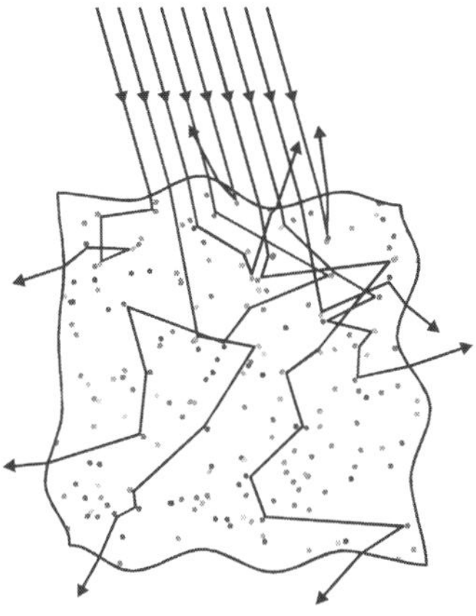

Figure 2.21. Schematic drawing showing scattering within a medium, shown here without absorption, i.e. all incident light is re-emitted as part of the reflected or transmitted component.

The difference between clear and scattering (also referred to as 'turbid') media can be shown with a simple demonstration using a glass of water. Clean water is clear and light passes through it without being scattered. If a small amount of milk is mixed with the water, the fat droplets of the milk act as scattering centres so that the light rays entering the liquid lose their directionality and are scattered. A laser beam from a laser pointer shone into the glass appears to 'light up' almost the whole glass.

Human tissue is highly scattering for wavelengths in the red and the near-infrared, which can be demonstrated when placing the tip of a finger over the exit aperture of a red wavelength laser pointer: the light shines through, i.e. is partially transmitted, but not as a beam, rather the whole finger tip is lit up (and one is at risk of being referred to as 'ET', the alien who wants to call home).

Depending on the ratio of the size of the scattering particle and the wavelength of the light, scattering can have a strong forward directionality, and can also vary with the wavelength. (With particles smaller than 1/10 of the wavelength, for example, the level of scattering decreases with the fourth power of the wavelength, so that small wavelengths are scattered more heavily, which is the reason for a blue sky and a red sunset). If the 'scattering strength' does not vary with depth, the absorption coefficient in equation (2.20) can be adopted to include the part of the radiation which is scattered out of the beam (and is then referred to collectively as the extinction or attenuation coefficient, as used in chapter 5 for the evaluation of beam power losses due to the atmosphere). For more detailed discussion on scattering in tissue the reader is referred to the literature listed at the end of the chapter [5].

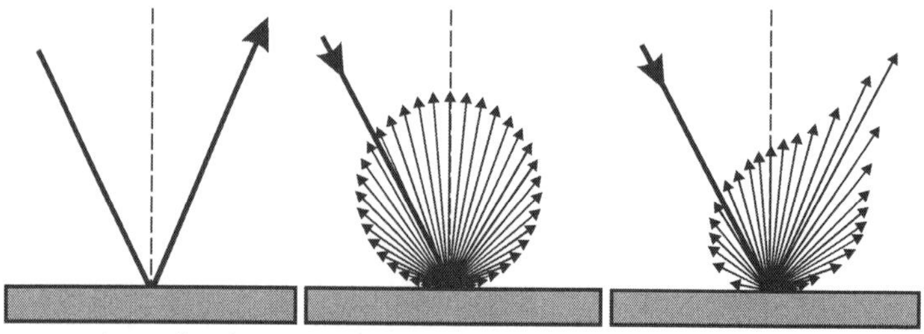

Figure 2.22. Specular and diffuse scattering, as well as mixed scattering (i.e. diffuse with some degree of specular reflection).

2.7.3 Diffuse reflection—surface scattering

The previous section has discussed volume scattering, i.e. the loss of directionality of a beam within turbid media. Scattering also occurs when radiation is reflected from a rough or matt surface. A rough surface can be regarded as a collection of small surface sections that each have different orientations so that different small sections of the beam experience different reflection angles and the beam is 'broken up' upon reflection, i.e. is scattered. Such a reflection is then referred to as diffuse reflection, in contrast to a specular reflection where the beam characteristics are conserved (but the beam is changed in direction). It is important to note that the type of reflection, i.e. diffuse or specular, can be very different for widely-separated wavelengths. For example, surfaces which are diffusely reflecting in visible light (such as a brushed or sandblasted matt steel) can be rather specularly reflecting for radiation from a CO_2 laser at a wavelength of 10.6 μm. This wavelength is larger than the scale of the structure of the 'matt' surface. This might have an effect on the hazard area, since for diffuse reflections (if the beam diameter at the reflecting surface is not too large) the inverse square law applies to the dependence of the irradiance on distance (see also section 5.5). For many rough or matt metallic surfaces, the reflection is mixed in nature, i.e. basically diffuse with some preferred scattering in the direction of the specular reflection (see figure 2.22).

2.8 Measurement instruments and detectors

In this section we explain relevant properties and limitations of detectors and instruments for the measurement of laser power or energy. This review is intended to help in the selection of the appropriate equipment for a given measurement task, and also to identify common pitfalls.

2.8.1 Parameters and uncertainty

Defined below are the main parameters usually used in describing the properties of detectors and radiometers. Where available, the definitions contained in the international standard on the performance of laser radiometers, IEC 61040 [6] are adopted. This international standard defines minimum requirements for laser power and energy measuring instruments for most of the parameters that are listed in this section, and also gives information on appropriate tests of these parameters. In addition it defines accuracy classes for such instruments, such as 'Class 10' or 'Class 2' which characterizes the maximum uncertainty of the detector or instrument. However, in practice, not many manufacturers classify their instruments using the accuracy classes defined in the standard.

Most laser radiation detectors nowadays are part of a complete instrument that includes a display unit ('indicator' or 'readout', often also referred to as the radiometer), so that a number of different detectors (or probes) can be connected to the display unit.

Calibration, responsivity

Laser power or energy cannot be measured directly as one could measure length, but only via some other physical effect that is related to laser power or energy and that can itself be directly measured, such a photocurrent induced in a photodiode. The calibration of a radiometer provides a link between the physical parameter that is to be measured, i.e. the laser power in watts or the laser energy per pulse in joules, to the parameter actually measured by the detector, such as photocurrent. Therefore, a calibration factor must be used, such as W A^{-1} (watts per ampere), which is the factor by which the measured photocurrent (in amperes) must be multiplied in order to obtain the desired radiant power in watts. An equivalent number, namely the responsivity, is defined as the quotient of the detector input signal (the incident optical power, measured in units of watts) and the response signal of the detector, usually an electrical potential difference measured in volts. The responsivity of a detector is high when a small optical power level induces a large response signal. The calibration factor is the reciprocal of the responsivity.

Active area

The active area describes the area of the detector that is sensitive to laser radiation. Some manufacturers specify an active area uniformity, a parameter that characterizes how much the responsivity varies across the active area. It is helpful when the active area is large enough in respect to the prescribed aperture so that there is no need for an additional lens which will introduce some losses.

FOV, acceptance angle, cone angle

The part of the space that is 'seen' by the detector, usually measured as a plane angle or solid angle is referred to as the FOV, acceptance angle or cone angle (see section 2.4). This geometrical parameter of the detector is often not specified by the manufacturer, as typical laser radiometers are used to measure single laser sources where the source is smaller than the field-of-view of the detector, and so the FOV of the detector does not influence the measurement. Some detectors feature a screw-in hood (Gershun type, see figure 2.11) that is used to reduce the FOV in order to minimize the signal arising from optical sources other the one that is to be measured. Such a tube, however, is not sufficient when a well defined FOV is required, as for eye safety measurements of large sources (see section 3.2).

Spectral response or spectral range

The spectral response, or spectral range defines the wavelength range within which the detector is to be used. Where the detector or radiometer is calibrated for only one wavelength, the spectral response is defined only at that wavelength (for instance, at 1.06 μm only). Thermal detectors usually have a rather broad and flat spectral response, while photodiodes have a limited spectral response which is not flat. Where the response depends on the wavelength, appropriate calibration or correction factors for different wavelengths may be provided by the manufacturer. For detectors with a strong wavelength dependence of the responsivity, such as silicon diodes, responsivity values are usually provided as a function of wavelength, often in steps of 5 nm within the range of sensitivity of the diode and the wavelength of the radiation which is to be measured needs to be input in the control unit of the radiometer.

Uncertainty

The uncertainty characterizes the degree of 'ignorance' or 'doubt' that is associated with the measurement. Since the exact (true) measurement value cannot be known, the value measured with the radiometer will deviate from the true measurement value. The manufacturer should specify the basic calibration uncertainty and should also note the distribution and confidence level for which the uncertainty is specified. For instance, when an uncertainty is specified as $+1$ mW, -1 mW, or more often in relative terms such as $\pm 2\%$ of the measured value based on a rectangular distribution, then this means that the true measurement value is assumed to lie within a range of $+1$ mW, -1 mW around the measured value and should never lie outside of this range. An example for such a rectangular distribution is shown in figure 2.23. The probability that the true measurement value lies somewhere in the specified range is constant, i.e. it is assumed that it is just as likely that the true measurement value lies at the measured value or at the border of the range.

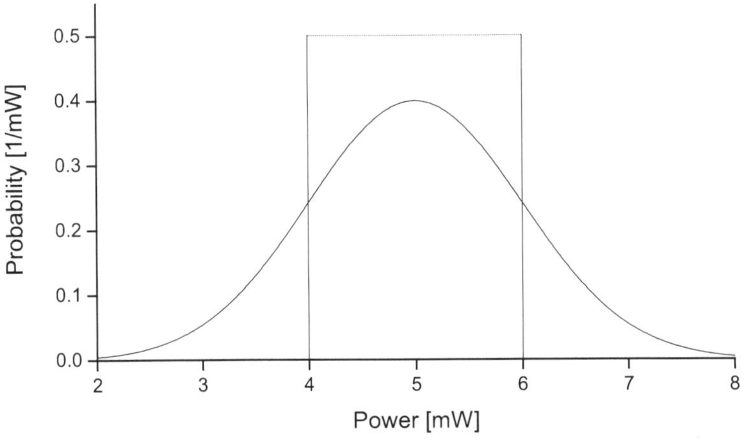

Figure 2.23. Example of a Gaussian and a rectangular distribution, both of which are often used to characterize uncertainties. The Gaussian distribution was plotted for a standard uncertainty (standard deviation) of ±1 mW.

Often, a Gaussian rather than rectangular distribution is assumed for the probability that the true measurement value lies some distance from the measured value. Following the Gaussian distribution (also shown in figure 2.23) there is a finite probability that the true measurement value is quite far away from the measured value. Therefore, with the assumption of a Gaussian distribution, the uncertainty has to be specified with some level of confidence that the true measured value is actually within the specified uncertainty. For instance, the uncertainty could be specified as standard deviation, also referred to as 'standard uncertainty' [6]. The standard uncertainty is associated with a confidence level of 68%. The probability that the true measurement value actually lies in the specified uncertainty range (that might be, for example, ±1 mW) is then 68% and the probability that it lies outside this range is 32%. When this level of confidence is not sufficient, an expanded uncertainty can be used that encompasses a wider interval and therefore a higher probability that the true value lies within the stated uncertainty. In recent years, the term 'coverage factor', k, has been standardized as the multiplication factor for the standard uncertainty, so that $k = 1$ refers to the standard uncertainty (for instance ±1 mW). For a coverage factor of $k = 2$, the uncertainty becomes, for instance, ±2 mW with a correspondingly higher level of confidence that the true measurement values lies within the specified range, i.e. for $k = 2$, the probability that the measured value lies inside the specified range is 95% (two standard deviations for a Gaussian distribution). A standard uncertainty can also be defined as a representative value for a rectangular distribution, and this is defined as $a/\sqrt{3}$, where a is half the width of the rectangular interval, in the example above 1 mW, so that the standard uncertainty in this case would be

±0.58 mW. The numerical value of the standard or expanded uncertainty is also often expressed as ratio with respect to the measured value, such as ±5%.

The basic calibration uncertainty of the radiometer needs to be combined with additional factors arising from other uncertainties associated with the radiometer. These can include the wavelength dependence of the responsivity or the active area uniformity and uncertainties associated with the radiation source or as statistically determined from a number of measurements. These uncertainties are usually combined as the root-sum-square, i.e. the individual uncertainties are squared, summed and then the square root of the sum gives the combined uncertainty.

While it is good practice and may also be considered as a legal requirement that a measured laser power level has to be below a certain limit (such as an allowable emission limit for a laser product class) including the uncertainty, it is the opinion of the authors that it is sufficient for laser safety purposes, considering the biological nature of the limits and also the safety factors in the limits, to account for the standard uncertainty only, and higher confidence levels (a coverage factor larger than $k = 1$) should not be required.

For a more detailed treatise on uncertainty see for instance ISO Guide 'GUM' [7].

Response time

When irradiation of the detector commences, for instance after opening the laser shutter or switching the laser on, for thermal radiometers, and in particular for thermopiles, the readout steadily increases from the initial value (the value without radiation, usually zero or close to zero) until the detector reaches steady-state conditions. The response time constant is usually defined as the time it takes for a radiometer output to rise from the initial value to 63% $(1 - 1/e)$, or sometimes to 90%, of its final value during irradiation with a constant power level. The response time for thermopiles is typically from one to several seconds, but can be longer for heavily insulated detectors. In practice, after commencement of the irradiation of the detector, the readout is observed and the measurement value is noted when one is satisfied that it has reached a stable condition.

Linearity

Within the range of application, i.e. within the rated maximum and minimum power or energy levels (see below), the radiometer should be sufficiently linear. Perfect linearity means that when the signal is increased by a given factor, the display (readout) increases by the same factor.

Temperature coefficient

The temperature coefficient defines the change of the readout with changing environmental temperature. A typical value for thermopiles is 0.1 mW °C^{-1}.

Maximum ratings

Every detector usually has a specified maximum rating for both maximum power and maximum irradiance or, in the case of energy detectors, for maximum energy per pulse and peak pulse irradiance, and sometimes also for maximum average power. If these values are exceeded, the detector response may no longer be linear and the detector can also be damaged. (Some manufacturers give separate maximum ratings for the range within which the detector is linear and for the power level above which damage of the detector can be expected.) Photodiode detectors such as silicon photodiodes have a rather low maximum rating that leads to saturation of the detector, sometimes at only a few milliwatts (although the value for damage is somewhat higher). Some manufacturers of thermal detectors provide metal plates that are covered with the same material as the detector to allow the user to test for possible damage (either surface ablation or some other change to the detector appearance), using the laser that is to be measured.

Noise equivalent power or energy

The noise equivalent power (or noise equivalent energy for energy meters) is defined as that power level (or energy per pulse) incident on the detector that produces a signal equal to the noise level. For this level of power (or energy) the signal to noise ratio is therefore equal to one. Put simply, the noise equivalent power or energy is a figure of merit indicating the lowest range of radiation that can be measured with the equipment. For reasonably accurate measurements, the incident power or energy should be sufficiently above the noise equivalent level in order to 'stand out' from the noise. For example, a power probe that can handle a maximum of 10 W could have a noise equivalent power level of 0.1 mW. The minimum level of radiation that should be measured with such a system is a few mW, i.e. a factor of at least 20 or 30 above the noise equivalent power level. Obviously, such a detector should not be used for measuring submilliwatt power levels, and an inexperienced user might mistake the displayed noise for the laser power.

Zero drift

The zero drift characterizes the extent of the drift of the readout signal while there is no laser radiation incident on the detector. The zero drift parameter is often specified, for instance, as '<0.5 mW after 30 min of operation'. This figure applies provided that the environmental conditions do not change, which might not be the case if there are changes in ambient air temperature, draughts or hot objects in the field-of-view, circumstances that can all affect thermopiles. It is, of course, good measurement practice to observe the signal after cessation of the irradiation to check that it returns back to zero level.

Polarization

Some detectors may have different responsivities for different directions of polarization. Polarization refers to the direction of oscillation of the electric component of the electromagnetic wave. A laser beam can have random polarization, i.e. the direction of oscillation changes randomly, or it might have a well-defined direction. Usually, black matt surfaces that are used for thermal detectors do not exhibit a dependence on polarization, but other materials especially when reflections at oblique angles are involved, might.

2.8.2 Types of radiometers

There are three basic and common types of laser radiometers: thermopiles, pyroelectric detectors and photodiodes. Other types such as photomultipliers may be used in spectroradiometers, but are generally not used to measure laser radiation.

Thermopiles

Thermopiles are thermal detectors typically used to measure radiant power of cw radiation or average power of pulsed radiation of medium to high power levels. Thermopiles derive their name from thermocouples that are positioned so that they act in series and accurately measure a temperature difference between the detector surface and a heat sink within the detector. The detector surface is covered with a non-reflecting coating. and should ideally exhibit little wavelength dependence. When laser radiation is incident on the detector surface, the radiation that is absorbed heats up the detector and an electrical potential difference is established in the thermocouples.

The wavelength dependence of the calibration is usually not very strong, so that one calibration factor is applicable for a wide wavelength range, from the ultraviolet to the far infrared. Some models use a volume absorber, where the radiation is not absorbed by the surface coating of the detector but it penetrates to some extent into the detector material so that the incident radiation is distributed over a larger volume and the maximum power level with which the detector can cope is higher (for a given power level, the induced temperature increase is lower). These volume absorber models, however, often feature a stronger wavelength dependence of the responsivity, so that a model that is calibrated for the visible and the near infrared can be more than 50% in error for measurements in the far infrared (for CO_2 radiation, for example). The broad spectral sensitivity is an advantage when it comes to measuring a wide variety of laser wavelengths, but it also makes the thermopile detector sensitive to other heat sources. The more sensitive models can measure the infrared radiation emitted from the skin (for instance the hand) and may also be influenced by the hot housing of the laser itself. Thermopiles are also highly sensitive to differences in temperature of the detector housing (between the sides and the back of the detector). When the housing is touched by the hand or an object having a different temperature to the housing, or when there is a draught that cools the housing, the measurement can be strongly affected.

The lowest power that can typically be measured with a thermopile is a few milliwatts. For special configurations with thermally insulating hoods covering the detector, one can measure down to levels somewhat less than a milliwatt, but the detector then exhibits an even longer response time (before indicating the true measurement) of up to a few minutes. Due to the response time of at least a few seconds, thermopiles are well suited to measure average power of repetitively pulsed lasers. When the pulse repetition rate is known, the energy per pulse can be calculated from the average power level measured with the thermopile. For low repetition rates (pulses further apart than the rise and fall time of the detector), thermopiles can also be designed to function as energy meters, by integrating the detector current over time. These detectors are usually intended for rather high pulse energy values of the order of somewhat less than one joule up to about 30 J.

For higher power levels, thermopile detectors need to be cooled either by forced convection (by a fan at the back of the heat sink) for power levels from about 50 W to about 200 W, and by water cooling for higher powers of up to 3 kW.

Some manufacturers provide a resistant heating element within the detector that can be used for calibration. With known electrical voltage and current through the heating element, one can calculate the electrical power and, with a correction factor to account for lack of optical losses for resistance heating in comparison to heating by laser radiation, this electrical power can be compared to the power displayed by the radiometer.

Pyroelectric detectors

Pyroelectric detectors are made from crystals where the polarization is strongly dependent on temperature, i.e. a temperature change causes a change in polarization and therefore an external potential difference. A pyroelectric detector does not respond to constant input power, only to signals (and temperature levels) that change with time. This effect can be compared to a piezoelectric crystal where a pressure difference at the crystal is associated with an electric potential difference across the crystal. Consequently, pyroelectric detectors are generally used to measure energy per pulse for pulsed sources, but may also feature an in-built chopper so that the power of cw sources can be measured. Such a pyroelectric, chopper-based power meter has some advantages over thermopiles, as it typically has a faster response and a lower noise equivalent power than comparable thermopiles. It also has fewer problems with draughts and with temperature variations of the housing. For most models using a chopper, the chopper may be detached and placed close to the laser source so that only the source itself is modulated and not the ambient light, thus reducing errors due to ambient lighting. Pyroelectric detectors designed to measure power by chopping the signal cannot be used to measure average power levels of pulsed sources.

When designed as a detector to measure energy per pulse, the maximum pulse duration and the maximum pulse repetition rate are critical parameters, as they have to be traded off against each other. The issue is mainly one of the amplifier and readout, as a higher maximum repetition rate means that a smaller maximum pulse duration can be measured. Typical general maximum pulse durations for pyroelectric detectors are of the order of 1 ms. Pyroelectric detectors might be affected by sudden noises such as claps or bangs (and some pulsed lasers, such as excimer lasers, may produce noise levels high enough to affect the measurement) and by vibrations.

Being a thermal detector, the surface coating of the active area can be highly absorbing, with little variation over a wide spectral range from the ultraviolet to the far infrared.

Photodiodes

Photodiodes are semiconductor devices where photons create electron-hole pairs resulting in an electrical current when the holes and electrons migrate within the n and p regions of the semiconductor. The photodiode with the widest use is the silicon photodiode, and the following discussion concentrates on this type. The silicon photodiode is sensitive in the near ultraviolet, visible and near infrared up to about 1100 nm; for higher wavelength ranges, other types, such as GaAs photodiodes are in use. The main advantage of photodiodes is the much higher sensitivity when compared to thermal detectors. Noise equivalent powers or energies can be less than 1 pW for detectors designed as power meters and less than 1 pJ per pulse for energy meters. Photodiodes are typically also quite fast and can, with the appropriate driving conditions, react within nanoseconds. As such, photodiodes can be also linked to an oscilloscope to determine pulse duration. A photodiode that is calibrated for radiative power and is linked to a digital storage oscilloscope or to a dedicated storage instrument linked to a PC can be a powerful radiometer for evaluating repetitively pulsed sources (especially non-constant pulse patterns). Using automatic processing, the data which is stored as 'power as function of time' can be analysed automatically to obtain the energy per pulse and the pulse duration for the pulses within the storage window (usually about 0.5 s).

Silicon photodiodes have a wide range of linearity, usually of the order of seven magnitudes, but, due to the high sensitivity, they saturate at laser powers of a few milliwatts if not specifically designed for higher powers (such as by placing a diffuser or filter in front of the photodiode).

It is also typical for photodiodes that the responsivity exhibits a strong wavelength dependence. Silicon diodes, for instance, have their highest sensitivity in the near infrared between 800 nm and 1000 nm, with a relatively steep decline of the responsivity for higher wavelengths and very little responsivity beyond 1100 nm. In modern instruments, the varying calibration factor is stored within the instrument and one can input the wavelength of the radiation so that the instrument automatically adopts the correct calibration factor. It should be noted that some manufacturers who provide a spectral responsivity calibration for silicon diodes really only calibrate the detector at one wavelength point, while the other points of the curve are not specifically calibrated for each detector but are derived from the typical relative wavelength dependence of the responsivity of the detectors. Some detectors feature a filter in front of the detector that to a good extent compensates for the wavelength dependence of the responsivity of silicon, i.e. it flattens the response. This is particularly useful if radiation that consists of several wavelengths is to be measured, such as for an eye safety analysis of laser displays or laser light shows.

References

[1] CIE 1987 *International Lighting Vocabulary* (Vienna: CIE)

[2] ICNIRP 1997 Guidelines on limits of exposure to broadband incoherent optical radiation (0.38 to 3 μm) *Health Phys.* **77** 539–55

[3] IEC TR 60825-9 1999 *Safety of Laser Products—Part 9: Compilation of Maximum Permissible Exposure to Incoherent Optical Radiation* (Geneva: IEC)

[4] CIE 1983 *The Basis of Physical Photonetry* CIE 18.2 (Vienna: CIE)

[5] Welch A J and Gemert M J C (ed) 1995 *Optical–Thermal Response of Laser-Irradiated Tissue* (New York: Plenum)

[6] IEC 61040 1990 *Power and Energy Measuring Detectors, Instruments and Equipment for Laser Radiation* (Geneva: IEC)

[7] ISO GUM 1995 *Guide to the Expression of Uncertainty in Measurement* (Geneva: ISO) also published by ENV 13005 1999 (Brussels: CEN)

Chapter 3

Laser radiation hazards

3.1 Introduction

Humans are exposed to optical radiation on a daily basis without suffering eye or skin injury, in fact the optical radiation from the Sun is necessary to sustain life on Earth. Humans need some ultraviolet radiation to produce vitamin D, they need the visible light to see and need infrared radiation for warmth. Since all hot surfaces emit infrared radiation, it is interesting to note that people also represent a source of optical radiation with a peak wavelength of about 10 μm and a total emitted power of about 100 W.

However, too much of anything is harmful, and this of course also applies to light. Damage of the skin and the eyes by exposure to sunlight is well known, in particular sunburn, skin cancer and snowblindness (photokeratitis, an inflammation of the cornea and conjunctiva) induced by the solar ultraviolet radiation, as well as retinal damage from staring at the Sun, for instance during eclipses. When sunlight is focused with a loupe (magnifying glass) it also presents a fire hazard. Due to the special nature of laser light, far less power is sufficient to produce serious injuries than when compared to conventional light sources, as already discussed in chapter 1. For the case of an exposure to the beam, the comparison of the level of exposure with international exposure limits (MPEs) for the eye and the skin indicates whether the exposure is safe or potentially hazardous.

In this chapter, we review anatomical aspects of the skin and the eye relevant for injuries induced by optical radiation, animal experiments as the basis to establish exposure limits, the MPEs for different wavelength ranges and exposure duration domains, as well as aspects related to the determination of levels of exposure to be compared to the MPEs. As the product emission limits (AEL) for Class 1, Class 2 and Class 3R are directly derived from the MPEs for the eye, the dependence of these AELs on wavelength, emission duration and angular subtense of the apparent source will not be repeated in the chapter on classification.

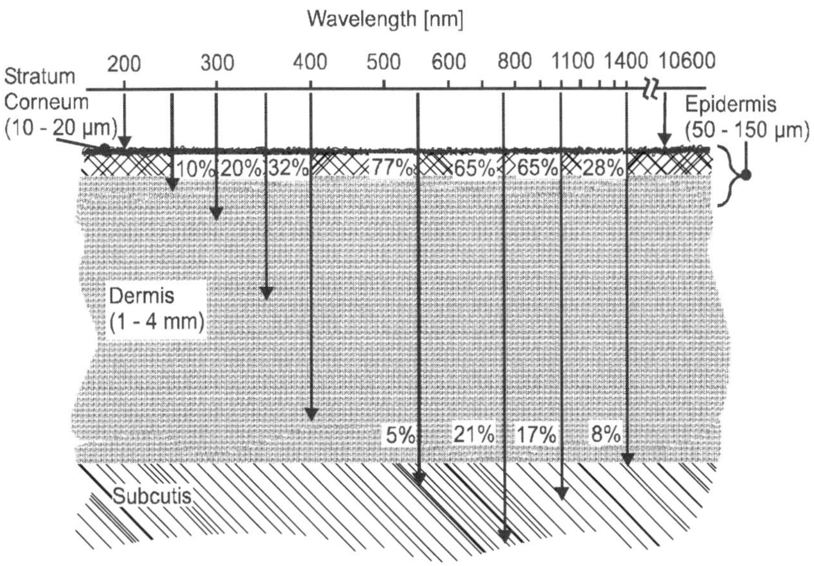

Figure 3.1. Schematic cross section through the human skin. The percentage figures for selected wavelengths indicate fractions of absorbed radiation (adopted from Sliney and Wolbarsth [1]).

A list of common misunderstandings and misinterpretations related to MPE analysis and laser safety classification is contained in the glossary at the end of the book.

3.2 The human skin

The outermost layer of the human body, the skin, besides other functions, has a protective purpose. The skin and underlying tissue absorb optical radiation (in contrast to other parts of the electromagnetic spectrum such as radio waves where absorption is negligible), thus the skin is at risk when exposed to excessive levels of radiation.

The human skin consists of three layers, as schematically shown in figure 3.1. In the outermost layer of living cells, the epidermis with a typical thickness of 50–150 μm, new skin cells are formed regularly so that the epidermis renews itself about every four weeks and the surface layer of the epidermis is made up of dead skin cells. This outermost surface layer is called the stratum corneum and it has a thickness of about 10–20 μm. In the epidermis, keratin, a fibrous kind of protein is formed. This makes up hair and nails and gives the skin its toughness. Tanning also occurs in the epidermis, a process where melanin is formed and distributed within the epidermis including the stratum corneum.

Melanin acts as a protective pigment as it inactivates the free radicals formed by high energy photons of UV radiation. Below the epidermis is the dermis, with a thickness of 1–4 mm, which is mainly a supportive layer, containing collagen and elastin for strength and elasticity, as well as blood vessels and nerve cells. The innermost layer of the skin is the subcutaneous layer, which is mostly fatty tissue and serves as shock absorber and thermal insulator. The thickness of the subcutaneous layer varies strongly from body region to body region and also between individuals.

The optical properties of the skin are strongly wavelength dependent: in the far UV (where 'far' is meant in respect to the visible), the radiation is absorbed mainly in the stratum corneum. With increasing wavelength the penetration depth also increases (i.e. the absorption coefficient decreases), up to the near infrared, where at around 800 nm a maximum is reached, with a penetration depth (the 1/e level, see section 2.7.1) of about 1 cm. At longer wavelengths, the penetration depth decreases with a wavelength dependence that is very close to water. There is a strong (water) absorption peak at a wavelength of about 3 μm and the absorption is very high for radiation beyond about 5 μm, so that radiation is again absorbed in the stratum corneum.

3.3 The human eye

The optical function of the eye is to form an image on the retina from visible light which is incident on the cornea and is emitted from light sources or (diffusely) reflected from objects. The main optical parts can be compared to those of a camera. Figure 3.2 shows a drawing of the horizontal cross section of the right eye.

The cornea and the lens form an image on the retina in the same way as a camera lens forms an image on a film or a digital camera forms an image on the CCD chip. The retina, which can actually be considered as an extension of the brain, is not only made up by the well-known light sensitive cells, called rods and cones. On top of the rods and cones, there are several neural layers, i.e. nerve cells, that already perform rudimentary image analysis of the signals provided by the underlying rods and cones (see figure 3.3).

The neural layers on top of the layer of rods and cones are together referred to as sensory retina. The outermost part of the retina, i.e. below the rods and cones is the retina pigment epithelium (RPE). It is interesting to note that the pigment layer in the RPE is the most absorbing layer in the retina, as the sensory retina is mostly transparent, only absorbing about 5% of the visible light. The RPE has the same antiscattering function as the antihalation backing in photographic film by absorbing light which has not been absorbed in the photosensitive layers. As the melanin pigment layer of the RPE is only about 5 μm thick, very little power is needed to produce a large temperature rise in the RPE. For near IR wavelengths, the radiation is partially transmitted through the RPE and is absorbed in the

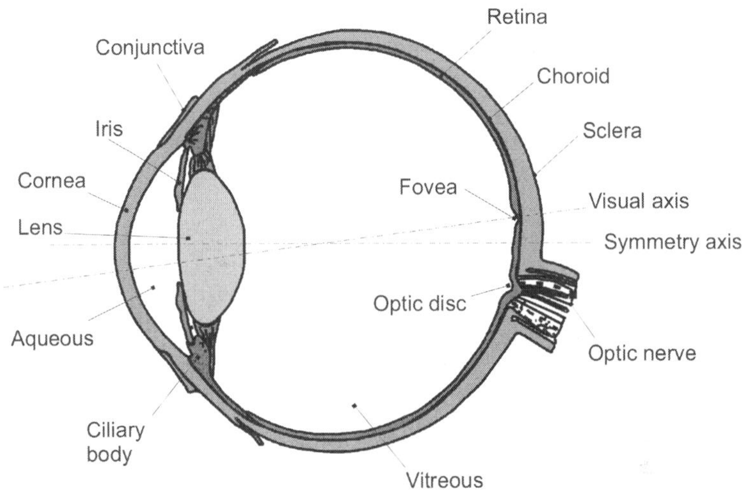

Figure 3.2. Schematic horizontal cross section of a human eye. Central sharp vision is only possible in the central part of the retina, the fovea, an area with densely packed cones. It is also interesting to note that the main refractive power of the eye is provided by the cornea, while the lens is mainly needed to adapt between distant and close objects.

choroid, where the absorption volume is much larger. In addition, for long term exposure, the blood flow through the choroid helps to limit the temperature rise.

Rods and cones are not distributed evenly over the retina. Cones are concentrated in a small central region of the retina, the fovea and the surrounding macula, which is responsible for high acuity vision, while there are relatively few cones in the surrounding retinal parts for peripheral vision which is mainly made up from rods. The rods are responsible for seeing in low light level conditions and can only see in 'black and white', the cones need more light to operate but they are responsible for colour vision. In a cross section of the eye (figure 3.4), the fovea is noticeable as a pit where the neural layer thins out and the thickness of the retina is basically made up by the layer of closely packed cones. In a top view of the retina in the back of the eye (the fundus) such as shown in figure 3.5, the diameter of the fovea is about 200 μm and it is surrounded by an area which contains yellow pigment and therefore has the name macula lutea (or for short, macula), with a diameter of about 2.5 mm.

It is the fovea, and to some degree the macula, that provides central high acuity vision, while peripheral vision has lower acuity due to the lower density of cones and the lower image quality off-axis. Our visual system, through eye movements, positions the high acuity area of the retina such that it centres on the region of interest of the field-of-view. Thus, we do not notice that only a small part of the retina can provide high acuity vision. The subordinate function

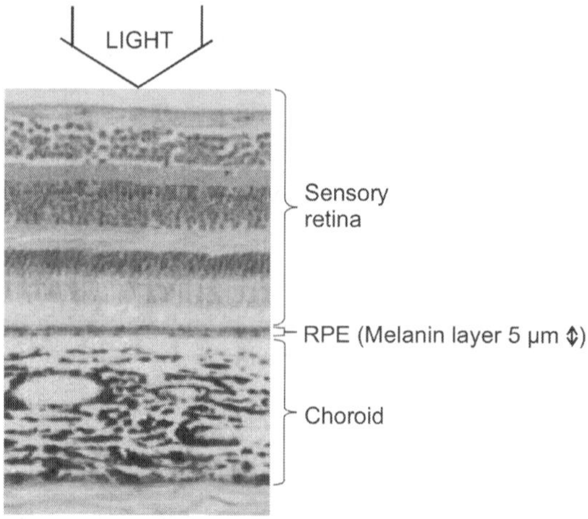

Figure 3.3. Cross section through the retina (light would be incident from the top). Note that the melanin layer in the retinal pigment epithelium (RPE) which in the visible wavelength range absorbs practically all of the incident optical radiation is only about 5 μm thick. The choroid is the layer below the RPE which contains the choroidal blood vessels. (Image kindly made available by David J Lund, US Army Medical Research Detachment of the Walter Reed Army Institute of Research, Brooks City-Base TX.)

of the peripheral retina becomes apparent when there are blind spots that are not even noticed. Everybody has at least one blind spot where there are no rods and cones, and that is where the blood vessel and nerve bundles pass into the eye. This region of the retina is called the optic disk and is seen as a bright spot on the left in figure 3.4 with a number of retinal blood vessels emanating from it. Usually, one does not notice the blind spot, as it is compensated for by the 'image processing' which occurs in the brain. While the damage to the optic disk would have a strong effect on vision, it is actually less susceptible to damage from laser radiation, it is relatively weakly absorbing and any impinging optical radiation is spread over a large volume with a relatively low temperature rise resulting.

With regard to the focusing (imaging) elements of the eye, it is interesting to note that the main refractive power of the eye is associated with the cornea, while the lens of the eye is mainly needed to adjust for imaging of objects at different distances. There is a fast 'autofocus' mechanism built into the optical system of the human eye: whatever we look at (i.e. whatever is the object of regard to which we direct our gaze), the sharpest possible image is obtained by adjusting the power (the focal length) of the lens so that a given object is imaged onto the retina. The focal length of the eye can be adjusted as the lens is elastic, and to look at close objects it is made thicker by contraction of the surrounding muscles. When we look at an object that is more than about 6 m away, the muscles can

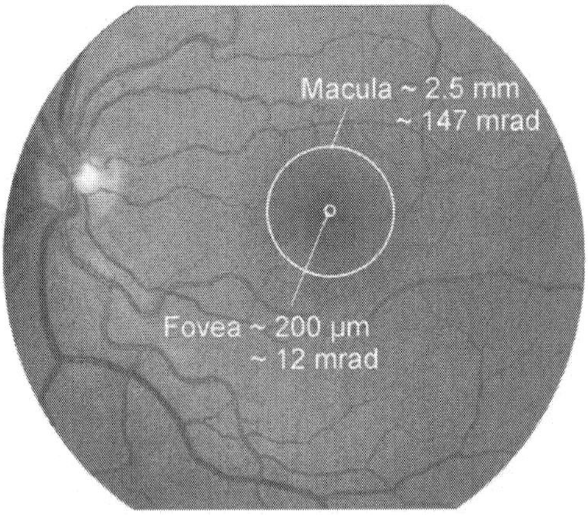

Figure 3.4. Fundus image of the human eye. The region with the highest density of cones and the where the neural layer is thinnest is referred to as fovea, with a diameter of about 200 μm. The foveal pit is surrounded by an area with relatively high density of cones referred to as macula with a diameter of about 2.5 mm. Fundus image kindly made available by Andre Akers, US Army Medical Research Detachment of the Walter Reed Army Institute of Research, Brooks City-Base TX.

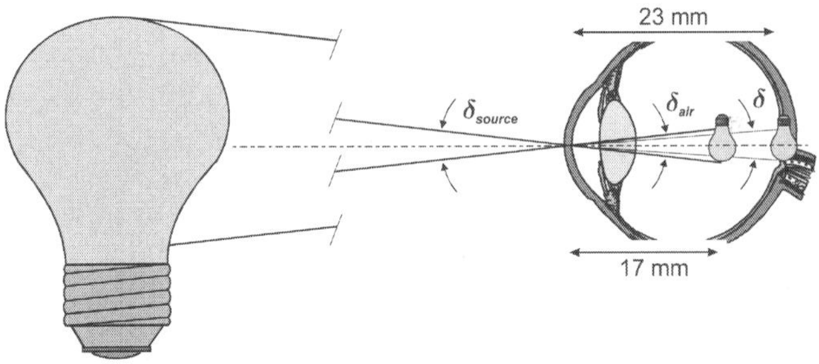

Figure 3.5. Schematical drawing of the angles subtended the image when the optical parameters of the eye are reduced to air-equivalent values (δ_{air}) in comparison to the angle in the real eye, δ, which is smaller than the value in air. The distance from the cornea (actually, the corresponding principle plane) to the retina in the real (normal) eye is 23 mm rather than the value of 17 mm that is used for the simplified model where the lens and the 'camera' are in air.

be considered to be relaxed and this state is generally referred to as 'the relaxed eye' (however, for small pupil diameters of 2 mm, due to the increased depth of focus, an object that might be as close as 2.3 m can be seen sharply even when the eye is accommodated (focused) for infinity). When the eye is relaxed, the combined focal length of the optical system is such that parallel rays that come from infinity are imaged onto the retina. The process of adjusting the focal length of the lens is referred to as 'accommodation'. The closest distance at which the eye can accommodate is limited by how much the lens can be thickened. The shortest focal length of the lens can be associated with the closest point that can be imaged onto the retina (the closest distance of accommodation), and this is therefore referred to as the near point. The near point varies with age and is closer for children and moves outwards with age. If the distance from the eye to the source is less than the near point of the eye, a blurred image will result. In laser safety, for an MPE evaluation, a reference near point viewing position is generally defined as 10 cm. While there might be people who have a near point that is less than 10 cm, such as myopes, the reference distance of 10 cm is still considered an sufficiently cautious (conservative) assumption for the closest viewing distance by the international laser safety committee of the IEC (the US laser safety user guideline published by ANSI even refers to a distance of 20 cm).

When discussing the optics of the eye, it is useful to introduce the refractive power of a lens (or lens system) that describes how 'strong' it is in terms of refracting light rays. The refractive power of a lens that is surrounded by air is defined as 1/(focal length in m), so that a power of 1 D is 1/1 m and a power of 10 D is equivalent to a focal length of 0.1 m $=$ 10 cm. The refractive power, or simply power, is measured in units of dioptre, D. It is also noted that the refractive power of a lens system does not depend on the medium around the lens system, but the associated focal length does. The general definition of the refractive power is the ratio of the refractive index of the material to the focal length that is valid in that material (as will be important for the eye, as the retina is not contained in air).

There is a common misconception that the lens of the eye is the main refractive element (i.e. one reads statements such as 'the lens of the eye focuses the laser beam'). However, this is not accurate, as the total refractive power of the relaxed eye (taken as an average value of the adult human eye) equals 59 D to which the cornea contributes a refractive power of 43 D but only 16 D is attributable to the lens [2]. A refractive power of 59 D corresponds to a focal length in air of 17 mm and therefore a simplistic eye model for the relaxed eye consists of a lens having a 17 mm focal length and a camera (CCD) array placed 17 mm behind the lens (or more accurately 17 mm behind the corresponding principal plane of the lens). This simplified concept is also used in the graphics of this book where the rays do not follow the correct path in the eye but rather the path that they would follow were all the refractive power due to the front surface of the cornea and the eye were filled with air. When this concept of the simplified single lens in air is applied to the near point, then an object distance (near point) of 10 cm would correspond to a focal length of 14.5 mm (this can be calculated with the lens formula and taking 10 cm as the object distance and 17 mm as the image distance). Consequently, in order to simulate the accommodation range from infinity to 10 cm, the air-equivalent focal length of the eye would change from 17 mm to 14.5 mm, respectively. For the natural eye that is not filled with air, the refractive index of the vitreous of 1.34 results in a higher focal length of the cornea-lens system on the image side: on the air side, the focal length of the relaxed eye is 17 mm, on the side towards the retina, the relaxed-eye focal length equals 1.34 times 17 mm, that is 23 mm. Correspondingly, in the average human eye the distance from the retina to the cornea (or more accurately to the corresponding principal plane) is not 17 mm but 23 mm. For a given source (as schematically shown in figure 3.5) due to this longer focal length and longer image distance, the angle subtended by the actual image on the retina δ is smaller than the angle subtended by the image of the object in air δ_{air} and therefore also smaller than the angle subtended by the source δ_{source}.

In the (real) eye, the image has the same diameter as in the air-model eye, but it is not at a distance of 17 mm behind the cornea but rather at a distance of 23 mm. For simplification, the optical parameters of the eye are often reduced to air-equivalent values so that the focal length of the eye is the same for the object and the image side, and the angle subtended by the object is the same as the angle subtended by the image. In the remainder of this book, we also simplify the discussion to this air-equivalent model eye.

The amount of light that enters the system and therefore the brightness of the image is regulated by the iris that is located in front of the lens, in much the same way as the aperture in a camera. The iris is a muscle layer which forms the coloured part of the eye, and may be blue, green, grey, brown or black. The hole in the centre of the iris is called the pupil, which changes its diameter from about 1.5–2 mm under bright light condition to a maximum of about 8 mm at very low light levels. (For completeness it is noted that the diameter of the pupil is here understood as the diameter of the pupil as seen from outside the eye, which is actually a magnified image of the pupil, and is referred to in optics as the entrance pupil).

The eyelid has several functions, the main ones being to continuously wash the front of the eye and to protect it from harm. For exposure to optical radiation, it acts as a protective ‘shutter’ which intentionally or unintentionally closes and protects the eye from prolonged exposure to bright light which might otherwise be harmful (see discussion on aversion response, section 3.9.4).

Inherent in the vision process are eye movements, and that is one way in that the eye differs from a photographic camera. Eye movements result in wander of the image of the source over the retina, thereby distributing the energy incident on the retina over an area larger than the image, which is especially effective for small irradiated spots such as typical for laser exposures. As eye movements mediate the level of hazard presented by a stationary retinal image, the extent and time domain of eye movements are important for exposure limits for exposure durations which are longer than the time domain of the eye movement. For exposure durations shorter than the eye movements, i.e. for pulsed exposures, the retina remains still in relation to the image of the source (like when using a flash to photograph a fast moving object). Generally, the longer the exposure (i.e. the viewing) duration, the larger the eye movements will become in terms of extent of the retinal area over which the image wanders. Experiments with trained observers showed [3] that fixation of a point target is possible for a limited period of time only, and for an observation of the point target over 100 s, the image moves over a retinal area of about 300 μm in the horizontal direction by 150 μm in the vertical direction when the head was stabilized by resting on a chin rest, and over an area of about twice the horizontal direction when the head was free (but the observer still fixated the target). Obviously, the latter case, where head movements add to the image wander, is the relevant one for safety relevant exposure situations. For viewing times longer than about 30–60 s, additional to the above-mentioned fixational eye movements, behavioural eye movements determined by the visual task produce even larger eye movements.

3.4 The concept of exposure limits (MPE)

Exposure limits are generally defined in the field of health and safety to separate exposure levels which may be hazardous from those which are generally considered safe. In laser safety, the level of exposure of the eye or the skin that can be considered the theoretical border between safe and potentially harmful is called the 'maximum permissible exposure', or MPE. The MPE values (or short 'MPEs') are set by the International Commission on Non-Ionizing Radiation Protection, ICNIRP (ICNIRP uses the term exposure limit, EL, instead of MPE) [4], in close liason and harmonization with the laser safety standardization committees of IEC and ANSI, who publish the MPEs in the respective laser safety standards IEC 60825-1 and ANSI Z136.1. Therefore, the exposure limit values published by ICNIRP and in the IEC and ANSI laser safety standard are identical, with the exception of a few special cases that will be discussed in the respective section of this chapter. There are two additional international documents that list laser exposure limits, namely 'The use of lasers in the workplace—a practical guide' published by the International Labour Office [5] and 'Environmental Health Criteria 23—Lasers and Optical Radiation' published by the World Health Organization [6]. Both documents have been prepared jointly with ICNIRP and therefore also reproduce the ICNIRP exposure limits. However, the documents have not been updated since ICNIRP published revisions of laser exposure limits in 2000 and are therefore partially outdated (the WHO publication actually predates the first publication of ICNIRP exposure limit guidelines in Health Physics and can be seen as a predecessor that served as the scientific rationale for the development of the INCIRP guidelines. In the US, the American Conference of Governmental Industrial Hygienists also publishes exposure limits (referred to as Threshold Limit Values, TLV®). While ACGIH and ICNIRP do not have an official liaison, both sets of exposure limit values are based on the same pool of experimental data and therefore the exposure limits are also mostly identical. Some other countries such as Russia and Poland have previously published their own exposure limits but are now also adopting ICNIRP exposure limits. As a

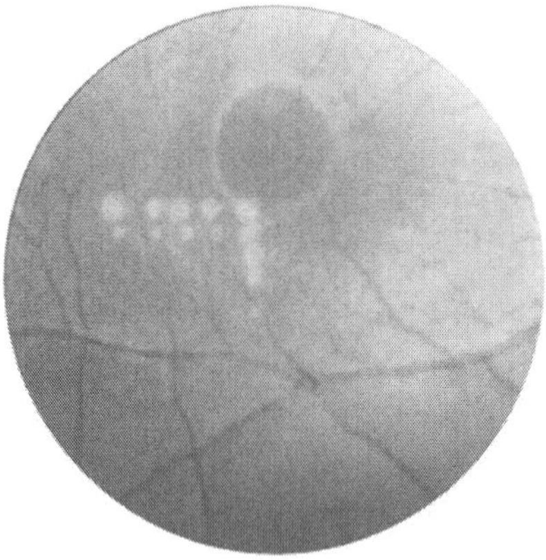

Figure 3.6. Fundus photograph of a monkey retina that was exposed to laser pulses with varying energy per pulse in a grid-like geometry. The outermost line of larger exposures on the left and bottom are marker lesions with higher power, the grid shows some sites which developed lesions and other sites where no lesions have developed. The wavelength, pulse duration and spot size on the retina was the same for all exposures, the only variation was in the energy per pulse. The exposure site was chosen to be at the rim of the macula (the larger round area) to cause the minimum of vision impairment to the animal. (Fundus image kindly made available by David J Lund, US Army Medical Research Detachment of the Walter Reed Army Institute of Research, Brooks City-Base TX.)

legal practice, setting of exposure limits is part of the sovereign right of a country, i.e. there is no obligation of a country to adopt exposure limits as defined by a commission such as the ICNIRP in their national legal framework. In practice, however, for laser safety and the safety of optical radiation, ICNIRP exposure limits are usually adopted on a national level. In this book we refer to the MPE values contained in IEC 60825-1 and note any differences to the values published by ICNIRP and ANSI where they exist.

For practical application, for instance in laser safety assessments or for calculating hazard zones, the MPEs are generally used as firm border between safe and potentially hazardous exposures, but one should be aware that the biological background of the MPEs is not such a firm and exact border between 'safe' and 'hazardous'. The MPEs are mostly derived from animal experiments and a limited number of exposures of human volunteers. In such threshold experiments, a number of exposures are delivered to the eye or the skin for a given laser wavelength, pulse duration and spot size at the target site (for instance the retina) while varying the energy per pulse or the cw laser power. A typical grid of exposures used to determine laser MPEs is shown in figure 3.6.

After each exposure, the exposed site is examined to see whether or not it develops a noticeable change or a detectable lesion (an injury). In simplified terms, the threshold for damage is that level of exposure where exposures above the threshold leads to an injury (or more generally to some defined effect), and exposure to below that value does not induce an injury. In laser safety, even only barely detectable lesions are considered as a positively detected effect, and therefore the lesions which are detected at the threshold exposure are called minimal visible lesion, MVL, or, if examination is performed with an ophthalmoscope for ocular exposures, they are called minimal ophthalmoscopically visible lesion, MOVL. Often other examination techniques are additionally used, such as fluorescence angiography or electron microscopy. As the exposed site is either damaged or potentially otherwise altered, one site can only be exposed once with a given pulse energy or power, and therefore a range of sites per animal and also a number of animals have to be exposed to determine the threshold. Generally it is found that there is not a sharply defined threshold exposure value below which no injuries occur and above which all exposures lead to damage. Rather, when a number of exposures are performed with a certain given exposure level, it is found that some of the exposures lead to lesions, while others do not result in a lesion. By dividing the number of exposures where a lesion was detected with the total number of exposures for the given exposure level, a percentage figure for the 'response' is calculated for each exposure level (the 'dose'). Such a response percentage number is determined for a range of discrete exposure levels (but always with the same wavelength, exposure duration and spot size) thereby constituting a dose–response curve. The spread of threshold values characterizes the spread of sensitivities for the different exposure sites and animals, as well as some level of uncertainty [7]. The percentage axis of the dose–response curve can be transformed into probability units of 'probits' after a statistical analysis developed for toxicology [8, 9] which makes a straight line out of the curve shown in figure 3.7, and therefore the dose–response plot is also often called 'probit plot'. The exposure dose at which 50% of the exposures lead to a lesion is called 'Effective Dose 50%' or ED-50 (see figure 3.7). The ED-50 is generally chosen and referenced as representative point of the dose-response curve and is also referred to as the 'threshold', even though there is a finite probability for damage at exposure energies somewhat below the ED-50.

The committees of ICNIRP in cooperation with ANSI and IEC laser safety committees examine these kind of data and set a human exposure limit (MPE) well below the ED-50 to ensure that there is negligible, i.e. practically zero probability for a detectable minimal lesion for exposures at the MPE (and obviously also for exposure levels that lie below the MPE) for all but extreme cases of exposure or sensitivity as discussed in the following paragraph. The factor between the ED-50 and the MPE for a given wavelength, exposure duration and spot size is usually referred to as safety factor, although it could be also seen as a scaling factor in terms of risk and probability for injury. The safety factor is often in the range of 10, however, depending on the wavelength and exposure duration it can also be much larger, as the MPEs often are simplified in terms of dependence on wavelength and pulse duration when compared to the variation of

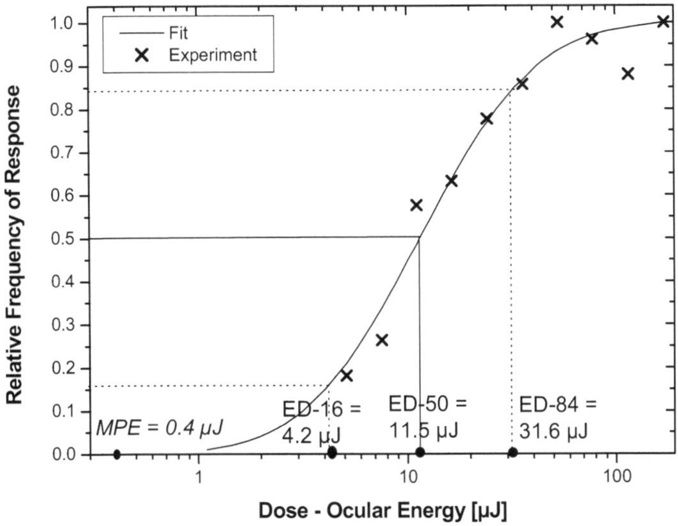

Figure 3.7. A typical dose–response curve as obtained in threshold experiments (here for a 850 nm laser, 180 ns pulse duration, minimal retinal spot size). For a given laser wavelength, pulse duration and retinal spot size the energy per pulse is varied, resulting in a distribution of percentages of exposures which lead to a lesion (in this example a total of 191 exposures in four animals). The ED-50 is the energy where on the average every second exposure (50%) leads to an injury. The MPE is set well below the ED-50 to insure that there is practically zero risk when exposed at the MPE level (in order to compare the ED-50 in terms of energy, the MPE as marked in the plot was multiplied with the area of a 7 mm diameter averaging aperture, see also section 3.6.6).

the ED-50. In some cases, where there is less variability and small experimental uncertainty, such as for corneal photochemical injury in the UV range, the safety factor can be less than 10.

The MPEs are defined such that exposure to levels below the MPE is considered low enough so that adverse health effects are prevented. Only for rare cases of hypersensitivity, for instance after contact or ingestion of photosensitizing agents, or for some kind of diseases which are linked with hypersensitivity to light, can exposure to levels below the MPE lead to adverse effects on the eye and the skin such as erythema (skin reddening). For an individual who is hypersensitive to ultraviolet and visible radiation, exposure to natural optical radiation is the main concern. Therefore, this kind of hypersensitivity is generally not an issue for laser radiation where exposure is very limited in terms of numbers of individuals exposed and duration of exposure as compared to natural optical radiation. Where the lens of the eye has been removed (in cataract surgery) and not replaced by an artificial lens, the transmittance of ultraviolet radiation to the retina increases such that for these so called 'aphakic' eyes, the retina can be photochemically damaged by ultraviolet radiation. However, nowadays, aphakic eyes are extremely rare as artificial lenses that absorb UV radiation are used as replacements for the natural lens. As the definition of current laser MPEs took account of eye movements, especially for exposure durations longer than a few seconds, as well as of pupil constriction for long-term exposure to bright light, the laser MPEs might not be sufficiently low for conditions where these normal mechanisms are suppressed. Examples are found in the medical and research field where during therapy or diagnosis eye movements can be reduced (for instance in anaesthesia) or pupils can be artificially dilated. Another extreme case where the MPEs do not apply is when a visible laser beam is focused onto one spot on the cornea for some time. In this case, the exposure of the retina might be below the MPE but the cornea could still be heated and damaged. These kinds of exposures need to be evaluated according to special exposure limits, and ANSI Z136.1 contains guidelines for exposure limits that may be used in these cases.

3.4.1 Exposures above the MPE

Following the discussion on the derivation of the MPEs it can be appreciated that an exposure to levels only slightly above the MPE usually does result in an injury, but the probability for an injury increases with exposure levels that are above the MPE. This gradual transition between 'MPE just about exceeded' and 'exposure leads to certain injury' is the background of the definition and understanding of Class 3R lasers where the MPEs for the eye can be exceeded but the likelihood for injury for the case of non-intentional, accidental (short time) exposure is small, so that these types of lasers can be used quite safely by trained personnel without the need for eye protection. However, even if (as a consequence of the safety factor between actual threshold for injury and MPE) exposure somewhat above the MPE might not lead to injury, the MPE values form the basis for laser hazard analysis that sets out to judge if an exposure can be considered safe or it is to be treated as potentially leading to an injury. The committees that deal with the definition of MPEs base their work mostly on the review of dose response curves and, by taking account of the uncertainty associated with the data, set the MPEs at levels of negligible risk. At the same time, it is not possible to judge the probability that an injury actually occurs nor to judge the severity of the injury for a given exposure level above the MPE. The amount by which the MPE is exceeded cannot therefore be used for probabilistic risk analysis where the probability for injury and the severity of the injury needs to be quantified. While it is generally correct that for a given wavelength, exposure duration and spot size, the injury will be more serious for exposure levels well above the threshold (also referred to as 'suprathreshold') than exposures which are just sufficient to lead to an injury (referred to as 'at threshold'), it is not possible to generally scale the level of severity as a function of by what factor the MPE is exceeded—both because the scaling factor between the threshold and MPE is not defined and known and because the severity depends on a great number of factors. Arguments for acceptable exposure levels to laser radiation above the MPE for the eye as a result of risk analysis are therefore strongly discouraged by the authors. However, for exposure of the skin above the skin MPEs, in practice, a risk analysis in terms of probability of exposure and severity of harm is adopted for some applications as a reasoning for less critical control measures of potential skin injuries. For instance, for surgical applications where the laser is guided by hand, traditionally, training of the surgeon is considered as appropriate control measure and the remaining risk for skin injury acceptable (and for instance comparable to cutting oneself with the scalpel) where otherwise skin protection (heavy gloves) would have to be worn which is obviously not practical. Formal or probabilistic risk analysis in the sense of a negligible probability that ocular exposure with optical viewing instruments occurs might be applicable for cases where the MPE is not exceeded for the naked eye but is exceeded when exposure with optical viewing instruments occurs. With the exception of Class 3R lasers, where the IEC laser safety committee judged brief non-intentional and accidental exposures as not very likely to produce an injury, exposure levels of the eye above the MPE should be treated as an unacceptable hazard, both in terms of the probability that actual injury occurs following an exposure as well as in terms of the severity of the injury. This also obviates any difficulty in trying to assign some category of probability (likely, probable, etc) or some level of severity to a given exposure level above the MPE.

At the same time as we caution against treating exposure levels above the MPE as acceptable in the framework of some risk analysis of a given laser application, we would like to emphasize that any level of hazard analysis and training should include a corresponding level of discussion and understanding of the nature of the potential injuries which may occur if the MPEs are exceeded, and this book also attempts to provide a corresponding overview.

3.5 Laser–tissue interaction

The MPEs for the skin and particularly for the eye depend in a rather complicated way on wavelength, exposure duration and, if the retina is at risk, on the size of the irradiated retinal spot. For an understanding of these dependencies as well as of the nature and severity of injuries from laser exposure above the MPEs it is beneficiary to review basics of how laser radiation interacts with tissue.

Generally, laser radiation needs to be absorbed in tissue to have any effect. The location and level of absorption of laser radiation in tissue, especially in ocular tissue, is strongly wavelength dependent, therefore different parts of the skin and the eye are at risk from exposure to radiation at different wavelength. Once the radiation is absorbed, then the type and nature of interaction of optical radiation with tissue can be distinguished to be thermal, photochemical or thermomechanical. For a given level of radiant exposure it depends mainly on the wavelength and on the interaction duration which type of interaction dominates.

The discussion in this section on laser–tissue interaction is kept at a rather basic level and the interested reader is referred to the textbooks listed at the end of the chapter for further details.

3.5.1 General optical absorption characteristics

There is a general principle of photobiology that radiation needs to be absorbed to have an effect. (The terms reflectance, transmittance, absorption coefficient and scattering were introduced in section 2.7.) When discussing absorption, the *absorption depth* (which can also be referred to as the *penetration depth*) is a useful parameter. The concept of absorption depth does not mean, however, that the radiation penetrates without loss as far as the respective depth and is then abruptly absorbed with no radiation transmitted beyond this point. It is rather that the level of radiation exponentially decreases with increasing depth. (At five times the penetration depth, when it is usually defined as the depth at which the level of radiation has fallen to a fraction of 1/e of the surface level, there is 0.7% of the level at $x = 0$ remaining.) Theoretically, there is no finite depth beyond which no further penetration of radiation into the tissue occurs. A dermatologist from Germany, Professor H Meffert, amuses his students with the thought that 'it is not dark in the stomach'. However, in practice, a certain level of exposure is necessary to induce a discernible therapeutic or hazardous effect and this limits the depth of affected tissue.

Absorption in tissue is dominated by water for wavelengths in the red and infrared, while in the visible the absorption is mainly determined by haemoglobin if the tissue is supplied with blood or by pigments, mainly melanin. In the ultraviolet the absorption is determined mainly by the absorption of proteins including, if present, melanin. The wavelength dependence of the absorption coefficient for blood vessels and for water is shown in figure 3.8.

As human tissue is generally not clear and homogenous, radiation with wavelengths which are not strongly absorbed, i.e. which penetrate into tissue, experiences strong scattering in the tissue. The cornea and the lens are, of course, an exception in that they are clear and highly transparent in the visible.

3.5.2 Types of interaction

The interaction effects of laser radiation with matter can be grouped into *photochemical*, *thermal* and *thermomechanical* processes, where the latter is often further subdivided into photoablation and photodisruption.

When optical radiation is absorbed by matter (which might be a gas, a liquid or a solid), the basic effect is that those atoms and molecules that absorb the radiation become excited. This excitation energy subsequently either dissipates as thermal energy and heats up the matter, or it can lead to chemical reactions

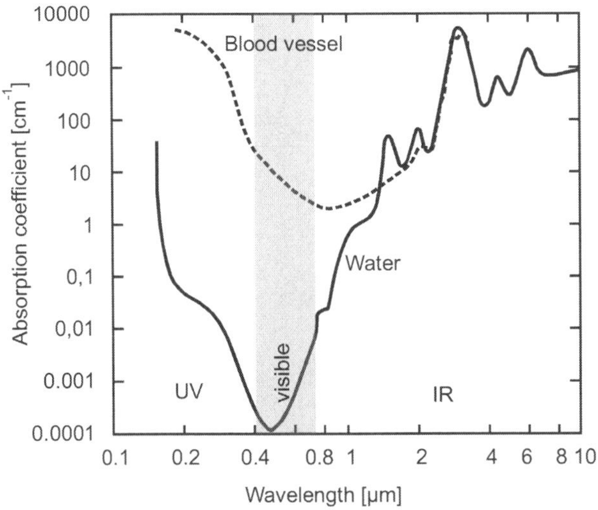

Figure 3.8. Wavelength dependence of the absorption of optical radiation in human tissue. In the UV, absorption is mainly governed by proteins, in the blue and green, the absorption is governed by haemoglobin (or if present by pigments) and for wavelengths in the red and infrared, by water (adopted from Berlien [10]).

such as ionization of excited atoms or bond breakage in excited molecules. These two types of interaction also apply to interaction of laser radiation with the tissues of the eye and the skin which may subsequently lead to injury. When the temperature of the tissue is increased above a critical temperature, proteins are denaturated and *thermally* induced damage occurs. If temperatures above 100 °C are induced, water in the tissue begins to boil and further temperature increases lead to a carbonization of the tissue. In the ultraviolet and blue end of the visible spectrum, *photochemical* damage can occur at irradiances that are not sufficient to cause damaging thermal effects. At these shorter wavelengths photon energies are sufficiently high to create toxic radicals in the cells or even to cause direct damage to macromolecules of cells such as to the DNA. Both thermal and photochemical injuries have distinctive dependencies on wavelength, exposure duration and retinal spot size that are discussed below. In terms of causing an injury, thermal and photochemical interaction can be considered as competing mechanisms.

Thermomechanical effects occur when the tissue is heated very rapidly, locally inducing rapid thermal expansion which lead to mechanical shockwaves. For very short pulses, a plasma may be formed that has very high temperatures and that also leads to secondary damaging effects.

Besides the influence of the wavelength, it depends mainly on the exposure or interaction duration and the (peak) irradiance which type of interaction

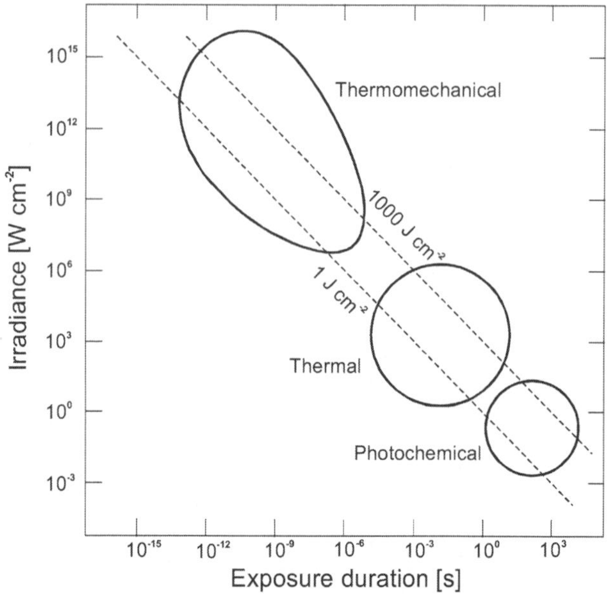

Figure 3.9. The type of interaction of laser radiation with tissue will be determined by the interaction duration on the one hand and the irradiance on the other. For instance, photochemical interaction mechanisms are dominant for low irradiance, long-term exposure, while thermomechanical effects only occur for short pulse, high irradiance exposure. Modified from Niemz [11].

applies or is the dominating one. While both parameters vary over about 16 orders of magnitude, the corresponding radiant exposure values, calculated by multiplication of the pulse duration with the peak irradiance, vary relatively little and lie in the range 1–1000 J cm^{-2} (as indicated by the diagonals in figure 3.9). The approximate areas of exposure (pulse) duration and irradiance for the different interaction mechanisms are plotted in figure 3.9.

3.5.2.1 Photochemical injury

Photochemical interaction encompasses all effects of light on matter where the light directly induces chemical changes within the matter. In nature, photochemical effects are ubiquitous, as photosynthesis as well as the vision process and tanning of the skin are photochemical processes. In medicine, psoriasis and billirubin deficiency in new-born infants is treated with light exposure, and light is used in photodynamic therapy for retinal degeneration and some forms of cancer. The chemical changes induced by light in tissue may not all be of a positive nature, however. The high-energy photons of ultraviolet radiation can cause sunburn in the skin, and in the extreme even skin cancer can develop as a consequence of extensive exposure. The eye can also be injured by photochemical effects: the cornea and lens from radiation in the ultraviolet and the retina from visible radiation.

There are a number of dependencies and properties that are typical of photochemical interactions. These are summarized in the following from the viewpoint of photochemical damage of the skin and the eye.

It is typical for photochemically induced injury that it depends on the number of absorbed photons (per unit area on the surface or per unit volume, equivalent to radiant exposure or energy density, respectively) whether injury occurs, but it does not (within a certain timeframe) depend on the time taken to deliver these photons. A short duration exposure with a high irradiance has the same effect as a long-term exposure with a correspondingly smaller irradiance. Photochemical damage has a cumulative, additive nature. Also individual exposure doses with some interval of non-exposure in-between add up over a period of several hours to produce the same effect as one continuous exposure having the same total radiant exposure. This dependence of the effect on the dose is referred to in photobiology as the 'Bunsen–Roscoe' law of reciprocity.

As a consequence of this additive nature, the exposure limit to prevent photochemical damage is expressed as a constant radiant exposure value that does not depend on the exposure duration. For instance in the far ultraviolet range, the same radiant exposure value is valid for exposure durations from 1 ns to 8 h.

It might be puzzling that a certain number of photons is necessary for injury to occur, while a single photon is capable of bond breakage or chemical change. While these single photon interactions may create oxidative substances or may even lead to cell death, a greater number of cell deaths is needed to constitute an injury. When this threshold is not reached, damage to the tissue is prevented by repair mechanisms, which also counteract the additive effect of photochemical processes. The additivity, i.e. the reciprocity law, is therefore only valid for exposure durations shorter than the time needed for repair mechanisms to become effective. This time frame is several hours, for damage to the skin or the cornea of the eye somewhat longer than for photochemical damage to the retina. It is also typical for photochemical damage that it needs some time to develop subsequent to the exposure that exceeded the threshold for damage. That is, the injury does not develop immediately after the threshold is exceeded, but only some time later, the delay being typically several hours.

Additionally to this typical dose additivity of photochemical injury, photochemical injury has a very strong wavelength dependence. Since the photon energy increases with decreasing wavelength, smaller wavelengths are typically more effective in producing a given effect than longer wavelengths. In other words, the tissue is generally more sensitive to exposure with shorter wavelengths than with longer wavelengths. This varying effectiveness or sensitivity for different wavelengths is expressed by an action spectrum, a concept which is used regularly for broadband sources and is discussed in detail in section 2.6.4. For laser sources, the strongly varying sensitivity is accounted for by the wavelength dependence of the exposure limits. Towards longer wavelengths, the decrease of sensitivity is exponential, limiting the range of photochemical damage to an upper wavelength range of about 400 nm for the cornea, lens and skin, and to about 550 nm for the retina. Towards lower wavelengths, the photochemical damage is limited in the skin by the absorption of radiation within the outermost layer of dead cells and for the retina due to the increasing absorption of the radiation by the lens and the cornea for wavelengths below about 400 nm.

The third property of photochemical damage that distinguishes it from thermal damage is the lack of dependence of the effect on the size of the exposed area or the irradiation spot size. Thermal interaction depends on the size of the exposed area or spot size as larger spots lack the radial thermal conduction component and for a given irradiance, larger spot sizes result in a higher tissue temperature than smaller spot sizes. Whether injury is induced photochemically depends on the radiant exposure only and does not depend on the extent of the exposed area of sensitive tissue (while the severity of injury is certainly affected by the area of the damaged tissue).

In those wavelength ranges where radiation can induce photochemical damage, it is in 'competition' with thermally induced damage. The process that produces an injury with the lower radiant exposure level for a given wavelength, exposure duration and spot size is the critical effect. Photochemical damage mainly occurs for exposure to continuous wave lasers or repeated exposures to irradiance levels that are too small to cause a temperature rise sufficient to cause thermal damage. Thermal effects dominate for short time exposures, especially for single pulses or bursts of pulses. In order for there to be sufficient energy to cause a photochemical effect for short exposure durations, the irradiance would need to be so high that thermal damage would occur before the dose for photochemical damage is reached.

We can summarize the properties of photochemical effects as given below.

- Exposure limit is given as radiant exposure or time integrated radiance.
- Individual exposures have to be added up within several hour's time frame.
- Strong wavelength dependence expressed by action spectrum (lower wavelengths being typically more effective).
- No irradiated spot size dependence.
- Injury takes some time to develop (in the order of an hour).
- Mostly relevant for longer exposure durations (about 10 s and longer).

3.5.2.2 Thermal injury

For a given irradiance level at the surface of the tissue, the level of hazard due to thermal interaction with the skin or the eye will depend mainly on the optical absorptivity of the radiation for the wavelength under consideration.

The absorptivity (section 2.7) plays a major role in the level of temperature increase: when the absorptivity is low, the absorption depth is large and the radiation incident on the surface is distributed into the depth of the tissue and therefore over a larger volume. This larger volume will not reach such a high temperature as would occur if the same quantity of radiation were absorbed in a smaller volume, i.e. where the power density (W m^{-3}) would be higher. For a given irradiance at the surface, the energy density relevant for the temperature increase is to a first approximation given by dividing the irradiance (W m^{-2}) by the absorption depth (m) which is equivalent to a multiplication by the absorption coefficient (m^{-1}). Consequently, for a given irradiance level, the magnitude of the temperature increase and hence the potential to induce thermal injury will depend directly on the absorption depth and inversely on the absorption coefficient. Since the absorption coefficient of the skin and of the components of the eye is strongly wavelength dependent, so will the hazard potential of a given irradiance depend strongly on the wavelength. For instance, the wavelength dependence of the injury thresholds and the exposure limits for the cornea from infrared radiation closely follow the wavelength dependence of the absorption coefficient of water for that wavelength range, as is discussed in section 3.13.

Whether thermal damage occurs or how severe the injury is doesn't only depend on the level of temperature elevation of the tissue but also on the duration, i.e. on the temperature 'history'. Heating of tissue can be roughly grouped into three temporal regimes: for exposure durations longer than a few seconds, thermal conduction into surrounding tissue and cooling by the blood establishes a constant temperature profile that no longer depends on exposure duration and is merely proportional to the incident irradiance. As the critical temperature for thermal damage is lower when the temperature is elevated for a longer period of time, the injury threshold decreases somewhat with exposure duration. However, thermal exposure limits are nonetheless defined as constant beyond some duration (such as 10 s for the skin) as the safety factor between actual injury and MPE is somewhat decreased in respect to the safety factor for short exposures, but it is still sufficient even for prolonged exposure durations, especially when one considers body movements. Consequently, in this time regime exposure limits are given as constant irradiance values with no time dependence. For the lower end of the time scale, when the pulse duration is shorter than the characteristic thermal diffusion time, then during the pulse the heat diffusion out of the heated volume is negligible, i.e. the heat is confined to the heated volume during the pulse duration and this condition is therefore referred to as 'thermal confinement'. For the thermal confinement condition, the maximum temperature reached in the tissue does not depend on the pulse duration, but only on the amount of energy (per unit area for surface absorption or per unit volume for deeper penetration) delivered into the tissue. In this regime, therefore, exposure limits are defined in terms of constant radiant exposure values with no time dependence. The thermal diffusion time t_{conf} depends on both the thermal diffusivity D_{therm} (in units of $m^2\ s^{-1}$) and the square of the penetration depth δ and can be approximated by [12]:

$$t_{\text{conf}} = \frac{\delta^2}{4D_{\text{therm}}}. \tag{3.1}$$

For a given tissue with a certain thermal diffusivity (often close to the value for water at 37 °C of 1.5×10^{-7} m^2 s^{-1}), the maximum pulse duration for which thermal confinement applies therefore depends strongly on the absorption depth which in turn depends on the wavelength. For deep penetration depths a large volume is heated and the diffusion of heat from the whole volume into the surrounding non-heated volume is correspondingly slower than for a very shallow absorption depth where the heat can be drained much faster and the temperature decrease within the heated volume occurs much more rapidly too. This dependence of thermal confinement on absorption depth is also well reflected in the MPEs: for large absorption depths in the near infrared region of 1500–1800 nm, the thermal confinement condition extends to many seconds and the MPE is given as constant radiant exposure value for an exposure duration up to ten seconds.

Between the thermal confinement condition for short pulses and the long-term steady-state conditions for very long exposure durations, there is a regime where the MPEs are given as a function of exposure duration. This dependence varies as $t^{0.25}$ ($t^{1/4}$) for corneal and skin damage and $t^{0.75}$ ($t^{3/4}$) for retinal damage. When these radiant exposure values are transformed into irradiance values by dividing by the exposure (pulse) duration, the dependence on pulse duration becomes $t^{-0.75}$ ($t^{-3/4}$) and $t^{-0.25}$ ($t^{-1/4}$), respectively. This dependence on the inverse of the pulse (exposure) duration results in higher allowed peak irradiance levels for shorter exposure durations. If heating and temperature elevation occurs only for short periods, the tissue can withstand higher temperatures without being damaged. The longer the temperature elevation lasts, the lower the maximum allowed temperature is to prevent permanent injury. This can also be experienced in everyday life: when one touches a hot surface such as an iron or a hot plate in the kitchen, if the exposure lasts only for a fraction of a second, relatively high surface temperatures (comparable to high irradiance levels) do not lead to a burn. If the contact to the surface lasts longer, then the temperature of the surface has to be correspondingly lower to prevent skin burns.

3.5.2.3 Thermomechanical injury

When tissue is exposed to pulses with durations in the nanosecond regime, the tissue is heated up very rapidly, causing thermal expansion of the tissue and subsequent stress waves which propagate through the tissue. In terms of eye safety, acoustic transients and especially shockwaves which result in mechanical damage of tissue are of major concern, as in the retina they lead to the rupture of tissue and possibly of blood vessels (haemorrhage). On the surface of a tissue such as the cornea for UV or far-IR wavelengths, thermomechanical effects lead to ablation of material, i.e. the heated material is rapidly removed from the bulk of the material which may otherwise be hardly affected by the process. Pulses having durations of less than one nanosecond are referred to as ultrashort pulses. In this time regime, peak irradiance values and corresponding electric field strengths are so high that they can lead to optical breakdown of the material. In this process, material is ionized and a plasma forms, inducing acoustic transients or even shockwave generation and ablation. This process is generally referred to as photodisruption, being regularly used in medicine for shattering stones in the urinary tract or for removing fibrous tissue growth which frequently forms in the eye following cataract surgery.

3.6 MPE evaluation and measurement concept

The injury thresholds for the eye and the skin differ, especially in the retinal hazard wavelength region and consequently there is one set of MPEs for ocular exposure and another for the skin. The MPE values for the eye and the skin are specified in units of J m^{-2} or W m^{-2} and depend on the exposure duration, the laser wavelength (which can have a widely varying 'effectiveness' in causing injury) and for the eye may also depend on the image size on the retina, especially for the thermal retinal hazard. This image size is expressed in terms of the angular subtense of the apparent source and is further discussed in section 3.12.1. For longer exposure durations the maximum safe exposure level is generally lower than for shorter exposure durations, but there are also wavelength and exposure-duration regions where the MPE does not depend on the duration of exposure.

3.6.1 Limiting aperture and angle of acceptance

The MPEs for the eye are specified at the cornea i.e. the focusing properties of the eye and the pupil size are accounted for in the derivation of the MPEs in the retinal hazard region. For a discussion on the temporal relationship between irradiance and radiant exposure we refer to chapter 2.1, here, in the discussion on MPEs we often treat the quantities of irradiance and radiant exposure as equivalent (both for the MPE value and the exposure level that is compared with the MPE) with the understanding that they are correctly transformed via the dimension of time. In order to evaluate the potential hazard of exposure to optical radiation, the level of exposure at a given location and for a given exposure scenario (which includes the exposure duration) need to be assessed, and this value is then compared to the MPE for the given exposure duration, wavelength and, if applicable, the angular subtense of the apparent source. It is important to note that the level of exposure, i.e. the irradiance or radiant exposure needs to be averaged over a limiting aperture having a diameter as specified in the laser safety standards (see table 3.1). The radiometric concept of averaging the irradiance or the radiant exposure over the limiting aperture is discussed in section 2.2.2. Due to its importance it is stressed again that the irradiance level that is to be compared to the MPE is to be averaged

Table 3.1. Limiting apertures to be used for averaging the irradiance or radiant exposure for comparison with the MPEs for the eye or the skin.

	Limiting aperture diameter for:	
Spectral region	Eye	Skin
180–400 nm	1 mm	3.5 mm
≥400–1400 nm	7 mm	3.5 mm
≥1.4–100 μm	1 mm for $t \leq 0.35$ s $1.5t^{3/8}$ mm for $0.35 \text{ s} < t < 10$ s 3.5 mm for $t \geq 10$ s	3.5 mm
≥0.1–1 mm	11 mm	11 mm

over the area of the limiting aperture in all cases, i.e. even when the laser beam itself is much smaller than the limiting aperture and the physical (non-averaged) irradiance within the beam is much higher than the averaged value. It is the (lower) averaged value (that could be considered a biophysically *effective* value) that is to be compared to the MPE, not the actual physical irradiance within the beam. As will be discussed in section 3.6.6, the radiometric concept of averaging the irradiance over a limiting aperture with a specified diameter is equivalent to determining the power or energy that passes through the respective aperture. This can be a simpler interpretation, especially for comparing the exposure level to the MPE values that relate to damage of the retina, as the power that passes through an aperture of 7 mm diameter is equivalent to the power that would pass through the dilated pupil of the eye.

The size of the averaging aperture is linked to biophysical factors such as the optical properties of the eye, body movements, scattering and heat conduction. For the retinal hazard region, the diameter of 7 mm corresponds to the maximum diameter of a dark-adapted pupil. This constitutes a worst-case assumption, as in a bright environment the usual pupil size is rather 2–3 mm in diameter. However, an adjustment of the laser MPEs for smaller pupil sizes in bright environments is not permitted (the smaller pupil size is already figured into the MPE values for photochemical retinal damage). For exposure to IR radiation of the skin and eye lasting for several seconds, the involuntary movement of the eyes and the body as well as heat conduction will average an irradiance profile over an area of at least several square millimetres, even if the irradiated body part is kept intentionally still. Near-infrared radiation penetrates relatively deeply into skin and due to scattering, the irradiance profile is averaged over corresponding dimensions. For wavelengths larger than 0.1 mm, an aperture size of 11 mm is specified, as smaller apertures would lead to inaccurate measurements because of diffraction effects.

While the limiting aperture relates to averaging of the irradiance in order to determine the biophysically effective irradiance level at the cornea, the second measurement parameter that can affect the determination of the exposure level for a given exposure location, the acceptance angles relate to limiting the power that is received from the source. The concept of the acceptance angle is discussed in section 2.2.2, where two methods were introduced that can be used to define the angle of acceptance: placing the field aperture at the source (which needs to be accessible) and imaging the (apparent) source with a lens onto the field aperture. It is noted here that the acceptance angle can limit the part of the (apparent) source that is seen by the detector, i.e. only the proportion of the power that is emitted from within the acceptance angle (and that arrives at the detector) is measured. Thus, the acceptance angle blocks or masks out part of the source if the source is larger as the acceptance angle. For a source that is smaller than the maximum acceptance angle, the measured power (or irradiance when averaged over the aperture area) is not affected, as shown for example in figure 3.10(*a*). However, if the source (that might also be a multiple source or an array) is larger than the acceptance angle, as shown in figure 3.10(*b*) then the power (or irradiance) that is measured through the limiting aperture is decreased when compared to the value measured with an angle of acceptance that encompasses the whole source. If the source is larger than the maximum acceptance angle, the use of an open acceptance angle would result in overestimating the hazard. In figure 3.10, in terms of irradiance, for comparison with the MPE, the effective irradiance at the location of the limiting aperture is that proportion of the total irradiance that is due to the single emitter that is within the angle of acceptance, and not the actual irradiance that arises from all three emitters. Thus, the angle of acceptance can be understood as excluding from consideration any radiation that arises from outside the angle of acceptance. When a lens is used to define the angle of acceptance it is important to note that the apparent source needs to be imaged onto the field stop.

The laser safety standards specify the following values for the maximum acceptance angle.

For measurement of the exposure level for comparison with the *retinal thermal limits*, the angle of acceptance should be limited to a maximum of 100 mrad, i.e.

$$\gamma_{\mathrm{th}} = 100 \text{ mrad.}$$

For measurement of the exposure level for comparison with the *retinal photochemical limits*, the angle of acceptance should be set to a value that depends on the exposure duration t:

$$\text{for } t < 100 \text{ s} \qquad \gamma_{\mathrm{ph}} = 11 \text{ mrad}$$

$$\text{for } t > 100 \text{ s but } t < 10\,000 \text{ s} \qquad \gamma_{\mathrm{ph}} = 1.1\sqrt{t}$$

$$\text{where } t \text{ is in seconds and } \gamma_{\mathrm{ph}} \text{ is in mrad}$$

$$\text{for } t > 10\,000 \text{ s} \qquad \gamma_{\mathrm{ph}} = 110 \text{ mrad.}$$

The biophysical background to these specifications of the angle of acceptance for the thermal and photochemical limit evaluation is discussed in

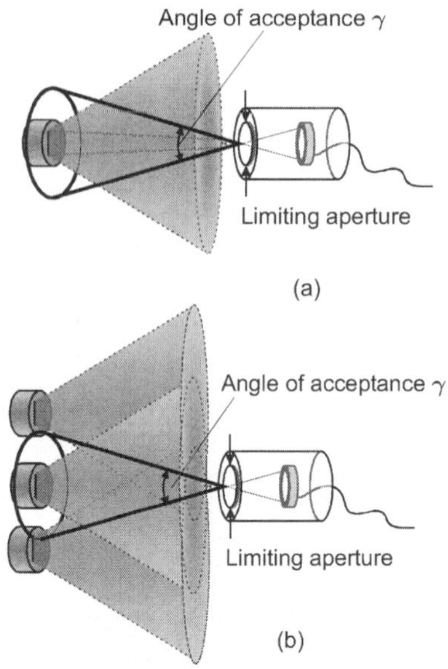

Figure 3.10. When the angular subtense of the apparent source is smaller than the angle of acceptance of the detector, the power passing through the limiting aperture (or the irradiance at the limiting aperture) does not depend on the size of the angle of acceptance (*a*). In (*b*) part of the source is outside of the angle of acceptance (i.e. the full source is larger than the angle of acceptance) and so the angle of acceptance blocks radiation from the other two emitters.

section 3.12. Just as the averaging process of the irradiance over the limiting aperture for beam diameters smaller than the limiting aperture resulted in a biophysical effective value that was smaller than the actual physical irradiance, so does the use of a maximum acceptance angle result in a smaller exposure level when the angular subtense of the source is larger than the maximum acceptance angle, and thus constitutes a biophysically effective value.

For determination of the exposure level to be compared to limits other than the retinal thermal and retinal photochemical limits, the acceptance angle should not be limited, i.e. an open acceptance angle (field-of-view), i.e. that is at least as large as the source, should be used.

3.6.2 Exposure location and exposure duration

The basic concept of an MPE evaluation is to answer the question 'is a certain exposure to laser radiation safe (to the eye or the skin)?' The word *certain* indicates here that an MPE analysis is performed for a specific scenario of exposure in terms of parameters of the laser radiation (wavelength, pulse pattern, beam power, etc) but also in terms of exposure location (at a certain distance from the laser, where the laser beam has a certain diameter to produce a given (averaged) irradiance at the location under evaluation). Another parameter that comes into play is the exposure duration, as the MPE values mostly depend on exposure duration. In other words the question 'is the exposure safe?' is answered for a certain choice of exposure duration—an exposure level that is safe for a short exposure duration might not be safe for a longer exposure duration (see sections 3.7.1 and 3.9.4 for typical exposure durations that are used for skin-MPE or eye-MPE evaluations, respectively). Consequently, for the safety evaluation of a specific scenario where exposure to laser radiation may occur, the location of the (potential) exposure with regard to the laser source, and the parameter 'exposure duration', both need to be specified. Since a safety evaluation is often carried out for scenarios where exposure to laser radiation is not intentional but rather accidental, and since the exposure location and exposure duration depend on human behaviour, these parameters can rarely be determined in an exact way but need to be estimated. Such an estimation can be done for varying degrees of risk: one could assume worst case exposure scenarios such as exposure at the location in the beam that has the highest hazard level (as further discussed in section 3.12.2) and prolonged intentional exposure or one could base the evaluation on typical exposure scenarios in terms of location and exposure duration (the topic of risk analysis is discussed further in chapter 7). The choice of exposure duration for a safety analysis should be based on the maximum *anticipated* exposure duration. While there are typical (maximum) anticipated exposure duration values recommended in the laser safety standards (and are also discussed in section 3.7.1 for the skin and in section 3.9.4 for the eye), these values are not mandatory and it might well be justified to use a shorter exposure duration (for instance if the issue is that a fibre from a Ho:YAG laser could break and move through the room, causing an exposure duration of less than 1 s rather than 10 s). Also regarding the location of exposure, the choice should be made following an analysis of the installation and use, and following the chosen level of risk: the choice of location for the safety analysis is often the closest location that is accessible for a given installation, for instance at some viewing window, or at some distance off the floor (such as 2 m) for a laser mounted on the ceiling. For the case of exposure to UV radiation, for instance from stray light emitted by an Excimer laser set-up, where the exposure is additive over the whole day, rather than assuming a simplified worst-case exposure at close distance over 8 hours, it might be more realistic to assign varying exposure durations to the different potential exposure locations, such as x seconds at position a(with irradiance level E_a), y seconds at position b (with irradiance level E_b), etc, and then calculate the total radiant exposure with $x \cdot E_a + y \cdot E_b$.

The three basic steps of an MPE analysis can be summarized as follows:

- Determine MPE.
  - Determine wavelength(s) of radiation.

  - Determine maximum anticipated exposure duration (either use typical recommended values or chose values according to scenario).
  - If relevant, determine angular subtense of the source (for retinal thermal limits) at the position of exposure (i.e. at the position where the level of exposure, see below, is determined) or for far-IR exposure of the skin, the area of the exposed skin.
  - If relevant, determine pulse pattern (pulse duration and repetition rate).
  - Calculate MPE value.
- Determine level of exposure.
  - Measure or calculate the level of radiant exposure or irradiance (the two quantities are related via the quantity of time, i.e. via pulse duration or exposure duration) at the position of evaluation. Exposure levels have to be averaged over the limiting aperture and for extended sources, the measurement FOV might be relevant.
- Compare level of exposure with MPE.

This procedure of determining the MPE, assessing the exposure level and comparing this with the MPE applies for exposure to cw radiation and to exposure to single pulses. In the case of exposure to multiple pulses, more than one criterion applies and the procedure has to be carried out for each criterion (such that each single pulse as well as any combination of pulses is below the respective MPE), as discussed in further detail in the following sections of this chapter. Similarly, for wavelengths in the retinal hazard region, the above straightforward procedure applies only to single-element sources. For non-homogeneous sources or sources that consist of multiple elements, the procedure of comparing the exposure level that stems from a certain source with the corresponding MPE might have to be done for each element as well as for different combinations of elements, as further discussed in section 3.12.5.6.

Since for this basic MPE analysis the level of exposure at a given location is compared to the respective MPE value, due to the dependence of the exposure level (and in some cases also of the MPE values) on the distance to the source, the analysis could be extended to answer the question ‘beyond which distance is the exposure to a given laser beam safe?’ The answer to this question is the so-called NOHD, the nominal ocular hazard distance: within the NOHD, the MPE is exceeded, outside of the NOHD the exposure level is below the MPE. It follows that to determine the NOHD one needs to measure (move the radiometer within the beam) or calculate the location in the beam where the level of exposure is equal to the MPE. As the MPE may depend on exposure duration, different choices of maximum anticipated exposure duration might lead to different NOHD values. An additional issue that might come into play in an MPE analysis is the potential use of optical instruments: in that respect above questions could be either asked for the naked, unaided eye (‘is the given exposure safe for the naked eye?’) but also for potential use of eye loupes or binoculars (‘is the given exposure safe when exposure occurs with eye loupes or with telescopes?’) as optical instruments can result in an increase of the level of hazard, i.e. can increase the exposure level above the MPE where the exposure of the naked eye is below the MPE. These issues are discussed in more detailed in chapter 5.

3.6.3 Representation of MPE values

The MPEs for the skin and the eye are presented in tabular form in IEC 60825-1 where a row represents a certain wavelength range and the columns represent ranges of exposure duration. The overall wavelength range extends from 180 nm to 1 mm (below 180 nm absorption in air becomes significant, but if there is concern of exposure to, for instance, 157 nm radiation from a F_2 laser exists, then the exposure limits specified at 180 nm can be used). The range of exposure duration extends from 10^{-13} s (100 fs) to 3×10^4 s (i.e. 30 000 s, about 8 h). In the laser exposure limit guidelines published by ICNIRP and in the ANSI laser safety standard, the MPEs are given in a list rather than a two dimensional table, but the values are equivalent.

In IEC 60825-1, the definition for the wavelength range is such that 'λ_1 to λ_2' means $\lambda_1 \leq \lambda < \lambda_2$, e.g. in the table, the wavelength range 400–700 nm includes the exact wavelength of 400 nm. As laser wavelengths will almost never be at these precise border values, this is more an issue of satisfying the mathematically keen mind rather than a practical one. For a given wavelength and exposure duration, the cells in the table contain the different MPE values, as shown in an example in figure 3.11.

To simplify the presentation of the MPE values, several dependencies of the MPE values on wavelength, exposure duration or on the angular subtense of the apparent source are represented by correction factors C_1 to C_7 (in the ANSI standard, the correction factors have subscript letters so that for instance the factor C_5 of IEC 60825-1 is referred to as C_P (P for pulses) the ANSI document—see the appendix for a table of all correction factors and both IEC and ANSI denominations). These correction factors are dimensionless, and it is important to note that where they are a function of the wavelength λ (such as $C_4 = 10^{0.002(\lambda-700)}$) the wavelength is to be specified in nm in the IEC document while the formulae for the correction factors are given for wavelengths measured in μm in the ANSI document. Where they are a function of the angular subtense of the apparent source α, (such as $C_6 = \alpha/\alpha_{\min}$) α has to be specified in mrad, and where they are a function of the exposure duration, t must be specified in seconds. There are also two parameters T_1 and T_2 that have the dimension of time, i.e. they are characteristic times: T_1 is the dividing line between thermal and photochemical limits in the ultraviolet wavelength range (in IEC 60852-1) or in the visible wavelength range (i.e. retinal limits) in ANSI Z136.1, and depends on the wavelength λ, while T_2 characterizes the time when eye movements become dominant in terms of exposed retinal area, and depends on α. (In the ANSI laser safety standard, the ultraviolet limits are represented as dual limits without a dividing line, so that T_1 is not used there and is used

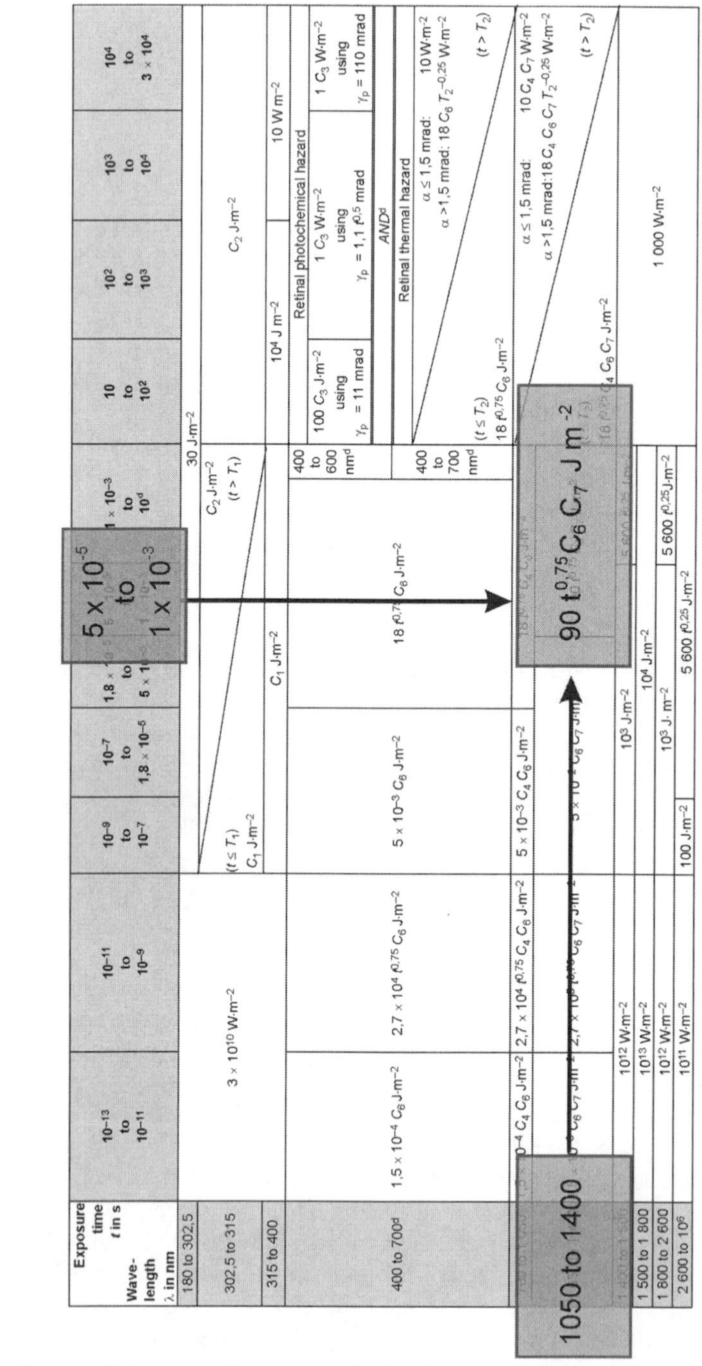

Figure 3.11. Example of a cell containing an MPE for the eye, where the line is chosen following the wavelength and the column is chosen following the exposure duration.

instead in the simplified MPE table for small sources as the dividing line between thermal and photochemical retinal limits.) All correction factors are discussed in more detailed in the following sections of this chapter.

Some cells feature a diagonal split, namely in the wavelength range 302.5–315 nm for exposure durations between 1 ns and 10 s, and for retinal long term exposure limits, where it depends on the chosen exposure duration which one of the two limits is to be applied. These cases are discussed in more detail in the section on MPEs in the UV wavelength range and in the section on retinal limits. For the visible wavelength range for exposure durations of longer than 1 s, dual limits are defined for the eye, namely thermal and photochemical limits: any given exposure level has to be below both MPEs, and it depends on the wavelength, exposure duration and the angular subtense of the apparent source which one of the two MPE values is the more critical one, i.e. the lower one.

For a given wavelength, the MPE values as a function of exposure duration are continuous in the sense that at the border between two cells within on line, when both values are evaluated for the exposure durations that applies to the cell border; the MPE value at the left of the border is the same (disregarding some rounding errors) as the value at the right of the border. Due to possible minor differences at the border, for the 'mathematical' treatment, it is always prudent to adopt the MPE value with the lower value when evaluating an MPE for an exposure duration that corresponds with a border value.

The reference to the temporal dependence of the MPE values is somewhat ambiguous: in this book, we use the term *exposure duration* to refer to the temporal dependence of the MPEs (as do the ICNIRP and ANSI documents), while in the IEC laser safety standard, the term *exposure time* is used. We see the term 'exposure duration' as a general term to denote the temporal parameter of the MPE values, and it needs to be distinguished from the *maximum anticipated* exposure duration. The difference in terminology becomes obvious when it comes to the evaluation of exposures to pulsed radiation: the *maximum anticipated* exposure duration used in the MPE analysis represents the overall maximum exposure duration to the pulse train, for instance the typical values of 0.25 s when the radiation is in the visible range, or 10 s when it is in infrared, but additionally to the overall exposure duration the exposure to individual pulses and subgroups of pulses need to be evaluated too. (The MPE tables are basically specified for single exposures, i.e. the values apply to the exposure to one single pulse, and there are additional evaluation requirements for multiple exposures.) For the evaluation of a single pulse, the temporal dependence of the MPEs could be referred to as pulse duration rather than the exposure duration; as for exposure to one single pulse, the pulse duration would be equivalent to the exposure duration. Seen rather mathematically, the MPE values can be considered to be dependent on the parameter t, where t can be referred to as the exposure duration that can range from the duration of a single pulse up to the maximum anticipated exposure duration. For a general and complete hazard analysis it is necessary to consider any values in between, for instance to evaluate groupings of pulses.

3.6.4 Summary and overview of dependencies

We would like to come back to the underlying question of a hazard analysis, and that is 'is the exposure level E_{exp} below the MPE?', or in a more mathematical form,

$$\text{is } E_{\text{exp}} < \text{MPE?}$$

From the discussion in the previous sections it is clear that this is not really such a simple answer when one considers all the dependencies and attempts to perform a hazards analysis without too many worst-case simplifying assumptions. In the following, we would like to summarize all the dependences and interdependencies of the MPEs and of the issues related to the determination of the exposure level, as well as simplifications. This section is followed by a section that discusses issues related to the location of measurement.

MPE values

The MPE values for the skin and the eye in principle depend on exposure duration and wavelength, although there are wavelength and exposure duration regions where MPE values can be constant, i.e. either not depend on exposure duration or on wavelength. (It is obvious, however, that when an MPE value that is specified in terms of irradiance that does not depend on time, i.e. is constant with respect to the exposure duration, then a recalculation of this value into a radiant exposure value introduces a linear dependence on the exposure duration.) In addition to the general dependence on exposure duration and wavelength, the MPE values that relate to the retinal thermal hazard depend on the angular subtense of the apparent source α. In the far-infrared wavelength range (wavelengths above 1400 nm), for exposure durations above 10 s, the MPEs for the skin depend on the irradiated area on the body, i.e. the beam diameter at the location of the hazard analysis.

It is also worth noting that in the ultraviolet and in the far-IR wavelength range the MPE values for the skin are the same as those for the eye. However, because there are different limiting apertures and maximum angle of acceptance defined for the assessment of the exposure level, it may be the case that for the same location in the beam and the same exposure duration, different exposure levels (one that represents the biophysically effective value for the skin and the other for the eye) are compared to the same MPE level. Similarly, for the retina in the visible wavelength range, dual limits are defined where the exposure level to be compared to the photochemical MPE may be lower than the one to be compared to the thermal limits as the maximum angle of acceptance for the photochemical limit evaluation is less than the one for the thermal limit evaluation for most exposure durations.

The above aspects can be summarized as shown in table 3.2.

Table 3.2. Summary of dependencies of MPEs.

Subsets of MPEs for different types of hazard	MPE-UV-thermal (same for eye and skin) MPE-UV-photochemical (same for eye and skin) MPE-retinal-thermal MPE-retinal-photochemical (visible only) MPE-far IR (same for eye and skin)
General dependence	Wavelength Exposure duration
Special dependence	MPE-retinal-thermal: α MPE-far IR for skin: spot diameter

Table 3.3. Summary of the dependencies of the measurement parameters limiting aperture and maximum acceptance angle that can affect the level of exposure E_{exp} for a given beam and a given exposure location in the beam.

Limiting aperture for irradiance averaging (equivalent to power through aperture)	Different for *eye* and *skin* (i.e. depends on the type of MPE to which exposure is compared)
	For eye in far-IR depends on exposure duration t_{exp}
Angle of acceptance	Different for *retinal thermal* and *photochemical*
	For retinal photochemical depends on exposure duration t_{exp}

Exposure level E_{exp}

As mentioned in the previous sections, the value of the exposure level that is to be compared to the different types of MPEs (see table 3.2) may be different since the limiting aperture and/or the acceptance angle may be different. For instance, for a given laser beam and a given location in the beam (and therefore a given physical irradiance level at that location), the biophysical effective exposure level can be different for different types of MPEs with which the exposure level is to be compared, and also for different exposure durations (when the limiting aperture or the maximum acceptance angle depend on exposure duration). The dependencies are summarized in table 3.3.

The different 'types' of dependence of the exposure level E_{exp} are summarized as follows.

- E_{exp} depends generally on location z in the beam (measured along the beam axis) as the power measured through a given limiting aperture and from angles within the acceptance angle depends on location

$$E_{\text{exp}}(z).$$

- E_{exp} for the same location z within the beam may be different for different 'types' of MPE, as for different types of MPE a different limiting aperture and maximum angle of acceptance may be specified. The question is $E_{\text{exp}} <$ MPE? for a given laser beam and a given location in the beam and should in such a case more accurately be written as (in the case of a beam in the visible wavelength range)

$$E_{\text{exp-skin}}(z) < \text{MPE-skin?}$$
$$E_{\text{exp-retinal-thermal}}(z) < \text{MPE-retinal-thermal?}$$
$$E_{\text{exp-retinal-ph.chem}}(z) < \text{MPE-retinal-photochemical?}$$

for a given location z in the beam.
- E_{exp} for the same location within the beam and for a given 'type' of MPE may depend on the exposure duration, as the diameter of the limiting aperture for the eye in the far-IR depends on the exposure duration, and the maximum acceptance angle γ_{ph} for the photochemical retinal exposure evaluation also depends on the exposure duration, so that, for example, for a beam in the visible wavelength range in the above 'question'

$$\text{is } E_{\text{exp-retinal-ph.chem}}(z, t_{\text{exp}}) < \text{MPE-retinal-photochemical?}$$

the exposure level that is to be compared to the retinal photochemical MPE could depend on the exposure duration, t_{exp}. This time dependence reflects the fact that the exposure is more hazardous when it lasts longer—in this case, the time dependence is contained in the measurement conditions that affect the effective exposure level and not in the exposure limits: instead of decreasing the MPE for longer exposure durations, the biophysically effective exposure level $E_{\text{exp-retinal-ph.chem}}(z, t_{\text{exp}})$ is increased.

It follows that depending on the beam and the problem at hand it can be important to compare 'matching' exposure levels and MPE values. Both the MPE and the exposure level may depend on exposure duration (so that it is important that both the exposure level and the MPE are evaluated for the same exposure duration) and different exposure levels need to be compared to different types of MPE values (for instance when dual limits are defined).

Hazard evaluation—simple worst case

The above dependencies of both the MPE and the effective exposure level on a number of parameters can make a safety evaluation of potential exposure to laser radiation an elaborate exercise. A hazard analysis, however, can be greatly simplified by a number of worst-case assumptions. The trade-off for the simplification is that the level of hazard as determined with the worst-case assumptions can be overestimated by a very large margin, up to a factor of 1000 or more in the extreme. The simplification can also be done to varying degrees, i.e. some parameters and dependencies can be taken as worst case while others could be accounted for more realistically. Simplifications certainly can save a lot of effort when the exposure level is below the MPE even for a simplified worst-case analysis. An overview of the parameters that can be either taken as worst-case simplified values or accounted for with all consequent dependencies is given in table 3.4. In some cases, some of the simplifying worst-case assumptions have no effect on the effective exposure level and thus can simplify the measurement or analysis but do not lead to an overestimation of the hazard. An example is concern about the measurement location when the beam has a small diameter and low divergence, so that it would fully pass through the aperture over a large distance. Other examples are the limitation of the acceptance angle for sources that are smaller than the acceptance angle, the use of a limiting (averaging) aperture for beams that have a large diameter at the location of evaluation and a homogenous irradiance profile across the aperture, and the use of an aperture stop for power measurements when the beam is much smaller than the aperture stop.

3.6.5 Evaluation and measurement position

We would like to complement the discussion on MPE evaluation by aspects relating to the choice of evaluation location (that is, the measurement position in the beam). The evaluation location along the beam (i.e. somewhere between the exit port of the laser product and at a great distance from the laser product) may be at specific position in the beam for which the question '$E_{\mathrm{exp}} <$ MPE?' needs to be answered. More often, the question will not be asked for a specific location but in general terms, i.e. is the MPE is exceeded anywhere along the beam?

Instead of actually performing an MPE analysis (i.e. by comparing the exposure level to the MPE) for all locations along the beam it is helpful to know the most hazardous exposure position (which in this book is abbreviated to MHP): if the exposure level at the MHP is less than the MPE, then the MPE is not exceeded anywhere else in the beam.

In the simplest case, for a diverging beam where the apparent source can be treated as a 'point' source even at a short distance from the source, the MPE value does not depend on the location along the beam. (For extended sources, the angular subtense of the apparent source depends on the exposure position, and, generally, the angular subtense of the apparent source decreases with distance to the source.) For retinal thermal hazards, the closest evaluation position which needs to be chosen in an MPE analysis is 10 cm (100 mm) from the apparent source, i.e. if there is a beam waist, then at 10 cm from the beam waist. The beam

Table 3.4. Worst-case assumptions and simplifications that can save a lot of effort but that can lead to gross overestimation of the hazard.

Maximize E_{exp}	Location (closest to source or beam waist)
	Unlimited acceptance angle (receive power from full source, not only from within maximum acceptance angle)
	Maximum irradiance at given location (do not average irradiance over limiting aperture)
	Total power (not limited by a limiting aperture) can for instance be done for LEDs where total power is given in the manufacturer's specification. Note that power averaged over a limiting aperture to calculate irradiance and compare to the MPE is equivalent to comparing the power through a limiting aperture with the MPE multiplied by the area of the limiting aperture (which is the AEL for Class 1 and Class 1M)
Minimize MPE	$C_6 = 1$
	Use the maximum exposure duration (typically 30 000 s but can in effect be lower when the MPE is constant beyond a given exposure duration, such as for far-IR from 10 s onwards)
	For a broader spectrum such as emitted from an LED that should be treated as a multi-wavelength exposure: evaluate the MPE for the lowest wavelength of the emission spectrum (generally, the MPEs are lower for lower wavelengths)

waist can also be a virtual one, i.e. inside the laser cavity or even 'behind' the laser. If the beam waist is further behind the exit port of the laser than 10 cm, then the appropriate closest evaluation location is at the exit port. The distance of 10 cm is derived from a conservative assumption of the near accommodation point of the eye, and it is not necessary to evaluate closer positions than 10 cm as the retinal image would be blurred. It follows that for these kind of beams, for retinal thermal hazard evaluation the most hazardous position is at a distance of 10 cm, and if the MPE is not exceeded there, then it is not exceeded at any other point along the beam. For the case that the MPE is exceeded at a distance of 10 cm from the apparent source, there will be a distance from the laser where the beam has expanded so much that the exposure level falls below the MPE. This distance can then be referred to as the 'hazard distance'. When this distance is based on the MPE for the eye, it is usually referred to as the nominal ocular hazard distance, NOHD. For evaluation of other hazards above the retinal thermal hazard, the most hazardous position is often the beam waist.

The analysis of a laser beam that is not simply diverging from a 'point source' can be somewhat more involved, as is discussed in other sections of this chapter. Examples of more complicated sources include extended sources (i.e. those that are larger than a 'point' or small source at short distances) or multiple sources (for example laser diode arrays). For such sources, the most hazardous position can be some distance away from the apparent source (i.e. greater than 10 cm). For example, the exposure level can remain constant when moving away from the source if the beam diameter is smaller than the limiting aperture, or the exposure level might even increase with increasing distance from the source in the case of converging beams or beams from an array that are pointed towards a point, while the angular subtense of the apparent source (and therefore the MPE) would decrease with increasing distance from the apparent source. In general, the most hazardous position for a given beam is defined as the location at which the ratio of the exposure level to the MPE is at its maximum. In a more mathematical form

$$\text{Most hazardous position (MHP): where } E_{\text{exp}}/\text{MPE is max.}$$

This concept is further discussed for retinal thermal limits in section 3.12.2.

When the MPE is exceeded at the most hazardous position, then there will a distance from the laser beyond which the exposure level is below the MPE, and this is the concept of hazard distance, or for the eye, the NOHD, as introduced in section 3.6.2. In a more mathematical form

$$\text{Hazard distance: where } E_{\text{exp}} = \text{MPE}.$$

For the case of beams that are not diverging from a 'point' source, it might be the case that there is a region close to the laser where the MPE is not exceeded, followed by a region where the exposure level exceeds the MPE (because the angular subtense of the apparent source and therefore the MPE decreases with distance while the exposure level remains constant) and beyond that a third region where the exposure level falls below the MPE again (because the laser beam has become larger than the limiting aperture). In this case two locations in the beam would exist where the exposure level is equal to the MPE. In such a case, the general definition of the hazard distance would be the *furthest* distance from the laser where the exposure level falls below the MPE. It is also noted that the hazard distance (the NOHD) is usually defined as the distance from the laser product, i.e. as the distance from the exit port of the laser, which is not necessarily the same as the distance from the position of the apparent source.

When the MPE is not exceeded at the most hazardous position, then exposure to the beam is safe at any position and there is no hazard distance (no NOHD for the case of ocular MPEs) associated with the laser product. (Such a product is then sometimes referred to as having a hazard distance of zero.)

To relate the dependencies of the biophysically effective exposure level, as an example we consider both the hazard distance and the MHP. Due to the dependencies summarized in the previous sections, for one and the same laser beam there might a different hazard distance for the skin than for the eye (NOHD), and regarding the NOHD, this might be different for the retinal and the photochemical hazard, and it also might be different for different exposure durations (both because of the dependence of the MPE as well as the E_{exp} on exposure duration). In such cases, the longest NOHD is the applicable one.

3.6.6 Background to the concept of dosimetry

For a discussion of the background of the dosimetry concept for the laser MPEs, specific wavelength ranges and consequently the type of tissue at risk need to be distinguished. Generally, the location of the injury corresponds to the location of the (main) absorption of the radiation at the particular wavelength. For optical radiation hazards for the eye, for wavelengths in the ultraviolet and the infrared above 1400 nm the tissues at risk are the cornea and the lens of the eye, and in the wavelength range 400–1400 nm, the retina. Besides the special role of the eye both in terms of providing vision and of susceptibility to laser damage, the skin as the outermost protective surface layer of the body is the general absorber over the full wavelength range of what is referred to as optical radiation. It follows naturally, that for MPEs that relate to surface absorption, i.e. MPEs of the skin and MPEs for the eye in wavelength ranges where the frontal parts of the eye (the cornea and the lens) are at risk, the MPEs are defined in terms of irradiance at the skin or irradiance at the cornea. However, even for wavelengths where it is the retina that is at risk, these 'retinal' MPEs are defined at the cornea, i.e. the irradiance or radiant exposure that exists at the position of the cornea of the eye is compared to the respective MPE value (that is also given in units of irradiance or radiant exposure). As the cornea and the lens of the eye images the radiation onto the retina, the 'natural' quantity to characterize retinal exposure and exposure limits would be radiance (see section 2.5), especially for extended sources, i.e. sources that form a non-point imaged on the retina. It is not the irradiance (profile) at the cornea that determines the retinal spot size, i.e. the retinal irradiance profile, but rather the emission (exitance) profile of the source that is imaged onto the retina. Just as with any other lens and conventional light sources such as lamps, when the lens is used to image the source, it is the emission profile of the source that determines the image and not the profile on the lens. Radiance has the advantage that it is proportional to the irradiance level at the retina. Consequently, radiance was used in previous editions of the international laser safety standard for extended source limits, and is used for retinal MPEs for broadband incoherent radiation. In the current version of the laser safety standard, however, an alternative concept is used, where the exposure level and the MPE are both defined in terms of *corneal* irradiance (as averaged over a 7 mm aperture) and the extent of the retinal spot is figured into the retinal thermal MPEs via a factor termed C_6 that depends on the retinal spot size (see section 3.12.1 for a detailed discussion of the apparent source). For a correct application of this alternative concept, for extended sources it is important to apply limiting measurement FOV as defined for thermal and photochemical retinal limits. Specifying the MPEs as irradiance at the cornea has the advantage that for well collimated laser beams that can be assumed to be point sources, C_6 and the limiting FOV can be neglected, and the evaluation thereby greatly simplified.

The averaging of the exposure level over the limiting aperture that is to be done for comparison with retinal limits can also be understood on the basis of the derivation of the MPEs from exposure experiments. For the retina, experimental ED-50 and threshold values are usually not reported in terms of the irradiance at the cornea but in terms of power or energy which is incident on the cornea and enters the pupil. This quantity is referred to either as intraocular energy, IOE, or as total intraocular energy, TIE. The energy or power incident on the retina is lower than the power incident on the surface of the eye due to reflection and absorption losses in the ocular media in front of the retina. However, for experimental studies and for the definition of MPEs, it is not necessary to actually characterize the power or energy that is incident on the retina. What is necessary for a complete characterization of retinal exposure, however, is the diameter of the irradiated spot, which can then be related to the angular subtense of the apparent source. As discussed in section 3.6, a safety factor, or scaling factor is introduced when setting MPE values based on experimental ED-50 values. For example, the experimentally determined ED-50 value for damage from pulses with duration in the range of 10 ns to 10 μs in the visible wavelength range is typically in the range of 2 μJ for minimal spot sizes. For derivation of the MPE, this value is decreased by a factor of typically 10 or more, and then this reduced value is divided by the area of the assumed pupil diameter of 7 mm (that is also the diameter of the limiting aperture). In our example, the MPE would be calculated as 0.2 μJ divided by 3.85×10^{-5} m^2, to result in the MPE that can be found in IEC 60825-1, namely 5×10^{-3} J m^{-2}. The reader who is familiar with the concept of deriving the AEL for Class 1 and Class 1M from the MPE, will have noted that the original limit value specified in terms of intraocular energy (i.e. before division by the area) of 0.2 μJ, is identical to the AEL of Class 1 and Class 1M. This is not surprising since the AEL is derived from the MPE by multiplying the MPE value with the area of the corresponding limiting aperture, as is discussed in section 4.2.1.

In the wavelength range where retinal damage applies, i.e. from 400–1400 nm, because it is the retinal irradiance which determines the hazard level of a given exposure, it would seem more straightforward and easier to understand if the retinal MPE values were defined in terms of 'power (or energy) passing through the 7 mm pupil'. The MPE values in that wavelength range would then be identical to the AEL for Class 1 and Class 1M. Rather than compare a corneal irradiance that is averaged over 7 mm with an MPE that is defined in terms of corneal irradiance it would seem more appropriate and consistent to refer to the power that enters the eye (and is then incident on the retina). For wavelengths outside the retinal hazard region, MPEs and exposure levels could still be defined in terms of irradiance and radiant exposure.

3.7 Injury to the skin

While the primary concern of laser safety is to prevent injury to the eye, laser radiation can also result in injuries of the skin. The skin can generally tolerate higher levels of radiation than the eye, especially in the wavelength range 400–

1400 nm, where for the eye the radiation is focused onto the retina and for irradiation of the skin obviously such a focusing does not take place.

Injuries of the skin from laser radiation can be categorized in a simplified way as either 'burn' or 'sunburn', where with 'burn' a thermal injury due to an increased temperature of the skin is meant, while 'sunburn' refers to photochemical damage of the skin that, however, is only possible from ultraviolet radiation. The term 'sunburn' should strictly be used only for exposure to solar radiation, a more general term would be 'photochemically induced erythema' (reddening of the skin). Skin injuries can vary greatly in severity depending on the extent of the area that is affected and by how much the injury threshold is exceeded. For thermal injury, exposure levels far above the threshold for mild burn can cause deep injuries that not only damage the skin but also underlying muscle tissue and even major blood vessels with potentially life threatening consequences. Exposures slightly above the threshold will lead to slightly inflammated (reddened) skin, that is also referred to as erythema. If this slight reddening occurs for small spot diameters of the order of a few millimetres it will in practice hardly be called an injury, even if the exposure as such is per definition above the threshold. Even if the power is somewhat above the threshold for reddening, when laser radiation is focused onto a tiny spot, the effect can be compared to a pin prick. However, exposure of a larger area of the body with the same irradiance can result in extreme injuries, whether from thermal burns or from photochemical damage. In practice, large area exposure of the skin to superthreshold irradiance levels is rather rare for thermal burn conditions (e.g. from CO_2 laser radiation) but occurs rather frequently for straylight exposure from high power UV lasers (e.g. Excimer lasers). Also, in the extreme, ultraviolet radiation might induce skin cancer: medical and epidemiological studies for solar radiation indicate that exposures well above the MPE (heavy sunburns) during the youth seem to increase the risk for melanocytic skin cancer, while chronic over exposure over a longer period of time during adulthood induce non-melanocytic skin cancer.

Generally, skin has a very good repair capacity for all but severe injuries (i.e. with the obvious exception of skin cancer).

The basic nature of thermal injury due to laser radiation is not different from burns that result from contact with hot surfaces, hot steam or liquid, or due to infrared radiation from red- or white-hot radiant heaters of grills. The main difference between these sources and lasers is that high-power, well-collimated laser beams can produce severe burns at a large distance from the actual laser source, as has happened for instance in an industrial setting where after service of the CO_2 laser materials processing machine a turning mirror of the beam guidance systems was not replaced and a number of workers were severely burned at the other end of a rather large manufacturing hall. A second difference stems from the strong wavelength dependence of the absorption depth of the radiation: the penetration depths are highest for near infrared radiation, such as to 810 nm wavelengths and also quite high for Nd:YAG laser radiation. With these wavelengths, thermal damage in the sense of denaturation of protein extends very deep (up to a centimetre or more) into tissue, and even for a focused beam, due to scattering, affects a relatively large volume of tissue. The nature of these types of injury can be compared to 'boiling' of the flesh, i.e. proteins are denaturated and the tissue colour becomes white, and the affected tissue subsequently is referred to as necrotic, i.e. the affected cells die. For wavelengths where radiation is absorbed in a thin surface layer such as for CO_2 lasers, high temperatures are induced that lead to carbonization (charring) of the tissue. There is practically no scattering but deeper layers of the tissue can be affected by 'burning off' of tissue within a fraction of a second—CO_2 lasers with powers between 30 W and 80 W are used in surgery as 'laser scalpels'. It is therefore not surprising that multi-kilowatt CO_2 lasers used for materials processing can result in serious and deep injuries.

It is often overlooked that there is quite a pronounced dependence of thermal damage MPEs on the size of the exposed area. For larger exposed areas, the MPE is decreased to account for reduced thermal conduction of heat away from the exposed site that is effective for small beam exposures. There are also lower MPEs to account for full body exposures (which is unlikely from laser radiation) to prevent heat stress, i.e. an increase of the body core temperature with corresponding potential effects of permanent damage to the brain, organs or heart failure.

Exposure to the ultraviolet (UV) radiation can result in 'sunburn'—however, as this effect is photochemical in nature, the term 'sunburn' is actually a misnomer. The basic effect of ultraviolet radiation from a UV laser source is no different than UV radiation from other sources such as the Sun, UV lamps or welding arcs (although the spectrum and radiant exposure levels for a given exposure have a strong influence on the actual effect of that exposure). It is typical for photochemical damage that the degree of damage (and also whether an effect is induced at all) depends on the radiant exposure, i.e. on the accumulated energy in terms of J m^{-2}, and a short duration exposure (seconds or less) with high irradiances has the same effect as a long-term exposure (hours) with a correspondingly smaller irradiance. It should be noted here however, that exposure to UV radiation from pulsed sources with very high irradiances, especially from Excimer lasers, can result in ablation of the top layers of the skin, as these high irradiances produce a thermal effect that in this case dominates over photochemical injury. Therefore, photochemical damage mainly occurs for exposure to continuous wave lasers or repeated exposures to irradiance levels (also from pulsed lasers) that are too small to cause a temperature rise sufficient to cause thermal damage. Also, for photochemical effects individual exposure doses add up over a period of several hours; for the skin, an additivity of exposures within up to about 8 h is assumed. Therefore, when working with UV lasers such as Excimer lasers, due to the photochemical additivity, even faint stray radiation or reflections, for instance from lenses and walls, can cause damage when exposure occurs for a few hours. While the hazard presented by the direct beam from such lasers is generally appreciated, the effect that exposure to faint stray radiation metres away from the beam or even via diffuse reflections of stray light from walls can have on the skin and the cornea (producing a very painful inflammation) is often underestimated. Also, not only UV lasers may present a 'sun burn' hazard; the plasma produced in high-power CO_2 laser beam welding also represents an intense source of UV laser radiation and when the unprotected skin is exposed in short distances to the plasma, serious photochemical damage would result within only a few minutes of exposure [13].

It is also typical for photochemical injury that the injury needs some time to develop, i.e. it takes several hours for mild erythema to develop when irradiation is discontinued at the time when the threshold is reached (for development of a mild sunburn from solar irradiation it takes about 6 h). When irradiation is continued after the threshold is reached, leading to an exposure several times above the threshold for a mild erythema, the response (erythema) develops sooner (for instance an irradiance level to UV-B radiation that results in an eight fold overexposure within one hour, erythema typically develops within that one hour). As is well known from experience with the Sun, a sunburn heals, depending on the severity, within a few days, but exposures to levels of a multiple above the threshold for mild erythema result in severe blistering with a potential of permanent pigmentary defects or scarring in the skin, or even delayed effects of melanoma skin cancer, a form of skin cancer with a high fatality rate. For the case that exposure to UV radiation occurs on a regular basis (chronic exposure over years) which might not have to be much above the threshold for light erythema, the effects on the skin are pigmentary changes (spots with high brownish pigment concentrations), loss of elasticity of the skin (photoageing) and in extreme cases carcinoma in the epithelial layer of the skin, that however, usually does not produce metastases (non-melanoma skin cancer) and does not have a high fatality rate unless it is left untreated.

3.7.1 Aversion response, typical exposure durations

Behavioural protective mechanisms, i.e. reflexes or cognitive reactions to a potentially harmful exposure that are designed to protect the body from an injury are also referred to as aversion responses.

Excessive heat is readily sensed by the skin, especially by the face. Exposure to optical radiation becomes painful when the skin temperature reaches about 44 °C, a temperature that is below temperatures that induce permanent damage for short term exposure durations. This strong pain response causes the exposed person to react within a relatively short time (that might be longer if the person is under the influence of drugs or alcohol or certain types of medication). Therefore, for exposure to visible and infrared wavelengths, the typical (conservative) exposure duration adopted for a skin hazard analysis is 10 s (see table 3.5). It is noted however, that the MPEs are expressed as constant irradiance for exposure durations longer than 10 s, therefore, an exposure that is below the MPE at 10 s exposure duration is also below the MPE for longer exposures. In other words, if

Table 3.5. Tabular overview of typical (default) exposure durations used for a safety analysis of exposure of the skin.

	Typical assumed exposure duration	Comment
UV 180–400 nm	Up to 8 h	Additivity (reciprocity)
VIS and IR 400 nm–1 mm	10 s	Induction of pain before damage occurs, aversion response limits exposure duration, constant MPE above 10 s

for a given irradiance level the skin is not burnt within 10 s, it is also not burnt for longer exposure durations. For example, it is a common experience when touching a hot object that it becomes apparent within a very few seconds whether it is likely to cause a burn or not.

Irradiance levels of exposures to ultraviolet radiation that may result in a 'sunburn' after some duration are generally below the level that would produce heat pain and therefore cannot be felt. Due to the delayed onset of the photochemical effect, there is no behavioural protective mechanism. For instance, if a given constant irradiance level (from stray light or a welding plasma) results in reaching the threshold for sunburn within 30 min, since it takes several hours for the skin to develop the inflammatory response, by the time the effect is observable, the threshold is exceeded several times. Since UV exposure cannot be felt (except for pulses with high irradiance levels which induce thermal effects) and since MPE values are defined as constant radiant exposure value for exposure durations up to 8 hours to reflect possible additivity of exposures, the typical maximum exposure duration that is adopted for a hazard analysis of an exposure is 30 000 s. Although there is no direct behavioural aversion response or protective mechanism for UV radiation, for repeated exposures at or somewhat above the threshold for erythema, the skin reacts with a number of protective mechanisms, such as production of protective pigments (tanning), and thickening of the skin. However, in laser safety, the general concern are acute exposures that are often well above the threshold and these kind of protective mechanisms of the skin do not apply in a laser safety analysis.

3.8 Skin MPE values

The current values of the skin MPEs as published in IEC 60825-1 (equivalent to ICNIRP and ANSI MPE values) are summarized in table 3.6. Please note that for the determination of an irradiance or radiant exposure level which is to be

Table 3.6. MPE values for the skin for single pulse exposures (in Edition 1.2 of IEC 60825-1 as published in 2001, the MPE table for the skin had formatting errors in the wavelength range 400–1400 nm—below are the correct values).

Wavelength	Exposure duration	MPE value
180–400 nm	1 ns–30 ks	Same as eye MPEs
400–1400 nm	1–100 ns	$200\ C_4$ J m^{-2}
	100 ns–10 s	$11\,000\ C_4 t^{0.25}$ J m^{-2}
	10 s–30 ks	$2000\ C_4$ W m^{-2}
1400 nm–1 mm	1 ns–10 s	Same as eye MPEs
	10 s–30 ks	Same as eye MPEs, i.e. 1000 W m^{-2}, but spot size dependence for spots larger than 0.01 m^{-2}

compared the MPE, limiting apertures are defined over which the irradiance or radiant exposure is to be averaged (see section 3.6.1).

In the UV wavelength range and for infrared wavelengths above 1400 nm, the skin MPEs are set identical to the MPEs for the eye, as in these wavelength ranges the optical absorption properties of the eye (mainly the cornea) are similar to the absorption properties of the skin. (Note, however, that the diameter of the limiting aperture can be larger than in the case of the eye so that the values characterizing the level of exposure that is to be compared to the MPE value will be different for small beam diameters.) While the skin is in some cases less susceptible to injury from laser radiation than the eye (especially in the UV wavelength range), setting the skin MPE to the same value as for the eye was done for simplicity. Therefore, these laser limits should not be applied to evaluate broadband radiation, as for instance the dose required to produce a minimal reddening (erythema) in the skin with far UV radiation is not less than about 150 J m^{-2}, even for fair skinned people, while the laser MPE equals only 30 J m^{-2}, a value that is derived from the peak of the sensitivity of the cornea to photokeratitis (inflammation) [14, 15]. However, it is prudent to have a higher safety factor defined for the skin MPEs to prevent possible delayed effects from exposure to levels below those causing a mild inflammation such as skin ageing and skin cancer.

In the wavelength range 400–1400 nm, where the skin MPEs are different to the eye MPEs, the factor C_4 represents the wavelength dependence of the absorption of the radiation in the skin, i.e. the MPE is lowest where the absorption depth is the shallowest, in the blue and green, while for deeper penetrating wavelengths in the red and infrared wavelength range, the MPE is correspondingly increased. The factor C_4 (termed C_A by ICNIRP and ANSI) is defined as $C_4 = 10^{0.002(\lambda-700)}$ in the wavelength range 700–1050 nm, equals 1 for

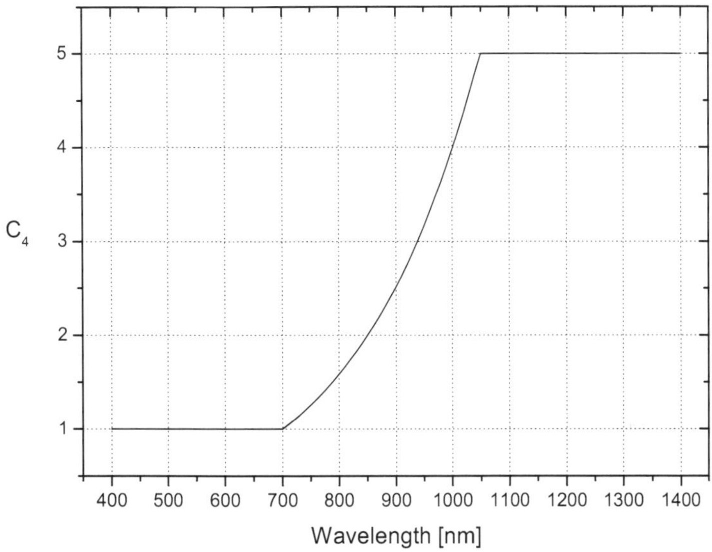

Figure 3.12. Wavelength dependence of the factor C_4 (C_A in the equivalent ANSI and ICNIRP documents) that very roughly accounts for variation in absorption depth of optical radiation.

wavelengths shorter than 700 nm and equals 5 for wavelengths above 1050 nm, as plotted in figure 3.12. C_4 roughly follows the wavelength dependence of the absorption depth of melanin, the primary absorbing pigment both in the skin and in the RPE of the eye. Compared to the wavelength dependence of photochemical damage as characterized by C_2 in the UV wavelength range and by C_3 for retinal photochemical damage, the wavelength dependence of the (thermal) skin MPEs in the range between 400 and 1400 nm is relatively weak.

Regarding temporal dependencies, the MPEs for the skin in the wavelength range 400–1400 nm are given as constant radiant exposure for exposure durations from 1–100 ns, as typical for short pulses with pulse durations below the thermal confinement time. For exposure durations longer than 100 ns up 10 s, the temporal dependence is $t^{0.25}$ (or $t^{1/4}$) with the MPE given as radiant exposure (see figure 3.13(*a*)). The logarithmic increase of the exposure limit with pulse duration when given as radiant exposure is rather 'shallow', as a 10 000 fold (10^4) increase in pulse duration leads to a 10 fold increase of the MPE. When recalculated to an irradiance value, the temporal dependence of the MPE becomes $t^{-0.75}$, i.e. the maximum permissible irradiance value decreases with increasing exposure duration corresponding to the decreased tolerance of the protein to elevated temperatures for longer times (see figure 3.13(*b*)). This temporal dependence of the irradiance specified MPEs means that there is a 1000 fold decrease of

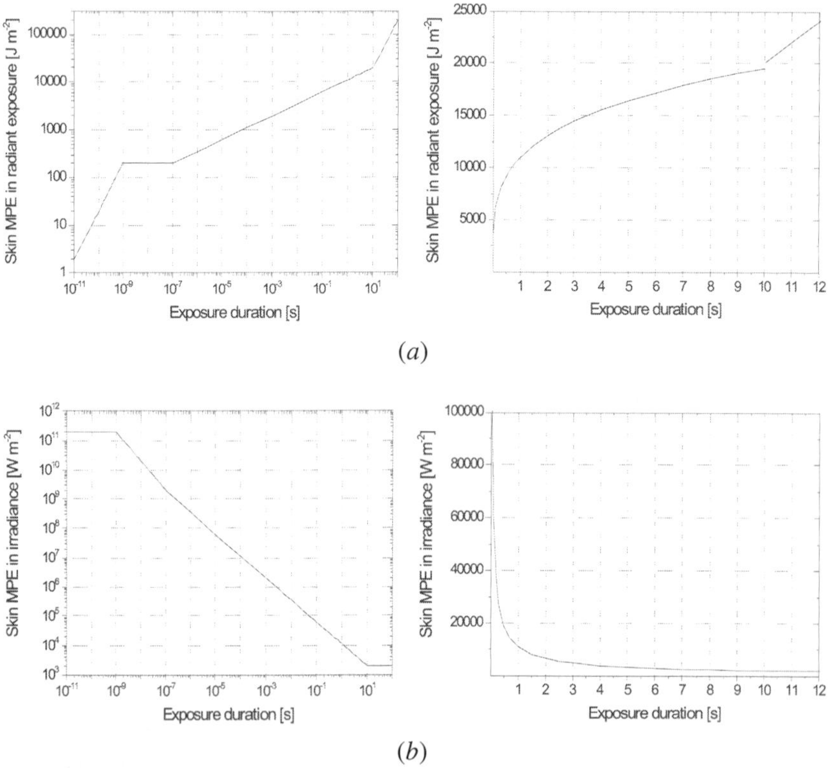

Figure 3.13. (*a*) Temporal dependence of the skin MPEs in the wavelength range 400–700 nm plotted as radiant exposure. On the left the full range of exposure duration is shown in a log–log plot and on the right the MPEs are shown in a linear plot. (*b*) As in (*a*) but data recalculated as irradiance values.

the MPE for a 10 000 fold increase in exposure (pulse) duration. For exposure durations longer than 10 s, a constant irradiance value is specified so that when a given level of irradiance is below the MPE for an exposure duration of 10 s, it is also considered safe for longer exposure durations. The temporal dependence of the skin MPEs is plotted in figure 3.13 for the wavelength range 400–700 nm. (For wavelengths between 700 and 1400 nm, the MPE values as shown can be corrected with the factor C_4). The radiant exposure plot is more appropriate to show the constant radiant exposure MPE value for pulse durations of 1–100 ns, while the irradiance plot is advantageous to show the constant irradiance MPE value for exposure durations longer than 10 s.

Larger exposed areas

It is often overlooked that the laser MPE value for the skin for the wavelengths above 1400 nm and for exposure durations longer than 10 s, which equals 1000 W m^{-2}, is applicable to exposure areas of up to 0.01 m^2 only (i.e. an area of 100 cm^2 or a circular area with a diameter of about 11 cm)—for larger exposed body areas, the MPE value decreases inversely proportional with the exposed area to a value of 100 W m^{-2} for an area of 0.1 m^2 (1000 cm^2 or a circular area with a diameter of 36 cm) or larger areas. This reduction very roughly takes account of the possibility of a hazardous heat load to a larger section of the body that might lead to thermal stress, i.e. an increase of the body's core temperature. However, for exposure to laser radiation this should rarely be an issue as typical accidental exposure scenarios would involve rather small area exposures and the sensation of heat would limit the exposure duration to below 10 s, where the MPE does not depend on the size of the exposed area.

Ultrashort pulses

As there is no animal study data on thresholds for skin damage available for ultrashort pulses, the MPEs for the skin are only defined to exposure durations of 1 ns and above. As a conservative approach, for exposure durations shorter than 1 ns, the MPE is kept at the irradiance level that applies at 1 ns. For instance, for visible wavelengths, the MPE at 1 ns equals 200 J m^{-2}, which corresponds to an irradiance value of 200×10^9 W m^{-2}. It is this (peak) irradiance value to which exposures with pulse durations below 1 ns should be limited, i.e. in terms of radiant exposure per pulse, the MPE decreases linearly with shorter pulse durations, so that at 10 fs (10^{-14} s), the MPE expressed as radiant exposure (per pulse) equals 0.002 J m^{-2}. As can be inferred from the known threshold values for the retina for ultrashort pulses and from theoretical considerations, such an evaluation will be conservative, i.e. the risk will be grossly overestimated.

Multiple pulses

For the case of exposure to multiple pulses, each pulse has to be below the MPE (the MPE is determined for the respective pulse duration), additionally to the requirement that the average irradiance needs to be below the MPE that is calculated for any averaging duration within the maximum anticipated exposure duration. It depends on the pulse pattern which of the two requirements is the more critical one, i.e. which of the two criteria limits the exposure level. The average irradiance criterion is based upon the establishment of a 'background' temperature that builds up from the repetitive exposure and that is proportional to the average irradiance. For a constant pulse pattern, the time over which the pulses are averaged does not influence the level of the averaged irradiance, however, for non-constant pulse patterns, the average irradiance requirement needs to be satisfied for all averaging durations within the selected maximum anticipated exposure duration. For instance, if the chosen exposure duration is 10 s and there are sections in the pulse train where pulses lie closer over a period of 1 s, then the irradiance needs to be averaged over this 1 s of closer-spaced pulses and the resulting irradiance value is compared to the MPE that is applicable to a 1 s exposure additionally to averaging over 10 s. Generally, every possible exposure within the chosen exposure duration needs to be below the corresponding MPE—both in terms of exposure duration as well as in terms of the 'position' of the exposure within the pulse train, i.e. for each pulse grouping in terms of number of pulses and position of the group within the possible output of the laser.

As an example of a general evaluation of pulsed exposure for exposure durations of 10 s and above (where the MPE value given in irradiance is the lowest) we calculate the average irradiance E_{aver} from the radiant exposure per pulse H and the repetition rate f as:

$$E_{\text{aver}} = Hf = E_{\text{peak}} t_{\text{pulse}} f \tag{3.2}$$

where the radiant exposure per pulse was replaced by the peak irradiance and the pulse duration t_{pulse}. For the example of a wavelength in the range 400–1400 nm, this average irradiance needs to be below the MPE of $2000C_4$ W m^{-2}, i.e. $E_{\text{aver}} < 2000C_4$ W m^{-2}. Subsequently it is possible to calculate the allowed peak irradiance in W m^{-2} for a given pulse duration in s and repetition rate in Hz as well as to combine the peak irradiance with the pulse duration to obtain a requirement for the radiant exposure per pulse (in J m^{-2}):

$$E_{\text{peak}} < \frac{2000C_4}{t_{\text{pulse}} f} \quad \text{or} \quad H < \frac{2000C_4}{f}. \tag{3.3}$$

When this average irradiance requirement is compared with the single pulse MPE for the different pulse duration regimes, a critical repetition rate can be identified. When the repetition rate is above the critical rate, the average irradiance criterion is the more critical one, when the repetition rate is slower than the critical rate, the single pulse criterion is the more critical one. Since the single pulse MPE, for pulse durations between 1 and 100 ns is a constant radiant exposure value of $200C_4$ J m^{-2}, a comparison with equation (3.2) shows that in this time regime the critical repetition rate is 10 Hz, while the critical repetition rate decreases further with the rate of $t^{-1/4}$ so that at 1 ms (a 10^4 fold increase in pulse duration) the critical rate equals 1 Hz as is also summarized in table 3.7. Therefore, in many practical cases, the average pulse criterion will limit the allowed exposure level, especially for longer pulse durations.

For completeness it is noted that above comparison of the single pulse and the average power criterion is equivalent to first calculating the MPE value for a 10 s exposure duration, i.e. 2000 W m^{-2} and the equivalent radiant exposure value of 20 000 J m^{-2}. The total radiant exposure of all pulses within 10 s ($10 \times f \times H$) needs to be below that value to fulfil the average irradiance criterion, and a division of 20 000 J m^{-2} by 10 s and f yields again the right-hand side formula of equation (3.2).

Table 3.7. Skin MPE values that apply to single pulses, calculated as radiant exposure values as well as irradiance values for different pulse durations. A comparison of these single pulse—MPE values with the average irradiance criterion for pulse trains shows that there is a critical repetition rate above which the average irradiance criterion will be the limiting one. The wavelength correction factor C_4 was left out in the table for simplicity so that the numbers are applicable for 400–700 nm only but may be corrected with C_4 for wavelengths up to 1400 nm.

	Pulse duration			
	1 ns	10 ns	100 ns	1 ms
Single pulse—MPE expressed as radiant exposure (J m^{-2})	200	200	200	2000
Single pulse—MPE expressed as irradiance (W m^{-2})	200×10^9	20×10^9	2×10^9	2×10^6
Critical repetition rate (Hz)	10	10	10	1

3.9 Injury to the eye

The nature and location of laser induced injuries of the eye is quite varied and depends mainly on the wavelength of the radiation and on the pulse length (exposure duration), as already discussed in section 3.5. Insofar as the wavelength dependence is concerned, the absorption characteristics of the different parts of the eye critically influence the location of the injury. In the ultraviolet region, the cornea or the lens is at risk from both thermal and photochemical injury; in the visible and near infrared wavelength range (up to wavelengths of about 1300–1400 nm), the retina is at risk mainly from thermal injury, but for shorter wavelengths in the visible band (i.e. at blue and to some extent green wavelengths), also from photochemical injury. In the far infrared (i.e. wavelengths beyond 1400 nm), the lens and the cornea are again at risk, but in this wavelength range only from thermal injury. The three main wavelength ranges in terms of the location of injury are schematically shown in figure 3.14. In this simplified introductory overview, 'thermal' does not include thermomechanical effects, as discussed in section 3.5, which for very short pulses are also possible in the ultraviolet wavelength range.

The ocular media in front of the retina, namely the vitreous, the lens, the aqueous and the cornea, are mainly transparent in the wavelength range of about 400–1400 nm which is therefore referred to as the retinal hazard region (although transparency is decreased for wavelengths above 1150 nm and minimal for wavelengths above 1350 nm). The wavelength dependence of the transmittance of the primate eye is shown in figure 3.15.

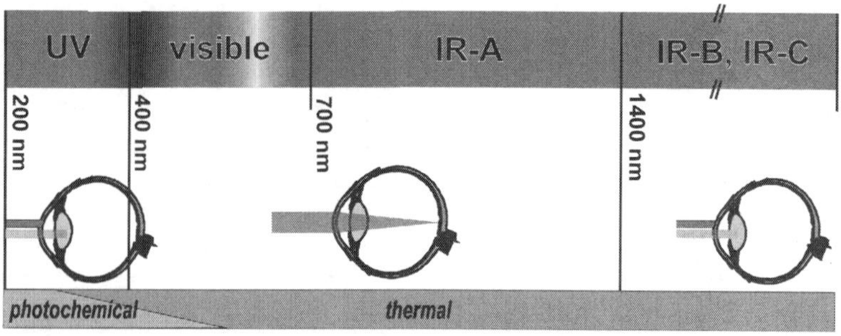

Figure 3.14. The part of the electromagnetic radiation spectrum of interest in laser safety. Ultraviolet, mid- and far-infrared radiation is absorbed by the cornea and lens of the eye, while radiation in the wavelength range 400–1400 nm is focused onto the retina, which results in very high irradiance levels on the retina even for small beam powers.

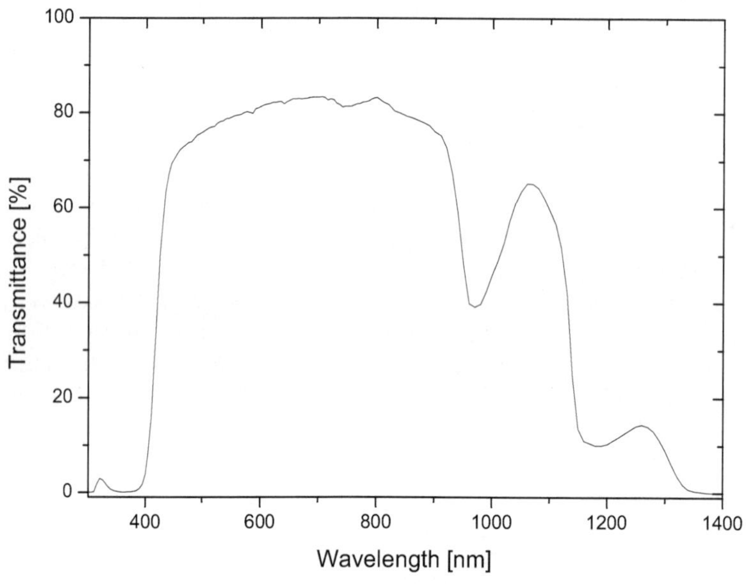

Figure 3.15. Combined transmittance of the ocular media in front of the retina of the young primate eye.

However, it is not only the transmittance of the ocular media alone that is responsible for the hazard to the retina, but also the special properties of laser radiation in terms of its focusing ability. The imaging properties of the eye for wavelengths between 400 and 1400 nm mean that remarkably low levels of laser radiation are sufficient to cause severe retinal injuries with corresponding

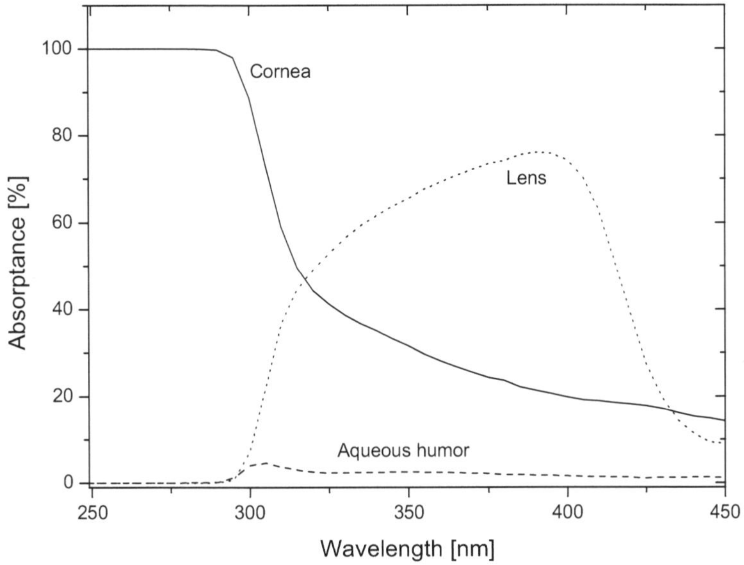

Figure 3.16. Absorptance of the frontal parts of the rhesus monkey eye—there are only minor differences to the absorption in the human eye. The absorptance of the lens depends heavily on the age and the data given here is applicable for a very young age with no yellowing of the lens.

detrimental effects on vision. Ultraviolet and far infrared radiation can result in severe injuries of the cornea and the lens, but radiation levels to produce those injuries are higher than those applicable for the retina. As such, there is no 'eye-safe' wavelength range, as the eye can be damaged at any wavelength, although the exposure levels necessary to cause injury can vary widely. (Wavelengths around 1500 nm are often erroneously called 'eye-safe' since the limits are much higher than in the retinal hazard area; these should rather be termed 'retina-safe'.)

3.9.1 Ultraviolet radiation

Across the ultraviolet wavelength range, the absorptance of the cornea and the lens is strongly wavelength dependent. In figure 3.16, the fraction of the radiation which is absorbed in the cornea, the lens and the aqueous humour is plotted as a function of wavelength.

In the far UV (UV-C, 180–280 nm), the absorption depth is very shallow and all the radiation is absorbed in the surface layers of the cornea. Correspondingly, exposure levels above the threshold for damage for short pulses can result in ablation of the cornea, for instance as observed for Excimer laser pulses with a typical pulse duration of about 20 ns and wavelengths of 193 nm or 248 nm. If the peak irradiance of those pulses is not high enough to produce photoablation, multiple exposures over a longer timeframe, for instance from stray-light, can still result in photochemical damage, i.e. inflammation of the cornea or the surrounding tissue, the conjunctiva. Such an inflammation (also referred to as photokeratoconjunctivitis) is also the relevant injury for exposure to cw sources and is well known from UV broadband sources such as from welding arcs, and can also occur from solar UV exposure, for instance from reflections from the snow, although irradiance levels from the Sun are much lower than from welding arcs at short distances or potentially possible from lasers. The inflammation is quite painful and can best be compared to the feeling of having sand in the eyes, but fortunately, if not very severe, heals within a few days without scars as the outer layers of the cornea are continuously renewed over a period of approximately two days. As is typical for photochemical damage, the injury takes several hours to develop and often it is only recognized the following morning after an exposure above the threshold has actually occurred. Such a photochemical effect may in practice be encountered when working with UV laser beams where eye protection is worn when working with or close to the beam, but it might be wrongfully thought that exposure some distance away from the beam or with the back to the beam is save. However, for the case that reflections from optical elements occur (such as 4% reflection from each surface of a lens) and/or diffuse reflections including those from walls, exposures to low irradiance levels over several hours can easily result in overexposure. With an MPE value of 30 J m^{-2} and an assumed exposure duration of 30 000 s (about 8 h), the allowed cw or average irradiance over 8 h would only be 10^{-3} W m^{-2}, which is equivalent to 0.1 μW cm^{-2}.

In the UV-B (280–315 nm) wavelength range, the radiation penetrates somewhat deeper into the eye and in this wavelength range both the cornea and the lens are at risk. Lens damage leads to a clouding of the lens which is also referred to as a cataract. The lens is less capable of repairing injury, and the reported studies of lenticular damage show that recovery only occurs following the lowest exposures that result in observable changes. Typically, exposure levels encountered in an accidental laser exposure are well above the threshold and damage to the lens is permanent, making cataract surgery and replacement by an artificial lens necessary.

In the UV-A (315–400 nm), the absorptance of the cornea is quite small and the lens is the predominant absorber. As the effectiveness of UV radiation to induce photochemical damage decreases with increasing wavelength (as the photon energy decreases), thermal damage of the cornea and the lens is the main concern at least from pulsed lasers.

3.9.2 Retinal damage

Between about 400 and 1400 nm, the parts of the eye in front of the retina transmit radiation (see figure 3.15) so that in this wavelength range, the retina is at risk. In very young eyes where the lens is still clear and transmittance is higher than

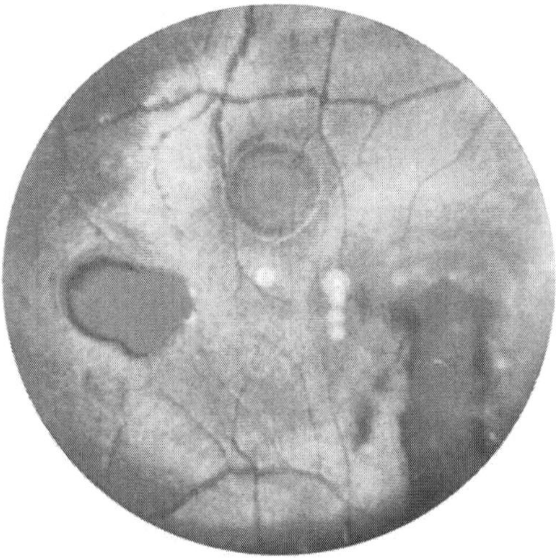

Figure 3.17. A range of injuries induced with a Nd:YAG laser. The white spots in the centre are thermal burns, i.e. coagulation of retinal layers. With larger energies, holes in the retina are produced which result either in bleeding into the vitreous, or the bleeding is contained in the layers of the retina, which results in functional loss in the affected area. (Photograph kindly made available by Joseph Zuclich, Northrop Grumman, TX.)

for adults, wavelengths of around 380 nm are transmitted to a limited degree to the retina, as can also be seen in figure 3.15. For prolonged exposure this may result in photochemical damage of the retina at values somewhat below the MPEs. While exposure to laser radiation at this kind of wavelength is generally rare and the exposure of infants might not be considered likely, there is some concern following the widespread use of LEDs emitting in the near UV wavelength range, for instance to check for counterfeit money. To meet this concern, the retinal limits might need to be extended down to 380 nm, in a similar way to the photochemical limit values for broadband radiation. Exposure to LEDs that emit over a bandwidth that might cover both the near UV and parts of the visible spectrum is more appropriately analysed by applying broadband incoherent radiation exposure limits [16, 17].

The fundus photograph shown in figure 3.17 shows several types of thermal and thermo-mechanical retinal injuries produced by a pulsed Nd:YAG laser. The white lesions result from thermally-induced denaturation of proteins and are characteristic of exposures that are longer than about 1 μs, up to an exposure duration of some seconds. For longer exposure durations to wavelengths in the blue and green part of the spectrum, photochemical injury can occur and this type of retinal injury visually appears similar to thermally produced injury. For these types of injuries, the area of the injury is in principal confined to the irradiated area. As the sensory retina (the nerve cell layers and the photoreceptors) is mainly transparent to optical radiation in the retinal hazard area, it is the melanin granules in the RPE that absorb most of the radiation in the visible (see also figure 3.3). In the near-infrared, the radiation penetrates into the choroid (which also contains lumps of melanin granules). Thermal injuries develop from exposure levels at threshold by heating of the melanin granules and subsequent thermal conduction to the segments of the photoreceptor cells closest to the RPE. Following thermal damage of the outer segments of the photoreceptor cell, the whole cell degenerates. For exposure levels far exceeding threshold levels (suprathreshold exposures) the elevated temperatures may well extend through all retinal layers and may also damage the nerve cell layers that are on top of the photoreceptors. Damage to these nerve cells may lead to large areas of vision loss when nerves are damaged that connect other areas of the retina to the optic nerve and that pass through the zone of the lesion. For laser pulses of duration less than 1 μs, rapid heating of the absorbing tissue causes thermomechanical injuries, possibly leading to haemorrhaging and damage extending well beyond the irradiated area, the flow of blood into the space between the retina and the vitreous greatly increasing the impairment of vision and the psychological impact of the injury on the victim. While the blood is absorbed after a few days, the extent and severity of such lesions is often greater than thermal or photochemical injuries. For instance, when the haemorrhage spreads within the layers of the retina, larger areas of the retina can be affected as the spreading blood could be trapped underneath a section of detached retina—such a lesion can be seen on the upper side of the fundus photograph shown in figure 3.17. To exemplify the effect of short pulse exposures with very high peak powers, we note that melanin granules can be brought up to temperatures of incandescence, liquid components can be converted to gas, bubbles may form around melanin granules and ionization may occur, i.e. the formation of a plasma. However, it is emphasized again that even minimal retinal lesions (at threshold) may result in permanent and serious vision loss if the damage is located in the foveal region of the retina.

As the retina as such does not have pain sensors, it has been speculated that thermally induced laser damage may go unnoticed, especially when it occurs in the periphery of the retina and not in the macula. Some victims of exposure to short pulsed laser radiation that induced thermomechanical damage have reported seeing a flash and hearing a popping sound, which might have been due to the small plasma created at the back of the eye. For exposure to blue or green wavelengths (or to a mixture of wavelengths, such as from the Sun, an arc lamp or an ultrabright white LED) for several (or many, depending on the brightness) seconds at exposure levels that may lead to photochemical damage, there is an initial bleaching effect similar to other temporary effects such as flash blindness. The actual permanent injury takes some hours to develop. If the lesion, even if only of minimal extent, is in the fovea or a larger lesion extending into the macula, loss of central vision is noticed as a central black (or sometimes white) region (referred to as a scotoma) in the visual field. Thus, vision in the peripheral (surrounding) field may be normal, but central sharp vision is lost and since peripheral vision is of low acuity, visual performance is greatly limited.

Retinal damage is permanent, i.e. damaged tissue areas do not form new functioning photoreceptors. However, it has been noticed in some cases of minimal lesions (close to the threshold) that vision does improve to some extent over the months following a laser accident. It is believed this may be due to sliding of viable retina into the damaged region. Medical treatment following an exposure to potentially hazardous levels of radiation should be carried out as soon after the accident as possible (certainly within a few hours) and be performed by a specialist ophthalmologist. Specific medical treatment for retinal injuries has not yet been established. Standard drugs are typically administered to prevent scarring of retinal tissue, but with varying success. An examination of the eye using fluorescein angiography within 24 h of exposure has largely only a legal value, as this is a sensitive technique which can detect even small lesions. Since fluorescein angiography can have serious side effects, it should never be used as a routine surveillance technique for monitoring the eyes of people working with lasers in the absence of other evidence that an injury may have occurred. For any kind of examination after a laser exposure that is suspected to have resulted in retinal damage, it is important that the ophthalmologist is aware that some small retinal irregularities, that many of us have and that do not impair vision, may look similar to a laser-induced lesion. These features would not be identified in a routine eye examination, but might be falsely identified as a laser lesion when the eye is examined more thoroughly following an exposure to laser radiation [18].

Laser eye accident statistics and analysis show that usually only one eye is hit by the laser beam, but there have been several cases where both eyes have been exposed and damaged, because the laser beam moved across both eyes, because the beam diameter was large enough to cover both eyes or because the person, having been exposed in one eye, then moved their head to get out of the beam and unintentionally exposed their other eye too! Due to eye movement, pulsed lasers may create more than one damage site on the retina of a single eye, which can be problematic as the scaring between the lesions may lead to tension in the retinal tissue.

3.9.3 Corneal damage from infrared radiation

Exposure to laser radiation in the infrared (IR) wavelength range outside the retinal hazard spectral region, i.e. between 1.4 and 1000 μm (1 mm), can result in corneal injury. Damage to the lens from infrared radiation is more likely to occur from chronic (i.e. long-term) exposure to broadband sources, and used to be well known in occupational medicine as glass-blower's cataract, following infrared exposure over many years. At irradiance levels necessary to acutely affect the lens (i.e. for short exposure periods), exposure levels at the cornea are high enough to also damage the cornea (in other words the cornea is damaged at an irradiance level lower than that necessary to acutely affect the lens). For the cornea, experimental threshold studies have identified thermal and, for shorter pulse durations, thermomechanical injury mechanisms. Accordingly, experimental injury thresholds and MPE values (as discussed below) generally follow the wavelength dependence of the penetration depth of radiation into the cornea.

The cornea heals within a few days of suffering superficial damage from exposure levels just above the threshold for injury, since the corneal epithelium is renewed every second day. However, accidental exposures to infrared radiation at wavelengths above 1400 nm are rarely superficial as the exposure levels that lead to injury, if exposure occurs, are usually well above the threshold. Although rare, corneal burns from CO_2 laser radiation have occurred.

3.9.4 Aversion response and typical exposure durations

As in the case of the skin, as discussed in section 3.7.1, nature has developed a number of protective mechanisms and aversion responses for the eye that limit the level of exposure and the exposure duration. However, it should be kept in mind that exposure levels well above the MPE can result in eye damage within a few milliseconds, far too short for aversion responses to take effect.

For exposure to ultraviolet radiation there is no natural protective mechanism that can be considered in a safety analysis to limit the exposure duration, since irradiance levels that are above the threshold for UV-induced photochemical damage are typically too low to be sensed as heat. Also, inflammation of the cornea typically takes a few hours to develop, and often develops overnight. Because of the additivity of UV radiation over a period of up to about 8 h, the typical exposure duration that is used in the safety analysis of UV radiation is 30 000 s (about 8 h). After approximately 8 h, the additivity of exposures breaks down, and repair mechanisms set in so that an exposure (irradiance) level that does not lead to the development of a minimal lesion after 8 hours will not result in any effect after a longer period of exposure.

Insofar as corneal thermal damage from infrared radiation is concerned, the cornea is very sensitive to heat and readily produces a sensation of pain. For a laser safety analysis of potential accidental exposure to far-infrared radiation, an exposure duration of 10 s is often used. This is because the MPE, when given in terms of irradiance, has its minimum value at 10 s and is then constant for longer exposure durations. When there is time to react to the exposure, i.e. when the irradiance level is such that it does not lead to an immediate injury (i.e. within a second or so), injury is prevented by reacting to the elevated temperature of the cornea and the pain that is sensed before actual damage occurs. 10 s is therefore a rather conservative assumption for potential accidental exposure durations.

Although the retina does not have pain receptors to prevent thermal injury, unlike the skin or cornea, there are a number of involuntary and intentional (cognitive) aversion responses when the human eye is exposed to bright light which all have the effect of either reducing the amount of light which enters the eye or limiting the exposure duration, thus reducing the potential for having a harmful effect on the retina. Constriction of the pupil, squinting or closing of the eyelids (the blink reflex), as well as moving the head or looking away, are all aversion responses which have developed mainly to protect the eyes from excessive sunlight, but are equally applicable to exposure from artificial, bright, visible light sources. Obviously those aversion responses only are induced by light which produces a corresponding stimulus, i.e. they only work for visible light. Infrared radiation that is transmitted through to the retina will not be seen and cannot be sensed, but may nevertheless injure the retina.

The laser safety committees have considered that a quarter of a second is an appropriate exposure duration for accidental exposure to visible light For this reason, exposures to visible light that do not exceed the MPE applicable to an exposure duration of a quarter of a second may be considered safe for accidental viewing. This is also the rationale for Class 2 as discussed in chapter 4.

A well-known protective response to exposure to bright light is the so called blink reflex, i.e. the involuntary closure of the eyelid following exposure to bright light. The blink reflex usually closes the eye within about 0.16 s after the onset of exposure. The blink reflex is a reaction to a sudden perceived threat to the eye, which includes being unexpectedly exposed to bright light. As such, the blink reflex is not a very strong reaction. If one expects that a bright light will appear in the field-of-view, or one looks intentionally at a bright source of light, the blink reflex does not operate. The blink reflex can also be impaired if people are under the influence of drugs (including alcohol). It is important to note, however, that the blink reflex is only one of a number of possible aversion responses, and if accidental, unexpected ocular exposure to a visible laser beam occurs, the person is usually startled and reacts accordingly (by turning their head and moving out of the beam). As such, Class 2 is not only based on the blink reflex, but on the general behaviour following exposure to bright light that can be expected to limit the exposure duration. It is also important to understand that while the exposure duration of 0.25 s was chosen to define Class 2 and results in a limit of 1 mW, somewhat longer exposure durations do not represent a significant hazard, as will be discussed in more detail in chapter 4. It is, however, possible to overcome the aversion responses to bright light and to intentionally stare into the beam for several seconds, just as people have stared into the Sun. In these cases, levels that would otherwise be safe for (short-duration) accidental exposure might lead to permanent retinal damage.

For infrared wavelengths in the retinal hazard region, i.e. for wavelengths in the range 700–1400 nm, the usually assumed exposure duration for an MPE analysis is 10 s, as it is not expected that somebody will stare fixatedly at a source for longer than 10 s that is not visible. (The MPEs are based on the minimal eye movements that would occur during intentional, deliberate viewing of the source. Where the source is within the eye's field-of-view but is not being stared at, larger eye movements will make the exposure less hazardous). There are, however,

Table 3.8. Typical exposure duration values used for eye MPE analysis.

UV 180–400 nm	Up to 8 h
VIS 400–700 nm	0.25 s or viewing duration for intentional viewing
IR-A 700–1400 nm	10 s (staring not assumed)
IR-B, IR-C > 1400 nm	10 s (based on heat aversion response)

exceptions to this that can occur when the source itself as opposed to the radiation emitted from it is being deliberately viewed (as when examining the end of a 'live' optical fibre, for example), or when another object adjacent to the infrared source is being viewed, such that the source remains stationary within the eye's field-of-view. In these cases exposure durations longer than 10 s may need to be assumed (although for small sources the MPE is constant for exposure durations longer than 10 s so that exposure durations longer than 10 s are not more critical than an exposure duration of 10 s).

The typical exposure durations used in an MPE analysis for exposure of the eye are summarized in table 3.8.

3.10 MPE values for the eye—also relevant to AEL values

In the following sections, the MPE values for the eye are discussed. We have split the discussion on MPE values that apply to the eye into three wavelength regions covering ultraviolet radiation, the retinal hazard region and infrared radiation at wavelengths greater than 1400 nm. The discussion of the dependence of the MPE values on wavelength, exposure duration and, for the retinal thermal limits, on the angular subtense of the apparent source, also apply directly to the AEL values for Class 1 and 1M, Class 2 and 2M, as well as to Class 3R (which is a factor of 5 above the AELs for Class 1 and Class 2). Where we refer to the parameter 'exposure duration' in this chapter on MPEs, for AELs the term 'emission duration' is used. While the exposure duration for an MPE analysis may be selected on the basis of the actual circumstances being considered, for classification (the comparison of the emission with the AEL), specific time base values are prescribed for each class that have to be used to calculate the AEL and assess the emission level. While the concept and terminology are different, the respective MPE and AEL values are numerically equivalent once the transformation of one into the other via the area of the limiting aperture is considered (see also section 4.2.1).

3.11 MPE values in the ultraviolet

The MPEs in the ultraviolet (UV) wavelength range are given in IEC 60825-1 as dual limits to protect the cornea and lens from both photochemical and thermal injury. The photochemical and thermal exposure limits have different dependencies on the exposure duration as well as on the wavelength, consequently it depends on the exposure duration and wavelength which of the two exposure limits is the critical (the lower) one.

The photochemical damage mechanism has a pronounced wavelength dependence which is followed by the laser MPEs only in a very simplified manner, as can be seen in figure 3.18. As is typical for photochemical interactions, the threshold for damage given in terms of radiant exposure (J m^{-2}) does not depend on the pulse duration (or exposure duration) over a very wide range from nanosecond pulse-durations to exposure periods lasting thousands of seconds. The 'photochemical' MPE when expressed in terms of radiant exposure therefore has no time dependence. (The limits referred to here as 'photochemical' do not exclusively protect against photochemical damage of the cornea, as discussed below, but the term is used to facilitate the discussion.) The wavelength dependence of the photochemical laser MPE is shown in figure 3.18 together with the ICNIRP exposure limit for broadband, incoherent radiation (natural solar radiation or from lamps) [14]. The ICNIRP limit follows more closely the experimental threshold values for inflammation of the cornea. Since photochemical threshold values for acute effects on the lens are generally above the thresholds for the cornea, only the thresholds for the cornea are relevant for the setting of the MPEs.

The photochemical laser limit of 30 J m^{-2} applies for all exposure durations above 1 ns in the wavelength range 180–302.5 nm. This value of 30 J m^{-2} is derived from the lowest broadband exposure limit, which occurs at a wavelength of 270 nm where the cornea has the highest sensitivity, as shown in figure 3.18. For wavelengths between 302.5 and 315 nm, the pronounced wavelength dependence of the photochemical limit is described by the factor C_2 (given by $10^{0.2(\lambda-295)}$), and this value also applies for all exposure durations between 1 ns and 30 000 s. For wavelengths above 315 nm, the 'photochemical' limit only applies to exposure durations longer than 10 s (for shorter exposure durations, a 'thermal' limit is defined, and is discussed later) and equals 10 000 J m^{-2} from 10–1000 s, and 10 W m^{-2} for exposure durations longer than 1000 s. Since for exposure durations longer than 1000 s, the exposure limit has a constant irradiance value of 10 W m^{-2}, the additivity of the exposures in terms of radiant exposure (dose) only applies up to 1000 s and not up to 30 000 s as in the case for shorter wavelengths. This long-term limit is not based on experimental data for corneal inflammation but is intended to protect the lens from long-term exposure effects to include both photochemical and a thermal component. The limits given in IEC 60825-1 are in this case less conservative (higher) than the limits specified in the ICNIRP guidelines and the ANSI laser safety standard Z136.1.

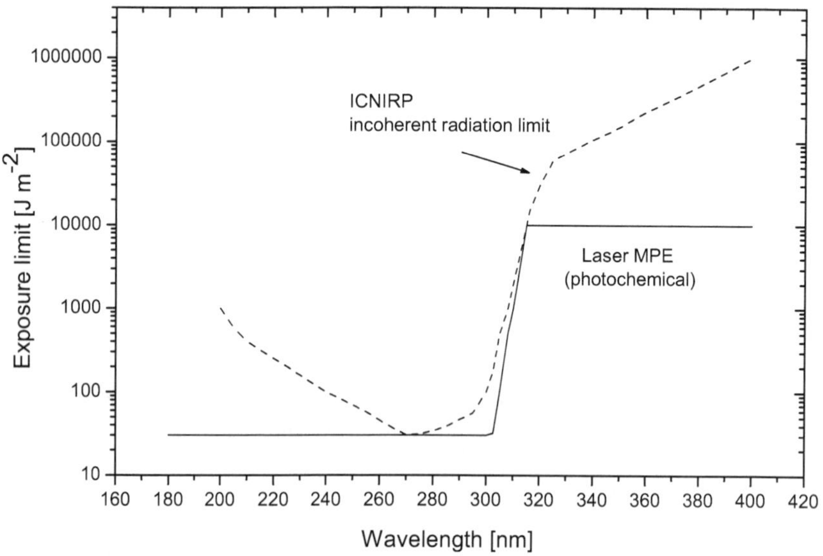

Figure 3.18. Comparison of laser and broadband photochemical MPE values.

In these two latter documents, the ocular MPE of 10 000 J m^{-2} is defined for the exposure duration range 10–30 000 s (not just up to 1000 s), with the result that for a potential exposure of 8 h duration, the irradiance needs to be limited to a value of 0.3 W m^{-2}, a factor of 30 lower than specified in the IEC document. With no experimental threshold data, both limits are based on an estimate of the natural solar UV exposure and epidemiological studies of the frequency of cataract occurrence in different countries. The approaches to this estimate and analysis are somewhat different in the different committees. The laser limits of IEC follow the broadband limit for UV-A as defined by the ACGIH [19], while the ANSI and ICNIRP laser limits follow the broadband limit for UV-A exposure as defined by ICNIRP [20].

A comparison of the photochemical broadband limits with the laser limits that are plotted in figure 3.18 shows that the laser limits are a rather crude simplification, and for wavelength ranges below about 250 nm and above 320 nm the difference to the broadband limits is up to a factor of 100. However, it should be noted that the laser limits also afford protection against pulsed exposure to submicrosecond pulses where photoablation might occur for wavelengths below 250 nm (as was shown with Excimer lasers), and for wavelengths above 320 nm they also protect against thermal long-term effects on the lens (these effects can be induced at exposure levels that are below the exposure limits for photochemical damage).

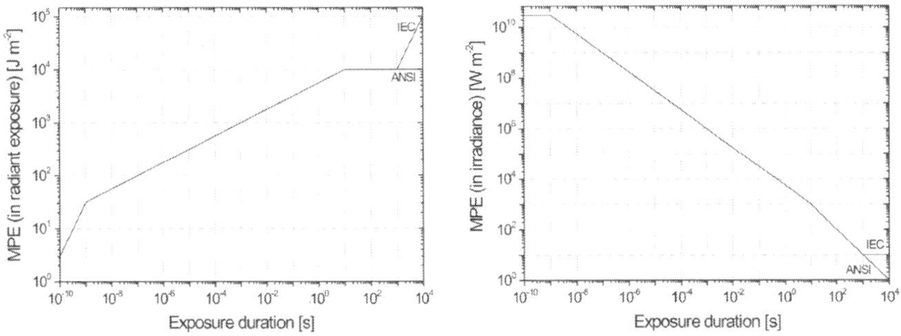

Figure 3.19. Function describing the thermal MPE for the wavelength range 302.5–400 nm as a function of pulse duration t. The middle section from 1 ns to 10 s is given by the time dependent factor C_1.

The 'thermal' MPEs are specified for wavelengths between 302.5 and 400 nm and for pulse durations between 1 ns and 10 s, and are defined as C_1 J m^{-2}, which is equal to $5600t^{0.25}$ J m^{-2}. (For 10^{-9} s the MPE equals about 30 J m^{-2} and for 10 s it equals about 10 000 J m^{-2}.) This function expresses the temporal dependence of the MPE for thermal damage to the cornea by pulsed near-UV radiation and is plotted in figure 3.19.

In the wavelength range 302.5–315 nm and for pulse durations between 1 ns and 10 s, the thermal limits overlap with the photochemical limits and therefore create dual limits. For exposure to single pulses it is possible to directly compare the two MPEs and calculate which one of the two is the lowest for a given wavelength and pulse duration. Therefore, in the IEC laser safety standard IEC 60825-1, the thermal and the photochemical MPE values are not explicitly presented as dual limits. Instead, a time $T_1 = 10^{0.8(\lambda-295)} \times 10^{-15}$ s is defined[1] as the boundary between the two sets of limits, and is shown in the MPE table as a diagonally split cell that contains the thermal limit in the lower left portion (longer wavelengths and shorter pulse durations) and the photochemical limit in the upper right portion (shorter wavelengths and longer pulse/exposure durations). When the exposure duration is less than the time T_1 the thermal MPE must be used; when the exposure duration is longer than T_1, then the photochemical MPE has to be used. The time T_1 is that pulse duration for which the two functions for the thermal and the photochemical limits are equal. T_1 therefore defines the diagonal in the MPE table as presented in IEC 60825-1 and the 'ridge' of the three dimensional presentation of the 'dual' MPEs as shown in figure 3.20. The temporal factor T_1 is simply obtained by equating the wavelength dependent

[1] In the ANSI laser safety standard Z136.1, the parameter T_1 is used to distinguish retinal thermal from retinal photochemical regions for small sources—the UV limits are presented as dual limits without defining a time for which one or the other is the more critical one.

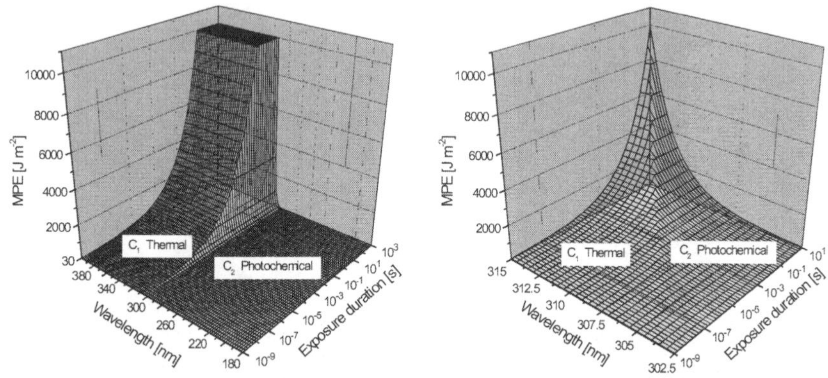

Figure 3.20. Three-dimensional plot of the UV laser MPEs for single pulse exposures. On the left-hand side, the full UV wavelength range is shown and exposure durations from 10^{-9}–1000 s, while on the right-hand side, the wavelength and exposure duration region where the thermal and photochemical limits 'meet' is enlarged. The photochemical limits exhibit a very strong wavelength dependence and change over more than two orders of magnitude within a wavelength range that is only 13 nm wide. The thermal limits exhibit a rather 'shallow' time dependence with the corresponding change from 30 J m^{-2} to 10 000 J m^{-2} for exposure durations of 1 ns to 10 s.

formula for the photochemical limit with the time dependent formula for the thermal limit and solving for the time.

The thermal limit of C_1 J m^{-2} applies also to the wavelength range 315–400 nm, for pulse (exposure) durations of 1 ns to 10 s, and connects to the value of 10 000 J m^{-2} that is valid for exposure durations between 10 and 1000 s. All the MPE values that apply for ultraviolet wavelengths are listed in table 3.9.

In both the ICNIRP laser exposure limit guidelines and the ANSI laser safety standard, the photochemical limit is listed as the primary limit, and the thermal limit is specified as an additional restriction for the full wavelength range 180–400 nm and the full exposure duration range of 1 ns to 30 000 s, which in fact constitutes the definition of dual limits as is done for the incoherent broadband limits and for the laser limits in the visible wavelength range.

3.11.1 Multiple pulses

The MPE values as given above apply to single continuous periods of exposures, i.e. to single pulses. Exposure to multiple pulses in the ultraviolet wavelength range needs to be evaluated according to the following special criteria.

Table 3.9. MPE values for the eye as given in IEC 60825-1 (in the ANSI laser safety standard and in the ICNIRP laser guidelines, the value of 10 000 J m^{-2} extends to 30 000 s).

Wavelength	Exposure duration t*	MPE value
180–302.5 nm	1 ns–30 000 s	30 J m^{-2}
302.5–315 nm	1 ns–30 000 s	If $t < T_1$: $5600t^{0.25}$ J m^{-2} (C_1 J m^{-2}) If $t > T_1$: $10^{0.2(\lambda-295)}$ J m^{-2} (C_2 J m^{-2}) where $T_1 = 10^{0.8(\lambda-295)} \times 10^{-15}$ s
315–400 nm	1 ns–10 s	$10^{0.2(\lambda-295)}$ J m^{-2} (C_2 J m^{-2})
	10–1000 s	10 000 J m^{-2}
	1000–30 000 s	10 W m^{-2}

* For exposure durations less than 1 ns, see the end of this section 3.11.2.

(1) Single pulse criterion

The first criterion is to directly apply the MPE value to each pulse (and it is therefore referred to as the 'single pulse criterion'); for a train of pulses, each individual pulse needs to be below the MPE, where the MPE is calculated for the respective pulse duration.

(2) Average irradiance criterion/additivity criterion

Additionally to the single pulse criterion, the average irradiance (averaged over some duration T_{av} up to the exposure duration) needs to be below the MPE that applies to the duration T_{av}. When the MPE is specified as a radiant exposure value that does not depend on exposure duration (as is the case for the photochemical limit) this rule transforms to the well-known dose, or additivity rule, as is shown later.

The average irradiance criterion needs to be fulfilled for all groupings of pulses within the pulse train and for all values of T_{av} between the pulse duration and the maximum anticipated exposure duration. For instance, if there are sections in the pulse trains where the pulses lie closer together or have a higher energy per pulse, then the irradiance averaged over that section of the pulse train will be higher than an irradiance averaged over longer or other sections of the pulse train. In general, the average irradiance E_{av} for some averaging duration T_{av} is calculated by summing all pulse radiant exposure H_{pulse} values within T_{av} and dividing the resulting total radiant exposure by T_{av}

$$E_{av} = \frac{\text{Sum of all } H_{pulse}}{T_{av}} \tag{3.4}$$

For constant pulse patterns (constant in terms of repetition rate f and radiant exposure per pulse), the average irradiance does not depend on the averaging duration and is calculated by multiplying the radiant exposure of a pulse with the repetition rate. The MPE is evaluated for the maximum anticipated exposure duration, since the MPE when specified in terms of irradiance either decreases with time or remains constant, and an evaluation at the maximum anticipated exposure duration produces the lowest MPE value. For the case where the MPE is specified in terms of a constant radiant exposure (i.e. independent of the exposure duration), such as for most of the photochemical UV limits, the average irradiance criterion transforms to what could be referred to as the additivity criterion: the total radiant exposure (the sum of individual radiant exposure values within the maximum anticipated exposure duration) needs to be below the radiant exposure MPE value that is valid for the maximum exposure duration. The equivalence of the average irradiance criterion with the additivity criterion can be seen when it is considered that the radiant exposure MPE value (such as 30 J m^{-2}) is converted into an equivalent irradiance MPE value by division with the exposure duration (in seconds). Since the average irradiance value (equation (3.4)) that is to be compared to the irradiance MPE is also derived by dividing by the exposure duration, the two divisions by time cancel out and the total radiant exposure (the sum of all H_{pulse}) that is incident on the cornea within the maximum anticipated exposure duration can be compared directly to the radiant exposure MPE value (of, for instance, 30 J m^{-2}). This form of the average multiple pulse requirement reflects the general additivity rule of individual exposures whenever the MPE is specified as a constant (non-time dependent) radiant exposure value, for photochemical limits and for thermal limits within the thermal confinement time (see also discussion on the basics of beam–tissue interaction, section 3.5). As an example of the additivity requirement, let us consider the MPE for the wavelength range 180–302.5 nm, which is 30 J m^{-2} for all exposure durations between 1 ns and 30 000 s. Assuming constant pulse energies, the total radiant exposure (to be compared to the MPE) is given by multiplying the radiant exposure of each pulse (H_{pulse}) by the number of pulses N within the anticipated exposure duration: $N \times H_{pulse} < 30$ J m^{-2}, which, by division with N can be recalculated into the requirement that the radiant exposure per pulse needs to be limited to 30 J m^{-2} divided by N.

In summary, the 'average irradiance criterion' applies to thermal MPEs for exposure durations longer than the thermal confinement time, and reflects the build up of a background temperature, due to repeated exposures, that is proportional to the average irradiance. The additivity criterion applies to photochemical MPEs to reflect the dose relationship typical for photochemical interaction, and also to thermal MPEs for exposure durations shorter than the thermal confinement time.

(3) $N^{-1/4}$ rule for thermal limits

In addition to the multiple pulse criteria that are currently defined in IEC 60825-1 as discussed above, the ANSI laser safety standard also requires that the radiant exposure value of each pulse needs to be below a reduced thermal single-pulse MPE that is calculated by multiplying the thermal MPE by the pulse reduction factor $N^{-1/4}$, where N is the number of pulses within the anticipated exposure duration. This criterion is well known for the thermal limits of the retina (and is discussed in more detailed in the section on retinal MPEs), and also applies to infrared thermal damage to the cornea. From an understanding of the biophysical background it can be surmised that the reduction of the exposure limit for multiple exposures also applies to thermal damage in the UV wavelength range. Depending on the repetition rate f, the pulse duration t_{pulse} and the maximum anticipated exposure duration $T_{\max}$, the $N^{-1/4}$ criterion can be more stringent (i.e. result in a more conservative MPE) than the average irradiance criterion described above. For constant pulse trains, it can be inferred that the $N^{-1/4}$ criterion is more stringent than the average irradiance criterion when the expression $T_{\max}^2 \times f^3 \times t_{\text{pulse}}$ is less than 1. When this additional criterion is applied to the MPEs as defined in the IEC document, the split between the photochemical and thermal limits as represented by T_1 is still valid, as the photochemical limit with the corresponding additivity criterion, for a given exposure duration, will always be more stringent than the $N^{-1/4}$ criterion. (However, the reduction of the thermal MPEs by the factor $N^{-1/4}$ leads to a discontinuity of the MPEs at T_1.) It should be stressed that the absence of the $N^{-1/4}$ rule for UV wavelengths in the current IEC document has little practical impact, since the MPEs are specified in terms of non-time dependent radiant exposure values for exposure durations of 10 s and above, and the additivity criterion is more conservative than the $N^{-1/4}$ criterion. It is only for shorter maximum anticipated exposure durations, such as 1 s, and for near UV wavelengths (where the thermal limit and not the photochemical limit applies) that the $N^{-1/4}$ criterion could be more stringent than the average irradiance (additivity) criterion, especially for high repetition rates. There are relatively few practical exposure situations where exposure durations less than 10 s would be used for an MPE analysis. (For classification, a time base of 30 000 s generally has to be used).

Example. Let us consider a set-up with a KrF Excimer laser that emits at a wavelength of 248 nm and has a repetition rate of 20 Hz. The pulse duration is 20 ns, the energy per pulse equals 0.1 J and the beam has a cross section of 1 cm^2. For simplicity, it is assumed that the energy is distributed evenly over the beam cross section to produce a radiant exposure of 0.1 J cm^{-2} or 1000 J m^{-2}. (Since the beam profile is assumed to be homogenous and the beam dimensions are much larger than the limiting (averaging) aperture, this value of radiant exposure can be compared directly with the MPE.) The beam is emitted horizontally across a table where a manually-adjusted optical set-up is being used for an experiment. Two exposure scenarios could be evaluated. The first is the possibility of exposure of the skin or the eye to the direct beam, for example when moving the head close to the beam while manipulating the optics, or because of a specular reflection of the beam. The corresponding radiant exposure equals 1000 J m^{-2} per pulse and this is far above the MPE for the skin and for the eye of 30 J m^{-2} even for exposure to a single pulse, let alone for a number of pulses, and adequate personal protective equipment is therefore needed. The second possibility is the case where the optics are well mounted and the risk of specular reflections is low, but exposure of the face to stray light from diffuse reflections (when eye protection is being worn) could still occur. An estimate of the diffusely reflected radiation at a distance of 1 m from the diffuse surface gives a radiant exposure per pulse of 0.03 J m^{-2}. (A procedure for calculating exposure levels arising from diffuse reflections is given in section 5.5.) Due to the additivity of the radiation (the MPE is specified in terms of constant radiant exposure), the MPE for the skin is reached after exposure to the diffuse reflection of 100 pulses, i.e. within 5 s for a repetition rate of 20 Hz. As the alignment procedure is likely to take longer than 5 s, protection of the skin during alignment or manipulation close to a diffuse reflection would also be necessary to prevent harm to the skin that can be quite severe for extended exposure durations. Following alignment, a screen or hood around the optical table could be closed when the experiment is conducted so that stray light from the experiment (that can also occur from reflections at lens surfaces) is shielded.

For comparison with the exposure to a wavelength in the far UV, let us assume that the laser is refilled with XeCl to emit at a wavelength of 308 nm and that the other emission characteristics remain as above. At a wavelength of 308 nm, the parameter T_1 needs to be evaluated to determine if the thermal or photochemical MPE applies for a single pulse. The parameter T_1 is equal to 25 μs, which is larger than the pulse duration of 20 ns, and so the thermal MPE applies, which is equal to 67 J m^{-2}. The exposure to a single pulse of the direct beam is again far above the MPE for a single pulse, and the multiple pulse criterion does not need to be evaluated. The exposure to diffuse reflections of 0.03 J m^{-2} per pulse is below the single pulse MPE and therefore has to be evaluated following the multiple pulse criteria. Since T_1 equals 25 μs, any realistic exposure duration to multiple pulses will be longer than that and consequently the photochemical limit applies. In the UV wavelength range, it is generally the most practical approach to analyse the photochemical limit first and to calculate the allowed exposure duration for a given level of average irradiance (this is equivalent to comparing the total radiant exposure with the photochemical MPE when specified in terms of radiant exposure). The photochemical limit for a wavelength of 308 nm equals 398 J m^{-2}, and division of the MPE by the radiant exposure per pulse gives the number of pulses for which the total radiant exposure would equal the MPE. This is equal to 13 266 pulses. With a repetition rate of 20 Hz, this gives an maximum safe exposure duration of 663 s or about 11 min. If the total exposure duration during one day does not exceed this maximum allowed exposure duration of 11 min (and it needs to be stressed that individual exposures are additive over the full day), protection of the face would not be considered necessary. Eye protection is still strongly recommended even if the MPE is not exceeded for exposure to the diffuse reflection, as the level of accepted risk from specular reflections or for a somewhat prolonged exposure is lower for the eye than for the skin.

Finally, we assume that the laser is filled with XeCl to emit at a wavelength of 355 nm and the emission characteristics otherwise remain as before. At a wavelength of 355 nm, the thermal limit applies up to 10 s. For direct exposure to the beam, the radiant exposure of a single pulse is still above the MPE of 67 m^{-2} (the same thermal MPE applies as for 308 nm). For exposure to the diffuse reflection, we can again calculate the allowed exposure duration for the given radiant exposure per pulse and repetition rate. We can first use the MPE that applies in the exposure duration range 10–1000 s, which for the UV-A wavelength-range equals 10 000 J m^{-2}. Due to the comparatively high MPE value for near UV radiation, the allowed exposure duration based on the MPE value would be 16 666 s or about $4\frac{1}{2}$ h. As this exposure duration is outside of the exposure duration range for which the MPE of 10 000 J m^{-2} is actually defined, we move into the regime of the MPE that is defined for exposure durations above 1000 s (and is a constant irradiance value of 10 W m^{-2}). The average irradiance that results from the diffuse reflections is calculated by multiplication of the radiant exposure per pulse by the repetition rate to obtain a value of 0.6 W m^{-2}, which is below the MPE and so the exposure to diffusely reflected radiation does not exceed the MPE, even for exposure durations of up to 8 hours. This is an example where the difference of the MPE values for the eye as specified in the ANSI document and in the IEC document does make a difference in the MPE evaluation. ANSI Z136.1 (and ICNIRP) specifies the 10 000 J m^{-2} limit up to 8 h, the value for the allowed exposure duration of $4\frac{1}{2}$ h as calculated above, remains in the exposure duration range for which the limit of 10 000 J m^{-2} is defined in the ANSI standard. While the average irradiance does not exceed the MPE as specified in the IEC document, under the ANSI definition of the MPEs the exposure duration should be limited to $4\frac{1}{2}$ h (which is, in practice, usually still large enough not to result in any practical restrictions).

3.11.2 Ultrashort pulses

For pulse durations less than 1 ns, no experimental injury threshold data are yet available in the UV wavelength range, and, as for the skin, exposure limits can be conservatively set using the constant irradiance value applicable for a pulse duration of 1 ns. For instance, from the MPE of 30 J m^{-2} that is valid down to 1 ns, the equivalent irradiance would be 30×10^9 W m^{-2}. For pulses shorter than 1 ns, the MPE expressed in terms of radiant exposure therefore decreases linearly with time, so that at 10^{-14} s, the limit for the radiant exposure per pulse equals 0.0003 J m^{-2}.

3.12 Retinal MPE values

For radiation in the wavelength range 400–1400 nm, the ocular MPEs relate to retinal injury. This wavelength range is therefore also referred to as the retinal hazard region. In this wavelength range, the ocular media in front of the retina are considered sufficiently transparent so that for a given level or irradiance at the cornea of the eye it is the retina that is the part of the eye that has the lowest injury threshold, i.e. that is the part of the eye that is at the greatest risk for injury. The retina is the most vulnerable part of the body to visible and near-infrared radiation, as laser radiation is typically focused onto a very small spot on the absorbing layers of the retinal tissue, producing a high retinal irradiance for relatively small beam powers that enter the eye. Consequently only a small amount of optical energy is sufficient to cause injury and the MPE values are correspondingly low.

For the discussion of retinal MPEs, it is important to keep the dosimetry concept as reviewed in section 3.6.6 in mind: by convention, the MPEs are defined at the position of the cornea and the level or irradiance or radiant exposure that is incident at cornea of the eye (averaged over a 7 mm diameter aperture) is the value that has to be compared to the MPE. Clearly, the actual level of irradiance or radiant exposure occurring at the retina will be very different from the value measured at the cornea, but this is taken into account in defining the MPE values. However, since the location of injury is the retina, it should be kept in mind that it is the irradiance and radiant exposure at the retina that is relevant for retinal injury, together, for the case of thermal damage, with the size of the irradiated retinal area.

It is one of the outstanding features of laser radiation that it can be focused to a tiny spot. This property can in a simplistic way be understood by envisaging a laser beam to consist of basically parallel rays (although even a very well collimated beam will not have totally parallel rays but the divergence and therefore the angles of the rays will be very small; for a more precise treatment of laser beams see chapter 5). When parallel rays fall on the eye, as schematically shown in figure 3.21(*a*) they are focused to a minimal spot on the retina, such as is the case for light that comes from a distant star and that, by the time it reaches the Earth, consists of parallel rays and also produces a minimal image on the retina (see chapter 1). The diameter of the smallest spot that it is possible to achieve at the retina is approximately 25 μm.

Instead of using retinal diameters in μm, the size of the retinal image or spot (at this stage of the discussion we assume a circular spot) is best characterized by the plane angle that the spot or image subtends as seen from the focusing elements of the eye, with the simplifying assumption that the eye is filled with air. This angle is measured in milliradian (mrad) and can be referred to as the angular subtense of the retinal spot (see section 2.3.1 for the definition of plane angle and 3.3 for a discussion of the optics of the eye). As can be seen in figure 3.21(*b*), the angle subtended by the retinal image (the angle to the right) is the same as the angle that is subtended by the source (the angle to the left). Therefore this

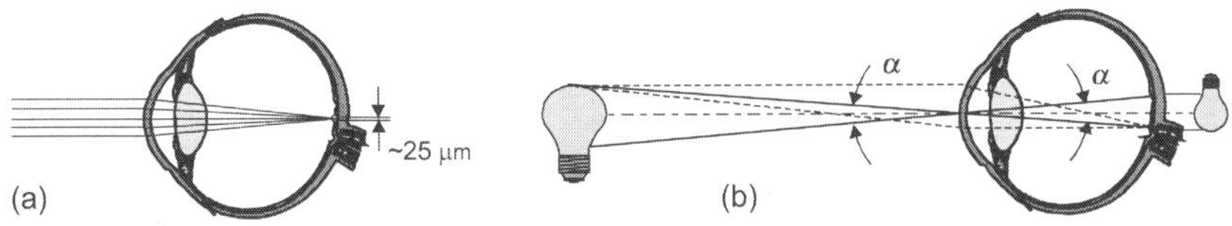

Figure 3.21. (*a*) A collimated laser beam can be considered to consist of practically parallel rays. Such parallel rays are focused to a tiny spot on the retina. Light from a 'conventional' source (*b*) forms a larger image on the retina. Retinal spot sizes can be characterized by the angle that the spot subtends as seen from the focusing elements of the eye. It is noted that in the figures, the ray paths through the eye are not correct but assume a thin lens in air at the position of the cornea that has the same refractive power as the imaging elements of the eye.

angle, having the symbol α, is referred to in the laser standards as the 'angular subtense of the apparent source', and not the angular subtense of the retinal spot (as this is difficult to determine and is actually not fully correct as the eye is not filled with air, as is discussed in section 3.3). However, it should be kept in mind that this angle characterizes the minimal spot size that can be obtained for a given laser beam and a given location of the eye, as will be discussed in more detail further below. The adjective 'apparent' indicates that for laser beams and LEDs with lens caps the size of the image on the retina is related to a virtual source of radiation (that for instance can lie a long distance behind the actual laser) and not to the actual physical source of radiation, i.e. the laser cavity or the LED chip. As such, the angular subtense of the apparent source should also not be confused with the divergence of the beam. That the divergence in the general case does not characterize the angular subtense of the source (or the image size on the retina) can be easily seen with the example of a star: the 'divergence' of the emitted radiation of a star is such that it covers the whole space of 360 degrees, but the star is still imaged as point source onto the retina. However, the directions of the diverging rays in the beam are roughly related to the location of the apparent source—in simple terms the location of the apparent source can be understood as the perceived origin (source) of the radiation, as schematically shown in figure 3.22. For a well-collimated beam (a beam with small divergence), the rays appear to originate from far behind the laser cavity and thus the apparent source is effectively located at infinity. For a highly divergent beam, the apparent source will be located roughly at the location from where the rays seem to diverge if their direction at the eye is extended back. It should be noted that this simple ray picture breaks down for exposures at or near the beam waist of a laser beam. In more technical and general terms (general as this applies to both incoherent radiation as well as to laser beams), it can be shown that the location of the apparent source is at the centre of curvature of the wavefront at the cornea (i.e. for a plane wavefront the location of the apparent source is at infinity and for a

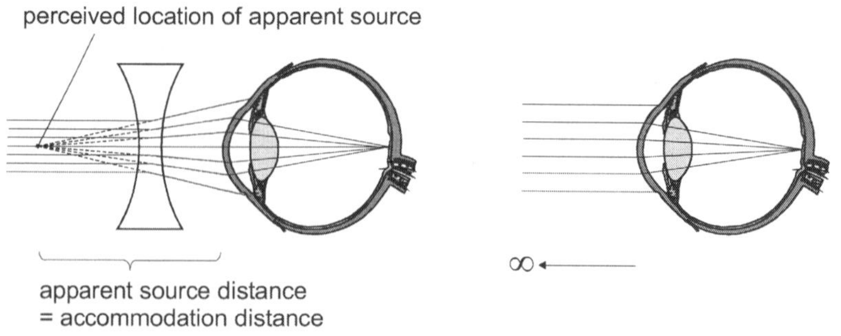

Figure 3.22. The location of the apparent source can be understood as the distance to which the eye accommodates to produce the smallest spot or image on the retina. As a simplified concept, the rays that are incident on the eye can be followed back to a (possibly imaginary) source. In the example shown on the left, the divergence is quite large and the apparent source is very close, on the right the beam is well collimated and the source is perceived as to lie at infinity, i.e. the eye is accommodated to infinity. The example on the left shows again that a large beam divergence does not mean that the retinal image spot (and α) is large too, as the rays will be imaged to a minimal spot if the original beam is well collimated and the lens is assumed to have negligible aberration.

spherical wavefront at the point of origin of the spherical wavefront as emitted from a point source).

The retinal thermal MPE values in their basic form are defined for the 'default' worst-case of a minimal retinal spot. Well-collimated beams and sources that are correspondingly small and/or far away produce such a minimal spot. The minimal spot corresponds to a minimal value of the angular subtense of the apparent source and this figure is referred to as $\alpha_{\rm min}$ and is equal to 1.5 mrad. For exposures where the retinal spot size is larger than this minimum value (referred to as 'viewing of *extended* sources'), the retinal thermal MPE values may be increased by a correction factor that has a maximum value of 66.6. The correction factor that affects this increase of the thermal retinal MPE values is referred to as C_6 in IEC 60825-1 (and $C_{\rm E}$ in the corresponding ANSI standard and in the ICNIRP guidelines). C_6 equals unity for minimal retinal spot sizes. Since it is often difficult and time consuming to determine the correct value of C_6, an MPE analysis (and classification) can be greatly simplified by neglecting the possibility of an extended image on the retina and by setting $C_6 = 1$. In most cases of collimated laser beams this worst case assumption is the correct value for C_6 anyway. Sources where C_6 may be larger as unity are LEDs, arrays and diffuse sources viewed from close distance. Even if a source is truly extended, if the analysis with an assumption of $C_6 = 1$ is satisfactory in terms of obtaining a certain classification or an exposure level below the MPE, then it is not necessary to invest time and effort to determine the correct value of α. However, even if C_6

is assumed to be unity, for classification and in some cases for an MPE analysis, the *location* of the apparent source can be relevant as measurement distances may be specified relative to the location of the apparent source, and not relative to the output aperture of the laser product.

The general concept of the apparent source is discussed first in this section, followed by a discussion of the variation of the thermal retinal MPE values with wavelength and with pulse (exposure) duration. Then the background and determination of α for the application of the retinal thermal limits is discussed, followed by a discussion of the photochemical retinal limits. Photochemical limits do not depend on the retinal image diameter, as it is assumed that the irradiated area is governed more by the extent of the eye movements than the extent of the actual image. However, measurements and calculations can be simplified when α is smaller than the angular extent that characterizes the eye movements (this angle γ_{ph} is defined as the limiting FOV (angle of acceptance) for measurements). At the end of this section on retinal limits, the treatment of multiple pulse exposures is discussed. Some examples are included in the following sections of this chapter, and more complete case studies at the end of chapter 4 are designed to not only provide practical examples of classification but also of the more general concept of MPE values and MPE analysis.

It is noted here that for a hazard analysis of optical radiation, as a worst case it is assumed that the eye accommodates to form the smallest angular subtense that is possible for the given laser beam and location in the beam even for near infrared radiation, i.e. even when the radiation cannot be seen. Consequently, the discussion in this section generally applies to radiation over the full retinal hazard range 400–1400 nm. This is a valid and prudent assumption, as is underlined by accidents that have occurred with Nd:YAG lasers in a laboratory setting. Although for Nd:YAG laser radiation at a wavelength of 1064 nm there is considerable chromatic aberration in the optical system of the eye, that serves to blur the retinal spot, accidental exposures often produce minimal spots. The chromatic aberration is unfortunately counteracted by an accommodation distance that is different from actual distance to the apparent origin of the radiation. This is because, in the laboratory, the eye is typically accommodated to look at objects close by (gauges, etc), but a collimated laser beam seems to originate from infinity.

3.12.1 Apparent source

The concept of the apparent source is used in optical radiation safety to characterize the most hazardous retinal image size. For a given power that enters the eye, the most hazardous exposure is obviously represented by the smallest retinal spot size. For this case of the minimal retinal spot the retinal thermal MPEs have their lowest values and the correction factor for extended sources, C_6, equals unity. For extended sources, the retinal thermal MPE values may be increased by a factor $C_6 = \alpha/\alpha_{min}$ so that the larger the irradiated spot on the retina (as characterized by the parameter 'angular subtense of the apparent source' α) is, the more power is allowed to enter the eye. The background and application of C_6 is discussed in section 3.12.5 while in this section, we discuss general aspects of the apparent source. As in the current edition of the international laser safety standards, measurement distances may be specified relative to the location of the apparent source, this discussion is relevant not only for retinal thermal limits for extended sources but also for the measurement geometry. Also, the background of photochemical limits and related measurement requirements can be better understood with an awareness of the concepts of retinal spot size and apparent source.

The smallest retinal spot size achievable with a well-collimated laser beam is of the order of 20–25 μm. Due to scattering in the eye, this value is somewhat larger than would be considered the smallest theoretically possible (diffraction limited) spot size. The actual size of the minimal spot is difficult to define, as it is made up of a central spot of the order of about 5–10 μm (which is akin to the diffraction limited profile) with somewhat higher irradiance, which is surrounded by an area ('skirt') of smaller irradiance, which, however, still contains an appreciable amount of the total power. The international laser safety standards refer to a source of radiation as a 'small source' when it is characterized by an angular subtense of less than 1.5 mrad (such as would be the case for an object with 1.5 mm diameter at a distance of 1 m from the eye, or an object of 0.15 mm at 10 cm distance from the eye). When we use the effective focal length of the standard relaxed eye in air of 17 mm to characterize the distance between the retina and the respective principal plane of the combined lens-corneal system, an angle of 1.5 mrad is equivalent to a retinal spot diameter of 25.5 μm. The angle of 1.5 mrad is referred to as the 'minimal angular subtense' and is denoted by $\alpha_{\rm min}$ (i.e. $\alpha_{\rm min}$ = 1.5 mrad). This angle characterizes the minimal spot that can be created by the optical system of the eye at the retina. Even when the source is characterized by an angle that is smaller than 1.5 mrad, the irradiated spot on the retina will not be smaller than the minimal spot size. It is noted that in previous editions of laser MPE guidelines and standards, $\alpha_{\rm min}$ was time dependent to characterize how eye movements can smear the image. However, this concept was not completely correct. Following the current and more appropriate definition, the minimal angular subtense $\alpha_{\rm min}$ characterizes the optical properties of the eye in terms of producing a minimal spot size, and this is irrespective of any eye movements.

For the cases where the spot that is produced on the retina is larger than $\alpha_{\rm min}$, the concept of the angular subtense of the apparent source, α, is used to characterize the extent of the spot on the retina. Also, the location of the apparent source can be relevant for a laser safety analysis. These issues are somewhat involved for laser beams and we discuss the simple case of a conventional light source first.

3.12.1.1 Simple sources of light, accommodation

For sources of incoherent broadband radiation such as a light bulb, a fluorescent tube or the Moon (which is actually a source of diffusely reflected radiation), it is obvious what the source (the object that emits or diffusely reflects the radiation) is and how large it is (although for the Moon, the actual size and distance is not obvious, but the subtended angle is), and also the location of the source is well defined. When we look at the object, the 'autofocus' system of the eye will adjust the focal length of the lens so that the optical radiation that is emitted or reflected from the object and that is intercepted by the eye is imaged onto the retina. The adjustment of the focal length of the lens of the eye is referred to as accommodation and is discussed in more detail in section 3.3. The distance to which the eye accommodates (i.e. the distance at which objects are imaged to form a sharp image on the retina) can be referred to as the accommodation distance. From an optics standpoint, accommodation means that the lens of the eye varies in thickness so that the focal length f of the cornea-lens system is adjusted until the accommodation distance D_{acc} coincides with the distance of the source to the eye D_{source}, and the image that is produced by the focusing elements of the eye lies at the position of the retina, i.e. the distance from the cornea to the retina D_{ret} coincides with the image distance D_{image}. When the eye is accommodated to image the source, these distances satisfy the lens equation

$$\frac{1}{D_{\mathrm{acc}}} + \frac{1}{D_{\mathrm{image}}} = \frac{1}{f} \qquad \text{and} \qquad D_{\mathrm{acc}} = D_{\mathrm{source}} \text{ as well as } D_{\mathrm{image}} = D_{\mathrm{ret}}. \tag{3.5}$$

Because of this process of imaging, rays that are emitted from one point of the object are recombined onto another point on the retina, thus forming a sharp image, as is indicated by some selected rays in figure 3.5 for the imaging of a light bulb. This sharp image is obviously needed for good vision as objects that are not imaged properly onto the retina appear blurred, which is what happens when, for instance, $D_{\mathrm{acc}} \neq D_{\mathrm{source}}$ (when the eye accommodates to some other distance than the source distance) or $D_{\mathrm{image}} \neq D_{\mathrm{ret}}$ (when the image is in front or behind the retina, as is the case for myopia and hyperopia, respectively). In the process of accommodation, the main criterion for adjusting the thickness of the lens and thereby the accommodation distance is to obtain sharp edges. For non-coherent sources, a sharp image of the source typically produces the desired edges and also represents the worst-case condition in terms of hazard to the retina, as it produces the smallest retinal spot size with the highest retinal irradiance. (When the image is blurred, the radiation that enters the eye is spread over a larger spot on the retina.) This is not always the case for laser beams, as is discussed in the case study on line lasers, where the image that is typically seen by an observer is a sharp line (that corresponds to a relaxed accommodation) but the worst case spot size is produced when the eye accommodates to the distance of the cylindrical lens that produces the line.

For the case that a simple source such as a light bulb is imaged onto the retina, the angular subtense of the source α is simply determined by dividing the diameter of the source by the distance of the source, and this angle also characterizes the angle that is subtended by the retinal image (for the air model of the eye). For instance, a frosted light bulb with a diameter of 5 cm (50 mm) positioned at a distance of 1 m from the eye subtends an angle of 50 mrad. The angular subtense of the retinal image is also 50 mrad, which corresponds to a retinal image diameter of 50 mrad $\times$ 17 mm $=$ 0.85 mm. In more general terms, the irradiance distribution of the image on the retina is directly related to the distribution of the emission (the exitance) that is radiated or reflected from the object in the direction of the eye. When the retinal irradiance is not constant and not defined by sharp edges, the techniques used to determine a laser beam diameter can be adopted to define α, as is discussed further below and in chapter 5. The treatment of non-circular or multiple sources is discussed in sections 3.12.5.5 and 3.12.5.6, respectively.

In summary, for a simple source such as a light bulb, the image formed at the retina is the image of the physical object that emits or reflects the radiation. Also, the accommodation distance, i.e. the distance at which the eye accommodates (focuses), coincides with the actual distance to the physical object. Consequently, the angular subtense of the retinal image is equal to the angular subtense of the physical object that emits or diffusely reflects radiation and this parameter (α) is for all distances to the source simply calculated by dividing the diameter of the source (assuming circular objects at this stage for simplicity) by the distance to the source. As a result, when the distance of the eye to the source is increased, the retinal image decreases steadily and linearly with increasing distance. Examples of what is referred to here as a 'simple' source and where the above description applies are:

- light bulbs without reflectors (either clear, when the filament is the physical source, or frosted, when the frosted glass, acting as a diffuse transmitter, becomes the optical source that is imaged);
- fluorescent tubes (the fluorescent coating is the source);
- bare LED chips (i.e. without a lens cap);
- any diffuse transmission or reflection including that of laser beams (see section 2.7.3 for a discussion on diffuse transmissions and reflections).

In more technical terms, it is common to these 'simple' sources that each point of the relevant source surface emits spherical waves, i.e. each point of the surface can be considered as a point source radiating in all directions into a hemisphere—what is termed a Lambertian radiator. Here, diffuse reflections are also considered as a 'source', as from each point of the diffuse reflection a spherical wave is emitted so that the rays coming from that point can be imaged to form a point in the image (even when the diffusely reflected radiation is monochromatic and originated from a laser beam that is incident on the diffuse surface). Furthermore, there are no intermediate or reflecting optics which could for instance magnify the source, as is discussed in the following paragraphs.

3.12.1.2 Collimated beams of incoherent light

For sources of incoherent radiation that employ a lens to produce a more or less collimated beam, such as search lights or low divergence LEDs, the emitting part (such as the filament of the light bulb or the chip of the LED) is located generally at the focal plane of the lens. As is schematically shown in figure 3.23, such a set-up collimates the radiation that is emitted from the lamp. Each point of the lamp emits rays in all directions that represent a spherical wave that are converted by the lens to parallel rays and a plane wavefront (equivalent to a spherical wave that originates at infinity). The beam still exhibits a certain degree of divergence (δ as shown in figure 3.23) that is related to the extent of the source, i.e. the smaller the source the smaller the divergence. For the theoretical ideal case as depicted in figure 3.23, the divergence of the beam is equal to the angle of the source subtended at the lens, i.e. the diameter of the source divided by the distance of the source to the lens (that is equal to the focal length of the lens). When viewing the source from within the beam, i.e. placing the eye within the beam and looking back into the source, when the eye is accommodated to infinity, the parallel rays from each point are recombined to form an image on the retina.

It is an interesting property of such a (perfectly) projected source that the angular subtense of the retinal image does not vary with viewing distance and is equal to δ, i.e. the angular subtense of the retinal image is equal to the divergence of the beam. For viewing distances beyond the focal point of the lens, the source appears magnified, i.e. the retinal image δ is larger as the physical dimension of the actual source of radiation α_{source} as shown in figure 3.23(*b*). When moving away from the projector or LED, the image of any physical object such as the projecting lens becomes smaller (ε_1 and ε_2) but the image of the emitting filament or chip remains the same up to the point (the flash distance) at which the image of the filament or chip fully fills the projecting lens (where $\delta = \varepsilon_1$) which is then termed 'flashed'. For viewing distances beyond the flash distance, the retinal image size is limited to the angle of the projecting lens.

As the eye is relaxed to form the image (the accommodation distance is infinity), the location of the source that is imaged appears to be at infinity. Therefore, in optical radiation safety, one does not refer to the source as such but to the *apparent* source, where the angular subtense of the apparent source α characterizes the retinal image size and the location of the apparent source is equal to the accommodation distance of the eye, i.e. the distance at which the eye accommodates to obtain the image with 'size' α.

For a real set-up, such as a search light or low divergence LED, a perfect projection will not be possible, since the lens will have aberrations and also the lamp will have some finite depth (extent along the optical axis), or the source may contain mirrors such as the cup in an LED, so that not all parts of the source

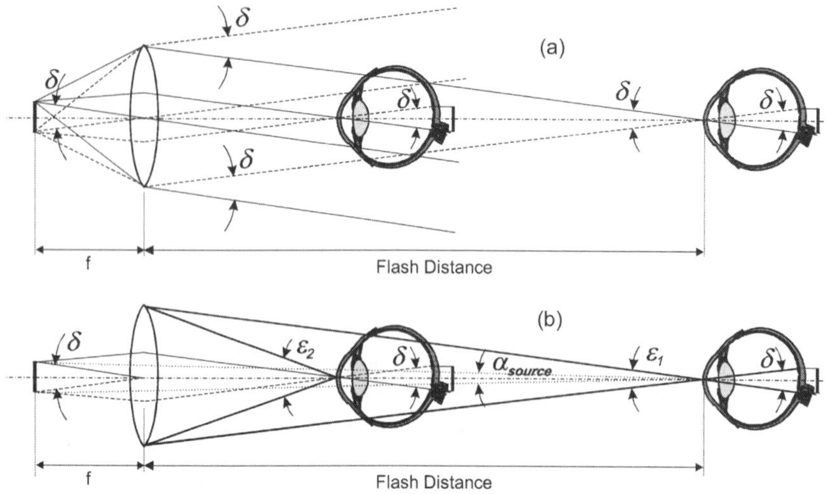

Figure 3.23. Configuration for search lights and low divergence LEDs where more or less collimated beams are obtained by placing the source at the focal plane of the projecting lens. The angular subtense of the retinal image is equal to the angular subtense of the source δ subtended from the lens and does not vary with viewing distance up to the flash distance when the lens subtends the same angle (ε_1) as the magnified lamp δ.

will be in the focal plane of the lens. Alternatively, instead of using a lens, it is possible to use a parabolic reflecting mirror which creates a combined source of the filament or LED chip at the centre of the image that is surrounded by the magnified image (which lie at different locations, and therefore at short distances of less than about 2 m cannot both produce simultaneously a sharp image on the retina). An analysis of such a real source is more complicated but it is still correct that the retinal image size varies little with distance to the projector and the apparent source is located some distance behind the actual source. Figure 3.24 shows images of the same type of LED where the lens cap was removed for the photograph on the left (i.e. the chip was not magnified) and with the usual lens cap on the right, that acts as projection lens. Both images were taken at the same distance to the LED but the focus of the camera was changed—for the image on the left-hand side, the focal distance was set to the actual position of the chip, for the photograph on the right-hand side, the focal setting was set some distance behind the chip.

The usage of the term 'the' apparent source can imply that for a given beam there is only one apparent source in terms of location and physical size. For instance, for a certain position of the eye x in the beam, one could determine α_x and the distance of the location of the apparent source to the eye, $D_{\mathrm{acc\_}x}$. From these two parameters it is possible to calculate the 'size' of the apparent source

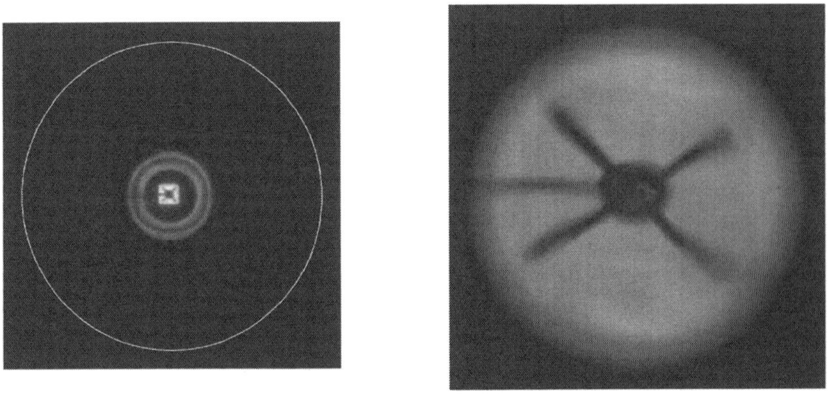

Figure 3.24. Images of a LED with and without a 'projecting' lens, taken at the same distance from the lamp. The lens-cap of the LED on the right magnifies the image of the chip such that it practically fills the lens (which is the same diameter as the LED, namely 3 mm as indicated by a white circle in the left image).

in mm (l_x) by $l_x = \alpha_x \times D_{\mathrm{acc\_}x}$, in the sense of the size of an imaginary object that would result in a retinal image angle α_x when placed at $D_{\mathrm{acc\_}x}$. However, it is not correct to consider this source size in mm as applicable for other viewing positions, i.e. it is not correct to calculate α for other viewing positions by dividing l_x with the distance to the previously determined location of the apparent source. The case of the projected incoherent source is one example where it is clearly seen that this approach is in error, as in the case of the projector the apparent source size does not change with distance (up to the flash distance). Clearly, the parameter α varies with distance in a different way as would be the case if there were an imaginary object representing the apparent source at a certain location. It is rather that α characterizes the angular subtense of the retinal image at a certain position of the eye within the beam and the corresponding value of α needs to be determined specifically for each position that is evaluated in an MPE analysis (where for that position, the power that passes through a 7 mm pupil is used to calculate the corneal irradiance and α is used to calculate the appropriate MPE value for that specific position). Experimentally, the value of α for a given beam and exposure position in the beam can be determined by placing a lens at the position of investigation to form an image of the source on a screen or a CCD array, just as the eye would form an image on the retina. We assume here for simplicity that the image is circular and has sharp borders that define the diameter of the image, so that the angle α is obtained by dividing the image diameter on the screen by the distance of the screen to the corresponding principal plane of the lens. As the focal length of the lens will be different from the focal length of the eye, and the distance from the screen to the lens will be different from the distance between the retina and the cornea. Also, the image size on the screen will be different from the image size on the retina. However, for such kinds of incoherent sources (where spherical waves are emitted from each point of the source) the image of the source also represents the most hazardous retinal irradiance pattern, and the size of this image can be characterized by the angle that the (magnified) source subtends. The image position and size can be determined by using geometric optics, i.e. with the lens formula (equation (3.5)) and by equating the image angle with the source angle, respectively.

At this point we would like to address two issues that often lead to confusion and misunderstandings. First, the use of the term 'intrabeam viewing'. With intrabeam viewing, we mean that the eye intercepts the beam, i.e. the beam is incident on the eye, and the direction of viewing is such that the source is within the visual field. Intrabeam viewing can also occur via specular reflections (i.e. via mirrors), as mirrors merely redirect the beam but do not otherwise change its characteristics. For intrabeam viewing it is typical that we image the source of radiation, such as the filament of the bulb in the search light, i.e. the rays that are emitted from a point of the surface of the filament and that are intercepted by the eye are recombined in one corresponding ('conjugate') point of the retinal image. This is in contrast to viewing diffusely reflected radiation, where in the process of diffuse reflection, individual rays are scattered and the 'information' about the original source of light is lost. In an optical sense, a diffuse reflection can be considered a secondary source of radiation from which radiation is emitted in all directions into a hemisphere, and consequently also the original beam geometry itself is lost. In the field of laser safety, the term 'intrabeam viewing' was historically (mis)used to refer to exposures that produce a minimal retinal spot size. This misuse arose since the 'typical' laser beam is well collimated and intrabeam exposure to such a beam invariably produces a minimal retinal spot size. The only type of exposure where such beams can be involved in producing an extended image is when these collimated beams are incident on a diffusing surface and the diffuse reflection (the beam diameter at the diffuse surface) would have sufficient size and is viewed from a close enough distance. In the pre-90s editions of the international laser safety standards and in other documents, there were exposure limits that applied to small sources that had the heading 'intrabeam viewing' and additionally to this, exposure limits (given in units of radiance) that applied for extended sources that had the heading 'viewing of diffuse reflections'. The usage of these terms with this 'historic' meaning can lead to confusion as, on the one hand, many sources including projectors and LEDs (and also some laser beams) can be viewed from within the beam but do not represent a small source, but on the other hand, diffuse reflections are only extended sources when formed by large diameter beams or when viewed at very close distances. (For example, a beam diameter of 1 mm at the scattering surface represents a small source for viewing distances larger than 67 cm.)

The second common area of confusion is to some extent related to the understanding of intrabeam viewing and to the viewing of diffuse reflections. The irradiance profile across the beam as it propagates through space needs to be kept conceptually separate from the irradiance profile at the position of the image. For intrabeam viewing, it is the emission profile of the source that is related to the retinal irradiance profile and not (at least not in the general sense) the irradiance profile of the beam as it propagates through space. For instance, the irradiance profile in the beam of a search light might be quite homogenous (which can be made visible by holding a diffuse surface in the beam), but the actual image is that of the filament (that can be made visible by placing a diffuse surface at the image plane behind a lens). In order words, if the beam as such falls directly on a surface (such as a piece of paper or the cornea of the eye) without being imaged by a lens, the irradiance profile at that surface is quite different from the irradiance profile of the image that is formed when that same beam passes through a lens and is then incident on a screen or the retina that is positioned in the image plane. There are cases where the beam profile of the propagating beam through space is closely related to profile of the image, as is for instance the case for a Gaussian beam (TEM_{00}) but this is rather the exception than the rule. In the general case, the rays that pass through any one point that does not lie in the image plane are coming from different points in the image and thus are 'mixed-up' in terms of creating an irradiance profile at a given cross section in the beam. (To be able to infer information on the object or the image, additionally to the irradiance profile, one would need information on the direction of the beams that pass through the given point—this is the background to holography where the hologram pattern is made up of interference patterns that are related to the direction of the wavefront). This difference of irradiance profile in the image plane to the irradiance profile at other positions in the beam is stressed since it is the irradiance profile in the image plane that is relevant for the characterization of the angular subtense of the apparent source and not the profile of the beam as it is emitted by the product or as it progresses through space.

3.12.1.3 Laser radiation

For hazard evaluation of intrabeam viewing of laser radiation, the concept of the angular subtense of the apparent source needs to be somewhat generalized, as a laser beam propagates differently to incoherent radiation. For instance, the rays that make up a laser beam do not form an 'image' behind a lens as do the rays that are emitted from conventional sources (as described above). For a laser safety analysis of a given laser beam and a given position of the eye in that laser beam, the general understanding of the angular subtense of the apparent source is such that the value of α characterizes the *smallest retinal spot that can be achieved at that exposure position*. That is, for a given position of the eye within a given beam (and therefore for a given power that enters the eye), it is assumed that the accommodation of the eye varies until the smallest retinal spot that is possible for the given laser beam and position in the beam is obtained. The accommodation range that is generally considered extends from infinity to the near accommodation point of 10 cm, which corresponds to a focal length of the eye in air of 14.5 mm and 17 mm, respectively. Even though very young people and myopic people can have an effective focal length of less than 14.5 mm and hyperopic people can have an effective focal length of more than 17 mm (so that even converging beams could be transformed to a minimal retinal spot), based on risk analysis of critical exposures occurring, the range 14.5–17 mm is considered generally sufficient.

The characterization of the angular subtense of the apparent source is simplest for high quality (well collimated) laser beams, i.e. low divergence beams that can be envisaged to consist of almost parallel rays: intrabeam exposure of the eye to such a laser beam produces a minimal retinal spot, i.e. $\alpha = \alpha_{\min}$ at all exposure positions within the beam. For this condition, the eye is relaxed, i.e. the accommodation distance tends toward infinity. From this simplest case it is also obvious that the apparent source size is not related to the beam diameter at the cornea or the diameter at the exit mirror: a well collimated laser beam produces a minimal image on the retina (i.e. one sees a small spot like from a star) independent of the beam diameter. For collimated laser radiation, the location of the apparent source is therefore also some distance 'behind' the laser. Seen from the standpoint of optics, such lasers exhibit a wave front that is close to plane and such a plane wave front is converted by the eye into a spherical wave front that converges towards a minimal (diffraction limited) retinal spot size.

However, lower quality (i.e. higher divergence and larger beam waist) laser beams can produce retinal spots that are larger than $\alpha_{\min}$ and thus may represent an extended source. In order to correctly determine α for such a laser beam, it is necessary to consider the concepts of beam diameter and propagation of a laser beam that are discussed further in chapter 5. Figure 3.25 schematically shows the envelope of a laser beam that is incident on the eye and is transformed by the cornea-lens system of the eye to result in some retinal spot size d_{ret} (measured in mm). For such a laser beam, the retinal spot size is mainly determined by the location and diameter of the beam waist in front of the eye and by the divergence of the beam. Even when there is no external beam waist (a beam waist outside of the laser cavity) and the emitted beam diverges from the exit aperture, a virtual beam waist can be assigned to that laser beam. This virtual beam waist will be located inside or some distance 'behind' the laser, which is indicated in figure 3.25 with dashed lines. For a given position of the eye in the beam, there is some variation of the retinal spot size possible for different accommodation distances (for different focal lengths of the lens), but again it is assumed that the focal length varies to produce the smallest possible spot size whose angular subtense is then termed 'angular subtense of the apparent source' α. (In this sense, α should not be used as a general symbol for the angular subtense of a retinal spot, but only for the *minimal* spot size that can be achieved by an appropriate focal length of the eye (for a given beam and location of the eye in the beam).

It is not possible to apply the techniques of geometrical optics to deduce the location and diameter of a secondary beam waist from the location and diameter of the primary beam waist in front of the lens, at least not for lens locations

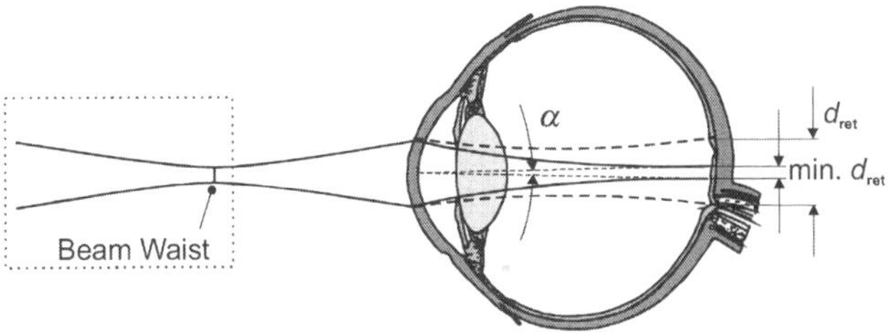

Figure 3.25. A laser beam with a beam waist some distance in front the eye is transformed by the focusing elements in the eye to result in a certain spot size on the retina. When this spot size, for a given beam and position in the beam, is minimized by variation of the focal length of the eye (indicated in the figure by two beam envelopes inside the eye as they would be formed by two different focal lengths of the eye), then the corresponding minimal angular subtense (calculated by $\min d_{\text{ret}}/17$ mm) gives α, the angular subtense of the apparent source.

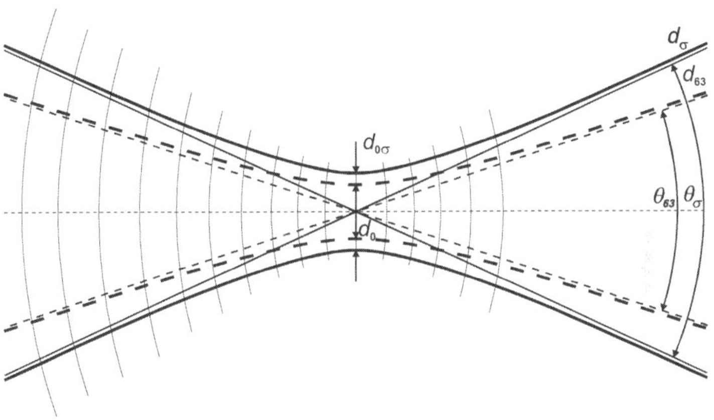

Figure 3.26. For one given beam, the envelope that represents the beam (and might be used to calculate the beam diameter at given positions along the propagation) depends on the choice of the criterion to determine and define the beam diameter.

that are relatively close to the primary beam waist. That geometrical optics is a simplification that cannot explain the shape of the beam waist is for instance seen when the rays in the far-field in figure 3.26 are extended back towards their apparent origin: they do cross the optical axis at the position of the beam waist (as such the location of the beam waist can be determined that way), but from simple geometrical optics standpoint of view they would create a focal spot with zero diameter. It is obvious that this cannot be correct.

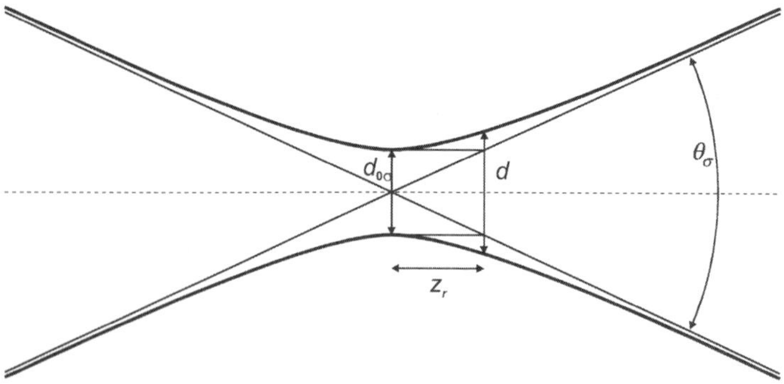

Figure 3.27. The Raleigh range is a figure of merit that characterizes the distance from the beam waist where the beam diameter increases only slightly with distance while outside the Raleigh range, the beam diameter roughly increases linearly as characterized by the far-field divergence. Large-diameter, low-divergence beams have large Raleigh ranges, small diameter, large-divergence beams have short Raleigh ranges.

Some basic principles of beam propagation are discussed in the following on the basis of Gaussian beam profiles. As discussed in chapter 5, the highest quality beam (the lowest possible divergence for a given wavelength and waist diameter) is a so-called TEM_{00} beam, which features a Gaussian beam profile all along the beam propagation including transformation by lenses. The diameter of the beam, however, varies along the direction of propagation in a very distinct way: at the beam waist the diameter is smallest and the beam increases in diameter on either side of the waist. This change of beam diameter is represented graphically and conceptually by the beam envelope. It is important to note that the shape of the envelope will change according to the criterion that is used to determine the representative point along the beam profile that is referred to as the 'radius' or 'diameter' (such as the second moment, or the 1/e irradiance criterion), as is shown in figure 3.27 for the beam envelope determined for both the 1/e (63% of total power) and the second moment criterion (for a Gaussian beam profile equivalent to the $1/e^2$ or the 87% of total power).

The shape of the envelope is basically hyperbolical. However, in the far-field, the beam envelope can be approximated well by straight lines that come from the position of the beam waist. In this region, the beam diameter linearly increases with distance with a degree that is characterized by the *divergence* θ (that is also often referred to as the *far-field divergence* to indicate that it has to be determined in the far-field to be accurate). It is helpful for the following discussion to note that one can associate a wavefront to the different positions along the beam. The wavefront of a Gaussian beam represents a section of a circle that is characterized by a centre (referred to as the *centre of curvature*, short for the centre of curvature of the wavefront) and the radius of curvature.

For a given point along the beam axis, it is obvious that the radius of curvature is equal to the distance of the centre of curvature to the respective point. It is a special characteristic of a laser beam that the wavefront in the beam waist is plane, corresponding to a centre of curvature that is at infinity. When one moves away from the beam waist, the centre of curvature moves closer to the position of the beam waist. In the far-field, the beam waist can be treated as the general centre of curvature. The extent of the near field can be characterized by a figure of merit that is referred to as the Raleigh range z_r. The Raleigh range extends from the position of the beam waist (with diameter d_0) to the distance where the cross section of the beam (the area) has doubled. Graphically, as shown in figure 3.27, this is equivalent to the position at which the straight lines that characterize the far-field divergence intersect with the beam waist diameter 'level'.

From simple geometrical considerations it follows that the Raleigh range is defined as

$$z_r = \frac{d_0}{\theta} \tag{3.6}$$

where z_r can be given in units of metre and θ in rad, or, more commonly, the beam waist diameter d_0 is given in mm and the divergence θ is given in mrad, which produces a Raleigh range that is given in metres. The Raleigh range characterizes the range over which the beam diameter is close to the value of the waist, so that a beam with large beam waist diameter and low divergence has a large Raleigh range. For instance, a beam with a waist diameter of 10 cm and a divergence of 1 mrad has a Raleigh range of 100 m, while a beam with a waist diameter of 1 mm and a divergence of 100 mrad has a Raleigh range 10 mm. As a rough simplification, for locations within the Raleigh range, the diameter of the beam increases quite slowly as one moves away from the waist, while outside of the Raleigh range the increase becomes linear with distance, as characterized by the far-field divergence. In the far-field, the beam behaves essentially like a section of a spherical wave that is centred at the beam waist.

A lens that is placed into the beam transforms the primary beam waist into another beam waist behind the lens. As a special case, when the beam waist happens to be located at the focal point in front of the lens, a secondary beam waist is formed at the focal point behind the lens, as is shown in figure 3.28. The second waist is not an *image* of the first waist; it is rather a second position along the beam at which the beam diameter has a minimum value. While in most optical systems we are concerned with forming an image and therefore utilize geometrical optics to analyse the behaviour of imaginary 'rays' as they pass through the system, in the case of laser beams we are usually concerned with the beam size. Although an object placed in the focal plane of a lens would give rise to an image at infinity and not in the second focal plane, the creation of a beam waist in the second focal plane of a lens as just described should not be construed as somehow contravening the laws of geometrical optics, nor of being directly concerned with imaging. Geometric optics and beam optics are simply different ways of analysing the propagation through space (usually filled with air)

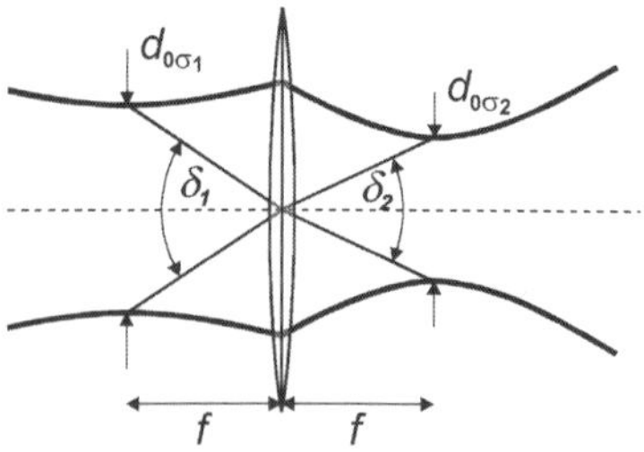

Figure 3.28. Transformation of a primary beam waist in a secondary beam waist, shown for the special case of the primary beam waist being located at the focal plane of the lens. Transformation of the beam by the lens results in the formation of a secondary beam waist which is also at the focal plane of the lens.

of optical radiation. (It is possible, incidentally, to use geometrical optics and conventional ray tracing techniques to demonstrate how waists can be formed.)

As can be seen in figure 3.28, when a beam with a certain beam waist position and diameter is transformed by a lens, the angle subtended by the primary beam waist, δ_1 is not the same as the angle subtended by the secondary beam waist δ_2, as would be the case if the two beam waists would be equivalent to 'source' and 'image' that would follow geometrical optics.

For intrabeam exposure of the eye to the radiation of a laser beam, the cornea and lens of the eye transform the beam envelope to form a secondary beam waist at some distance behind the cornea of the eye, although it is not generally the case that the secondary waist forms at the retina. For a given primary beam, the diameter and location of the secondary beam waist will to some extent depend on the focal length of the lens of the eye. The angular subtense of the source α is defined as the angular subtense of the smallest angular spot that is possible for a given beam and a given location in the beam. The angular subtense of the apparent source α can be characterized either by experimental techniques that basically use an artificial eye with a lens and a CCD array to simulate the focusing process inside the eye, or by model calculations of the beam diameter at the retina. There is a well-developed formalism available that describes the beam diameter as a function of propagation distance including transformation and formation of a secondary beam waist by a lens [21]. The propagation theory is based on Gaussian beams, but it has been shown that when the beam diameter is determined following the second moment method as described in an ISO standard [22], then the formalism also satisfactorily describes the propagation of non-Gaussian beams. A beam (at least a non-astigmatic beam) is fully characterized by the waist diameter, the waist position relative to the eye, and the far-field divergence of the beam. In the following, any reference to a beam diameter or divergence will imply the determination according to the second moment method. For Gaussian beam profiles (TEM_{00} beams), the second moment method yields the same results as the $1/e^2$ irradiance level criterion (or 84% power criterion).

By application of the beam propagation theory it is possible, with some simplifying assumptions, to calculate the minimal retinal spot diameter for a given beam and a given exposure position [23, 24]. While discussion of the detailed model is outside the scope of this book, in section 3.12.5.9 we present basic results that should help the general understanding of the issue.

3.12.2 General evaluation approach

The concept of a safety evaluation in the sense of comparing the exposure level (at different positions in the beam, or at the most hazardous position) to the MPE was introduced in section 3.6. Following the detailed discussion of the apparent source in the previous sections we now have the basis to further explore a generalized approach to a safety evaluation. More specific concepts that apply to the different types of MPE values are treated in the respective sections where the MPE values are discussed.

Such a general approach sets out to determine whether or not an exposure anywhere in the beam is above the appropriate MPE values for retinal exposure. As a drastically simplified version of the general approach, we can assume the worst case of a small source (set C_6 in the retinal thermal angle of acceptance MPE to unity) and measure with an unrestricted FOV retinal thermal angle of acceptance. (A smaller FOV as specified in some cases in the laser safety standard might reduce the power that is measured (as discussed in section 2.4) but it complicates the measurement set-up.) This choice of C_6 would produce the lowest MPE value and the open FOV would produce the largest measurement value that is to be compared to the MPE[2]. For such a simplified analysis, without having any more background information or knowledge about beam propagation in the eye, it is clear where the most hazardous location in the beam would be, and that is the position of the beam waist (as shown in figure 3.29), since there the power that enters the eye (or the irradiance at the cornea) is at its maximum. If the beam diverges from the exit window or exit aperture of the product outwards (see figure 3.29 on the right), i.e. if there is no external beam waist, then the most hazardous position (where the highest power would enter the eye) is at the position of closest access, i.e. at the exit window or aperture. In practical terms, the most simplified approach is to just look for the position in the beam where the measured power through a 7 mm aperture is maximized. If the MPE for $C_6 = 1$ is not exceeded at these worst case positions, then the MPE will not be exceeded anywhere else in the beam. We will later see that the results of the beam propagation model for the retinal thermal hazard allow us to move this worst case

[2] We note that it is actually the irradiance at the cornea, averaged over a 7 mm pupil, that is to be compared to the MPE. However, as is explained in section 3.6.6, this is equivalent to defining the MPE in terms of 'power or energy into the eye' and to compare this value with the power measured through a 7 mm pupil. In the discussion in this section, we follow the latter way of looking at it.

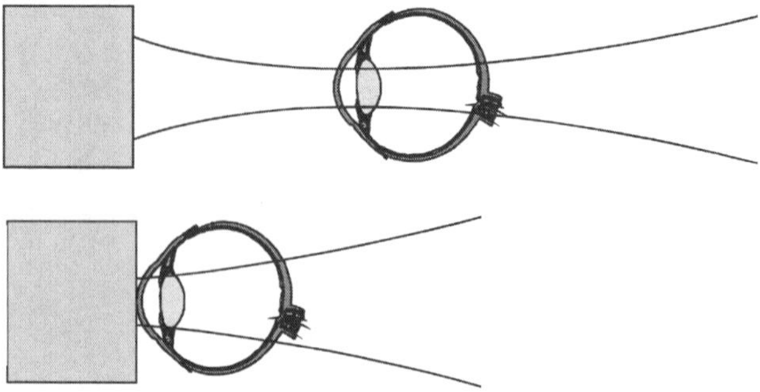

Figure 3.29. In a simplified worst-case analysis that assumes a small source ($C_6 = 1$ and an open FOV), the most hazardous exposure position is where the irradiance (the power into the eye) is maximized. This is the case either at the beam waist for an external beam waist (top) or at the exit aperture of the laser.

position to 10 cm from the beam waist, which for higher divergence beams (from roughly 70 mrad onwards) would result in a lower power value compared to the power value measured in the beam waist.

When a more complete (but potentially greatly involved) safety analysis is to be performed, the level of exposure needs to be compared with the corresponding MPE value for various positions along the beam, as is schematically shown in figure 3.30 for three positions. (In practice, the number of positions that are evaluated would generally need to be higher.) For each position, the power through a 7 mm aperture is determined (considering the appropriate limiting field-of-view (angle of acceptance) if relevant). For the power measurement to be compared to the photochemical limit, different measurement FOV are defined than those applicable to the thermal limits (as will be discussed in sections 3.12.5 and 3.12.6, respectively). Consequently, for wavelengths where the photochemical limit applies as well as the thermal limit, two different power measurements might be necessary, that may, depending on the FOV and the nature of the source, result in two different power values P_{ph_chem} and P_{therm}. P_{ph_chem} (or rather the irradiance averaged over a 7 mm aperture) can be directly compared to the retinal photochemical limit (which does not depend on α). As the retinal thermal MPE value depends on the angular subtense of the apparent source α, and the value of α depends on the position within the beam, α therefore needs to be characterized at each evaluation position, i.e. at each position at which P_{therm} was determined.

Finally, the power P_{therm} that was determined for each position (after division by the area of the 7 mm aperture to obtain the irradiance value) is compared to the MPE that is calculated with the value of α that corresponds to the measurement position. A complete analysis that sufficiently covers all potential exposure positions will also yield the most hazardous exposure position (MHP), which is

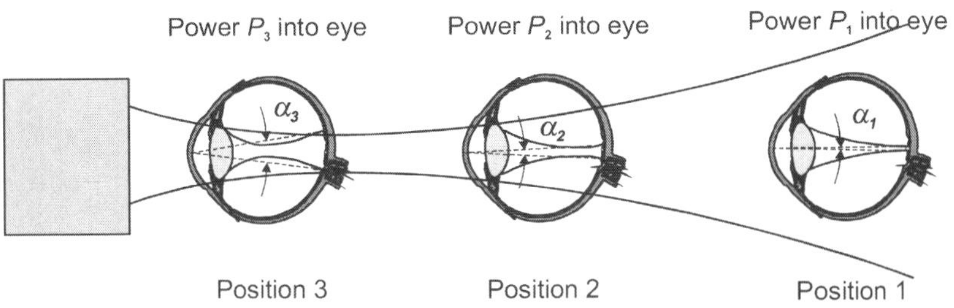

Figure 3.30. For the most general laser safety analysis, the power that enters the eye at different positions in the beam needs to be characterized and compared to the MPE. As the MPE values for the retinal thermal hazard depend on α, they are different for different exposure positions.

defined as the location at which the ratio of the exposure level (or power into the eye) to the MPE, P/MPE has the maximum value. (Note that there will be different MHPs for the thermal and the photochemical hazard. In the following we concentrate on the MHP for the thermal hazard, as this is more complex.)

If the most hazardous exposure position were known, the MPE analysis could be based on this single position, as the ratio between the exposure level and the MPE will be smaller for all other positions along the beam. In contrast to the simplified worst-case approach shown in figure 3.30, for the general case, predicting the most hazardous exposure position is not straightforward: while for a certain position more power would enter the eye than for another, it might well be that this position is less hazardous than the other if α is correspondingly larger than for the other position. For instance, when the exposure occurs right in the beam waist, i.e. when the cornea is at the position of the beam waist, it can be shown that α will be equal to the far-field beam divergence. When the distance to the beam waist increases, both the power that enters the eye as well as the angular subtense of the apparent source will decrease, but in a nonlinear fashion. The distance from the beam waist at which the MHP is located will depend on the beam divergence and the diameter of the beam waist.

For evaluation of the retinal thermal hazard it is possible to specify a simple procedure for the evaluation that is based on beam propagation as described in section 3.12.5.9.

3.12.3 Retinal thermal—wavelength dependence

In this section, we discuss the wavelength dependence of the retinal thermal limits. This wavelength dependence applies to retinal thermal limits in general and is basically the same for all exposure durations, i.e. the wavelength dependence is expressed by way of two correction factors that are multiplied to the basic MPEs (and these basic MPEs are a function of exposure duration). There

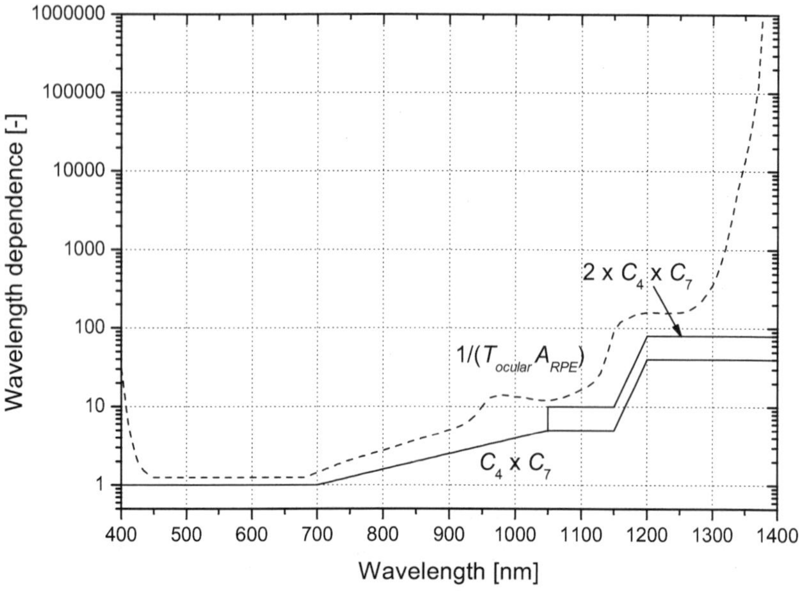

Figure 3.31. Wavelength dependence relative to the value as defined for the visible wavelength range. For exposure durations above 50 μs, the wavelength dependence is give by the factors C_4 and C_7 (that can increase the retinal thermal MPEs in the near infrared up to a factor of 80 as compared to the visible wavelength range), for pulse durations less than 18 μs there is an additional increase of a factor of 2 for wavelengths above 1050 nm so that the maximal increase is up to a factor of 160. The dashed line is calculated from the transmittance curve of the ocular media in front of the retina and the absorptance of the absorbing layer in the retina (data adopted from Jack Lund [25]).

is an exception to this 'general' wavelength dependence for wavelengths above 1050 nm and exposure durations less than 50 μs as is shown in figure 3.31.

To place this section into perspective we note that retinal photochemical MPE values (that are dual limits to the retinal thermal limits for exposure durations above 10 s and in the visible wavelength range) feature a different wavelength dependence which is discussed in section 3.12.6.

For the retinal thermal MPEs, the two wavelength correction factors C_4 and C_7 (in ICNIRP and ANSI C_A and C_C, respectively) are defined as follows:

$$C_4 = \begin{cases} 1 & \text{for } \lambda < 700 \text{ nm} \\ 10^{0.002(\lambda-700)} & \text{for } 700 \text{ nm} \le \lambda < 1050 \text{ nm} \\ 5 & \text{for } 1050 \text{ nm} \le \lambda < 1400 \text{ nm} \end{cases} \tag{3.7}$$

Table 3.10. MPE values for the retina for exposure durations less than 10 s for single exposures in the wavelength range 400–1400 nm. The values below are given for small sources and need to be multiplied with the factor C_6 for extended sources. (a) MPE values for the wavelength range 400–700 nm. For the wavelength range 700–1050 nm these values are multiplied with the factor C_4 so that the values for a wavelength of 1050 nm are a factor of 5 higher as the ones shown in the table. (b) MPE values for the wavelength range 1050–1400 nm.

(a)

Exposure duration t	MPE value
<100 fs	1.5×10^9 W m^{-2}
100 fs–10 ps	1.5×10^{-4} J m^{-2}
10 ps s–1 ns	$2.7 \times 10^4 t^{3/4}$ J m^{-2}
1 ns–18 μs (T_i)	0.005 J m^{-2}
18 μs–10 s (actually to T_2)	$18t^{3/4}$ J m^{-2}

(b)

Exposure duration t	MPE value
<100 fs	$1.5 \times 10^{10} C_7$ W m^{-2}
100 fs–10 ps	$1.5 \times 10^{-3} C_7$ J m^{-2}
10 ps s–1 ns	$2.7 \times 10^5 t^{3/4} C_7$ J m^{-2}
1 ns–50 μs (T_i)	$0.05C_7$ J m^{-2}
50 μs–10 s (actually to T_2)	$90t^{3/4} C_7$ J m^{-2}

$$C_7 = \begin{cases} 1 & \text{for } \lambda < 1150 \text{ nm} \\ 10^{0.018(\lambda-1150)} & \text{for } 1150 \text{ nm} \le \lambda < 1200 \text{ nm} \\ 8 & \text{for } 1200 \text{ nm} \le \lambda < 1400 \text{ nm.} \end{cases} \tag{3.8}$$

They are applied to the exposure limits for the eye for the wavelength range 400–1400 nm for exposure durations up to 10 s as given in table 3.10 (up to exposure durations of 10 s, there is only one retinal limit, and that is the retinal thermal limit). The time dependence of the retinal thermal limits is discussed in section 3.12.4.

The wavelength dependence of the retinal thermal limits for exposure durations above 10 s is expressed by multiplying the basic value that is defined for the visible wavelength range by both C_4 and C_7.

3.12.3.1 Background of correction factors

As discussed in section 3.9.2, for exposure levels at or only a small amount above the threshold for injury, it is the temperature increase in the absorbing layer of the retina, the melanin layer in the RPE, that leads to thermal damage of the photoreceptor cells. Consequently, the temperature increase (and the threshold for injury) depends on the energy arriving at the RPE (that may be reduced by transmission losses in front of the retina) and on the absorbed energy within the RPE. (For wavelengths in the near infrared some of the energy passes through the RPE into the choroid.) It follows that the temperature rise in the RPE is directly proportional to the transmittance of the pre-retinal ocular media and to the absorptance of the RPE. Both parameters exhibit a certain wavelength dependence (as is shown for the transmittance of the ocular media in figure 3.15) and therefore, for otherwise equal exposure conditions, radiation with different wavelength will lead to different temperature rises in the RPE. For instance, in the visible wavelength range, minimal losses in front of the retina and high absorption in the RPE will lead to comparatively high temperatures in the RPE. Since the tissue will be damaged when a certain critical temperature is exceeded, in order to prevent damage, the level of ocular exposure as defined at the cornea needs to be limited to lower values when the transmittance in the ocular media and when the absorptance in the RPE is high.

The correction factor C_4 as given above is derived from the change of absorptance of the melanin layer in the RPE and is also used in the skin MPEs where melanin also constitutes the main absorbing constituent (chromophore) in the visible and near infrared wavelength range. The correction factor C_7 characterizes, although only very roughly, the pronounced increase of the injury threshold level for near-infrared wavelengths where the water in the ocular media (especially in the vitreous) starts to absorb strongly, and only a small fraction reaches the retina. The energy is absorbed in front of the retina over a relatively large volume so that the resulting temperature increase in that volume is minimal. The two correction factors are shown in figure 3.31 together with the inverse of the transmittance of the ocular media in front of the retina T_{ocular} and the absorptance A_{RPE} of the RPE. (This curve was also corrected for aberrations in the eye which only play a minor role, and was adopted from work by Jack Lund [25].) The curve describes the wavelength dependence of experimental thresholds very well for all wavelengths within the retinal hazard region, so that it is also sometimes referred to as 'action spectrum for retinal thermal injury'. (However, the curves in figure 3.31 are both relative so that the distance between the two curves does *not* represent the absolute distance between the experimental threshold levels and the MPE!)

As can be seen in figure 3.31, the wavelength dependence of the MPEs does not completely follow the detailed spectral dependence of the thermal action spectrum and is constant for wavelengths greater than 1200 nm (the combination of C_4 and C_7 gives a factor of 80 for wavelengths above 1200 nm), resulting in a rather large safety factor for wavelengths above about 1250 nm. For wavelengths around about 1300 nm, the energy is absorbed essentially over the whole volume of the eye, and less than 10% of the radiation incident on the cornea reaches the retina. For wavelengths closer to 1400 nm, the lens and cornea can also

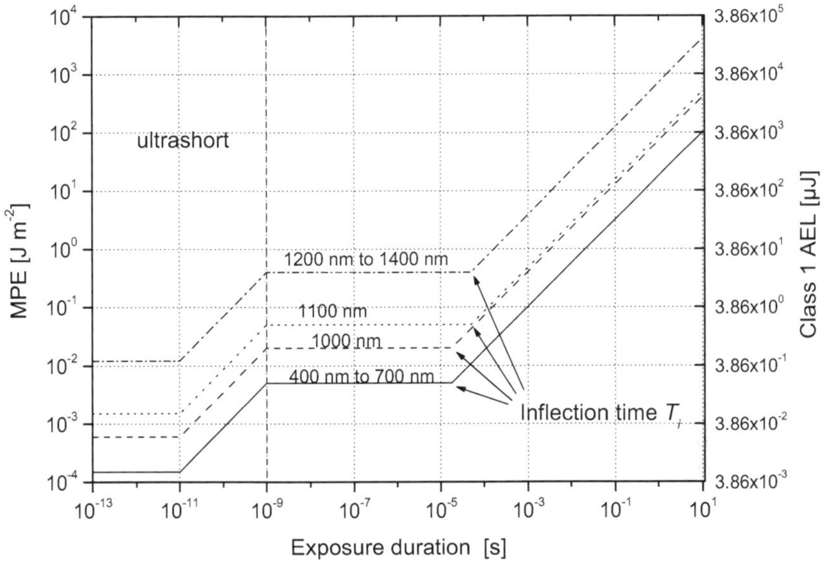

Figure 3.32. Retinal thermal MPE plotted as function of exposure duration for up to 10 s and for a range of wavelengths. The MPE values are given for $C_6 = 1$ and apply to single pulse exposures. The wavelength dependence follows the correction factors as discussed in the previous section.

be damaged, depending on the beam profile within the eye, but at much higher irradiance levels as characterized by the current retinal MPEs.

The discontinuity in the wavelength dependence at 1050 nm for pulse durations less than 50 μs comes from the difference of the dependence on exposure duration for wavelengths less than 1050 nm and above 1050 nm. For exposure durations down to 50 μs, the wavelength dependence is generally given by $C_4 \times C_7$. However, the MPE for wavelengths above 1050 nm is a constant radiant exposure value for an exposure duration of 50 μs, while for wavelengths below 1050 nm the MPE becomes constant only at 18 μs (see also figure 3.32 in the following section). Thus, for wavelengths above 1050 nm, the MPE stops decreasing with exposure duration at 50 μs, while for wavelengths below 1050 nm the MPE decreases with exposure duration further down to 18 μs, so that at 18 μs the discontinuity in the wavelength dependence at 1050 nm is a factor of 2 (from $50^{0.75}/18^{0.75}$), as shown in figure 3.31. The MPE values as specified in table 3.10 also show this wavelength dependence as the values given in (b) for pulse durations less than 50 μs are a factor of 10 higher than the corresponding values given in (a), while the factor C_4 that applies to the values given in (a) up to a wavelength of 1050 nm, is only a factor of 5 at 1050 nm.

Table 3.11. Inflection times for the different wavelength ranges in the visible and infrared. It should be noted that in wavelength regions where basically the whole eye is heated up (i.e. from 1500–1800 nm), the inflection time is as large as 10 s. Where absorption occurs superficially as in the far infrared, the inflection time is only 100 ns.

Wavelength	T_i
400 nm $\leq \lambda <$ 1050 nm	18 μs
1050 nm $\leq \lambda <$ 1400 nm	50 μs
1400 nm $\leq \lambda <$ 1500 nm	1 ms
1500 nm $\leq \lambda <$ 1800 nm	10 s
1800 nm $\leq \lambda <$ 2600 nm	1 ms
2600 nm $\leq \lambda \leq 10^6$ nm	100 ns

3.12.4 Retinal thermal—time dependence

The retinal thermal MPE values can be grouped into a number of distinct sections regarding the exposure duration. First, we discuss MPE values for exposure durations less than 10 s, as already listed in the previous section, and as shown in figure 3.32 for $C_6 = 1$ and for single pulse exposures (exposures to multiple pulses are discussed in section 3.12.8).

For exposure durations less than 1 ns, the regime is referred as 'ultrashort' and will be discussed further below. For exposure durations, t, above 1 ns, we see in table 3.11 in the previous section that the MPE values are specified as a constant radiant exposure (horizontal lines in the plot) up to a certain inflection time, referred to as T_i, beyond which the MPE values specified in terms of radiant exposure increase following a $t^{3/4}$ dependence. The biophysical background for the inflection time is related to the volume of tissue in which the radiation is absorbed and to the associated thermal diffusion time domain for this volume. The region of constant MPE (in J m^{-2}) up to the inflection time is associated to the 'confinement' of the heat within the absorbed volume for a certain time, which is discussed in more detail in section 3.2.2.2 as the concept of 'thermal confinement'. (There is also confinement of a pressure wave which is governed by the speed of sound in the tissue.) The thermal confinement applies to pulses which are shorter than the time it takes the heat to diffuse out of the volume where the radiation is absorbed. Consequently, for otherwise constant parameters, the peak temperature of the heated tissue does not depend on the pulse duration but only on the radiant exposure. For instance, outside of the retinal hazard region, the inflection time T_i is large for wavelengths between 1400 and 2600 nm, where the penetration depth into the eye is relatively large, and in particular for radiation from 1500–1800 nm where the laser radiation is absorbed over a large volume in the eye which includes the vitreous. In a simplified way, the inflection time characterizes the 'speed' or 'time-resolution' with which the tissue can react to changes of temperature. It is therefore also indicative of thermal confinement that pulses that are incident *within* the thermal confinement time (within T_i) are thermally added together, i.e. two pulses with radiant exposure H that are some time apart but both are incident within the thermal confinement time, produce practically the same temperature in the tissue as one pulse with $2H$ radiant exposure—the tissue does not have time to cool before the second pulse arrives. Thus, the second pulse, in terms of thermal load, just adds its energy to the first one. Therefore, for an MPE analysis, all the energy that is incident within T_i is added together and the total energy within T_i is compared to the (constant energy) MPE. Since the inflection time T_i is directly related to the volume that absorbs the radiation (large volumes take longer to react), T_i depends on the wavelength, as the optical absorption depth in the eye strongly depends on wavelength. The values of the inflection time T_i for wavelengths above 400 nm (i.e. not only limited to the thermal hazard region of 400–1400 nm) are summarized in table 3.11.

3.12.4.1 Ultrashort

Thermal confinement as such still applies to ultrashort exposure durations. However, in this time regime, additional affects come into play that reduce the MPE to below the value that applies for exposure durations above 1 ns. The exposure duration dependence of the MPE values follows a $t^{3/4}$ dependence from 1 ns down to exposure durations of 10 ps. For exposure durations less than 10 ps down to 100 fs the MPE is expressed as a constant radiant exposure. The range of constant radiant exposure MPE values between 100 fs and 10 ps is not reflected in the experimental threshold values, which tend to decrease steadily for pulse durations of less than about 10 ns, but less steeply than the MPE values between 1 ns and 10 ps. The region of constant MPE between 100 fs and 10 ps was basically intended as a simplification for practical MPE evaluations so that the pulse duration would not have to be determined. No experimental data are available for exposure to pulses with durations less than 100 fs so that the MPEs in that range are specified conservatively as a constant pulse peak irradiance. The value for the MPE expressed in terms of constant irradiance for exposure durations less than 100 fs is derived by dividing the MPE as given in J m^{-2} for 100 fs by 100 fs (10^{-13} s). If the limit for pulse durations less than 100 fs is expressed in terms of radiant exposure (in J m^{-2}), then it decreases linearly with decreasing exposure duration, while it can be expected that the actual experimental threshold will not decrease that strongly with decreasing pulse durations.

3.12.4.2 Long term

Exposure durations above 10 s can be considered as ‘long term’ exposure durations. In this time regime, eye movements play an important role in terms of the level of hazard that is represented by a certain power that enters the eye.

When the laser spot as such (with a given retinal diameter as characterized by α) is stationary, then eye movements cause the laser spot to move relative to the retina, which is similar to the effect of moving a photographic paper relative to a stationary beam. Due to eye movements, a given retinal location is exposed for a shorter duration than would be the case if the eye did not move. Even when we stare fixatedly at a spot (i.e. into the laser or at a LED) there are fast involuntary eye movements so that at the retina, the exposure trails a random path around the centre (which would usually be at the fovea) that covers a certain retinal area around the fovea. Thus, eye movements have the effect of increasing the exposed area in comparison to the actual laser spot size at the retina and consequently of decreasing the time-averaged irradiance at the retina. In the case of a stationary eye, the retinal irradiance would be calculated by dividing the power of the beam by the area of the stationary laser spot, but with eye movements, the average retinal irradiance is calculated by dividing the power of the beam by a representative retinal area arising from the effect of eye movements during the exposure duration. In effect, eye movements spread the power that is contained in the beam over a larger area of the retina, thereby decreasing the hazard[3]. The extent of eye movements, i.e. the diameter of the retinal area covered by the laser beam, increases with increasing exposure duration. For short exposure durations, the retina does not move during the exposure duration, similar to the effect of using a 'fast' (i.e. very short) shutter speed for taking a photograph of a moving object, and so for exposure to pulses, eye movements do not have any effect. The longer the exposure duration (the viewing duration), the larger (in terms of angle or retinal area) the eye movements tend to become.

For discussion of the long-term retinal thermal limits it is advantageous to re-calculate the MPE as given in table 3.10 for visible small sources ($18t^{3/4}$ J m^{-2}) in terms of irradiance, i.e. by division by t to obtain $18/\sqrt[4]{t}$ W m^{-2}. We see that this maximum permissible *irradiance* level decreases with increasing exposure duration, which is to be expected for a thermal damage mechanism: the longer the tissue is kept at a certain elevated temperature level, the lower this temperature level needs to be to induce damage. Eye movements, however, counteract the decrease of the MPE with longer exposure duration. This influence of the eye movements is reflected in the latest revision of the laser exposure limits by setting the MPE to a constant irradiance value for exposure durations longer than a certain time T_2, instead of further decreasing the MPE with increasing exposure duration. For the simplest case of small sources ($\alpha < \alpha_{\min}$), the retinal thermal MPE remains at a constant level of 10 W m^{-2} for exposure durations above 10 s, i.e. for small sources the break time T_2 equals 10 s. As shown in figure 3.33, for extended sources, the break time T_2 for the exposure duration at which the MPE is set to

[3] However, the hazard (the temperature rise) from a spot that moves across the retina is still greater than if the power were uniformly spread over the retinal area covered by the moving spot. This is also the reason why scanned patterns on the retina cannot be simply treated as extended sources, as is further discussed in the case study of the scanned laser, section 4.8.4.

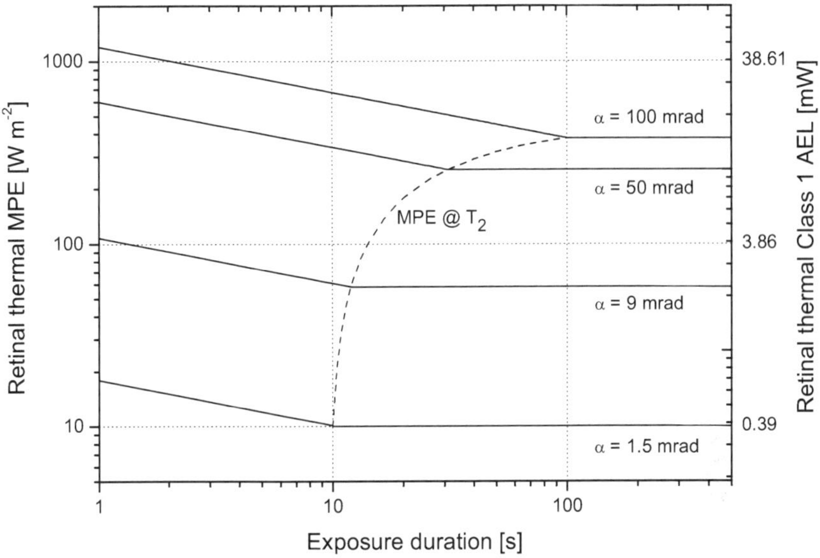

Figure 3.33. Retinal thermal MPE values decrease with increasing exposure duration up to point T_2 which depends on the angular subtense of the apparent source α. The scale on the right are the retinal thermal AEL values for Class 1 and Class 1M which are derived from the MPE values by multiplication with the area of the 7 mm limiting aperture.

constant irradiance levels increases with increasing α (i.e. with increasing retinal spot size).

The dependence of the break time T_2 on the angular subtense of the apparent source α is shown in figure 3.34 and is defined by

$$T_2 = 10 \times 10^{\frac{\alpha - 1.5}{98.5}} \tag{3.9}$$

where α needs to be expressed in units of mrad (and is limited to values between α_{min} and α_{max}) and T_2 is calculated in seconds.

The dependence of the break time T_2 on the angular subtense of the apparent source is based on the relationship of the beam spot size at the retina (as characterized by α) to the magnitude of eye movements. The eye movements only effectively increase the irradiated retinal area when the extent of the eye movements is larger than the beam spot size, as schematically shown in figure 3.35. In other words, the larger the retinal spot size is, the larger the eye movements need to be so that the increase of the irradiated area on the retina is notable and can sufficiently counteract the general decrease of the thermal MPE with increasing exposure duration. Since the extent of the eye movements increases with increasing exposure duration, and large eye movements are needed

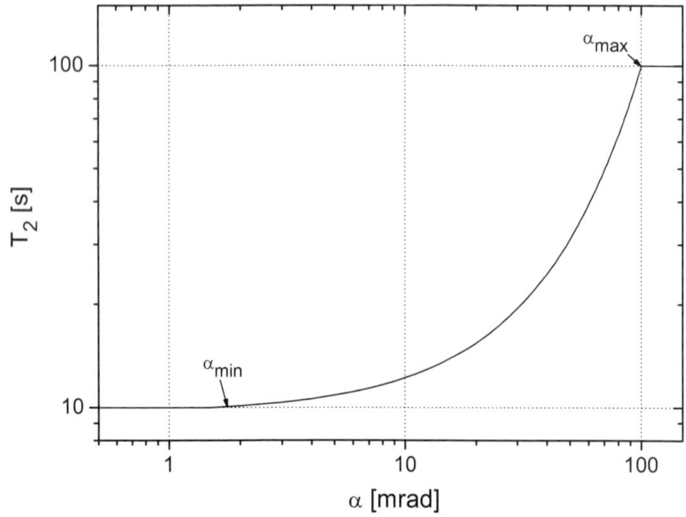

Figure 3.34. The break time T_2 increases with increasing angular subtense. For $\alpha < \alpha_{min}$, (i.e. for small sources) the break time equals 10 s and it is limited to 100 s for $\alpha < \alpha_{max}$.

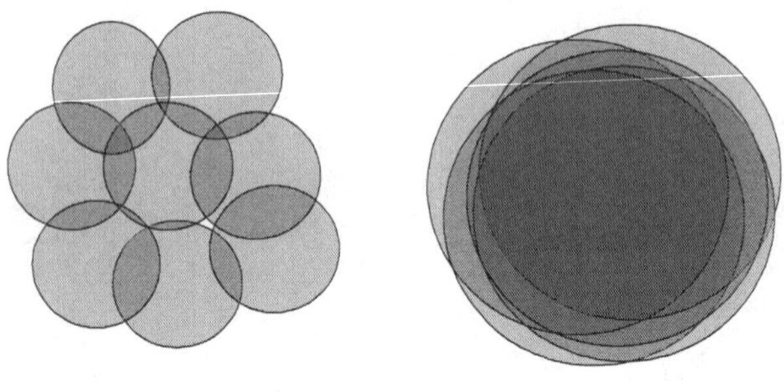

Figure 3.35. Simplified schematic drawing to show that the extent of eye movements needs to be large as compared to the retinal spot size, indicated by circles. Both cases are drawn for the same extent of eye movements, i.e. the same displacement of the centre of the circles. When the retinal spot size is large (left), the eye movements have little effect in terms of increasing the irradiated retinal to beyond the area of the beam spot size.

for large spot sizes to have an effect, the break time T_2 shifts to longer exposure durations for larger spot sizes.

The complete definition of the retinal thermal MPE values for the visible and for exposure durations above 10 s is: for $\alpha < \alpha_{min}$, 10 W m^{-2} (constant value for all exposure durations above 10 s); for $\alpha < \alpha_{min}$ for exposure durations below the break time T_2, $C_6 18 t^{3/4}$ J m^{-2} (the same as for exposure durations less than 10 s) which is equivalent to $C_6 18t^{-1/4}$ W m^{-2}, this value is applicable up to an exposure duration of T_2, beyond that the MPE is a constant value of $C_6 18T_2^{-1/4}$ W m^{-2}.

For wavelengths above 700 nm, all of the above values need to be multiplied with both wavelength correction factors C_4 and C_7, i.e. the MPE is the lowest in the visible wavelength range where it has a value that does not depend on wavelength, but for wavelengths in the infrared, the MPE is increased for increasing wavelengths as discussed in section 3.12.3.

As is generally the case, the worst case lowest MPE value is given for small sources, i.e. for $\alpha < \alpha_{\mathrm{min}}$ and $C_6 = 1$. The respective MPE value equals 10 W m^{-2} and thus constitutes the lowest MPE value for thermal damage of the retina (for all exposure durations and wavelengths). The corresponding AEL value for Class 1 and Class 1M, i.e. the MPE value multiplied by the area of the 7 mm limiting aperture, is 0.4 mW. It is only for short wavelength visible radiation that the photochemical MPE can be lower than this value (by a factor of 10, as discussed in section 3.12.7). While the retinal thermal long-term exposure limit also features C_6 as a multiplication factor, because of the varying break times T_2 the value of the minimum MPE that applies for exposure durations above T_2cannot simply be derived by multiplying 10 W m^{-2} by C_6. For example, for $\alpha = 100$ mrad, the minimum MPE value equals $66.6 \times 18 \times 100^{-1/4}$ W m^{-2} = 380 W m^{-2} (or the equivalent AEL Class 1 value of 14.6 mW) while 10 W m^{-2} times 66.6 would be almost double that value.

3.12.5 Retinal thermal—dependence on α

The concept of the apparent source and the angular subtense of the apparent source α was introduced in section 3.12.1. Here we discuss in more detail how the retinal thermal limit depends on the angular subtense of the apparent source and how the numerical value of α is determined.

3.12.5.1 Terminology

The angular subtense that characterizes the smallest spot size that can be optically achieved at the retina is referred to as the *'minimum angular subtense'* and has the symbol α_{min} and the numerical value of 1.5 mrad, i.e. $\alpha_{\mathrm{min}} = 1.5$ mrad. The minimum angular subtense α_{min} characterizes the minimum retinal spot size that can be obtained, i.e. even if the source (the 'object' that emits the radiation in the optical sense) would itself be characterized by an angular subtense of less than 1.5 mrad, the spot size on the retina would not be less than 1.5 mrad (due to diffraction and scattering). In previous editions of the laser safety standard, α_{min} was time dependent to characterize the influence of eye movements, however this concept was not fully consistent and eye movements are now better reflected by the temporal factor of T_2. There is also a '*maximum angular subtense*' with the symbol α_{max} and the numerical value of 100 mrad, i.e. $\alpha_{\mathrm{max}} = 100$ mrad. In contrast to α_{min}, the maximum angular subtense of 100 mrad (equivalent to angle in degrees of 5.7°) does not reflect an actual optical limitation of the retinal image size, as can be easily recognized when it is considered that the field-of-view (the section of the hemisphere in front of the eye that is visually perceived, i.e. that is imaged onto the retina) of the naked eye is certainly a lot larger than 5.7°. The maximum angular subtense of 100 mrad rather characterizes a break-point in the dependence of the retinal thermal hazard on the diameter of the retinal spot size, and for a correct evaluation it is important that the measurement angle of acceptance is limited to 100 mrad together with the limitation of α to 100 mrad, as will be discussed below.

Depending on the angular subtense of the apparent source, different terms are used: when the angular subtense of the apparent source is less than α_{min}, it is referred to as a '*small source*'. Sometimes such a source is also referred to as a 'point source', although the term point source is used in optics as a theoretical concept of a source having zero extent and is therefore not really equivalent to the term 'small source', as a small source can have a finite extent, namely up to an angular subtense of 1.5 mrad. When the angular subtense of the apparent source is between 1.5 mrad and 100 mrad, it is referred to as an '*intermediate*', when it is beyond 100 mrad, then it is referred to as a '*large source*'. The term '*extended source*' is also often used and generally refers to a source larger than 1.5 mrad (but is not limited to 100 mrad). (As was discussed in section 3.12.1.2, the term 'intrabeam viewing' is no longer used to mean viewing of a small source. Since the angular subtense of the apparent source depends on the exposure location, i.e. the location in the beam and the distance to the apparent source, these terms should really be accompanied by a specification of the reference viewing position to prevent confusion and misinterpretation. In the strict sense these terms only apply to a well-defined position in the beam, so we can say, for example, 'at position x, the apparent source represents an extended source'. More generally, the use of these terms can also be applied to the most hazardous position or to the position necessary for the classification of a product, so that 'extended source' means that it is an extended source at the most hazardous position while it might be the case that for exposure positions closer to the beam waist, the source could be a 'large source', and when the eye is sufficiently far away from the beam waist, it will become a small source.

3.12.5.2 Dependence on angular subtense

The retinal thermal exposure limits depend on the angular subtense of the apparent source by way of the factor C_6 (in the ANSI standard, this factor is referred to as C_E), which is defined as

$$C_6 = \frac{\alpha}{\alpha_{min}} = \frac{\alpha}{1.5 \text{ mrad}} \tag{3.10}$$

where α is given in units of mrad and is limited to values between α_{min} and α_{max}. If the actual angular subtense of the apparent source is less than 1.5 mrad, the

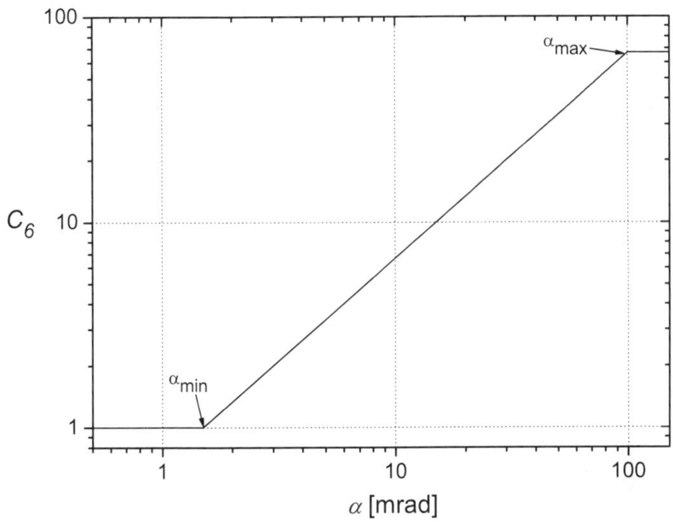

Figure 3.36. The factor C_6 increases the small source limit of the retinal thermal exposure limits that apply in the wavelength range 400–1400 nm as a function of the angular subtense of the apparent source α.

value of 1.5 mrad is assigned to α; if it is larger than 100 mrad, the value of 100 mrad is assigned to α. Consequently, C_6 can assume values between $C_6 = 1$ when $\alpha < 1.5$ mrad and $C_6 = 66.6$ when $\alpha > 100$ mrad. For sources larger than 100 mrad it is important to note that the angle of acceptance for measuring the exposure level that has to be compared to the MPE value must also be limited to 100 mrad. (The specification of C_E in the ICNIRP guidelines and in table 6 of the current version of the ANSI laser safety standard can be misinterpreted to mean that α is not limited to 100 mrad and C_E can increase beyond 66.6, which is not consistent with the limitation of the measurement angle of acceptance (also referred to as limiting cone angle) to 100 mrad, as discussed further below.) A plot of C_6 as function of α is shown in figure 3.36.

The factor C_6 can be seen as permitting a relaxation of the basic limit that is specified for small sources: for small sources, $C_6 = 1$ and can be neglected, while it increases the exposure limit for extended sources up to a factor of 66.6. For instance, the small source MPE value for visible radiation and an exposure duration of 0.25 s equals 1 mW when the MPE is multiplied with the area of the averaging aperture (so that the power that passes through a 7 mm aperture represents the power that enters the eye). For $\alpha < 1.5$ mrad, the limit remains at 1 mW, while for large sources, the exposure limit increases up to 66.6 mW.

3.12.5.3 Discussion of background of C_6 and associated measurements

The linear dependence of the experimental retinal threshold values on the retinal spot size is well reflected by experimental and theoretical studies for a number of wavelengths and pulse durations[4] and can be understood in simplified terms as follows. As already discussed (section 3.6.6), the retinal MPE values are specified at the position of the cornea and can also be understood as setting a limit to the power that is allowed to enter the eye through the pupil. It is obvious that when the power that enters the eye through the pupil is spread over a larger area on the retina, more power would be allowed to enter the eye than if that power were concentrated onto a minimal spot on the retina. Consequently, it is to be expected that the MPE would increase for increasing values of α. As a first guess, one might think that the retinal hazard (and therefore the MPE) would scale with the irradiance at the retina. If this were the case, the MPE would be expected to depend on α^2, as the retinal irradiance is proportional to the power that enters the eye divided by the area of the retinal image (and α characterizes the diameter of the retinal image). However, for thermal injury this relationship is reduced to a linear dependence on α, since for the same level of retinal irradiance, larger spot sizes are actually more hazardous than smaller ones. For small spot sizes, cooling to surrounding non-irradiated tissue is much more effective than for large images. For larger images, heat transfer from the centre of the image can only occur into the retina and not radially (sideways), as is schematically shown in figure 3.37. Consequently, for the same level of retinal irradiance, larger spots produce a higher temperature than smaller ones. This effect of higher temperatures for larger spot sizes reduces the exposure limit, since the effect of cooling means that the damage threshold does not depend simply on the retinal irradiance. One could understand the effect of the missing radial cooling as if the basic α^2 dependence (reflecting the area dependence of the irradiance when calculated from the power that enters the eye) is divided by a factor of α to reflect the higher temperatures (and decrease in the damage threshold) for larger spot sizes. This division reduces the square dependence of the MPE to a linear dependence as characterized by C_6. If the MPE were not given in terms of the irradiance level at the cornea but in terms of the irradiance level at the retina, then this MPE would actually depend on $1/\alpha$, i.e. the exposure limit would decrease with increasing spot sizes to reflect the fact that for the same level of irradiance, larger spots are more hazardous (have a lower exposure limit) than smaller ones. (The broadband retinal thermal limit is specified in terms of radiance, and as radiance is proportional to retinal irradiance, this limit is a function of $1/\alpha$).

[4] Recent research (presented at ILSC 2003) shows that for short pulse durations (less than roughly the thermal confinement time T_i), the damage threshold does not depend on the spot diameter but only retinal irradiance, which makes the current definition of C_6 overconservative for these exposure durations. This could be accounted for in the future by decreasing α_{max} and γ_{th} for short pulse durations or by defining C_6 such that for those short pulse durations it would depend on α^2.

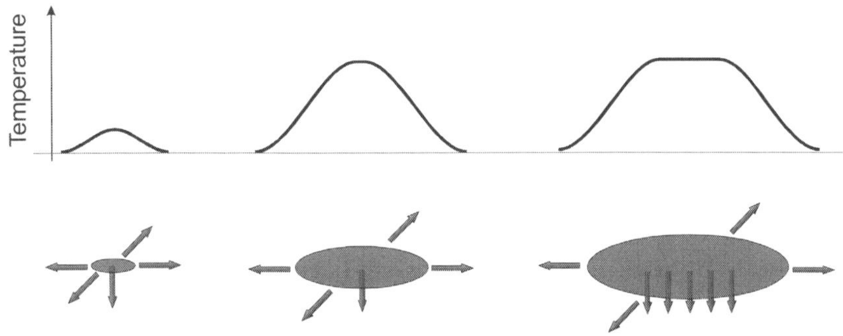

Figure 3.37. Heat transport from the centre of the retinal spot is much more effective for small spot sizes (left). For larger spot sizes (middle), heat from the centre can be conducted only into the depth of the retina (downwards in the figure), and for the same retinal irradiance the temperature profile in the centre of a large spot is higher as compared to a small image. For very large spot sizes (right), the temperature profile in the centre is no longer affected by radial cooling and therefore does no longer depend on the spot size diameter.

This dependence of the injury threshold on spot size diameter, however, does not apply to very large spot sizes since the temperature profile in the centre of the spot assumes a more or less flat profile, and this value is not affected by any further increase to the size of the actual image, as the edges that are cooled radially are too far away from the centre and heat drain into the depth of the retina involving choroidal blood flow is dominant. Consequently, for large sources ($\alpha > 100$ mrad), the injury threshold depends only on the irradiance at the retina and no longer on the diameter of the spot (or α). Instead of increasing the MPE proportional to α^2 for source sizes beyond 100 mrad, the power (or irradiance at the cornea) that is compared to the MPE is decreased in comparison to the actual power that enters the eye: the MPE is limited by limiting α to a maximum of $\alpha_{max} = 100$ mrad, but at the same time the measurement angle of acceptance (the FOV) for the determination of the exposure level is limited to $\gamma_{th} = 100$ mrad (the geometrical concept is shown in figure 3.38).

This approach has the advantage of being able to detect hot spots within the retinal circle defined by the 'diameter' of 100 mrad (1.7 mm), as can be understood by considering that the irradiance within that circle can be calculated by dividing the power that enters the eye by the area of the 100 mrad circle (we are neglecting absorption losses for this argument). Thus the retinal irradiance is actually *averaged* over this area as defined by a 100 mrad circle. Because of this averaging effect, non-uniform sources that feature hot spots smaller than 100 mrad need to be treated with smaller FOV values (over which the retinal irradiance is averaged) and correspondingly smaller α values as discussed in section 3.12.5.6. Since the radiation emitted from the source is imaged onto the

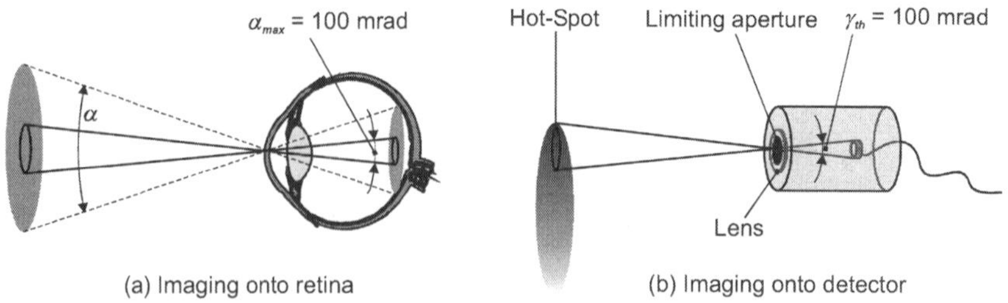

Figure 3.38. For sources that subtend an angle at the eye that is larger than 100 mrad, C_6 is limited to the maximum value of 66.6 but it is important to measure the exposure level with an angle of acceptance (γ_{th}) that is also 100 mrad. When a lens is used to produce a well-defined angle of acceptance then it is important that the apparent source is imaged onto the field stop. The source is to be scanned for hot spots to characterize the maximum irradiance at the retina (however, a non-uniform source with hot spots smaller than 100 mrad needs to be treated differently as discussed in section 3.12.5.6).

retina, it follows that the retinal irradiance profile (power incident per unit area on the retina) is directly proportional to the power emitted per unit area of the source (the 'exitance' profile) in the direction of the eye. This is equivalent to imaging the radiation of the source onto a detector where the diameter of the detector (or the field stop at the detector) determines the angle of acceptance of the radiometer (see also section 2.4). A radiometer that features an angle of acceptance of, for instance, 100 mrad is set up so that the sensitive part of the detector (or the field stop in front of the detector) subtends an angle of 100 mrad as seen from the imaging lens. Such a detector set-up therefore 'accepts' radiation that is emitted from the source from within the angle of acceptance, such as 100 mrad, and the power that passes through the limiting aperture is 'distributed' over the detector area, and is thus in terms of irradiance profile on the detector averaged over the area of the detector, or in terms of angular subtense, is averaged over 100 mrad. It is important to note that this approach only yields accurate results when the apparent source is imaged onto the field stop (or onto the detector if the area of the detector functions as a field stop). If the apparent source is not imaged properly, the radiation field at the field stop is blurred and the hazard would be underestimated. The proper position of the field stop relative to the imaging lens can therefore be found by adjusting the field stop (with the detector) relative to the lens to maximize the signal.

The reader who is familiar with the concept of radiance and radiance measurements will have recognized the above set-up as being equivalent to a radiance measurement where the radiance is averaged over the angle of acceptance of 100 mrad (see section 2.5 for an introduction to radiance). To obtain an irradiance value, the power that enters the detector is divided by the area of the 7 mm limiting aperture at the lens. This irradiance value is what has to be compared to the MPE that is specified in terms of irradiance at the cornea. (Similarly, the power value is what has to be compared to the AEL for those classes that are directly related to the ocular MPE values and that are specified in terms of 'power through an aperture'.) To obtain a radiance value, this irradiance level is further divided by the solid angle that corresponds to 100 mrad, i.e. by 7.9×10^{-3} sr. Consequently, the MPE for large sources can be expressed in terms of radiance (which does not depend on α) by dividing the large source irradiance-MPE (where $C_6 = 66.6$) by the solid angle of 7.9×10^{-3} sr. In other words, if $\text{MPE}_{\text{small}}$ is the retinal MPE value for small sources in units of W m^{-2}, i.e. the value as a function of wavelength and exposure duration but for $C_6 = 1$, then the radiance-MPE for large sources ($\text{MPE}_{\text{large\_rad}}$) in units of W m^{-2} sr^{-1} can be calculated by:

$$\text{MPE}_{\text{large\_rad}} = \text{MPE}_{\text{small}} 66.6/7.9 \times 10^{-3} = 8430\text{MPE}_{\text{small}}. \qquad (3.11)$$

(The value given in the ICNIRP guideline is rounded up to 8500.) However, it should be noted that in order for the radiance-MPE and the radiance measurement to be fully consistent with the irradiance-MPE and irradiance measurements as discussed above, the radiance measurement has to be performed with an angle of acceptance of 100 mrad in order to scan the source for hot spots. If the angle of acceptance for the performance of the radiance measurement (or calculation) were larger than 100 mrad, this could underestimate the hazard, by 'averaging out' any hot spots that were present. For a complete safety analysis of inhomogeneous, non-uniform sources (and therefore inhomogeneous irradiance profiles of the image at the retina or at the CCD array), the source should also be evaluated with smaller acceptance angles and correspondingly smaller values of α as described in section 3.12.5.6.

By using an angle of acceptance of 100 mrad, the power that enters the detector is limited to only that part of the radiation that is emitted by the source from within the angle of acceptance. For sources larger than 100 mrad, the power that enters the detector (and which is compared as the biophysically-effective value to the MPE) is lower than for a larger (or 'open') angle of acceptance. For the eye, the safety limit is specified on the basis of the retinal irradiance averaged over $\alpha_{\text{max}} = 100$ mrad, and thus both α and the angle of acceptance γ_{th} are limited to 100 mrad, but the actual size of the retinal spot and the actual power that enters the eye is not relevant. The lower hazard that is represented by a source where the power that enters the eye is spread over a circle larger than 100 mrad is thereby not accounted for by an increase of the exposure limit (which would be proportional to α^2) but rather by a decrease of the measured exposure level (proportional to α^2 for homogenous sources) that is compared to the MPE.

Only if the retinal irradiance level within the source image (i.e. within α) is constant is the ratio '*Power that enters eye from within FOV of 100 mrad/*$(\alpha_{\text{max}})^2$' the same as '*Total power that enters eye (i.e. power from within* α*)/*α^2', as only then the irradiance averaged over 100 mrad is the same as the irradiance averaged over the larger angle α. Consequently, only for homogenous sources with no hot spots is it valid to perform the measurement using an 'open' field-of-view that would measure the power from the total source that arrives at the eye, and to increase the small source MPE by a factor beyond 66.6 proportional to α^2, starting with 66.6 for $\alpha = \alpha_{\max}$, i.e. by

$$C_6^{\text{open}} = 66.6\frac{\alpha^2}{\alpha_{\max}^2} = \frac{\alpha_{\max}}{\alpha_{\min}}\frac{\alpha^2}{\alpha_{\max}^2} = \frac{\alpha^2}{\alpha_{\min}\alpha_{\max}}. \quad (3.12)$$

If there are 'brightness' hot spots smaller than 100 mrad in the source, then this method of 'increased C_6 and increased measurement level' (as compared to α limited to $\alpha_{\max}$ and the measurement limited to the FOV as is specified in IEC 60825-1) *underestimates* the hazard presented by the source. The right-hand side of equation (3.12) is given in the current ANSI laser safety standard and the ICNIRP guidelines, and it is not indicated clearly enough that this is only appropriate if the source is homogenous and more importantly if the measurement angle of acceptance is *not* limited to 100 mrad. If the above value of C_6 is used to increase the MPE beyond a C_6 of 66.6 (i.e. α would *not* be limited to 100 mrad) but at the same time the field-of-view for the measurement *is* limited to 100 mrad (as indicated elsewhere in the ANSI and ICNIRP documents), then this would *seriously underestimate* the hazard (proportional to $\alpha^2/\alpha_{\max}^2$). It is expected that this issue will be clarified in future editions of the ANSI and ICNIRP laser safety documents.

3.12.5.4 Defining the diameter of the source (or image)

So far we have assumed that the (apparent) source as imaged onto the retina results in a retinal irradiance profile that has sharp edges, i.e. a top-hat profile. These sharp edges define the diameter of the source and the angle subtended by the source at the eye, α, is calculated by division of the source diameter by the distance of the eye to the (apparent) source. (Generally the dimensions of the source cannot be measured directly and α is characterized by imaging the source onto a CCD array as discussed below). However, when the source emission (the radiant exitance) and therefore the retinal image is inhomogeneous and does not have sharply defined edges, a criterion is necessary to assign some lateral extent (diameter) to the irradiance distribution.

Although currently IEC 60825-1 does not define a specific criterion that should be used to determine a, we would like to argue that the appropriate and generally applicable procedure follows from the way multiple sources need to be analysed. The principle is to analyse a non-homogeneous source in respect to the most hazardous combination of power (or energy) contained within a certain part of the source and the angular subtense of that part of the source (the power is compared to the MPE, and the angular subtense is used to calculate C_6 which is part of the MPE). This principle is discussed in detail in section 3.12.5.6 on non-uniform and multiple sources. In the strict sense, all but a constant irradiance profile (a top-hat profile) has to be analysed as non-uniform source. For instance, the appropriate angular subtense of a Gaussian retinal irradiance profile, following this principle, is not defined on the basis of the 63% of the total power, d_{63} or the 1/e of the peak irradiance, as was argued before. When one applies the principle of maximizing the ratio of the power within a certain part of the source and the angular subtense of that part of the source, then it is found that the worst case diameter that should be used to determine α for a Gaussian profile contains 72% of the total power. Thus for a hazard analysis of a source that produces a Gaussian beam profile on the retina, following the general principle suggested here, 72% of the power that passes through the measurement aperture is compared to the MPE where C_6 is calculated from the corresponding diameter that contains 72% of the total power, which is somewhat larger than the 63% diameter (with the assumption that this diameter is larger than 1.5 mrad). This is quite different (and about 40% less conservative) to previously suggested approaches where the total power in the retinal image (provided it is smaller than 100 mrad) is compared to the MPE where C_6 is calculated with the 63% radius.

3.12.5.5 Non-circular sources

The parameter α as discussed so far applies in this simple form only to circular apparent sources. For non-circular apparent sources (retinal images), the angular subtense of the apparent source to be used in the calculation of C_6 is determined by the arithmetical mean of the angular subtense α_x that characterizes the source in one direction and α_y for the other direction (as shown in figure 3.39), i.e.

$$\alpha = \frac{\alpha_x + \alpha_y}{2} \tag{3.13}$$

where α_x and α_y are again limited to a minimum value of 1.5 mrad and to a maximum of 100 mrad. For instance, for a 'thin' and 'long' line image that can be produced by a line laser, which might be 'thinner' than 1.5 mrad, and 'longer' than 100 mrad, the corresponding parameters in equation (3.13) are set to 1.5 mrad and 100 mrad, respectively, giving a value for α of 50.8 mrad (although this is not the most hazardous viewing condition for a line laser as discussed in the case study, section 4.8.3). (Note that the reason why a line or 'fan-beam' laser produces a line image on the retina when the source is viewed directly at close range is that because the beam has two widely different—horizontal and vertical—values of divergence, there is no single apparent source. One point of origin is usually effectively at infinity and the other is close to the cylindrical lens that fans the beam out. The beam is asymmetric and the eye cannot focus simultaneously at both points of origin. In one or other direction the beam is always out of focus and 'smeared out', thereby producing a line image.)

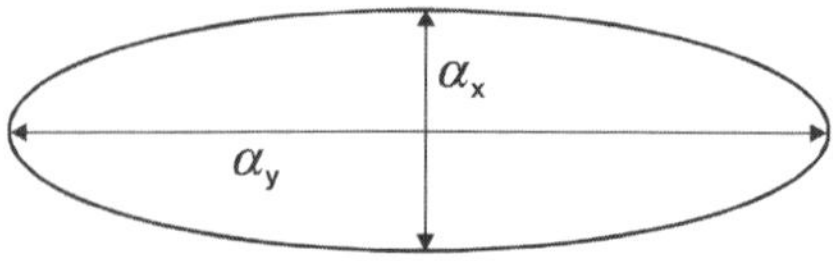

Figure 3.39. Example of a non-circular irradiance profile that characterizes the emission of the source (the angular subtense of the source) and the irradiance profile at the retina.

3.12.5.6 Non-uniform and multiple sources

Sources that consist of subsources such as arrays, bundles of fibres or other inhomogeneous, non-uniform sources that produce a non-uniform irradiance profile at the retina (or at the CCD array when imaged with a lens) need to be evaluated following what could be termed a ‘combinational’ approach: additional to the evaluation of the source as a whole, each subsource (or area of higher brightness, i.e. source ‘exitance’) needs to be evaluated as an individual source, as well as all combinations of subsources. Each subsource or combination is associated with a certain value of α, and for each subsource or combination, a certain power level (or corneal irradiance level) will be measured within the 7 mm limiting aperture. (Thus, radiation contributed from other subsources than the ones which are being evaluated is disregarded.) The ‘matching’ pairs of power values and MPE (depending on α) are subsequently compared, and exposure to the source can be characterized as ‘below’ the MPE when none of the subsources or combinations are above their respective MPE value. Regarding the spacing of subsources, the combinational approach is limited to minimum values of 1.5 mrad and to maximum values of 100 mrad, such that subsources smaller than 1.5 mrad are treated as one source and contributions to the exposure that arise from any emission that occurs outside the central 100 mrad region are disregarded. In other words, when spaces between sources are larger than 100 mrad, the sources are treated as completely independent and not evaluated in a combination (as the corresponding images on the retina can be considered thermally independent).

As an example, we consider an array of surface mounted non-capped LEDs as shown in figure 3.40: the exposure level from each individual LED needs to be below the MPE calculated with the angular subtense of the apparent source that applies to the single LED. In practice, the corresponding exposure level measurement can be performed by blocking off all other LEDs with some opaque material. As a next step the power that originates from two neighbouring LEDs is determined by blocking off all but these two neighbouring LEDs. The angular subtense is determined for this combination using equation (3.13) and the dimensions α_x and α_y of the combined source. It is noted that for the evaluation of combinations of individual sources, the actual physical spacing can be used to determine the angular subtense of the combination.

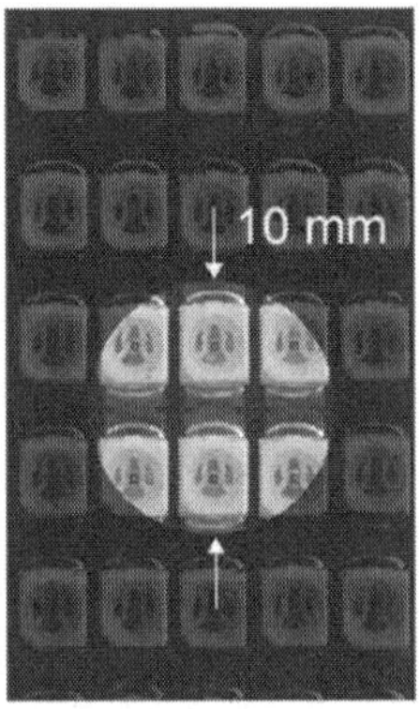

Figure 3.40. An array of surface mounted, non-capped LEDs. The maximum area of combination representing an angle of acceptance of 100 mrad at a distance of 10 cm is indicated by a circle.

Further analysis would have to be performed for ever increasing numbers of LEDs treated as a combined source, up to an area having a diameter of 1 cm, which corresponds to 100 mrad at the viewing distance of 10 cm. However, in many cases the analysis can be greatly simplified if the subsources are all identical and if the spacing of subsources is such that the angular subtense associated to two sources is more than double the angular subtense of one subsource. In such cases it is clear that the single individual source will be the critical case, i.e. independent of the arrangement and size of the array (and therefore of the total power emitted by the array), the only relevant exposure evaluation is that of a single LED. The exposure to a single LED will represent a higher hazard level (power divided by the MPE which is a function of α) than all combinations, since for two sources, the power that enters the eye cannot be more than doubled but the angular subtense for two subsources is more than doubled, thus decreasing the level of hazard for any combination of subsources.

For the case that low divergent beams are emitted from the individual subsources, care needs to taken to determine the most hazardous exposure distance to the source. For the case of surface mounted non-capped LEDs that are each practically Lambertian emitters, it is obvious that the worst case exposure distance will be 10 cm from the array. However, for individual sources that emit low divergence beams that are collimated or might even cross over at some distance from the array, a distance further from the source can represent the most hazardous position: when moving away from the multiple source, the angular subtense that characterizes the spacing of the individual subsources is reduced but the power that enters the eye might not be reduced to the same degree. In the case of crossing beams it might even increase, thus the level of hazard would increase with increasing distance from the source up to the most hazardous position. The general dependence of the hazard level on exposure location was discussed in more detail for a single low divergence beam in section 3.12.2 but has the same background that with increasing distance the power that enters the eye decreases more slowly (or not at all) than the angular subtense of the source. The case of low divergence multiple sources that point in the same direction raises the complexity level of the evaluation, since for each exposure location, different combinations of subsources (producing different 'matching' pairs of power and MPE) could be the critical one.

As a more general description of the 'combinational' approach we suggest the following general scheme. This involves a varying size angle of acceptance that for each size is scanned across the source and the detected highest exposure level is compared to the corresponding MPE that is evaluated with α equal to the angle of acceptance. The varying measurement angle of acceptance (size of the field-of-view, FOV) could be produced by a rectangular aperture (such as made up from four blades that can be moved in respect to each other) that can be varied in size and shape and that is placed on top of the array or multiple source or that is realized digitally when an array image is available. The evaluation procedure begins with a square aperture that subtends 1.5 mrad $\times$ 1.5 mrad from the evaluation distance (where the power measurement is to be carried out through a 7 mm limiting aperture that is centred with respect to the position of the blocking aperture). By moving the blocking aperture together with the measurement radiometer relative to the source, the whole source is scanned with this minimal FOV and the maximum power value would be recorded. This maximum power value can then be compared to the MPE that is evaluated for a minimal source ($C_6 = 1$). As a next step, the minimal FOV is maintained in one direction but extended in the other to form a slit having a width of 1.5 mrad and a step-wise increased length up to a length that subtends 100 mrad. Again for each size of FOV (each slit length) the whole source is scanned (the slit is scanned transversely as well as rotated) and the maximum power recorded, and compared to the MPE that is applicable to a value of α as calculated using equation (3.13), where α_x and α_y are the length and height, respectively, of the rectangular blocking aperture. This procedure of adjusting the blocking of the aperture and searching for the highest power passing through the respective FOV is continued until either the full source is encompassed or until the aperture reaches 100 mrad $\times$ 100 mrad (i.e. where the complete source is larger than 100 mrad $\times$ 100 mrad). While this approach can in practice be simplified to evaluate only a selection of aperture shapes and sizes, it helps in understanding the basic intention of the combinational approach: the exposure that originates from each part of the source should be below the corresponding MPE in order for the exposure to the source to be in the general sense referred to as 'below the MPE'. (In the case of classification, the accessible emission as measured through the applicable apertures has to be below the relevant AEL.) In the case of inhomogeneous sources that do not consist of distinctive, easily distinguishable and well arranged subsources, the description of this theoretical approach can be helpful in developing an appropriate evaluation strategy.

Blocking off parts of the complete source as just described is a way of analysing subsources which are accessible. An alternative way of carrying out such an analysis is by imaging the source onto a CCD array from a laser beam analyser. (It is important to have a CCD array that is linear in terms of incident irradiance and output signal, and to have a good background subtraction algorithm.) The variation of the apertures and the determination of the relative power originating from a subsource can be done using the image data in conjunction with appropriate analysing software. As an example we show the irradiance profile of the image of an LED in figure 3.41 where the central chip has a higher irradiance and the surrounding ring of radiation stems from a reflector cup. The image is directly proportional in terms of size to the angular subtense of the apparent source (and the image on the retina) and in terms of pixel signal is directly proportional to the local radiance of the source and the local irradiance at the retina. Therefore, relative power contributions from partial sources can be conveniently evaluated with beam analysis software. The signal of each pixel is proportional to the irradiance at that pixel, and the signals within an area can be summed up and this sum is then proportional to the power that is contributed by this part of the image (representing part of the apparent source) to the total power. Thereby subsources can be evaluated by calculating the ratio of signal within an area to the diameter (or an appropriately averaged extent for non-circular areas) of the area. This ratio is indicative of the 'level of hazard'. In the example of figure 3.41, when the total power incident on the array (all signals added up) is assigned a value of 1, then the relative power that is contributed to the total power by the central square chip is 0.13, i.e. 13% (and therefore 87% of the power that would enter the eye or the radiometer comes from the reflecting ring-cup around the chip). Additional to the partial relative power of the subsource, a diameter representative of the extent of the subsources is derived for the subsource by maximizing the ratio of the power within the FOV to the diameter of the FOV. This diameter (given for example in units of 'pixel') can also be taken relative to the second moment diameter of the complete source (i.e. chip plus ring), so that for the example of figure 3.41 the chip diameter represents 25% of the diameter of the complete source. The relative hazard level of the chip as a subsource can be calculated with the ratio of (relative) power and (relative) size. With a partial power that is 13% of the total power and a source size that is 25% of the total source size, the chip as a subsource has a hazard level half that of the total source. Thus, this kind of 'image analysis' yields the part of the source that represents the highest hazard level in terms of power and angular subtense of the (sub)source. By using the same limiting aperture for imaging of the source onto the CCD array as for the power measurement, and by placing the imaging lens with the aperture at the same place as the power measurement, a calibration factor for the CCD camera signal can be calculated by dividing the total signal for the complete source by the measured power. When the signals are multiplied with

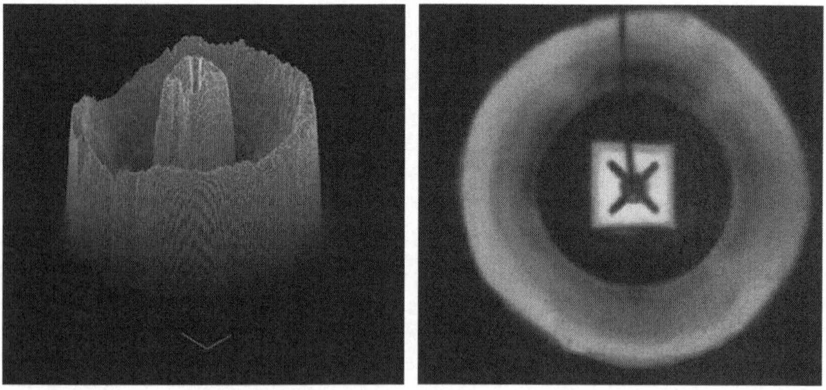

Figure 3.41. Three-dimensional (3D) plot of irradiance associated to the each pixel in the image plane of an infrared LED as measured with a CCD array (left). The irradiance profile in the image (shown as 2D plot on the right) can be associated to a radiance, or brightness profile of the source. The central chip produces a higher irradiance level at the image (and therefore at the retina) than the surrounding reflecting cup. However, the chip only represents 13% of the total power incident on the CCD array while 87% of the total power are associated to the reflecting cup. Such a source should be evaluated as a non-uniform source.

the calibration factor, the image analysis can be performed in terms of absolute radiative power. Furthermore, pixel diameters can be calibrated in terms of the angle subtended from the lens, so that from these measurements and evaluations both the relevant angular subtense of the apparent source α and the partial power for the subsource with the highest hazard level can be calculated, and the power (divided by the area of the 7 mm limiting aperture) compared to the respective MPE.

3.12.5.7 Measurement of α

Currently there is no standardized method for measuring α. The actual physical extent of a source can only very rarely be used to determine α, for instance only if the source is very simple and sharply defined such as a diffusing plate that is irradiated from the back and that has some aperture close to the plate to produce sharp edges, or for a multiple source where each individual source is assumed to be less than $\alpha_{\rm min}$ and the physical spacing of the elements of the array can be used to determine the angular subtense of combinations of elements as discussed above. Generally, the extent of the retinal image has to be characterized by using a lens to mimic the imaging properties of the eye and some screen or detection array to represent the retina. The most general and ideal set-up would be a lens that can

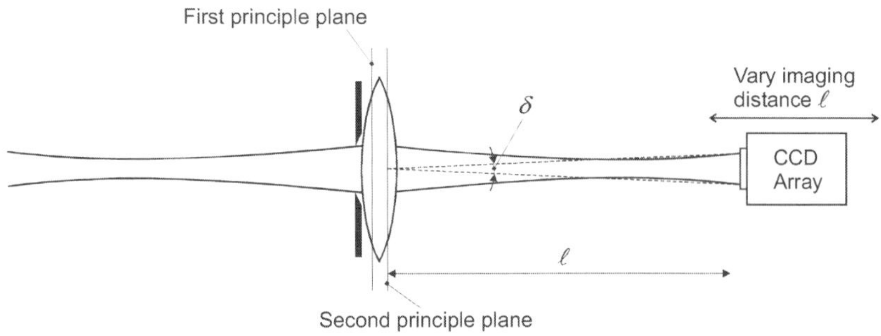

Figure 3.42. Set up for the determination of the angle α subtended by the apparent source at the exposure location represented by the lens (i.e. during the procedure, the lens is kept in place relative to the laser). The CCD array is moved to determine the spot with the minimum angular subtense.

change its focal length between 14 mm and 17 mm, however, this is currently not available.

The approach recommended by the authors is to use a lens and a CCD array at a variable distance behind the lens. The position of the lens corresponds to the position at which α is determined (and at which also the exposure level needs to be determined). It was shown [24] that for Gaussian beams, it is not necessary to use an artificial eye in the sense of having the exact geometric proportions and distances (with a fixed retinal distance but a variable focal length of the lens) of the eye, but any focal length lens can be used as long as α is determined by *minimizing the angular subtense* that the spot at the CCD array subtends. This minimum angle is then equivalent to what it would be for an artificial eye. It is important to note that by varying the distance between the lens and the array, the smallest angle (defined as the spot diameter divided by the distance of the CCD array to the lens) is to be determined, which is not necessarily produced by the smallest spot on the CCD array. In other words, one is not looking for the smallest spot in terms of mm, but for the smallest angle, which might be obtained for a spot that is somewhat larger as the minimal spot but is somewhat further away resulting in a smaller α as the minimal spot, as shown schematically in figure 3.42.

In practice, the procedure could be as follows: a lens with a focal length f (which can be chosen for the measurement condition at hand) is placed at the position in the beam at which α is to be determined. All the components of the measurement need to be centred with respect to the axis of the laser beam. An aperture with a diameter of 7 mm should be placed in front of the lens to simulate a dark-adapted pupil of the eye for highly astigmatic beams (line lasers). However, for other sources, the aperture should be larger than 7 mm (by a factor that reflects the longer focal length and larger image that is obtained with the experimental set-up). The lens diameter should be at least double the size of the aperture to minimize spherical aberration.

As an image 'detector', a CCD array (without focusing optics) or another matrix based detector is placed at the focal position behind the lens and moved away from the lens while 'observing' the size of the irradiated spot on the screen. (It is intentional here that the term 'image' is not used.) The spot diameter should be determined as described in the previous sections. When the observation of the spot as a function of the lens-array distance shows that the spot is minimal, then the corresponding angular subtense is determined, but the lens-array distance is increased beyond that point to see if the angular subtense of the spot further decreases with distance. This is particularly important if the beam waist of the beam that is to be evaluated is close to the focal plane of the transforming lens. If the focal plane of the lens is outside of the Raleigh range of the test beam, then it is sufficient to look for the smallest spot on the CCD array as this spot will then have the smallest angular subtense. The minimal value of the angular subtense of the spot on the CCD array as subtended from the lens, δ_{min} is used as the angular subtense of the apparent source α. Once the lens-array distance l that produces the minimal angular subtense is determined, this distance can be used to calculate the location of the apparent source D_{acc} for the respective exposure position with the lens equation (3.14). (The exposure position relative to the beam is given by the location of the lens; the measurement of power for the exposure level measurement would have to be determined at that position for 'matching' pairs of exposure level and MPE as a function of α).

$$D_{acc} = \frac{l \cdot f}{l - f}. \tag{3.14}$$

Although it is not necessary to know the location of the apparent source when the evaluation follows the general principle laid out in section 3.12.2, it might still be an interesting parameter that can help in understanding the evaluation, and can also be used to determine the correct position of the field stop when a lens is used to define the angle of acceptance by imaging the source onto the field stop.

It is important to note that the range of movement of the array with respect to the lens is limited by the range of accommodation that needs to be simulated. To place the array at the focal plane of the lens corresponds to the case of the relaxed eye, as the retina is located at the focal length of the eye for accommodation to infinity. However, positions of the array within the focal length of the lens should also be used to determine α, to account for both hyperopic (long-sighted) vision and for a focal depth of the eye that might be larger than used for the experimental set-up. Placing the array inside of the focal length of the lens can produce the minimal angular subtense for the case of a converging beam. For a well-collimated beam, the smallest angular subtense will be detected close to the focal plane of the lens while for a diverging beam, the smallest subtense will be detected some distance away from the focal plane. The furthest point of placement of the array from the lens, l_{max}, is given by the restriction of the optical object distance to a near point of 10 cm. By using the lens equation it follows that

$$l_{\max} = \frac{f \times 100}{100 - f} \tag{3.15}$$

where the distances are specified in mm. As a focal length of more than 100 mm would result in a negative value for the placement of the array, the practical choice of lenses is restricted to focal lengths of less than 100 mm. For instance, if a lens with a focal length of 80 mm is chosen, $l_{\max}$ equals 400 mm (i.e. 40 cm).

For non-circular sources, for each lens-array distance, the image beam diameter values need to be determined in the two principal axis and the arithmetic mean value is used as the effective diameter of the spot.

In the case of scanning beams, the extent of the scan across the retina (or as detected with an imaging lens) cannot be directly used to determine α, as is discussed in case study 4.8.4.

3.12.5.8 General approach versus current standard specification

In the previous sections we have discussed the dependencies of the retinal thermal value on exposure duration, wavelength and angular subtense of the apparent source, as well as related measurement criteria for the determination of the effective exposure level that has to be compared to the MPE. The general evaluation approach (the comparison of the exposure level to the MPE) as described in section 3.12.2 is rather time-consuming, but the advantage of such a general approach (where all possible exposure locations are analysed and for each position the minimal retinal spot size is characterized) is that the *location* of the apparent source is actually not relevant. As long as α is the correct value for each evaluation position, it does not matter where the eye accommodates in order to produce this value of α. The location of the apparent source only becomes relevant when it is used to define the position at which the MPE analysis (or for classification the comparison of level of exposure with the AEL values) is to be performed. This is the current practice in the international laser safety standards. Compared to the general approach, the definition of such a single evaluation position is of course a significant simplification in terms of the number of power and α measurements. However, it has the drawback that the location of the apparent source needs to be determined. For the definition of the evaluation position it is important that this single evaluation position is in fact also the position of the greatest hazard so that when the exposure level is below the MPE at that single position, it will not exceed the MPE at any other position within the beam. The current laser safety standards assume that the worst-case exposure position is 10 cm from the apparent source, and for the case of a laser beam with an accessible beam waist, 10 cm from the beam waist (i.e. in the current standards, the location of the beam waist can be treated as the location of the apparent source). Consequently, both the exposure level (the power through a 7 mm pupil) and α are to be determined at a distance of 10 cm from the beam waist. The basis for this specification is the assumption of a near point of 10 cm, i.e. 10 cm would be the closest position at which the eye would be able to image the apparent source. Closer distances would result in higher powers entering the eye but a blurred, less hazardous image would result. For incoherent sources this is certainly a valid assumption, and a viewing distance of 10 cm would certainly correspond to the MHP when it is assumed that the near point can be as close as 10 cm. For the case of laser beams, a theoretical analysis for thermal retinal hazards showed that this is a good approximation for beam divergences larger than about 100 mrad but underestimates the hazard for lower divergences. This can be understood since for low divergence beams (with reasonably small beam waist diameters), essentially the total beam power enters the 7 mm pupil of the eye even at some distance from the beam waist. The angular subtense of the apparent source is largest at the beam waist (where it is equal to the divergence of the beam) and decreases as the distance to the beam waist increases. Consequently, when moving away from the beam waist, the power that enters the eye stays more or less constant, but with further distance α becomes smaller, so that the level of hazard increases with increasing distance from the beam waist. It is only at the distance where either α becomes smaller than α_{min} or where the beam diameter becomes so large that a sufficient amount of power is lost at the pupil, that the level of hazard does not further decrease with distance. It follows that the higher the divergence of the laser beam, the closer the MHP moves towards the beam waist. It was shown with the help of the beam propagation model that the MHP does not move closer to the beam waist than 10 cm. Therefore, for a beam divergence value of about 100 mrad and larger, the most hazardous position is at 10 cm from the beam waist. This is the consequence of the assumption of a near point of the eye of 10 cm.

Although the current editions of the laser safety standards generally specify an evaluation position of 10 cm from the beam waist, it is clear that the prudent evaluation position for medium to low divergence values should be at the MHP, which for the retinal thermal hazard might be further from the beam waist than 10 cm. While at the correct MHP, the power that is measured through a 7 mm aperture will be not much less than the level measured at 10 cm, the value of α at the MHP will be smaller than the value at 10 cm. For instance, for a beam with a divergence of 10 mrad, the MHP will be about 1 m from the beam waist and for a beam waist diameter of 3 mm α at the MHP will be about 2 mrad, while the same beam at a distance of 10 cm would result in an α of about 21 mrad, a factor of 10 higher than at the MHP, while the power through a 7 mm pupil at the MHP is only about 35% less than the total power that would have been measured at a distance of 10 cm. It can be expected that future editions of the international laser safety standards will take account of this concept of the MHP for medium divergence beams. The development of such a specification is discussed in the following section.

3.12.5.9 Exposure level and α at most hazardous position

For beams that do not exhibit a notable degree of astigmatism (i.e. beams that have a close to circular beam profile), beam propagation models can be used to calculate the most hazardous position as a function of beam divergence and beam waist diameter. There are two approaches to determine the exposure level and α at the most hazardous position: (i) by direct measurement of the power that passes through a 7 mm pupil located at the MHP and by measurement of the angular subtense of the apparent source α with a lens and a CCD array to simulate the eye (as described in section 3.12.5.7), or (ii) by characterizing the beam divergence, θ_σ, and the waist diameter, $d_{0\sigma}$ (with the method of the second moment) and using formulae to calculate the power that passes through a 7 mm pupil and for α at the MHP. The formulae for the latter approach were derived for Gaussian beams with the assumption of aberration free optics and neglect of diffraction effects at the pupil. It therefore constitutes a worst-case approach and might result in overcritical results. As both approaches specify the evaluation position directly as a certain MHP relative to the beam waist and not relative to some (abstract) location of the apparent source, it is not necessary to characterize the location of the apparent source. However, since the location of the apparent source is associated with the accommodation distance of the eye (the distance at which the eye focuses), the model provides the information that is necessary to calculate the location of the apparent source. In the beam propagation model, the focal length of the lens of the eye is varied to obtain the smallest retinal spot size. The focal length that produces the smallest spot for a given position of the eye within the beam can be used to calculate an object distance with the lens equation (the 'image' distance equals 17 mm). The calculated object distance is equivalent to the accommodation distance and therefore also the location of the apparent source in the sense that when a virtual object is placed at the location of the apparent source and the angular subtense of this virtual object is equal to α, the image formed at the retina will have the same dimension as the retinal spot that is formed by the laser beam. It has been shown by beam propagation modelling that the location of the apparent source is at the centre of curvature of the wavefront that is incident on the eye, and the angular subtense α that characterizes the smallest spot on the retina is equal to the angular subtense α of the beam diameter at the location of the apparent source, as is schematically shown in figure 3.43. As in the far-field (outside of the Raleigh range), the centre of curvature is close to the position of the beam waist, for exposure in the far-field, α will be approximately equal to the beam waist diameter divided by the distance to the beam waist.

While a theoretical beam propagation analysis can be very helpful, it contains a number of worst case assumptions that may lead to an overestimation of the hazard for beams with non-Gaussian beam profiles. But for multiple small sources it might also represent an underestimation of the hazard. If it is possible to perform accurate experimental characterization of α, then this might be the

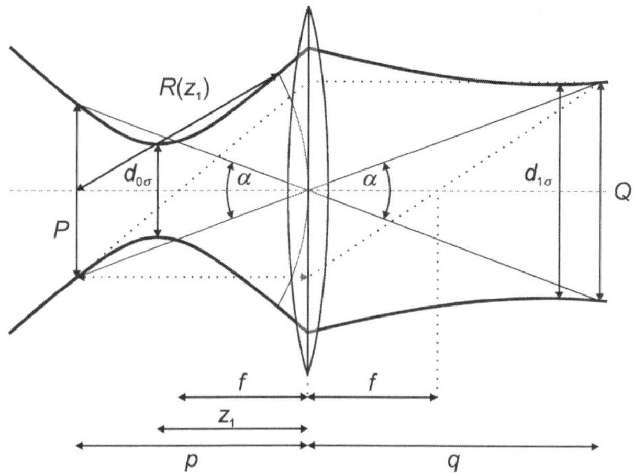

Figure 3.43. The location of the centre of curvature of the wavefront at the position where the beam is incident on the eye can be associated to the properties of the location of the apparent source.

preferred way especially for higher order (lower quality) laser beams. For the direct measurement of exposure level and α for beams that deviate not too much from a Gaussian one, the criteria for the measurement location, i.e. the MHP, can be derived from the beam propagation model calculations. This simplified approach is valid only for beam waist diameters of up to about 6 mm and beam divergence values of up to about 500 mrad.

If the divergence of the beam is larger than 92 mrad, the MHP is given as 10 cm from the beam waist. If the divergence of the beam is less than 92 mrad, then the MHP is the location in the beam where the power that is measured through a 7 mm aperture has fallen to 72% of the total power (i.e. beyond that point, less than 72% of the total power pass through the aperture) or where the value of α falls to below 1.5 mrad, whichever distance is closer to the beam waist. In practice, the following procedure can be used (as represented in the flow chart in figure 3.44). With the 7 mm aperture (ideally using a radiometer that has a sensitive detector diameter of 7 mm) one starts at the beam waist and notes the total beam power there. For this total power measurement, it is not necessary to know the accurate position of the beam waist. One then moves outwards until the detected power is 0.72 of the total power value. The distance of this position from the beam waist is in the following given the symbol l_{72}. (It can be shown that at this position, the (second moment) beam diameter of the beam is 8.8 mm, i.e. the most hazardous exposure occurs at that position in the beam where the beam diameter equals 8.8 mm.) If l_{72} is less than 10 cm, as will be the case for beam divergence values of more than about 100 mrad, one is allowed to move out to 10 cm from the beam waist to measure the power and α there. (For a high accuracy analysis it would be necessary to determine the location of the beam waist following a second moment method, however, it might also be sufficient to determine the position of the beam waist with simpler methods.) If l_{72} is more than 10 cm, one measures α at the determined position (the measurement of α is described in section 3.12.5.7). If the determined α is larger than 1.5 mrad, the measurements are completed and the power through the 7 mm aperture determined at that position can be used to calculate the irradiance level that is compared to the MPE which is calculated with the value of α. For the case of low divergence beams it is possible that the value of α determined at the 72%-position is less than 1.5 mrad, and in that case the most hazardous position is closer to the beam waist. In most cases, the most hazardous position is where $\alpha = \alpha_{\text{min}}$ and the fractional power there is less than 72%. However, instead of experimentally finding that position to measure the power at this position, it is recommended to simply use the total power for the following comparison with the MPE (or for larger beam waist diameters, the power that passes through the 7 mm aperture located at the beam waist). This will not be much more conservative than determining the position where $\alpha = \alpha_{\text{min}}$ and measuring the power through a 7 mm aperture there, as typically almost the total beam power will pass through the aperture anyway.

The following figures show the results of the beam propagation modelling for the retinal thermal hazard. We also give simplified formulae for the fractional power that passes through a 7 mm aperture for the assumption of a Gaussian beam profile and for the angular subtense of the apparent source α at the most hazardous position. The model is based on well established beam propagation theory that describes the transformation of a Gaussian laser beam by a lens for beam divergence values of up to about 500 mrad. By specifying the input parameters of the model, the beam divergence θ_σ and the beam waist diameter $d_{0\sigma}$, using the second-moment method, the model is also applicable to non-Gaussian beams. However, the model and concept introduced below cannot be used for multiple or non-uniform sources or for astigmatic beams. For the model, the eye consists of a lens with a variable focal length, a 7 mm aperture as the pupil and the retina located 17 mm behind the lens. For a given divergence and beam waist diameter, the most hazardous position is defined as the position at which the level of the thermal retinal hazard has its maximum value. This can be obtained by varying the focal length of the lens between 17 mm and 14.5 mm. The level of thermal retinal hazard is defined as the fraction of the power that passes through the 7 mm aperture (for a total power of unity) divided by the retinal beam spot diameter.

The calculations show that the accommodation position of the eye is at or close to the beam waist, which is to be expected since the limitations on beam waist diameter and divergence result in a Raleigh range of less than 10 cm, so that the most hazardous positions are in the far-field and the centre of curvature is close to the beam waist position. With this far-field restriction, the results of

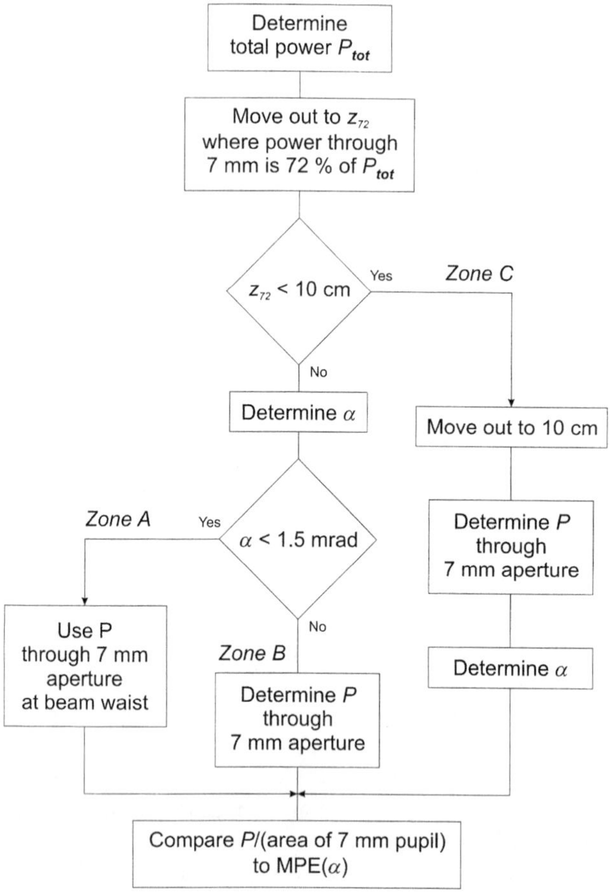

Figure 3.44. Flow diagram representation of the proposed practical method for determining the exposure level and the angular subtense of the apparent source at the MHP for an evaluation of a Gaussian circular symmetrical beam (i.e. with a low degree of astigmatism) with respect to the thermal retinal hazard, i.e. for wavelengths in the wavelength range 400–1400 nm. When the beam diameter is larger than 6 mm, additional evaluations are necessary.

the beam propagation model can be easily understood and interpreted with the assumption that the beam waist is the apparent source in terms of location and size, and that the beam spread is equal to the divergence θ_σ (i.e. approximating the hyperbolic beam envelope by a cone that originates at the beam waist, which is a good approximation outside of the Raleigh range). It is then possible to define simple equations for the most hazardous position z_{haz} in terms of distance to the beam waist, α and the fractional power through the 7 mm pupil at the MHP

(P_f). This approximates the results of the beam propagation model very well. In establishing these simplified equations, the only parameter that was adopted from the beam propagation model was the fraction of the power that defines the beam diameter for the simplified notion of representing the beam by a cone with sharp borders. The corresponding value was found to be 72% and based on this value, a comparison with the beam propagation model shows that all derived formulae provide a conservative result for the calculated parameters. This value of 72% defines the beam diameter so that for the simple understanding of the beam propagation model, the beam can be treated as having sharp borders that are located where 72% of the power pass through the aperture, as schematically shown in figure 3.45. This value lies between the 1/e (64%) and the 1/e^2 (87%) definition for beam diameter. It directly follows for a Gaussian beam profile that the second moment (87%) beam diameter at that position in the beam where 72% of the total beam power pass through a 7 mm aperture equals 8.8 mm (i.e. 87% would pass through a 8.8 mm aperture). For an exposure distance of 0.1 m (10 cm) from the beam waist, the (second moment) beam diameter of 8.8 mm is obtained when the (second moment) divergence equals 92 mrad.

The above-mentioned criteria (see also the flowchart) for the MHP correspond to three zones with the following properties, which can be also identified in figures 3.46–3.49.

Zone A—beams with very low divergence or very low beam waist diameter (or both). In this zone, at the most hazardous position, $\alpha < \alpha_{\min}$ and the zone border is basically defined by the condition of $\alpha = \alpha_{\min}$. The fraction that passes through the 7 mm aperture is close to the total beam power.

Zone B—beam divergence values up to a maximum of 92 mrad. The MHP is further from the beam waist than 10 cm. For the simplified formulae, the MHP is the distance from the beam waist where 72% fractional power passes through the aperture (see the plateau in figure 3.48). The angular subtense α at the MHP depends on the beam waist diameter and can be approximated by dividing the beam waist diameter (corrected by $\sqrt{2}$ for the 63% diameter top-hat assumption for the retinal profile) by the most hazardous position.

Zone C—beam divergences above 92 mrad. The most hazardous position is 10 cm from the beam waist, and α can be approximated by dividing the beam waist diameter by the distance of 10 cm (again corrected for the 63% diameter top-hat assumption).

As can be seen from figure 3.47, the angular subtense at the most hazardous positions for the beam divergence and beam waist diameter values plotted do not reach 100 mrad so that it is not necessary for the simplified procedure discussed above to limit the angle of acceptance to 100 mrad. For beam waist diameters more than about 6 mm, however, it is possible that the most hazardous position is closer to the beam waist than 10 cm. For beam divergence values larger than about 500 mrad and beam waist diameter values larger than about 6 mm it is not recommended to use the simplified approach as some assumptions of the model start to break down. For divergence values larger than 500 mrad, the

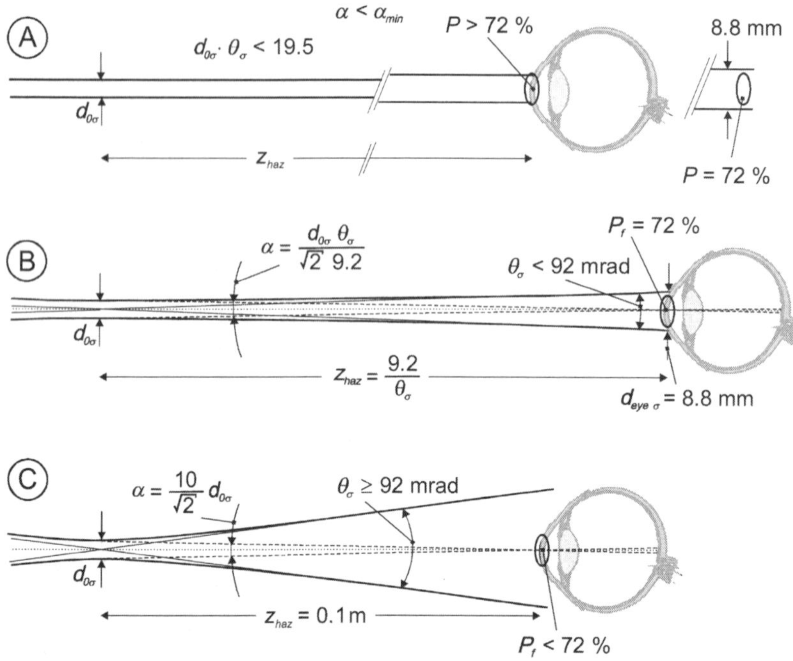

Figure 3.45. Schematical drawing of simple understanding of the most hazardous position for a beam with varying divergence. The beam waist diameter specified according to the second moment method is denoted by $d_{0\sigma}$. The three sections are further discussed in the text and are also represented in figure 3.49. The formulae are simplifications that are derived with the assumption that the beam waist is the apparent source and the beam spread is linear with the origin at the beam waist.

beam propagation theory that was used becomes inaccurate. For beam waist diameters larger than about 6 mm, exposure at or close to the beam waist can become the most hazardous position, so that the far-field assumptions on which the simplified formulae are based, do no longer apply. However, for beams with realistic divergence values, for these kind of beam diameters, the difference in power to a position of 10 cm from the beam waist is less than 1%.

It should be noted that the above formulae and evaluation scheme were developed for and can be directly applied only to close to Gaussian non-astigmatic beams and to stationary beams. For astigmatic beams, for instance for line lasers, or for scanning beams, the simple approach described above cannot be used and the general evaluation concept described in section 3.12.2 must be adopted. The case of line lasers and scanning lasers are treated as a case study in sections 4.8.3 and 4.8.4, respectively.

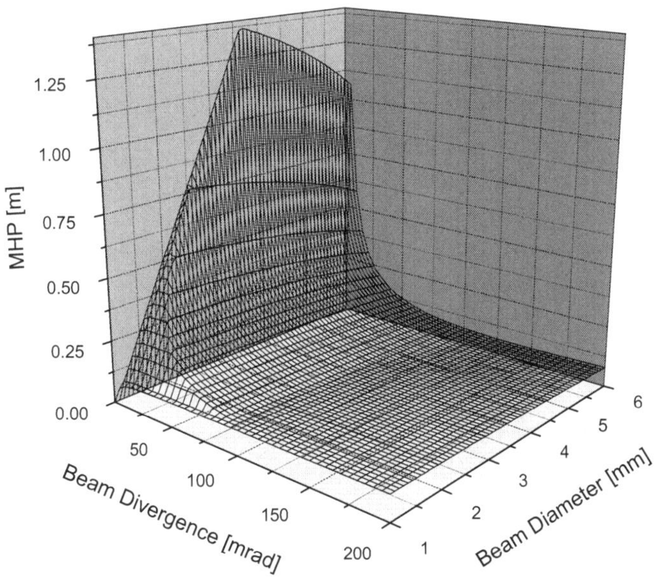

Figure 3.46. 3D plot of most hazardous position (MHP, here defined as a distance from the beam waist) as a function of the beam divergence and waist diameter. For divergence values larger than about 92 mrad, the MHP has a constant value of 10 cm, which is the assumed near point of the eye. For closer positions than 10 cm, the hazard is reduced as the spot size on the retina increases. For divergence values less than about 92 mrad, the MHP increases up to the point where at the most hazardous position, α equals about 1.5 mrad (see figure 3.47).

Note on assumptions and limitations of the beam propagation model

The beam propagation formalism is an exact solution for Gaussian beams in the paraxial approximation, i.e. it strictly only applies when the beam divergence is larger than about 500 mrad. The model does assume aberration-free optics and neglects diffraction effects at the pupil, which both result in worst case calculations when compared to the real case. In reality, relatively low aberration is obtained by reducing the pupil to 2 mm, but diffraction effects at the pupil can lead to an increase in the retinal spot size. A larger pupil would, for small beam diameters at the cornea, produce minimal diffraction but then heavy aberration results in a broadening of the minimal retinal spot size. For large beam diameters with the assumption of a 7 mm pupil, both aberration and diffraction effects act together to increase the spot in comparison to the worst case values that are produced by the model. The only assumption in the model which does not necessarily produce worst case values is the Gaussian profile; a top-hat distribution at the cornea, for the same beam diameter, would result in more

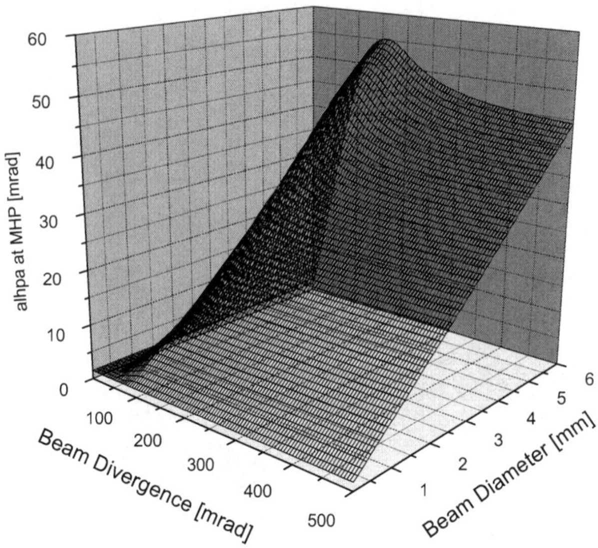

Figure 3.47. 3D plot of α (based on the 63% definition of α) at the MHP, as derived from the calculated retinal spot size. The three zones can be well distinguished, and the large tilted plane, representative of Zone C, shows that in this region the retinal spot size is basically directly proportional to the beam waist diameter and does not depend on the divergence, which can be understood as the MHP in this region is a constant value of 10 cm.

power entering the eye than a beam with a Gaussian profile. However, this difference in power can only become relevant for beam diameters at the cornea which are in the region of the pupil diameter, and then the effects of aberration and diffraction would overcompensate any non-worst case assumptions with regard to the power that enters the eye. Insofar as the distribution on the retina is concerned, calculations based on the assumption of a top-hat distribution show that the dependence of the parameters presented in the above figures are largely equivalent.

3.12.6 Retinal photochemical

For exposure durations above 10 s and wavelengths between 400 and 600 nm, photochemical limits are defined in addition to the thermal limits discussed in the previous sections. The photochemical and thermal damage mechanisms can be seen as a 'competing' mechanism in the true sense of the word, since the damage mechanism that first results in injury for a given exposure level is the critical damage mechanism in that particular circumstance. Consequently, in order that an exposure to a given laser beam at a given position in the beam and

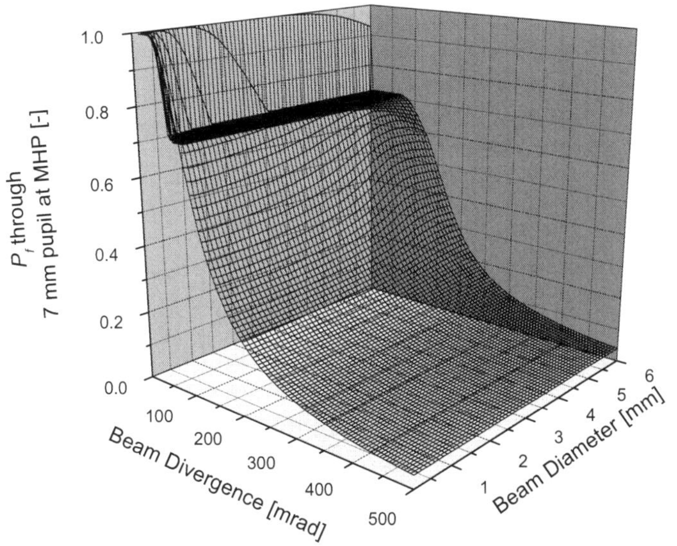

Figure 3.48. 3D plot of the fraction of the power through the 7 mm aperture that is located at the MHP. The plateau at the level of 72% of Zone B can be clearly seen. In Zone A, the fractional power is larger than 72% as there the MHP is given by the location in the beam where $\alpha = \alpha_{\text{min}}$ and this distance is closer to the beam waist than the distance defined by the 72% fraction. In Zone C, the fraction of power into the 7 mm pupil is approximately proportional to the beam divergence, as there the MHP is at 10 cm from the beam waist.

for a given duration to be safe, the exposure level needs to be below the MPE for both the thermal and the photochemical case. It has already been noted that for extended sources, the biophysically-effective exposure level that has to be compared to the photochemical MPE can be smaller than the exposure level that has to be compared to the thermal MPE, as the maximum angle of acceptance for the photochemical case is smaller than the one for an evaluation of the retinal thermal hazard. Thus, it is not sufficient or generally applicable to compare the two MPEs in order to determine which is the more critical one, it is rather that one needs to compare hazard levels, i.e. the ratio of the exposure level for the two cases divided by the respective MPE. The photochemical limits are compared to retinal thermal limits in more detail in section 3.12.7.

The retinal photochemical MPE values are given in table 3.12. The basic limits are applicable for the wavelength range 400–450 nm, and are increased for longer wavelengths by the wavelength correction factor C_3, which is discussed in the following section. The time dependence and the associated maximum angle of acceptance are discussed in section 3.12.6.2.

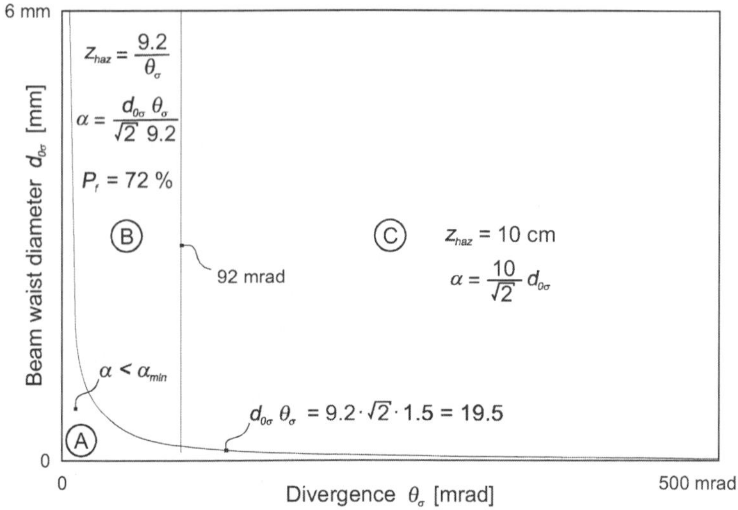

Figure 3.49. 2D plot representation of the three zones and corresponding formulae.

Table 3.12. Retinal photochemical MPE values and associated values for the angle of acceptance that are relevant for the determination of the effective exposure level when the apparent source is larger than the angle of acceptance. The angles are given both in terms of plane angle and solid angle. The values given as a solid angle were calculated by using $\Gamma = (\pi/4)\gamma^2$.

		Associated angle of acceptance	
Exposure duration t	MPE value	Plane angle	Solid angle
10–100 s	$100C_3$ J m^{-2}	$\gamma_{ph} = 11$ mrad	$\Gamma_{ph} = 10^{-4}$ sr
100–10 000 s	$1C_3$ W m^{-2}	$\gamma_{ph} = 1.1\sqrt{t}$ mrad	$\Gamma_{ph} = 10^{-6} \times t$ sr
$t > 10\,000$ s		$\gamma_{ph} = 110$ mrad	$\Gamma_{ph} = 10^{-2}$ sr

3.12.6.1 Wavelength dependence

As is typical for photochemical limits, the photochemical retinal limit strongly depends on wavelength, as is shown in figure 3.50, where the retinal photochemical laser limits are compared to the ICNIRP photochemical limits for incoherent broadband radiation (the 'blue light hazard'). It can be seen that for simplicity, the wavelength dependence of the laser MPEs are described by a logarithmic function which results in a somewhat decreased safety factor for some wavelengths.

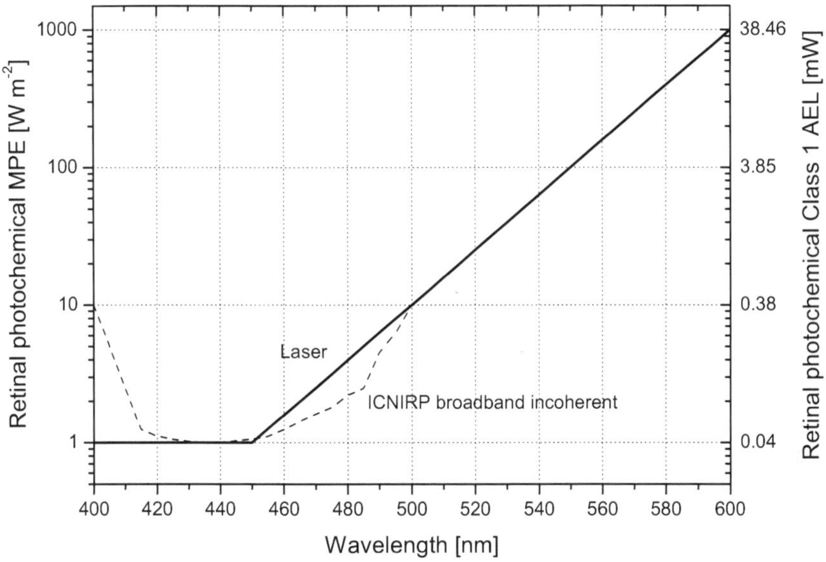

Figure 3.50. The retinal photochemical MPE (left axis) increase strongly with wavelength (values shown are valid for exposure durations larger than 100 s). The values given at the left-hand ordinate are also equal to the wavelength correction factor C_3, as the basic MPE value equals 1 W m^{-2}.

The wavelength correction factor C_3 (in ICNIRP and ANSI called C_B, where 'B' stands for 'blue light hazard') increases the retinal photochemical limit exponentially, so that it changes from the lowest value of 1 W m^{-2} valid in the wavelength range 400–450 nm up to a value of 1000 W m^{-2} for a wavelength of 600 nm. Consequently, the retinal photochemical MPE is the critical MPE when compared to the retinal thermal hazard for wavelengths in the blue and green end of the visible spectrum. In the yellow and red, the retinal thermal hazard is usually the more critical one.

3.12.6.2 *Time dependence, angle of acceptance*

The time dependence of the retinal photochemical MPE is closely linked to the angle of acceptance that is used for the determination of the exposure level in order to compare this with the MPE. The retinal photochemical limit for laser radiation is derived from the broadband photochemical limit, which is given as a constant time-integrated radiance (in units of J m^{-2} sr^{-1}) for all exposure durations between 10 and 10 000 s, as is further discussed in section 3.12.6.4. The values specified for the laser limits are obtained by multiplication by the specified limiting angle of acceptance, and therefore the time dependence of the laser limit is due to the time dependence of the limiting angle of acceptance. The values specified for the angle of acceptance reflect the angular extent of the eye movements that move the laser spot across the retina (see the discussion on eye movements in section 3.12.4 for the retinal thermal exposure limits). The dependence of the limiting angle of acceptance γ_{ph} on exposure duration (table 3.12) is intended to roughly reflect the dependence of the extent of the eye movements. The maximum angle of acceptance γ_{ph} is a constant value of 11 mrad for exposure durations between 10 and 100 s, where the MPE is also a constant value of 100 J m^{-2}, and increases with the square root of the exposure duration to a value of 110 mrad at an exposure duration of 100 s.

As is typical for photochemical processes, the effect depends only on the cumulated dose and not on the exposure duration (within a certain time frame), and consequently the MPE is given as a constant radiant exposure of $100C_3$ J m^{-2} for exposure durations up to 100 s. This means that for cw radiation, the maximum permissible *irradiance* level is given by $100C_3$ $1/t$ W m^{-2} and decreases linearly with exposure duration, varying from 10 W m^{-2} for an exposure duration of 10 s to the 1 W m^{-2} for an exposure duration of 100 s. When separate exposures occur within 100 s, the correct way of performing an analysis is to sum the radiant exposure values of the individual exposures within the exposure duration and compare the sum to the MPE value of 100 J m^{-2}. If there is continuous exposure but the irradiance level of the exposure varies, then the appropriate treatment is to integrate the irradiance level over time within 100 s to obtain the total radiant exposure value that is subsequently compared to the MPE value. An alternative method is to determine the average irradiance (averaged over 100 s) and multiply the average irradiance by 100 s to determine the total radiant exposure within 100 s. For exposure durations greater than 100 s, the retinal photochemical MPE assumes a constant irradiance value of 1 W m^{-2}, but it should be noted that there is a time dependence in the angle of acceptance specified for the determination of the exposure level. Non-constant irradiance levels or pulsed exposure need to be averaged before being compared to the MPE of 1 W m^{-2}. It is therefore not necessary to limit the peak irradiance to the MPE, but only the averaged value. (This can be understood on the basis of the underlying dose relationship.) However, for non-uniform exposure levels or non-uniform pulse patterns it is important to note that averaging durations as low as 100 s need to be considered in order to determine the maximum average irradiance. It is not permissible to only average over the maximum anticipated exposure duration (or time base in case of classification), as this may be longer than 100 s. Evaluation of pulsed exposure or emission is further discussed in section 3.12.8.

For an evaluation of a non-uniform or pulsed exposure where the averaging duration needs to be varied, the possible influence of the angle of acceptance that is defined as a function of the exposure duration (which for non-uniform or pulsed exposure is the averaging duration) should be considered, i.e. for an averaging duration of 100 s, the specified angle of acceptance is 11 mrad and this increases for longer averaging durations. In terms of additivity of exposures, exposure to small ('point') sources only needs to be considered up to 100 s, i.e. if one exposure occurs per 100 s, the individual exposures can be compared individually to the MPE for 100 s. However, where the angular subtense of the source is larger than 11 mrad, longer exposure durations have to be treated as additive, since for longer exposure durations larger values for the angle of acceptance lead to larger measured values. The analysis can be simplified by using the angle of acceptance that is specified for the maximum anticipated exposure duration (for instance 110 mrad for 10 000 s and longer) for the evaluation with an averaging duration of 100 s. However, the specified limiting angle of acceptance only has an effect, i.e. needs to be considered, if the angular subtense of the apparent source (α) is larger than the specified angle. For smaller sources, as also discussed in section 2.4, the angle of acceptance does not have an effect on the measured value since the complete source is 'seen' by the detector. For this case it is also not necessary to account for the angle of acceptance in the measurement—it is possible to determine the exposure level with a usual radiometer that has an 'open' and undefined angle of acceptance. It is only when α is larger than γ_{ph} that the determined exposure level (power measured through the 7 mm limiting aperture divided by the area of the limiting aperture) is larger if the maximum angle of acceptance is not accounted for in the measurement. In other words, the use of an unlimited angle of acceptance for extended sources would result in an overcritical exposure value, since the *effective* value correctly determined using the maximum angle of acceptance as specified in table 3.12 would be lower. The increase of γ_{ph} with exposure duration has the following effect on the determination of the effective exposure level. Let us assume a source that has some angular extent α which is larger than 11 mrad, say 55 mrad. When an exposure to the source is evaluated with the assumption of an exposure duration between 10 and 100 s, then the angle of acceptance should be 11 mrad. Since the source is larger than this, limiting the angle of acceptance of the measurements means that only part of the radiation that is incident on the limiting aperture from the whole source contributes to the effective exposure level, i.e. only that proportion of the total irradiance that has its origin in the part of the source that is within the angle of acceptance is measured. The source needs to be scanned for hot spots, i.e. the field-of-view of the radiometer has to be pointed to different parts of the source to maximize the effective irradiance value. To get some information on the degree of reduction of the exposure level due to the limited angle of acceptance, one can take the squared ratio of γ_{ph} to α, which, if the source were homogenous (i.e. does not contain hot spots), is the ratio of the irradiance measured with an unlimited angle of acceptance to the irradiance measured with a properly limited angle of acceptance. (The square comes from taking the ratio of the areas of the detector and the image at the detector plane.) For our example, this would mean that due to the specified maximum angle of acceptance of 11 mrad, the effective irradiance is a factor of $11^2/55^2$, i.e. 1/25 or 4% of the value that is obtained with an unlimited angle of acceptance (and in the extreme, i.e. for a source size of 110 mrad or larger, the possible difference between the exposure level determined with 11 mrad and with 110 mrad is a factor of 100). However, due to the increase of the limiting angle of acceptance with exposure duration, this ratio decreases steadily for longer exposure duration. In our example, the angle of acceptance is equal to 55 mrad for an exposure duration of 2500 s (42 min). For exposure durations longer than this, the full source is measured and the value of the angle of acceptance is no longer relevant, both in terms of what is specified as the maximum value γ_{ph} and in terms of what is used for the measurement (as long as the angle of acceptance of the radiometer is larger than α).

In summary, retinal photochemical MPEs are relevant only for long-term (usually intentional) exposure to short wavelength visible radiation. For exposure durations of 100 s and longer, the limit in the wavelength range 400–450 nm equals 1 W m^{-2} (the corresponding AEL for Class 1 and Class 1M equals 0.04 mW), and this value is increased to 1000 W m^{-2} for a wavelength of 600 nm. A time-dependent angle of acceptance is associated with the limits that have to be used to determine the exposure level. For sources which are larger than the specified angle of acceptance, the effective exposure level that is compared to the MPE is smaller than the actual exposure level at the measurement position.

3.12.6.3 Practical hazard analysis

In the previous section we have discussed the possible influence of the angle of acceptance on the determination of the effective level of exposure that is compared to the photochemical MPE value. For 'typical' well-collimated laser beams, the angle of acceptance does not affect the measured value, as such lasers represent a 'point' or small source, where the source is smaller than the angle of acceptance even for exposure positions close to the source. For these lasers, the worst case exposure position is also easily identified; it is simply the position in the beam with the smallest beam diameter. This is at the exit port when the emitted beam is diverging from the laser or at the beam waist when the emitted beam is converging to an external beam waist some distance away from the device. If the beam diameter is smaller than the 7 mm limiting aperture over part of its length, then the 'level of hazard' does not vary over the length of beam that is smaller than 7 mm. Also, the measurement distance relative to the apparent source (which is usually the beam waist) should not be limited to a minimum distance of 10 cm (100 mm). While it is correct that the image would be blurred (α becomes larger) for distances less than the assumed near point of 10 cm, for the photochemical hazard the limit is in any case based on the 'smearing out' of the image due to eye movements. For beam diameters larger than the 7 mm aperture, moving closer to the apparent source than 10 cm increases the power that enters the eye (and that is measured through the 7 mm aperture). However, the increase of the irradiated retinal area at distances closer than the near-point might be less than the extent of the eye movements, and the hazard is consequently increased. For an exposure at the beam waist (i.e. when the cornea of the eye is at the position of the beam waist), the angular subtense of the retinal spot is the same as the beam divergence, as the refractive power of the eye does not have an effect. (This is also referred to as 'Maxwellian viewing'.) The retinal spot cannot become larger than this and it follows that the beam waist is the worst case exposure position for beams with a divergence less than the maximum angle of acceptance γ_{ph} (i.e. less than 11 mrad for exposure durations up to 100 s and less than 110 mrad for exposure durations of 10 000 s and above). If for these types of lasers the beam waist diameter is less than the limiting aperture, there will be a range of (almost) equal hazard level extending from the position of the beam waist over which practically the full power is measured.

For exposure at the beam waist to beams having divergences larger than γ_{ph}, the specified angle of acceptance results in a smaller effective exposure level so that an exposure some distance away from the beam could be more hazardous. Beam propagation models developed at Seibersdorf research in Austria show that the location of the maximum hazard level strongly depends on the retinal irradiance profile (which is the 'image' of the projected exitance profile of the source) so that it is not possible to give a simple rule as it was developed for the retinal thermal hazard. For beams with a divergence larger than the maximum angle of acceptance γ_{ph}, (i.e. the angle of acceptance affects the effective exposure level) the hazard evaluation in the general case has to be done experimentally. For the photochemical limit, the experimental evaluation is somewhat simpler than for the thermal hazard, as the photochemical MPE does not depend on the value of α and thus the same MPE is valid for all exposure locations. Thus, the procedure for a general hazard evaluation is to vary the distance of the detector to the laser product to maximize the measured exposure level which is compared to the MPE. What complicates the general approach is the limiting angle of acceptance that can be realized, as describe in section 2.4, with two different optical set-ups. For sources that are accessible, such as diffuse transmissions or reflections, or arrays, it is possible to place the field stop at the source to define the angle of acceptance. If this approach is taken, the diameter of the field stop needs to be changed depending on the measurement distance to obtain the correct angle of acceptance for each measurement distance. If the apparent source is not accessible which is the general case for laser beams, then a lens is needed to image the apparent source onto the field stop in front of the detector. For this set-up it is necessary that for each exposure position, the lens-field stop distance is adjusted so that the detected power or energy value is maximized, which is the condition for imaging the apparent source (where apparent source is to be understood as an imaginary source that produces the smallest retinal image or spot size). It is not sufficient to use a fixed set-up where the imaging lens is at some fixed distance relative to the field stop.

For non-homogeneous or multiple sources, it is not necessary to consider smaller angles of acceptance than the prescribed values (as was the case for the thermal hazard), but, for a given exposure distance relative to the laser device, the detector set-up with the imaging lens needs to be pointed to different regions of the source and also needs to be moved normal to the beam axis to detect possible hot spots. It might even be the case that that different parts of the source have different apparent locations, especially for sources that consist of a collimating reflector around the emitting part of the source, such as search lights and LEDs. The reflected light is perceived to come from further behind the actual source, so that for a complete analysis of the source it might be necessary (depending on the size of the source and the evaluation distance) to adjust the imaging distance in the imaging radiometer set-up when the radiometer is pointed at different parts of the source.

In terms of the dependence of the hazard level on the distance to the beam waist, the general behaviour is that if the beam divergence is larger than γ_{ph} there will be some range around the beam waist where the apparent source is larger than γ_{ph}. In this range it might be the case that the level of hazard does not depend on distance, which is the result of the radiance theorem. When the distance to the beam waist is increased, the angular subtense of the apparent source decreases so that at some distance from the beam waist the angular subtense of the apparent source will be equal to γ_{ph}. Exposures beyond that point will be less hazardous when the beam width is not much less than 7 mm, as the full source is within the angle of acceptance, but the power that enters the eye (or the 7 mm aperture) decreases steadily with distance from the beam waist.

3.12.6.4 Background to the derivation of the limit

The photochemical hazard does not depend on the diameter of the irradiated area (as does the thermal hazard) as long as the irradiance remains constant (i.e. if the power that enters the eye is doubled and the irradiated area is doubled too, then the irradiance and the hazard level remains the same). Since the retinal irradiance arising from an extended source is directly related to the radiance of the source being viewed (and not directly related to the irradiance produced by the source at the cornea), the basic photochemical limit is best specified in terms of radiance, as is done for incoherent broadband radiation. The laser MPE for retinal photochemical injury is directly derived from the exposure limit for broadband incoherent radiation as defined by ICNIRP [16] (but also in an ACGIH [19] and IEC [26] document) as a time-integrated radiance of 10^6 J m^{-2} sr^{-1} for exposure durations up to 10 000 s, and as a constant radiance level of 100 W m^{-2} sr^{-1} for exposure durations above 10 000 s. (For the broadband limits, the wavelength dependence of the limit is defined by an action spectrum used to weight the effective exposure level). The definition of the broadband MPE as a constant time-integrated radiance (or 'dose') up to exposure durations of 10 000 s reflects the cumulative biophysical effect of the exposure (the additivity), which is valid for several hours. The laser MPE values are calculated by multiplying the broadband limits by the angle of acceptance as given in table 3.12. The time dependence of the angle of acceptance very roughly characterizes the time dependence of the extent of the eye movements. The time-integrated radiance MPE value can be converted to a radiance value by dividing it by the exposure duration t:

$$10^6\ \mathrm{J\ m^{-2}\ sr^{-1}} = 10^6 t^{-1}\ \mathrm{W\ m^{-2}\ sr^{-1}}.$$

The time dependence that comes from the dose relationship is balanced by the linear time dependence of the maximum angle of acceptance in terms of solid angle, so that the multiplication by the measurement angle of acceptance results in a constant irradiance value for the MPE:

$$10^6 t^{-1}\ \mathrm{W\ m^{-2}\ sr^{-1}} \times 10^{-6} \times t\ \mathrm{sr} = 1\ \mathrm{W\ m^{-2}}.$$

Just as the eye movements lead to levelling out of the retinal thermal MPEs beyond a certain exposure duration, so do the eye movements in the case of the retinal photochemical limits compensate for the additive effect of the photochemical interaction and transform a basic dose MPE into an MPE value that is given as a constant irradiance. While for the retinal thermal limits the dependence of the extent of the eye movements was reflected in the break time T_2 that increased for larger spot sizes, in the case of the photochemical limits this dependence of the extent of the eye movements on exposure duration is directly represented by the definition of the limiting acceptance angle. While the value of 11 mrad for exposure durations less than 100 s is closely linked to the extent of retinal lesion areas that occur when people stare at welding arcs, for longer exposure durations the value for the maximum angle of acceptance becomes somewhat arbitrary in terms of characterizing 'real life' eye movements. For instance, for exposure durations up to 10 000 s (about 3 h), staring continuously at a single point without looking elsewhere is difficult to imagine, and a value of 110 mrad for a retinal area covered by eye movements for exposure durations that long is certainly a conservative value (and was intended to be a number close to $\alpha_{\max}$ for the retinal thermal value, but has in fact a completely different background to $\alpha_{\max}$).

The effect of the eye movements as represented by the angle of the acceptance can best be understood for radiance measurements. (Taking an irradiance measurement with a well-defined angle of acceptance and comparing this value with an MPE derived by multiplication of a radiance-MPE value by the angle of acceptance is identical to carrying out a radiance measurement averaged over the same angle of acceptance and comparing this value to the radiance-MPE.) The angle of acceptance for the measurement is obtained by imaging the source onto the field stop in front of the detector (see figure 2.13). The field stop defines the angle of acceptance, which for radiance measurements serves to average the radiance over the angle of acceptance. This is equivalent to averaging the irradiance profile at the image plane over the field stop. (A limiting aperture of 7 mm at the position of the imaging lens represents the pupil, and the irradiance profile at the position of the cornea is averaged over this limiting aperture.) Since imaging the source onto the field stop is equivalent to imaging the source onto the retina, averaging over the field stop is equivalent to averaging the retinal irradiance over a certain retinal area (both the 'size' of the field stop and the retinal area are best described in terms of the angle subtended by the area). In evaluating the photochemical hazard, it is the eye movements that lead to the averaging, i.e. an angle subtended by a laser spot on the retina (or by the field stop in front of the detector) that is smaller than the angle subtended by the area resulting from eye movements (or smaller than the field stop) will be distributed over the respective area on the retina to produce an average irradiance which is less than the true irradiance in the spot. For the photochemical hazard it is appropriate to average the irradiance both in terms of time and area, as it is the total radiant exposure which is the relevant quantity. For instance, when a beam with 1 mW power is incident for 100 s, the energy delivered equals 0.1 J. When the beam is stationary and has a cross-sectional area at the target of 1 cm^2, it has delivered a radiant exposure of 0.1 J cm^{-2}. However, the same radiant exposure is delivered in the 100 s by a smaller beam that scans homogeneously across the area of 1 cm^2. The radiant exposure value can also be obtained by first averaging the irradiance over the area of 1 cm^2 to result in 1 mW cm^{-2} and then multiplying this irradiance by 100 s. As long as the beam is smaller than 1 cm^{-2}, the actual size of the beam does not influence the averaged irradiance or radiant exposure. Although the local and momentary irradiance is much higher for the smaller beam, the number of photons delivered during the 100 s per 1 cm^2 is the same as for the larger stationary beam. Since the photochemical interaction depends on the total number of photons incident per unit area of tissue, both cases produce the same biophysical effect. This is in contrast to the thermal hazard where the eye movements, while having the effect of decreasing the hazard in comparison to the non-moving case, produce a local temperature rise that is higher for the moving small spot than for a non-moving larger spot. This is why thermal limits cannot be averaged over the angle of acceptance that describes the eye movements.

Since the basic photochemical MPE is given in terms of time-integrated radiance, it might be tempting to make use of the principle of the invariance of radiance (also referred to as 'conservation of brightness', or 'radiance theorem', see also section 2.5) for a photochemical MPE analysis. However, this should be done with caution, as this principle is only correct for the real physical radiance and cannot be generally applied to the biophysically effective radiance value determined by averaging over the angle of acceptance. This can be understood when one considers that for the determination of the effective radiance, the irradiance level at the image plane (i.e. at the retina or at the field stop in front of the detector) is averaged (spread out) over a certain area related to the angle of acceptance, and that radiance is directly related to irradiance in the image plane. While the *physical* radiance will *not* depend on the distance to the source, the *effective* radiance, if the source is smaller than the averaging angle of acceptance, will. The physical irradiance in the image plane (and therefore the physical radiance) remains constant with decreasing distance from the source, as the increasing power level that enters the imaging system (through the limiting aperture) is compensated by a larger image size to produce an invariant irradiance in the image plane. However, when the irradiance in the image plane is averaged over a certain area, then the effective retinal irradiance (and therefore the effective radiance) increases at closer distances, as more power enters the eye but the area over which this power is spread out (averaged) does not change with distance (as long as the image of the source is smaller than the averaging area). Consequently, the hazard increases when one moves closer to the source up to the point where the angular subtense of the apparent source α is equal to the averaging angle of acceptance γ_{ph}. For closer exposure distances, the hazard does not increase, as α becomes larger than γ_{ph} and the law of conservation of radiance again applies (provided that there are no hot spots in the source). The radiance theorem is not only often applied for issues related to distance but also to optical instruments. In this form the theorem states that radiance cannot be changed by optical elements (except when decreased by reflection or absorption losses). Again, this is only correct for the true physical radiance. For instance, a telescope does not increase the physical radiance of the source (or the physical irradiance at the image plane), provided the whole area of the input optics is filled, as the higher power that enters the eye through the telescope is compensated by an equal increase of the size of the image. However, in terms of effective radiance (or effective irradiance in the image plane), when the magnified image is smaller than the averaging area, the averaged area is the same with and without the optical instrument but a lot more power enters the eye through the telescope, thereby increasing the hazard in respect to the naked eye.

3.12.7 Comparison of thermal and photochemical retinal limits

In the wavelength range 400–600 nm and for exposure durations longer than 10 s, both photochemical and thermal MPE values need to be considered in an MPE analysis. Depending on the wavelength, exposure duration and angular subtense of the apparent source, one of the two MPEs will be the more critical one. For a comparison of the photochemical and thermal limits it is important to note that the measurement requirements (the angle of acceptance) specified for the determination of the exposure level are different for the two hazards. The (maximum) angle of acceptance for the determination of the exposure level used for comparison with the thermal limit equals 100 mrad, while the specified angle of acceptance for the photochemical case, γ_{ph}, equals 11 mrad for exposure durations up to 100 s, and for longer exposure durations increases with exposure duration up to a maximum of 110 mrad. The different angles of acceptance can lead to different effective exposure levels when the angular subtense of the source is larger than the angle of acceptance. For instance, let us assume a source that at the evaluation distance has an angular subtense of 55 mrad (see also the example in section 3.12.5.2) and an exposure duration of 100 s, so that $\gamma_{ph} = 11$ mrad. For the determination of the exposure level for the thermal case, the full source is measured, which could for example produce an irradiance at the evaluation position of 100 W m^{-2}. The effective exposure level for the photochemical case, due to the smaller angle of acceptance, would be only 4 W m^{-2}. Depending on the wavelength it might well be the case that the photochemical MPE is lower than the retinal thermal MPE (so that a direct comparison of the MPEs would lead to the erroneous impression that the photochemical case is more critical). However, the much smaller photochemically-effective exposure level could mean that the thermal MPE is exceeded while the photochemical MPE is not. Therefore, in the general case, it is best to compare the thermal and photochemical 'hazard ratios', that is the ratio of the effective exposure level to the corresponding MPE.

However, for small cw sources where $C_6 = 1$, a comparison of the two competing MPE values is simple and applicable for all exposure durations, as the angle of acceptance is always larger than the source, i.e. the full source is measured and the two exposure levels are the same. The retinal thermal MPE for this case is simply a constant value of 10 W m^{-2} for all wavelengths in the visible wavelength range and for all exposure durations longer than 10 s. The photochemical MPE value for the wavelength range 400–450 nm for an exposure duration of 10 s is also 10 W m^{-2}, i.e. for this wavelength range and exposure duration, the two limits are identical (the treatment of repetitive exposures is different, however, as described in section 3.12.8). In terms of the time dependence of the photochemical limit when expressed in terms of irradiance, the MPE value decreases to a value of 1 W m^{-2} for an exposure duration of 100 s, and remains constant for exposure durations longer than 100 s. For wavelengths beyond 450 nm, the photochemical limit has a pronounced wavelength dependence expressed by the factor C_3, which increases the MPE by a factor of 10 for a wavelength of 500 nm, a factor of 100 for a wavelength of 550 nm and 1000 for a wavelength of 600 nm. It follows that for small sources, for the wavelength range 400–450 nm, the photochemical limit is the lower one for all exposure durations (for 10 s the two limits are identical). For wavelengths of 500 nm, the photochemical limit for exposure durations longer than 100 s is 10 W m^{-2}, the same as the thermal limit. Therefore, for wavelengths above 500 nm, for small sources, the thermal limit is the more critical one for all exposure durations, as shown in figure 3.51.

For the wavelength range between 450 and 500 nm it is possible to calculate a critical exposure duration T_{crit} (as a function of wavelength) which expresses the exposure duration beyond which the photochemical MPE is lower than the thermal MPE. This critical wavelength is referred to as T_1 in the ICNIRP guideline and in the ANSI laser safety standard (in the IEC laser safety standard, T_1 is used for an equivalent purpose in the UV wavelength range, and no specific function is given for the more critical MPE). The corresponding formula can be found by equating the retinal thermal limit of 10 W m^{-2} with the photochemical limit and solving for the exposure duration:

$$T_{\text{crit}} = 10 \times C_3 = 10 \times 10^{0.02(\lambda-450)} \tag{3.16}$$

where λ is expressed in nm and T_{crit} results in seconds.

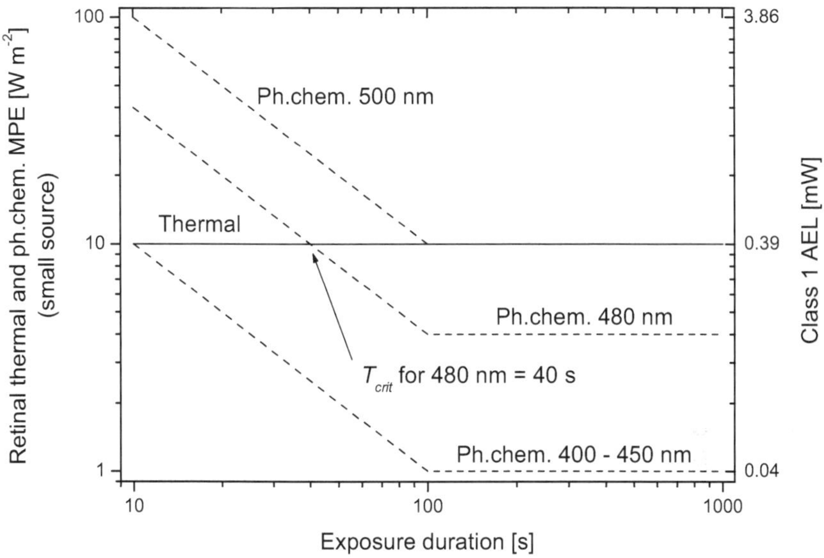

Figure 3.51. Comparison of the thermal and photochemical retinal MPE for small sources ($\alpha \leq 1.5$ mrad). In the wavelength range between 450 and 500 nm it depends on the chosen exposure duration which one of the two MPEs is the more critical one.

More generally, MPEs for extended sources (where the different maximum angle of acceptance may play a role) can be compared with the assumption of a homogeneous source profile (leading to a homogeneous retinal irradiance profile). In this case the reduction of the exposure level due to the angle of acceptance for the photochemical limit γ_{ph} that decreases the exposure level, can be replaced by a corresponding increase of the photochemical MPE value (equivalent to the increase of C_6 beyond 66.6, see equation (3.12)). The factor to increase the photochemical MPE value for the case that the angular subtense of the source α is larger than the specified angle of acceptance γ_{ph} equals $\alpha^2/\gamma_{\text{ph}}^2$ (with the condition that the source is circular and homogeneous).

When the impact of the smaller angle of acceptance for the photochemical limit is taken into account it can be seen that the photochemical retinal limit, when extended to below 10 s, can never be more critical than the retinal thermal limit. Therefore, the statement in the current version of the international standard IEC 60825-1 that the photochemical limit is to be applied as dual limit to the retinal thermal limit does not seem to be necessary.

The potential of optical instruments to increase the hazard is discussed in more detail in the subsequent chapters, but it is interesting to note here that the relevant retinal damage mechanism for staring into the (noon-time) Sun for many seconds with the naked eye is the photochemical one, because the blue part of the solar spectrum causes photoretinitis, a photochemical injury (which is why it is safe to view the sunset, as the blue light is scattered and the Sun appears red and not white as at noon). During a solar eclipse, the exposed area of the retina is reduced, but not the retinal irradiance in that area, and therefore viewing an eclipse without protection can just as well produce retinal photochemical damage as starting into the full Sun. When a telescope or higher power binocular is used to view the Sun, then thermal damage is produced within a few seconds or less. Although the retinal irradiance remains the same because of the radiance theorem, the image size is increased and for equal irradiance but larger image size, cooling is less effective than for the smaller image of the Sun produced with the naked eye, and the thermal damage mechanism is the critical one (i.e. it leads to damage before the photochemical mechanism does).

3.12.8 Multiple pulses in the retinal hazard region

The retinal MPE values discussed in the previous sections apply to single exposures (single pulses or continuous periods of constant exposure) only. For evaluation of an exposure to a pulse train or of multiple exposures, the following criteria need to be considered. For an exposure to be 'below the MPE', all criteria need to be satisfied. However, we will show that it is rarely necessary to evaluate the pulse train against all possible criteria, as one criterion will generally be more critical than others, or in some cases will lead to identical restrictions.

For an evaluation of non-uniform pulse patterns (i.e. pulse patterns where not all pulses have the same energy, pulse duration and spacing), it is important that not only the complete pulse pattern within the maximum anticipated exposure duration (or for classification, within the time base) is evaluated, but also any subgroup of pulses. Currently, this is not stated explicitly enough in the laser safety standards and is therefore easily overlooked. However, it is clear from a practical point of view that this is necessary and important, as both the maximum anticipated exposure duration and the time base for classification are *maximum* durations, and an actual exposure to the radiation can occur at any instant during the pulse train and can obviously last for a shorter duration than the maximum anticipated one. For uniform pulse trains, the maximum anticipated exposure duration will generally be the most critical one, as MPE values, when expressed in terms of irradiance, generally decrease with increasing exposure duration. However, for non-uniform pulse patterns, this is not generally the case, as an exposure to a group of pulses with higher pulse energy can be above the safety limits while the evaluation for the maximum exposure duration could be below the safety limits. Therefore, in the general case, it is important to vary the 'evaluation duration' T_{eval} between the pulse duration of the shortest pulse and the maximum evaluation duration. Additionally it is important to vary the 'position' of this 'evaluation window' across the complete pulse pattern to simulate exposure at different times.

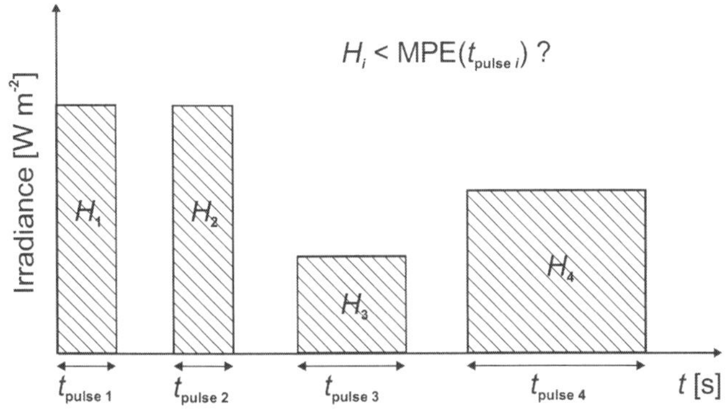

Figure 3.52. Schematic representation of the single pulse criterion. The radiant exposure of each pulse H_i is compared to the single pulse MPE that is applicable for the respective pulse duration $t_{\text{pulse\_}i}$.

In the following we first discuss the pulse criteria as they are currently given in the laser safety standards. However, we will also show that appropriate application of one of the three criteria, for evaluation against the thermal limit, covers all necessary evaluations. For wavelengths and exposure durations where the photochemical limit becomes relevant, the average irradiance needs to be evaluated additionally.

3.12.8.1 Single pulse criterion

The radiant exposure of every single pulse in the pulse train needs to be below the MPE for the pulse (i.e. the MPE evaluated for the pulse duration of each pulse). The principle is shown schematically in figure 3.52.

3.12.8.2 Average irradiance criterion

The irradiance averaged over evaluation durations up to the maximum considered exposure duration (for instance 0.25, 100 or 10 000 s), needs to be less than the MPE for the respective evaluation duration. While evaluation of this criterion for the maximum considered exposure duration (or for classifications, for the time base) will be the critical condition for uniform pulse patterns, for non-uniform pulse patterns it is important that any grouping of pulses and shorter exposure durations be considered too, so that the irradiance E_{aver} averaged over a shorter duration T_{eval} is compared to the MPE that applies to this shorter exposure duration. For varying pulse energies and repetition rates, averaging durations shorter than the maximum anticipated exposure duration (or the time base for classification) will generate a higher average irradiance, and it depends both on the pulse train as well as on the time dependence of the MPE which combination

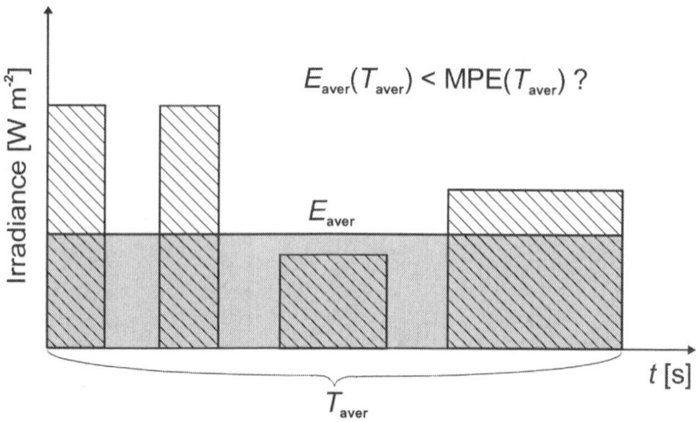

Figure 3.53. Schematic representation of the average irradiance criterion. For irregular pulse patterns, the irradiance needs to be averaged over varying groupings of pulses and the averaged irradiance needs to be below the MPE that is applicable for the averaging duration.

of pulses within the pulse train is the more critical one for the average irradiance criterion. For regular pulse patterns and constant pulse energies, the average irradiance does not depend on exposure duration and the most critical exposure duration is the one which produces the lowest MPE, where the MPE needs to be specified in terms of irradiance (however, since the MPE values either decrease with exposure duration or remain at a constant irradiance level, the maximum exposure duration is generally the critical one). The concept is schematically shown in figure 3.53.

For the thermal limits, this criterion relates to a 'background' temperature rise which results from multiple exposures and which builds up over the exposure duration. The temperature increase is directly proportional to the average irradiance, but for irregular pulse shapes may depend on the actual exposure duration and the section of the pulse train to which the exposure occurs. For the photochemical limits this criterion reflects the additivity of individual exposures over time. For instance, the photochemical MPE is given as constant radiant exposure value for exposure durations between 10 and 100 s. The 'additivity' or 'dose dependence' rule to add up the energy within the maximum anticipated exposure duration (or the time base for classification) is equivalent to the average irradiance criterion, as the average irradiance is calculated by dividing the total energy that is contained within the maximum anticipated exposure duration (up to 100 s) by that exposure duration, and the transformation of the radiant exposure-MPE into an irradiance-MPE is also via division by the same exposure duration. For evaluation of a source that has an angular subtense above 11 mrad, and where the maximum anticipated exposure duration is longer than 100 s, the impact of the angle of acceptance needs to be taken into account. While the MPE as such is specified as a constant irradiance value, the effective exposure level for extended sources may increase with exposure duration since the angle of acceptance increases with exposure duration. For a uniform pulse pattern, the most critical condition in this case is the angle of acceptance that is specified for the maximum anticipated exposure duration. For a non-uniform pulse pattern it is important to consider averaging durations (T_{eval}) as short as 100 s for the determination of the exposure level, while the angle of acceptance for the evaluation may be chosen according to the respective value of T_{eval} (see also the discussion on the photochemical limit, section 3.12.6.2).

3.12.8.3 Additivity criterion

The additivity criterion can be specified in two forms, the well known '$N^{-1/4}$ rule', or the recently introduced total-on-time pulse (TOTP) rule. Since the factor $N^{-1/4}$ is used to reduce the single pulse MPE, it is also sometimes referred to as the 'reduced pulse criterion'. We propose the term 'additivity criterion' here as a general term that applies to both forms of the rule, since we would like to argue that the TOTP 'version' is at least as important as the $N^{-1/4}$ rule, and both characterize the additivity of the thermal hazard for multiple exposures.

The $N^{-1/4}$ rule for uniform pulse patterns

For uniform pulse trains, the $N^{-1/4}$ rule takes the following form. The radiant exposure per pulse needs to be less than the thermal MPE that is applicable for the pulse duration and that is reduced by a factor C_5, where $C_5 = N^{-1/4}(N^{-0.25})$ and N is the number of pulses within the maximum anticipated exposure duration, or within T_2 for the case that T_2 is less than the maximum anticipated exposure duration. (In ICNIRP and ANSI, the pulse MPE reduction factor is referred to as C_P.) The upper exposure duration limit for the determination of the number of pulses N is T_2, since eye movements will result in the exposure of different sites on the retina for exposure durations above T_2 (see also discussion of T_2 in section 3.12.4). The concept is shown for the case of a constant pulse pattern in figure 3.54.

For uniform pulse patterns it is clear that this requirement will always be more critical than the single pulse criterion.

Example. As a simple example we consider a pulse train with a repetition rate of 80 Hz and a maximum anticipated exposure duration (or time base for classification) of 0.25 s. This results in the number of pulses $N = 20$. The MPE reduction factor equals $C_5 = N^{-1/4} = 0.47$. Let us assume emission in the visible wavelength range with a pulse duration of 1 μs for which the MPE (with the assumption of $C_6 = 1$) for exposure to a single pulse equals 5 mJ m^{-2} (and the Class 1 AEL equals 0.2 μJ). For exposure to the pulse train of 20 pulses, the MPE is reduced by multiplication by C_5 to about half of these values. So that the additivity criterion for exposure to pulse trains is satisfied, the pulse

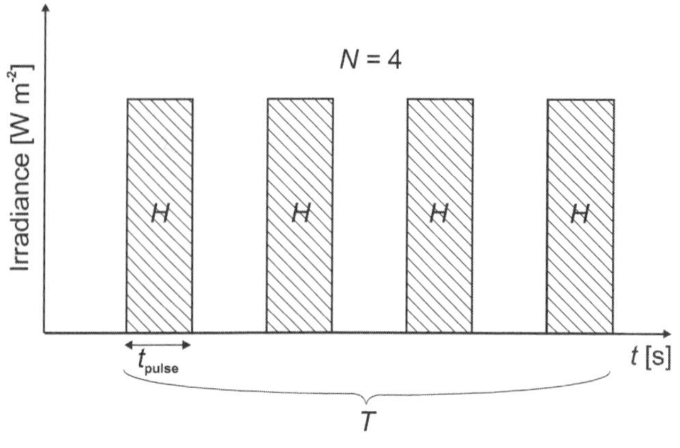

Figure 3.54. Schematic representation of the reduced pulse criterion for uniform pulse patterns. The radiant exposure per pulse H is compared to the single pulse MPE that is reduced by a factor $N^{-1/4}$ (the criterion is often referred to as the 'N to minus one quarter rule'). Here, T stands for either the maximum anticipated exposure duration (for classification, the time base) or the time factor T_2 which is a function of α, whichever is the lower value. (For high repetition rates, the analysis is somewhat modified as discussed below).

radiant exposure needs to be limited to this lower value. The reduced MPE equals 2.9 J m^{-2} and the AEL for Class 1 and Class 1M equals 0.09 μJ.

Thus, depending on the repetition rate and exposure duration, the factor $N^{-1/4}$ can severely reduce the allowed exposure level (or for classification, the energy per pulse). The dependence of the factor on N is shown in figure 3.55 where it can be seen that for a factor of 10 000 in pulse number, the reduction factor is decreased by a factor of 10, as is expected for an exponent of $-\frac{1}{4}$.

We have previously pointed out that the assumption of $\alpha = 1.5$ mrad ($C_6 = 1$) is the worst case assumption which alleviates the necessity to characterize the angular subtense of the apparent source α since it produces the lowest limits. At first glance, it might seem that this is not an appropriate assumption for the case of multiple exposures, since for $\alpha > 1.5$ mrad, T_2 increases from 10 s to up to a value of 100 s for $\alpha \geq 100$ mrad. Consequently, for an anticipated maximum exposure duration of 100 s (or a time base of 100 s for classification) the number of pulses N would be up to a factor of 10 higher than for the assumption that $\alpha = 1.5$ mrad. However, the reduction of the MPE by C_5 for higher pulse numbers N needs to be considered in relation to the increase of the MPE with C_6, so that it is easily seen that the assumption of $\alpha = 1.5$ mrad is still the general worst-case assumption.

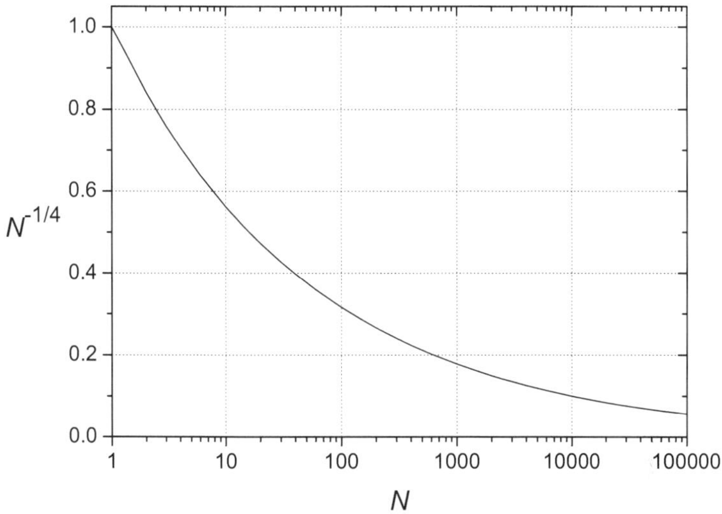

Figure 3.55. The reduction factor $C_5 = N^{-1/4}$ that reduced single pulse MPEs as a function of the number of pulses N that are contained in the exposure.

Biophysical background to additivity criterion

The reduction of the MPE for the evaluation of repeated exposures reflects the experimental finding that ocular tissue is more sensitive to repeated exposures. That is, repetitive exposure to a given radiant exposure per pulse can produce lesions even if the exposure per pulse is below the single pulse threshold (and the repetition rate is sufficiently low not to cause thermal damage due to the steady-state temperature rise, which is covered by the average irradiance criterion[5]). The biophysical background relates to a certain additivity of the individual exposures in terms of producing an injury. Single exposures at exposure levels below the (single pulse) MPE might produce damages on the cellular level which do not lead to an actual injury (in the body, cells die on a regular basis 'from natural causes' and are continuously replaced, and a certain level of 'insult' on the cellular level is repaired without causing a lesion). However, when the exposure is repeated, these subthreshold insults add up and at some stage can produce an actual injury. Thus, there is an additivity effect of multiple exposures although the additivity is not as strong as for photochemical damage where the effect depends on the dose

[5] It is a common misunderstanding that the $N^{-1/4}$ rule is based on the increase of the retinal 'background' or 'steady-state' temperature which results from repeated exposures that have a repetition rate high enough so that the tissue does not cool between pulses and the 'background' temperature builds up (until it reaches a steady-state temperature). The $N^{-1/4}$ rule is *not* related to the background temperature increase, as is obvious since it does not depend on the repetition rate. It is rather that the increase of the background temperature from multiple exposures is the basis for the average irradiance criterion.

(i.e. on the total number of photons and not on the time it takes to deliver these photons). For the thermal damage mechanism, the additivity is less pronounced, which is reflected in the time dependence of the MPE values and also results in an elegant alternative method for the $N^{-1/4}$ criterion, i.e. the TOTP criterion, as discussed below.

The $N^{-1/4}$ rule for non-uniform pulse patterns

For non-uniform pulse patterns where the pulse energy or the pulse spacing varies (but the pulse duration is constant), the *average* radiant exposure may be used for comparison with the reduced MPE. However, this is only an appropriate analysis if different evaluation durations (averaging durations) are considered, not only the maximum one. That is, exposure durations less than the maximum one need to be considered, so that exposure to *groups* of pulses within the maximum considered exposure duration (or the time base for classification) are evaluated in the same way as the complete pulse train. This is shown with the example of figure 3.56, where the pulse train consists of four larger pulses and four smaller ones, and the smaller pulses have a radiant exposure per pulse equal to H and the larger ones of three times H. (It is assumed here that the repetition rate is lower than a critical value of 55.6 kHz for wavelengths up to 1050 nm as discussed in the next subsection.) The evaluation of exposure to all pulses will result in a reduction factor for the single pulse MPE of $8^{-0.25} = 0.6$, and the corresponding *average* radiant exposure equals $2 \times H$. The MPE reduction factor for the group of four large pulses is $4^{-0.25} = 0.7$, while the radiant exposure per pulse equals $3 \times H$. It can be easily seen that the two different evaluation durations lead to different restrictions on the 'allowed' radiant exposure, when the reduction factors for the MPE values are compared to the respective radiant exposure values: 0.6–$2 \times H$ and 0.7–$3 \times H$. The MPE as such is lower for the full pulse train, but the comparison of the larger pulse radiant exposure for the first four pulses with the higher MPE that applies to exposure to this group of pulses is still more critical (by about 30%) than the comparison of the lower MPE with the average radiant exposure value.

When there are sections in the pulse pattern that feature a group of identical pulses, then the evaluation need not be performed for subgroups within these groups. However, for general irregular patterns, the evaluation principle as described can be applied down to groups of two pulses, or even down to one pulse so that this rule would then also include the single pulse evaluation described in section 3.12.8.1, where $N = 1$ and each pulse has to be below the MPE that applies to single pulse exposure.

For irregular pulse patterns where the pulse durations vary, the authors recommend the use of the alternative form of the $N^{-1/4}$ rule, the total-on-time pulse (TOTP) rule as described below. For varying pulse widths (with the restriction that the pulse duration needs to be larger than T_i and the repetition rate is low enough so that no grouping is necessary, see next paragraph), the $N^{-1/4}$

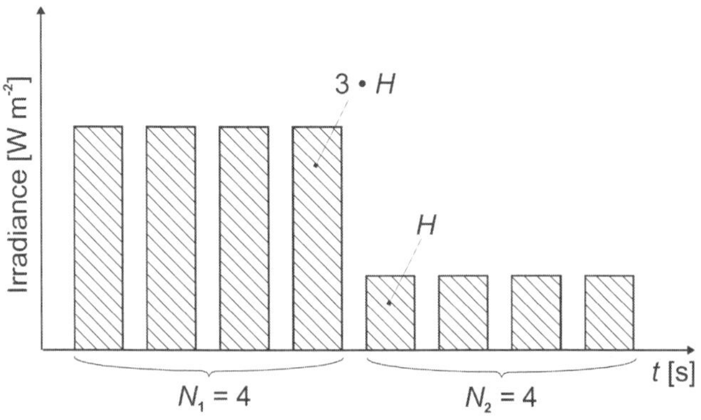

Figure 3.56. Example for a non-uniform pulse train where exposure to the first group of pulses is the critical case.

rule can be made equivalent to the TOTP rule, (which is the more general rule for varying pulse patterns), by comparing the average radiant exposure within the evaluation duration to the reduced MPE, where the MPE is evaluated for the average pulse duration

$$\frac{\sum H_i}{N} < 18\left(\frac{\sum t_{\text{pulse\_}i}}{N}\right)^{0.75} N^{-0.25}. \tag{3.17}$$

It is noted that the N's in this equation actually cancel out and the formula directly represents the TOTP rule as discussed below.

Grouping of pulses (high repetition rate)

The discussion of the $N^{-1/4}$ rule so far assumed that the pulses are spaced far enough so that not more than one pulse would lie within a time frame of T_i, the thermal confinement time. For the determination of N, if the repetition rate is high enough so that multiple pulses occur within a time frame of T_i, the pulse energies within T_i need to be added since they all contribute to producing a temperature increase. (See section 3.12.4 for a discussion on the thermal confinement time T_i and the biophysical background of the grouping of pulses.) That is, the radiant exposure within T_i is summed up and the group of pulses within T_i is counted as one pulse for the determination of N. The total radiant exposure within T_i is subsequently compared to the MPE that applies to the exposure duration of T_i and is reduced by a factor of $N^{-1/4}$ (where N is less than the actual number of pulses and can therefore be considered an 'effective' number of pulses). For the retinal hazard region, T_i is either 18 μs (for wavelengths between 400 and 1050 nm) or 50 μs (for wavelengths between 1050 and 1400 nm). It follows that for constant repetition rates, pulse energies need to be added when the repetition rate

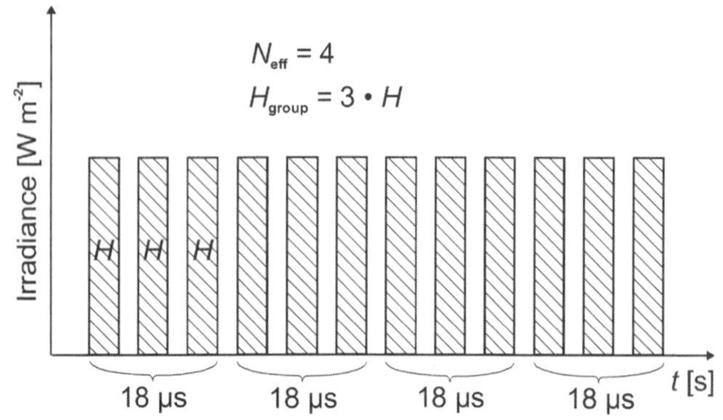

Figure 3.57. Schematic representation of the $N^{-1/4}$ rule when more than one pulse lies within the thermal confinement time T_i. For the example shown, $N = 4$ (i.e. there are four 18 μs 'groups') and the sum of the radiant exposure within the group H_{group} equals three times the radiant exposure of a single pulse.

is higher than 55.6 kHz for wavelengths up to 1050 nm and higher than 20 kHz for wavelengths above 1050 nm (which is simply calculated as the reciprocal of T_i). The principle of this grouping is shown schematically in figure 3.57.

In older versions of the laser guidelines and standards, this 'grouping requirement' was not explicitly stated but was implicitly contained in the definition of the MPE as a constant radiant exposure value for exposure durations of less then T_i. It should be noted that the grouping of pulses that leads to a reduced effective number of pulses N_{eff} is considerably more conservative than counting all pulses and comparing the radiant exposure of each single pulse to the reduced MPE. While in the second case, the MPE will be reduced to a lower value than in the first case (due to higher value of N, in the example shown in figure 3.57, the actual number of pulses equals three times N_{eff}), this reduction of the MPE, due to the exponent of -0.25, has a much smaller effect than the adding of the pulse energies (or radiant exposure values) for the grouping. (In the example shown in the figure, the radiant exposure 'per group' is three times the radiant exposure of a single pulse.)

In practice, especially for constant pulse patterns, it is helpful to introduce what could be referred to as a 'packing factor' PF which characterizes both the reduction of N due to the grouping and the increase of the radiant exposure of the group as compared to the single pulse radiant exposure value. The packing factor, as defined here, can also be seen as the 'number' of pulses within the thermal confinement time T_i (i.e. the number of pulses that are grouped together due to thermal confinement) but is not restricted to be an integer number. The packing factor is therefore calculated by multiplication of the repetition rate of the pulse

Table 3.13.

	MPE	AEL Class 1 and Class 1M
Limit for exposure duration < 18 μs (not reduced)	5 mJ m^{-2}	0.2 μJ
Reduction factor C_5	0.092	
Reduced limit for 18 μs 'packet'	0.46 mJ m^{-2}	0.018 μJ
Packing factor—number of pulses within 18 μs	1.8	
Reduced limit for single pulse	0.26 mJ m^{-2}	0.010 μJ

pattern f (in hertz) with the time T_i (in seconds):

$$PF = f \cdot T_i. \tag{3.18}$$

As T_i^{-1} is the critical repetition rate (e.g. 55.6 kHz), the packing factor also characterizes how much higher the repetition rate f of the pulse train is compared to the critical rate, i.e. the packing factor is also the ratio of the repetition rate f to the critical repetition rate.

The packing factor can be used to calculate N_{eff} and the radiant exposure per group H_{group} from the actual number of pulses N within T and from the radiant exposure per pulse H, respectively:

$$N_{\text{eff}} = \frac{N}{PF} \qquad H_{\text{group}} = H \cdot PF \tag{3.19}$$

where PF and therefore N_{eff} will generally not be an integer number.

Example. The above example of this section can be adopted for the grouping of pulses, when the repetition rate is increased to a value of 100 kHz, leading to a number of pulses within 0.25 s of 25 000. This repetition rate is above the critical value of 55.6 kHz so that more than one pulse lies within the thermal confinement time of 18 μs for the wavelength range under consideration. The packing factor equals $100/55.6 = 1.8$ so that the effective number of pulses within the exposure duration reduces from 25 000 to 13 888. The MPE reduction factor is $C_5 = N^{-1/4} = 13\,888^{-1/4} = 0.092$ to result in the MPE of 0.46 mJ m^{-2}. The AEL for Class 1 and Class 1M equals 0.018 μJ. However, this MPE (or AEL) does not apply to a single pulse in this case, but limits the radiant exposure (or energy) per 18 μs. Since 1.8 pulses (the packing factor) are contained within the time window of 18 μs, the restriction for the 'allowed' radiant exposure *per pulse* is really the MPE (or the AEL) divided by the packing factor, so that each pulse is limited to a value of 0.26 mJ m^{-2} and 0.01 μJ, respectively. The numbers are summarized in table 3.13.

Comparison with average irradiance criterion

It is interesting to note that for pulse patterns with repetition rates *above* the critical value for the grouping of pulses, the average irradiance criterion (applied to the thermal MPEs) is identical to the grouping criterion as just discussed. This is reasonable, since pulses that are within the thermal confinement time are thermally 'smeared out', and when the pulse train is continuous, i.e. one smeared-out group follows the other without interruption, the thermal impact is equivalent to a continuous exposure to the average irradiance. The equivalence, however, can also be shown mathematically, since for the irradiance criterion, the decrease of the thermal MPE when specified in terms of irradiance with exposure duration (exponent of $-\frac{1}{4}$) is identical to the decrease of the single pulse MPE with $N^{-1/4}$. For a practical evaluation, one can choose which one of the criteria is used, although it is a valuable 'check' to evaluate both criteria as they need to result in the same ratio of exposure level to MPE.

It can also be shown that for repetition rates *less* than the critical value for grouping (e.g. less than 55.6 kHz for wavelengths up to 1050 nm), the $N^{-1/4}$ rule is generally more critical than the average irradiance criterion. The mathematical treatment shows that both criteria lead to equal restrictions when the sum of all pulse durations within the evaluation period is equal to the evaluation period, which is the case for a continuous exposure, and the average irradiance criterion would become the more critical one when the sum of all pulses is longer than the evaluation period, but of course this is not possible. The current editions of the laser safety standards generally require evaluation of all three criteria, although this is rarely necessary. We discuss this below by using an elegant proof in relation to the TOTP method.

Ultrashort pulses

The evaluation as described above generally also applies to multiple exposures to ultrashort pulses. However, for pulses with pulse durations less than 1 ns and repetition rates high enough (above 55.6 kHz for wavelengths up 1050 nm) so that thermal grouping occurs, the evaluation needs to be done with caution, as the MPE for pulse durations less than 1 ns is less than the MPE that applies to the thermal confinement time. Biophysically, nonlinear effects result in the reduction of the single pulse MPE for pulse durations less than 1 ns. However, it is obvious that thermal grouping of pulses within the thermal confinement time has to be considered also for pulses shorter than 1 ns, even though *additionally* to the thermal damage, nonlinear effects tend to decrease the single pulse limit. The appropriate treatment of high repetition rate exposures to ultrashort pulses is currently not sufficiently discussed in the laser safety standards, which is mainly due to the lack of experimental data. The appropriate approach should be to consider grouping of pulses (adding of pulse energies) within the thermal confinement time, as described in the previous paragraphs, as well as to evaluate the single pulse without grouping, i.e. to compare each single pulse radiant exposure to the single pulse MPE. The current version of the international laser safety standard IEC 60825-1 implies that grouping of pulses does not have to be considered for an evaluation, but this will surely be corrected in future editions of the standard.

TOTP rule

In the 2001 revision of the laser safety standard IEC 60825-1, an alternative method to the $N^{-1/4}$ rule is given, which is referred to as the TOTP rule. For uniform pulse patterns, the total-on-time, TOT, is simply the sum of all pulse durations within T_2 or within the anticipated exposure duration (or for classification, the time base), whichever is smaller. For non-uniform pulse patterns, the criteria needs to be applied also to shorter evaluation durations, in the extreme down to two pulses, so that the TOT is the sum of all pulse duration within a certain evaluation duration T_{eval}.

For the determination of the TOT, pulses with pulse durations less than T_i are assigned the duration T_i. For non-uniform pulse patterns, if more than one pulse occurs within the duration T_i, these pulse groups are assigned pulse durations of T_i. For uniform pulse patterns, this procedure means the TOT simply becomes equal to the maximum anticipated exposure duration (or to the time base for classification) if the repetition rate is higher than the critical value for thermal grouping (i.e. T_i^{-1}, e.g. 55.6 kHz for wavelengths up to 1050 nm).

The MPE is determined for the 'exposure duration' of TOT and the sum of all pulse radiant exposures within the evaluation duration is compared to this MPE. The 'mathematical' representation of the TOTP rule for visible wavelengths is:

$$\text{is the sum of radiant exposure within } T_{\mathrm{eval}} < 18C_6\ \mathrm{TOT}^{0.75}?$$

Where TOT is the sum of all pulse durations within T_{eval} (considering the assignment of pulse duration T_i to pulses with pulse durations less than the thermal confinement time T_i and to pulse groups within T_i).

It can be shown that this rule is mathematically equivalent to the $N^{-1/4}$ rule as discussed above. The comparison as shown in the table below is done for pulses having non-constant radiant exposure values as well as for constant radiant exposures per pulse. For non-constant values, the average radiant exposure is used for the $N^{-1/4}$ rule (the average radiant exposure is calculated by dividing the sum of all radiant exposure values with the number of pulses N), see table 3.14.

The TOTP rule is the rule of choice for non-uniform pulse patterns, and for uniform pulse patterns provides a valuable way of confirming the $N^{-1/4}$ rule evaluation. The TOTP method also provides additional insight in the biophysical additivity of repeated thermal insults, as the hazard level for repeated exposures is equal to the hazard from a continuous exposure that lasts for TOT seconds. That is, the individual pulses can be envisaged to be moved together so that there is no pause between the pulses to form one long 'pulse' which is safe provided that

Table 3.14.

Rule	TOTP	$N^{-1/4}$
'Exposure duration' to determine MPE	$\mathrm{TOT} = t_{\mathrm{pulse}} \cdot N$	t_{pulse}
MPE (in units of J m^{-2})	$18\mathrm{TOT}^{0.75} C_6$	$18 t_{pulse}^{0.75} N^{-0.25} C_6$
Compare radiant exposure to MPE	Sum of $H < 18\mathrm{TOT}^{0.75} C_6$	(Sum of H) $N^{-1} < 18 t_{\mathrm{pulse}}^{0.75} N^{-0.25} C_6$
	Substitute TOT by $t_{\mathrm{pulse}} \times N$ in above	Bring N from left- to right-hand side
Criteria are identical	Sum of $H < 18 t_{\mathrm{pulse}}^{0.75} N^{0.75} C_6$	Sum of $H < 18 t_{\mathrm{pulse}}^{0.75} N^{0.75} C_6$
Proof for constant pulse radiant exposure values H (no averaging necessary)		
Compare radiant exposure to MPE	Sum of $H < 18\mathrm{TOT}^{0.75} C_6$	$H < 18 t_{\mathrm{pulse}}^{0.75} N^{-0.25} C_6$
	Substitute TOT by $t_{\mathrm{pulse}} \cdot N$, (Sum of H) by $N \cdot H$ and bring N to the right-hand side	
Criteria are identical	$H < 18 t_{\mathrm{pulse}}^{0.75} N^{-0.25} C_6$	

Table 3.15. Comparison of the averaging (first three lines) and the TOTP rule (last line). (Sum of H in T_{eval}) is the sum of all radiant exposures within the evaluation duration.

Average irradiance < MPE in units of W m^{-2} for averaging duration T_{eval}:	(Sum of H in T_{eval}) $T_{eval}^{-1} < 18T_{eval}^{-0.25}$ Bring T_{eval} to the right-hand side (Sum of H in T_{eval}) $< 18T_{eval}^{0.75}$
Sum of H in T_{eval} < MPE in units of J m^{-2} for TOT:	(Sum of H in T_{eval}) $< 18TOT^{0.75}$

the total energy of that long pulse is below the MPE for the corresponding (TOT) 'pulse duration'.

On a mathematical level, the equivalence is basically due to the time dependence of the retinal thermal MPE with the exponent of $\frac{3}{4}$ when the MPE (between T_i and T_2) is expressed in terms of radiant exposure and the $-\frac{1}{4}$ exponent for N.

Invariance of hazard for scanned or chopped sources

The dependence of the MPE as discussed in the previous paragraph also results in the invariance of the hazard for scanned or chopped cw beams when the scan rate or the speed of the chopper wheel is varied. Varying the scan rate (for instance the frequency of the scanning mirror) or the chopper speed results in a variation of the repetition rate of the pulsed exposure. The peak power of the pulses is related to the cw power of the beam and is in principle not affected by scan and chop rates. For increasing repetition rates, the pulse duration decreases; however, the number of pulses N increases by the same ratio. When the retinal thermal MPEs for pulses are expressed in terms of irradiance, then the exponent of the pulse duration is $-\frac{1}{4}$ so that a decrease of the pulse duration is cancelled out by the increase of N, when we use the $N^{-1/4}$ rule. Since the averaged irradiance is generally not affected by the chopping or scanning frequency (as the increase of repetition rate and decrease of pulse duration cancel out), when the frequency becomes so high that the pulses need to be grouped within T_i, the hazard is also not affected.

3.12.8.4 Summary

At the beginning of the summary we show in table 3.15 that the TOTP rule (and therefore the $N^{-1/4}$ rule) is generally more critical than the average irradiance criterion. From the restrictions expressed by the inequalities in table 3.15 it follows that the average irradiance criterion is more critical than the TOTP criterion if T_{eval} < TOT, however, this inequality can never be satisfied since

TOT is the sum of all pulse durations within T_{eval}. In the limit of repetition rates equal to the critical value for grouping of pulses due to thermal confinement, the TOT becomes equal to T_{eval}.

By applying both pulse criteria not only to the maximum anticipated exposure duration (or for classification, to the time base) but also to shorter exposure (emission) durations as described by T_{eval}, the above proof is also valid for non-uniform pulse patterns. When the evaluation duration T_{eval} is reduced so that single pulses are evaluated, the TOTP rule (or equivalently, the $N^{-1/4}$ rule) can even be interpreted to include the single pulse criterion as described in section 3.12.8.1.

In summary when the $N^{-1/4}$ rule and the alternative but equivalent form, the TOTP rule, is applied to varying evaluation durations including single pulses, for the evaluation of the *retinal thermal* hazard, other criteria need not be considered. For the wavelength and exposure duration range where the *retinal photochemical* limit is defined additionally to the thermal one, the average irradiance criterion as applied to the photochemical MPEs can be more critical than the evaluation of the thermal hazard (depending on the wavelength and angular subtense of the source).

For repetition rates higher than the critical value for grouping (e.g. 55.6 kHz for wavelengths up to 1050 nm), the average irradiance criterion is identical to the additivity criterion.

For scanned or chopped radiation of a cw beam, the hazard level (for a given exposure position in the beam) does not depend on the scanning or chopping rate.

3.13 MPE values in the far-infrared

For wavelengths above about 1400 nm, the absorptance of the ocular media in front of the retina becomes so high that the cornea becomes more sensitive to damage by optical radiation than the retina, i.e. the cornea is damaged at lower exposure levels than those necessary to cause retinal damage. Experimental threshold studies identify only thermal and, for short pulse durations, thermomechanical injury mechanisms. Accordingly, experimental injury thresholds and MPE values for exposure durations less than the thermal confinement time generally follow the wavelength dependence of the penetration depth of radiation into the cornea. Where water absorption is strong and absorption occurs in the uppermost layer of the cornea (i.e. for wavelengths above about 2600 nm), the MPEs are relatively low. When absorption occurs in a larger volume (which in the wavelength range 1500–1800 nm includes the vitreous), resulting temperature rises are comparatively small and exposure limits are higher. For large penetration depths and therefore large MPE values in terms of radiant exposure, the thermal confinement time as characterized by the inflection time T_i is also long (see discussion in section 3.12.3 on thermal confinement), as can be seen in table 3.16.

Table 3.16. Inflection times T_i for the different wavelength ranges for wavelengths above 1400 nm. In wavelength regions where basically the whole eye is heated up (i.e. 1500–1800 nm), the inflection time is 10 s. Where absorption is very superficial as in the far-infrared, the inflection time is only 100 ns. Also shown are the MPE values for exposure durations less than T_i and the reciprocal of T_i, the critical frequency for adding of pulses within T_i.

Wavelength	T_i	MPE	T_i^{-1}
1400 nm $\leq \lambda <$ 1500 nm	1 ms	1000 J m^{-2}	1 kHz
1500 nm $\leq \lambda <$ 1800 nm	10 s	10 000 J m^{-2}	0.1 Hz
1800 nm $\leq \lambda <$ 2600 nm	1 ms	1000 J m^{-2}	1 kHz
2600 nm $\leq \lambda \leq 10^6$ nm	100 ns	100 J m^{-2}	10 MHz

For exposure durations above 10 s, the MPE assumes a constant value of 1000 W m^{-2} for the full wavelength range 1400 nm to 1 mm. The constant irradiance MPE reflects that longer exposure durations than about 10 s do not result in a further increase of temperature, i.e. the steady-state temperature profile is established. For such long exposure durations that are beyond the thermal confinement time even for deeply penetrating wavelengths (T_i for the wavelength range 1500–1800 nm is 10 s), the effect of the wavelength dependence of the absorption depth is counteracted by thermal conduction, so that even for shallow absorption depths, deeper layers and larger volumes are heated by thermal conductivity.

The wavelength dependence of the MPE for pulse durations up to the inflection time is shown in figure 3.58 together with some experimental ED-50 data for corresponding pulse durations and the absorption curve for the cornea and for saline water adjusted to the experimental data [27]. It can be seen that the general wavelength dependence of the experimental data follow the wavelength dependence of the absorption depth very well. This can be related to a thermal damage mechanism, where the temperature rise scales directly with the inverse of the absorption depth, i.e. for a large absorption depth, the energy is distributed over a larger volume resulting in smaller temperature rises, which in turn is reflected by higher ED-50 values.

The dependence of the MPE on the exposure duration for exposure durations between T_i and 10 s is expressed as $5600t^{0.25}$ J m^{-2} and reflects the reduction of the threshold due to the reduction in thermal diffusion for shorter exposure durations. The MPE values for wavelengths above 1400 nm and for all exposure durations from 1 ns to 10 s are summarized in table 3.17. For exposure durations above 10 s, the MPE value for all wavelengths above 1400 nm equals 1000 W m^{-2}.

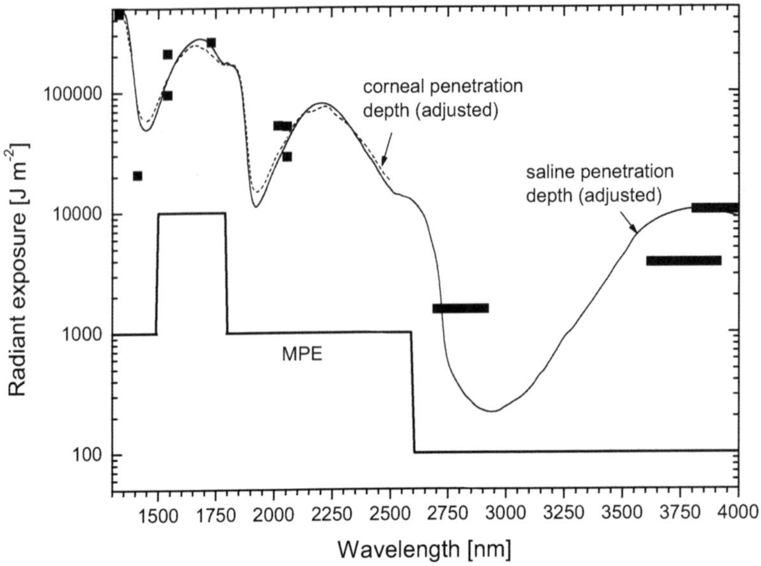

Figure 3.58. Wavelength dependence of the MPE for wavelengths above 1400 nm for pulse durations less than the inflection time T_i. Also shown are some experimental ED-50 data for the appropriate pulse duration range and measured penetration depths for corneal tissue (up to 2500 nm) and for saline solution. The penetration depth curves are adjusted by the eye to fit the experimental data.

Table 3.17. MPE values for infrared wavelengths greater than 1400 nm and pulse durations from 1 ns to 10 s.

Wavelength	Exposure duration	MPE
1400–1500 nm	1 ns–1 ms	1000 J m^{-2}
	1 ms–10 s	$5600t^{0.25}$ J m^{-2}
1500–1800 nm	1 ns–10 s	10 000 J m^{-2}
1800–2600 nm	1 ns–1 ms	1000 J m^{-2}
	1 ms–10 s	$5600t^{0.25}$ J m^{-2}
2600–10^6 nm	1–100 ns	100 J m^{-2}
	100 ns–10 s	$5600t^{0.25}$ J m^{-2}

As in the ultraviolet wavelength range, for pulse durations less than 1 ns, due to lack of experimental data, a conservative approach of limiting the peak pulse irradiance to the value that is derived from the MPE at 1 ns is followed.

For instance, for wavelengths between 1400 and 1500 nm, the MPE for exposure durations less than 1 ns equals 10^{12} W m^{-2}.

3.13.1 Multiple pulse exposures

Multiple exposures in the wavelength range 1400 nm to 1 mm need to be evaluated according to the three criteria that were already discussed in section 3.12.8 for the retinal hazard region. However, due to the different time dependence of the limits, the application of the rules are somewhat different for the wavelength range above 1400 nm. For wavelengths above 1400 nm, the number N for the $N^{-1/4}$ rule only needs to be calculated for a maximum duration of 10 s, i.e. N is the number of pulses (or the number of pulse groups when pulses are within T_i) within the maximum considered exposure duration (or time base for classification) if this is less than 10 s, or otherwise within 10 s. Due to the exponent of 0.25 for the exposure duration dependence of the far-IR MPEs, the TOTP rule is not an equivalent alternative to the $N^{-1/4}$ rule and cannot be used. Also, the comparisons regarding the critical condition do not apply. For evaluation of pulses in the wavelength range above 1400 nm against the ocular MPEs it is also interesting to note that the limiting aperture increases with exposure duration t from 1 mm for $t < 0.35$ s to 3.5 mm for $t \geq 10$ s, as discussed in section 3.6.1. For evaluation against the single pulses criterion and the $N^{-1/4}$ criterion, the limiting aperture is determined by the pulse duration (if the pulses are shorter than T_i, then by T_i), while for the average irradiance criterion, the limiting aperture is determined by the averaging duration. Therefore, depending on the beam diameter at the evaluation position, the exposure level that is determined for the average irradiance criterion may be up to a factor 12 smaller than the exposure that is determined for the two criteria that relate to single pulses.

For repetition rates lower than the critical one as also shown in table 3.16 (i.e. where no grouping due to thermal confinement needs to be considered), it depends on the evaluation duration T_{eval}, the pulse duration t_{pulse} and the repetition rate if the average irradiance or the additivity criterion is the more restrictive one. For uniform pulse trains, for the condition that the pulse duration is longer than T_i, it is possible to calculate the repetition rate f_{crit} above which the average criterion is more critical than the $N^{-1/4}$ rule:

$$f_{\text{crit}} = T_{\text{eval}}^{-4/6} t_{\text{pulse}}^{-4/12}. \tag{3.20}$$

For instance, for an evaluation duration of 10 s, which is the typical maximum anticipated exposure condition for wavelengths above 1400 nm, and for a pulse duration of 1 ms, the average irradiance criterion is the critical one when the repetition rate is higher than 2.2 Hz. For a pulse duration of 1 μs the critical repetition rate calculated with equation (3.20) equals 21.5 Hz, however, this pulse duration satisfies the condition for the applicability of equation (3.20), namely that pulse durations are longer than T_i, only for wavelengths above 2600 nm.

The rule for summing up the radiant exposure within T_i for the $N^{-1/4}$ rule naturally also applies to wavelengths above 1400 nm, which in the extreme, for wavelengths between 1500 and 1800 nm, means that the total radiant exposure within 10 s is added and compared to the exposure value of 10 000 J m^{-2} (which is equivalent to determining the average irradiance with an averaging duration of 10 s and comparing this value to the MPE of 1000 W m^{-2}). However, it is interesting to note that due to the steep decrease of the MPE with exposure duration when the MPE is specified in terms of irradiance (with an exponent of -0.75), the average irradiance criterion is more critical unless the evaluation is limited to very short maximum exposure durations. For uniform pulse patterns with repetition rates above the critical one for summing of pulses within T_i, it can be shown that the average irradiance criterion is the more critical one for evaluation durations longer than 31 ms for the case that $T_i = 1$ ms, and longer than 0.31 ms for the case that $T_i = 100$ ns. Unless the emission of the product is limited to these short durations and does not emit a second time within 10 s, the average irradiance criterion generally is the more critical one.

3.14 Multiple wavelength exposures

For the evaluation of coinciding exposure to laser radiation of more than one wavelength, the respective exposures have to be treated as additive when the same kind of tissue is at risk and are treated independently when different tissues are at risk. The current wording of IEC 60825-1 indicates that exposures also do not need to be treated as additive when the pulse durations differ by more than one order of magnitude, however this cannot be generally correct as, for instance, it is clear that exposures within the thermal confinement time as characterized by T_i or exposures in the UV need to be treated as additive even when the difference of pulse durations is a lot more than one order of magnitude. As adding of exposures is more conservative than treating them separately (comparing the separate exposure levels to the different MPE values), the authors strongly recommend that pulses are treated additively irrespective of the pulse duration.

Regarding the additivity rule for affecting the same kind of tissue, exposure to 532 nm and 1064 nm q-switched radiation will be treated as additive, as both affect the retina, while exposure to 2.1 μm and 1.064 μm radiation will be treated independently, as one affects the cornea and the other the retina. Consequently, the wavelength regimes can be tabulated in respect of being additive for exposure of the skin, marked with 'S', or for ocular exposure, marked with 'O' (table 3.18).

Where the wavelength ranges are shown as additive, the exposure levels have to be weighted by the respective MPE, i.e. the assessment for two wavelengths is satisfied when

$$\frac{\text{Exposure level}_1}{\text{MPE}_1} + \frac{\text{Exposure level}_2}{\text{MPE}_2} \text{ is smaller than 1.} \qquad (3.21)$$

Table 3.18. Additivity of exposure to different wavelengths as currently specified in IEC 60825-1, where additivity for the skin is marked with 'S' and additivity for ocular exposure is marked with 'O'.

Spectral region	UV-C and UV-B 180–315 nm	UV-A 315–400 nm	Visible and IR-A 400–1400 nm	IR-B and IR-C 1400 nm–1 mm
UV-C and UV-B 180–315 nm	O S			
UV-A 315–400 nm		O S	S	O S
Visible and IR-A 400–1400 nm		S	O S	S
IR-B and IR-C 1400 nm–1 mm		O S	S	O S

The concept can of course also be extended to more than two wavelengths and it can also be used to evaluate broadband exposure. For instance, when a spectral irradiance at a given position from the LED is given in units of W m^{-2} nm^{-1} determined over a certain wavelength increment, the above procedure is equivalent to weighting the data of the spectrum (the exposure level) by an action spectrum and summing up over the wavelength range. As the action spectrum is derived as the reciprocal curve of the wavelength dependence of the exposure limit and by normalization for the minimum exposure limit (as discussed in section 2.6.4) the above equation is mathematically equivalent to

$$\text{Exposure level}_1 \cdot S_1 + \text{Exposure level}_2 \cdot S_2 < \text{MPE}. \tag{3.22}$$

Instead of the action spectrum, for the laser MPE values, we have functions describing the wavelength dependence, such as C_3 for the retinal photochemical hazard. These functions can be treated as reciprocal values of the action spectra. In fact, the factor of C_3 was derived in that way from the action spectrum that is defined for broadband incoherent exposure limits.

While the concept as such is mathematically equivalent to the treatment of broadband sources, laser limits are not intended to be applied to broadband sources and have shortcomings especially when the spectrum of the source extends over more than one region of table 3.18. In such cases it is recommended that the broadband limits are applied to analyse the source additionally to the laser limits. For instance, in the ICNIRP broadband limits, there is an action spectrum for photochemical damage in the UV wavelength range that extends to 400 nm so that the whole UV wavelength range would be treated as additive (although with highly reduced effective exposure levels in the UV-A due to small action spectrum values), which would not be treated as additive and with the appropriate exposure limits if the laser limits were used. Similarly, the retinal photochemical damage action spectrum extends down to 380 nm and overlaps with the action spectrum for the UV wavelength range, which is not reflected in the procedure defined in the laser safety standard.

References

[1] Sliney D H and Wolbarsht M 1980 *Safety with Lasers and Other Optical Sources* (New York: Plenum)

[2] Smith G and Atchinson D A 1997 *The Eye and Visual Optical Instruments* (Cambridge: Cambridge University Press) p 777

[3] Ness J W *et al* 2000 Retinal image motion during deliberate fixation: implications to laser safety for long duration viewing *Health Phys.* **78** 131–41

[4] ICNIRP 1996 Guidelines on limits for laser radiation of wavelengths between 180 nm and 1000 μm *Health Phys.* **71** 804–19

ICNIRP 2000 Revision of guidelines on limits for laser radiation of wavelengths between 400 nm and 1.4 μm *Health Phys.* **79** 431–40

[5] ILO 1993 *The Use of Lasers in the Workplace—A Practical Guide* (Geneva: International Labour Office)

[6] WHO 1982 *Environmental Health Criteria 23—Lasers and Optical Radiation* (Geneva: WHO)

[7] Sliney D H, Mellerio J, Gabel V P and Schulmeister K 2002 What is the meaning of thresholds in laser injury experiments? implications for human exposure limits *Health Phys.* **82** 335–47

[8] Finney D J 1971 *Probit Analysis* 3rd edn (Cambridge: Cambridge University Press)

[9] Mush A 1996 Dose-time-effect-relationships *Toxicology: Principles and Practice* ed R J M Niesink, J deVries and M A Hollinger (Boca Raton, FL: Chemical Rubber Company)

[10] Helfmann J 1992 Nichtlineare Prozesse in Berlien Müller *Angewandte Lasermedizin* (Landsberg: Ecomed)

[11] Niemz M 2002 *Laser–Tissue Interactions* (Berlin: Springer)

[12] McKenzie A L 1990 Physics of thermal processes in laser–tissue interaction *Phys. Med. Biol.* **35** 1175–209

[13] Schulmeister K, Schmitzer Ch, Duftschmid K, Liedl G, Schröder K, Brusl H and Winker N 1997 Hazardous UV and blue-light emissions of CO_2 laser beam welding *Proc. Int. Laser Safety Conf.* (Orlando, FL: LIA)

[14] ICNIRP (IRPA) 1985 Guidelines on limits of exposure to ultraviolet radiation of wavelengths between 180 nm and 400 nm *Health Phys.* **49** 331–40

[15] Schwaiger M, Schulmeister K, Brusl H and Kindl P 2000 UV-hazard evaluation using different international guidelines and MEDs *Radiat. Protection Dosimetry* **91** 227–30

[16] ICNIRP 1997 Guidelines on limits of exposure to broadband incoherent optical radiation (0.38 to 3 μm) *Health Phys.* **77** 539–55

[17] CIE S 009/E:2002 *CIE Standard Photobiological Safety of Lamps and Lamp Systems* (Vienna: CIE)

[18] Mainster M A and Sliney D H 1997 But is it really light damage *Ophthalmology* **104** 179–80

[19] ACGIH 1997 *Threshold Limit Values for Chemical Substances and Physical Agents; Biological Exposure Indices* (Cincinnati: ACGIH)

[20] ICNIRP(IRPA) 1989 Proposed change to the IRPA 1985 guidelines on limits of exposure to ultraviolet radiation *Health Phys.* **56** 971–2

[21] Pedrotti F L and Pedrotti L S 1993 *Introduction to Optics* (Englewood Cliffs, NJ: Prentice-Hall)

[22] ISO 11146 2003 *Lasers and Laser-Related Equipment—Test Methods for Laser Widths, Divergence Angle and Beam Propagation Factor* (Geneva: ISO)

[23] Ward B A 2003 Measurement of laser and LED beams for prediction of angular subtense *Int. Laser Safety Conf. 2003 Conf. Proc.* (Orlando, FL: LIA)

[24] Galbiati E 2001 Evaluation of the apparent source in laser safety *J. Laser Appl.* **13** 141–9

[25] Lund D J 1999 The action spectrum for retinal thermal injury *Measurement of Optical Radiation Hazards* ed D H Sliney and R Matthes (München: ICNIRP, CIE) pp 209–28

[26] IEC TR 60825-9 1999 *Safety of Laser Products—Part 9: Compilation of Maximum Permissible Exposure to Incoherent Optical Radiation* (Geneva: IEC)

[27] Schulmeister K, Sliney D H, Mellerio J, Lund D J, Stuck B and Zuclich J 2002 Review of exposure limits and experimental data for corneal and lenticular damage from short pulsed UV and IR laser radiation *Proc. Laser Bioeffects Meeting* (Paris: CEA) pp 12-1 to 12-15

Chapter 4

Laser product classification

The core question of laser safety 'is a given exposure to laser radiation or the emission of a specific laser product safe?' sounds simple and the answer seems straightforward too: compare the exposure to the exposure limit (MPE) for the eye. However, an MPE analysis can be quite complicated especially for pulsed sources or for extended sources, as becomes apparent from the discussion in chapter 3, and it is not practical that somebody who buys a 0.5 mW laser pointer also performs an MPE analysis to determine if the product is hazardous for different exposure situations or not.

Therefore, the international laser safety classification scheme was set up to provide basic information regarding potential hazards to the eye or skin associated with a specific laser product. Based on the emission level of the product, the manufacturer has to assign the product to one of the safety classes. In the simplest case, such a laser classification scheme could consist of two classes: 'safe' and 'potentially hazardous'. However, such a simple distinction between safe and potentially hazardous is really not possible, since the level and nature of hazard presented by a product depends on a number of factors, such as exposure duration (a product that is safe for short accidental exposure might not be safe for intended viewing), viewing conditions (a product that is safe for exposure of the naked eye might not be safe when exposure occurs during use of optical viewing instruments such as binoculars or eye loupes) and one might also want to distinguish a product for which direct exposure of the eye is hazardous but exposure to radiation from diffuse reflections or exposures of the skin are safe, from a product that also represents a skin hazard and where diffuse reflections might be hazardous too. The current international laser safety classification scheme groups laser products into a number of safety classes that account for all of above multi-faceted hazard aspects. The advantage of this classification scheme is that products can be grouped into comparable hazard level classes and that for each class, 'default' user control measures appropriate to the hazard can be defined. The disadvantage, however, is that a rather complicated classification scheme results. However, since the classification as carried out by the manufacturer can generally not account for specific conditions of use and because there are some worst-case assumptions inherent in the laser safety classification scheme, it may well be that a product is safe for a specific application but it might be classified as potentially hazardous. For these cases the classification scheme may be critiqued as still too rough, or as too 'worst case', if one would not keep in mind that the 'default' user controls are also only necessary and appropriate when the 'worst case' hazard really exists and that more appropriate user controls should be defined following an application specific risk analysis.

The laser or LED product is anything from a simple device such as a laser pointer or an LED flashlight to a large machine that incorporates one or more laser devices. Classification applies to the complete system, including any accessories supplied with the equipment, at the highest level of integration. For example, where one or more lasers are incorporated into an industrial processing machine, it is the complete machine that constitutes the laser product and is subject to classification. In the case of a laser supplied with accessories, for example a low-level therapeutic laser for which a number of different output attachments are available, it is the most hazardous combination of the equipment with attachments that determines the product class.

The definition of the classes and the corresponding criteria that have to be fulfilled so that a product can be assigning a certain class are contained in the international laser safety standard IEC 60825-1. The European standardization organization has published an identical document that is referred to EN 60825-1. All European member countries are obliged to publish this European standard as identical national standard. The IEC standard is also adopted by practically all nations who publish a laser safety standard, including Australia, Japan and Canada. There are some limited differences in the current US user standard ANSI Z136.1 and in the current US manufacturers standard which is under responsibility of the CDRH and that are further discussed in section 4.5. It is noted that the discussion of the classes in this book is based on edition 1.2 (i.e. edition 1 plus the changes contained in amendment A2) of the standard IEC 60825-1 as published in the year 2001. In the following, whenever we refer to IEC 60825-1 without special further reference, we refer to the edition published in 2001 and to equivalent national standards.

The classification of a given laser and LED product generally has to be carried out by the manufacturer. As the manufacturer does not know who is going to buy his products and also often does not know the details of usage, the classification scheme is not based on an exposure analysis in terms of an *exposure* level to be compared to the MPE for the eye and skin as discussed in chapter 3. Rather, the classification is based on the radiation *emission* of the product: to determine the class of a laser or LED product, the energy or power passing through an aperture with a given diameter at a specified distance from the product is compared to a set of maximum allowed energy or power values for each class, referred to as AEL values, Accessible Emission Limit values. This concept is schematically shown in figure 4.1. However, not surprisingly, the AELs

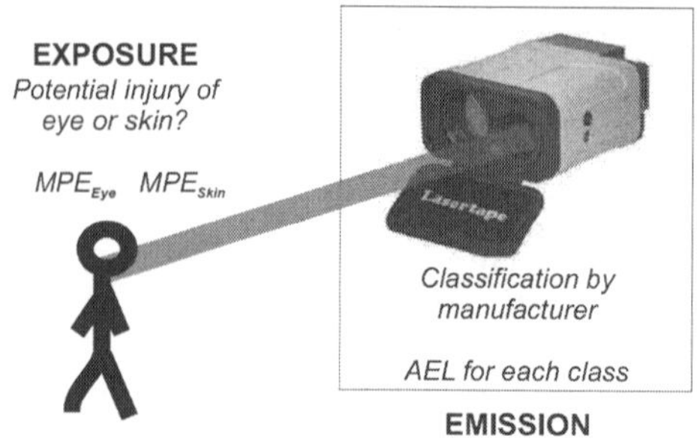

Figure 4.1. A general laser hazard analysis is based on comparing a certain *exposure* level (at a certain distance from the laser product) to the *exposure limits* (MPEs) of the eye or the skin that depend on the exposure duration. In contrast, the classification of a laser product is based on the emission of the product and this *emission* level is compared to *emission limits* that are defined for each laser class (AEL for Class 1, AEL for Class 2, etc). (Photograph of laser by Riegl Measurement Systems.)

for those safety classes that group lasers that are safe for ocular exposures, are derived directly from the ocular MPEs.

It is also noted here that the classification has a prescriptive nature: for classification, the manufacturer has to follow the detailed procedures defined in the standard. As the classification is based on worst-case assumptions of usage and exposure geometry, it might place products that are safe in their actual and specific use into a class which would indicate a certain level of hazard. Classification aids the hazard evaluation process as for instance for Class 1 products, the exposure will always be below the MPE. Class 3B laser emit radiation which is significantly above the MPE for eye, and for Class 4 also above the MPE for the skin, however, it depends on the beam geometry, set-up and application, if this grave hazard exists only close to the exit aperture, is enclosed by guarding or screens around the laser or if it extends over several kilometres. The classification is a useful guide for the potential maximum level of hazard, but one should not define controls strictly on the basis of the class without evaluating the real hazard for a given application, as this may otherwise lead to generalized controls that may be over-restrictive. It should be also noted that the laser safety class only refers to the hazard to the eye or skin from exposure to the direct beam or to exposure from diffuse reflections, it does not give any information about other hazards that may be presented by the laser product or the specific use of the product such as mechanical, electrical or chemical hazards (see chapter 6 for a discussion of these additional hazards). The role of the classification scheme for the definition of 'default' control measures and general principals of hazard and risk analysis that is performed to define application specific control is further discussed in chapter 7.

In this chapter, the meaning of the laser safety classes, the process of classification and requirements that have to be fulfilled by the product are discussed in detail. While the meaning of the classes and a basic understanding of the process of classification is important for the user, the detailed description of the tests that are to be performed for the classification is rather contained in the book to help manufacturers to apply the laser safety standard. However, when the laser product is changed by the user so that safety related aspects are affected, it is the duty of the user to ensure that the laser class still applies, or if it was changed, to have the product reclassified. A common example for the latter is the removal of parts of the protective housing or guards or of interlocks of the protective housing of an embedded Class 1 laser product. The level of risk that is associated with the various classes is discussed in greater detail in section 7.1.1. While section 7.1.1 is included to help in the process of risk analysis for a specific application and to select the appropriate user control measures, the discussion should be also relevant for the manufacturer who is required to provide information on the hazards associated to his product and on safe usage.

4.1 Overview

The meaning of the laser safety classes following the current version of IEC 60825-1 can be summarized as follows:

Class 1. No risk to eyes (including use of optical viewing instruments). No risk to skin. Lasers that are safe, including long-term direct intrabeam viewing. Also safe when exposure occurs while using optical viewing instruments (eye loupes or binoculars). Class 1 also includes high-power lasers that are fully enclosed so that no radiation is accessible during use (embedded laser product).

Class 1M. No risk to the naked eyes, no risk to skin. Lasers that are safe for the naked eye (unaided eye), including long-term direct intrabeam viewing. Eye injury may occur following exposure with one of the two categories of optical viewing instruments (eye loupes or binoculars).

Class 2. No risk to eyes for short time exposure (including use of optical viewing instruments). No risk to skin. Lasers emitting radiation in the visible wavelength range (400–700 nm). For such lasers, the aversion response to bright light (for instance the blink reflex) usually limits the duration of retinal exposure. Therefore, although the power is higher than for Class 1, they are considered safe for usual exposure situations, but are potentially hazardous for intentional staring into the laser beam. Safe levels are not exceeded when exposure occurs while optical viewing instruments are used. These lasers may, however, cause dazzle and flash blindness, that may present a hazard for instance when steering a vehicle or aircraft.

Class 2M. No risk to naked eyes for short time exposure. No risk to skin. Visible lasers that are safe for short time exposure only for the naked (unaided eye). Eye injury may occur following exposure with one of the two categories of optical viewing instruments (eye loupes or binoculars). They may present a dazzle hazard.

Class 3R. Low risk to eyes. No risk to skin. Low risk for eye injuries provided that only accidental exposure (short time exposure) occurs. Emission levels somewhat higher than for Class 2 (in case of visible emission) or for Class 1. Emission level is high enough to result in eye injury when intentional intrabeam viewing occurs. Intended for professional use (for instance for alignment) by trained personnel. Class 3R is defined only for wavelengths larger than 302.5 nm. They may present a dazzle hazard.

Class 3B. Medium to high risk to eyes. Low risk to skin. Lasers for which intrabeam exposure is hazardous, including accidental short time exposure, but for which the viewing of diffuse reflections is normally safe. Natural aversion response to localized heating prevents serious skin injury, or skin injury can only occur if beam is focused onto tiny spot, so that the effect can be compared to a pin-prick.

Class 4. High risk to eyes and skin. Lasers for which intrabeam viewing and skin exposure is hazardous and for which the viewing of diffuse reflections may be hazardous. These lasers also often represent fire hazard.

Note on nomenclature. 'M' in Class 1M and Class 2M is derived from Magnifying optical viewing instruments. 'R' in Class 3R is derived from Reduced or Relaxed, requirements: reduced requirements both for the manufacturer (e.g. no key switch and interlock connector required) and for the user (e.g. usually no eye protection necessary). The 'B' for Class 3B has historical reasons, as in the version of the standard before the 2001 edition, a Class 3A existed, which had a similar meaning to what is now Class 1M and Class 2M (see section 4.2.6).

It should be noted that for the above descriptions, whenever we use 'hazardous' or refer to high risk for injury, this hazard and risk only exists within the area around the laser where the corresponding MPE levels are exceeded, i.e. for exposure of the naked eye within the nominal ocular hazard distance (NOHD; see section 3.6.2) or, for Class 1M and 2M, within the extended NOHD (see section 5.4 for a discussion of the NOHD). It may well be that a particular (Class 3B or Class 4) laser product has a very short NOHD associated with it, so that for a particular installation or application, at the location where personnel could be irradiated, the MPE of the eye is not exceeded and eye protection is not necessary. Examples for such installations are scanning lasers or line lasers mounted at the ceiling of the manufacturing hall that project a pattern or line onto the work piece in the work area below. While the power level and scan pattern could be such that the exposure in the work area is below the MPE and safe, maintenance and service routines will need special consideration as exposure at closer distances, for instance when up on a ladder to clean the exit window, or the non-scanning beam might be hazardous. Also for Class 4 laser products, there is a NOHD associated to diffuse reflections, although quite limited in dimension. The characterization of the risk associated with a particular laser and application is the task of a risk assessment and is further discussed in chapter 7.

In terms of viewing or exposure conditions that might be hazardous for one class but are not hazardous for other classes, the classification scheme distinguishes between:

- exposure to diffuse reflection versus intrabeam (direct) exposure;
- short-term (accidental) versus long-term (intentional) viewing;
- exposure of the naked (unaided) eye versus exposure with optical viewing instruments.

These exposure and viewing conditions as considered in the classification scheme are discussed in the following.

4.1.1 Diffuse versus intrabeam (direct) viewing

Viewing of radiation reflected from diffuse target (or diffuse transmission for instance through translucent or 'frosted' glass) is far less hazardous than intrabeam viewing (direct exposure) of the same laser beam. For a more basic discussion of the nature of diffuse reflection or transmission (also closely linked to scattering) see section 2.7.3. With 'intrabeam exposure' we mean that the laser beam is incident at the surface of the eye, but it is not necessary for the eye to be looking directly at the laser source. (For radiation at wavelengths within the retinal hazard region, it is necessary for the source to be within the eye's field-of-view in order for the retina to be exposed.) For collimated laser beams, this viewing condition will produce a minimal retinal spot, irrespective of the position of the eye in the beam. This type of exposure, sometimes referred to as 'direct' exposure, also includes exposure via specular reflections, i.e. via a mirror, as the mirror merely redirects the beam. If exposure occurs via a specular reflection, one looks into the laser via the mirror. In contrast, when exposure from diffuse reflections occurs, the beam as incident on the rough surface is broken up and the power contained in the beam is scattered into a wide range of directions. Due to this scattering, depending on the distance to the scattering surface, the irradiance at the cornea of the eye is much less than from an exposure to the direct beam. Additionally, when the viewer is close to the scattering surface and the beam diameter at the surface is correspondingly large, then the diffuse reflection (for wavelengths in the retinal hazard area of 400–1400 nm) constitutes an extended source with correspondingly higher limit values.

It is noted here that historically, 'intrabeam viewing' was often used as a synonym for exposures to small (i.e. point) source in contrast to viewing of extended sources, that are typically mostly produced by diffuse reflections of laser beams (for instance, in the pre-1993 edition of the standard, then called IEC 825, there was one table of limits for point sources with the heading 'intrabeam viewing' and there was a separate table for extended sources given in units of radiance that would be typically be applied to assess diffuse reflections).

However, this usage of the term ‘intrabeam viewing’ to mean ‘exposure to radiation from small sources’ is discouraged, as on the one hand extended sources such as line lasers or LEDs can also be viewed directly and on the other hand, diffuse reflections are only extended sources for large diameter beams incident on the scattering surface or for very close viewing distances (for instance, a beam diameter of 1 mm at the scattering surface represents a small source ($\alpha \leq 1.5$ mrad) for viewing distances larger than 67 cm).

Regarding the meaning of the laser safety classes, Class 4 incorporates lasers with powers high enough so that even diffuse reflections might be hazardous, while viewing of diffuse reflections of Class 3B and lower power lasers is generally safe (with the exception of prolonged staring at diffuse reflections at very close distance for higher power Class 3B lasers, as discussed in section 7.1.1). Observation of diffuse reflections is for instance generally done to align lasers—should this be done with Class 3B lasers or even Class 4 lasers (outside of the NOHD for diffuse reflections), then it would have to be made sure that exposure to the direct beam cannot occur, for instance by enclosing the beam in a tube (up to a diffusing surface).

4.1.2 Viewing duration

MPE values generally decrease with increasing exposure duration (at least up to exposure duration of 10 s) to reflect that longer exposures are more hazardous than shorter ones. Correspondingly, long-term intentional viewing is more hazardous than short-time, accidental exposure. The maximum emission power levels that are allowed for Class 1 and Class 1M are set so low that even long-term exposure and intentional intrabeam viewing of the laser beam is safe.

For visible beams, for power levels associated with Class 2 and Class 2M lasers, natural aversion responses to bright light will usually limit the exposure duration. Aversion responses are further discussed in section 3.9.4. The time base used in the standard for short time, unintentional exposure is 0.25 s. Corresponding to the shorter exposure duration, higher power levels are allowed for Class 2 lasers than for visible Class 1 lasers with the same wavelength and source size. Class 2 lasers can be considered to be quite safe: while it is possible to purposely stare into the beam of a Class 2 or Class 2M laser, it is perceived as highly uncomfortable and can rarely be upheld for longer than a second—for the powers allowed for Class 2 lasers, due to the safety factor built into the limits, such an exposure can still be considered as safe. Besides natural aversion responses to bright light, the exposure duration that is typically associated with accidental exposure (in contrast to purposefully looking into the laser beam) is also limited due to movements of the beam relative to the eye: for accidental exposure, usually, either the laser beam or the head is not stationary, for instance when the laser beam moves across the eyes or when the beam is stationary as from a mounted laser but one unintentionally moves the head through the beam. The concept of Class 3R also relies on exposure durations that are associated with accidental exposure and that are limited in duration when compared to intentional viewing and intentional exposures. Since somewhat higher power than Class 2 are allowed for visible Class 3R, it is important for Class 3R lasers that intentional viewing does not occur and users should be aware of that residual risk for Class 3R lasers. For the case of visible Class 3B and Class 4 lasers, natural aversion responses to bright light are not sufficient to protect the eye as damage occurs almost instantly.

4.1.3 Naked (unaided) eye versus exposure with optical viewing instruments

The international classification scheme distinguishes between hazards to the naked eye and hazards that may arise for exposure with optical viewing instruments such as eye loupes (magnifying lenses used for close-up viewing) or telescopes and binoculars. Regular prescription glasses and contact lenses are not considered an optical viewing instrument, as they only correct vision. It is important to distinguish two groups of optical viewing instruments that increase the level of hazard of certain types of laser beams in different ways. Telescopes or binoculars can strongly increase the level of hazard of well-collimated large-diameter beams by collecting (with their large diameter input optics) and directing more energy onto the eye as would be the case for the naked (unaided) eye. Magnifiers or eye loupes increase the optical power of the eye and thus allow one to view a source closer than would be possible with the naked eye. When a source that emits a divergent beam is viewed with an eye loupe, more power is collected at the closer distance than compared to the unaided eye and thereby the hazard for such sources is increased by eye loupes. The potential increase of power levels that are incident on the eye due to optical viewing instruments is accounted for in the classification procedure by defining aperture diameters and measurement positions that simulate worst-case exposure geometries with optical viewing instruments. The concept of measurement distances and aperture diameters and the potential increase of the hazard when exposure with optical viewing instruments occurs is discussed in more detail in section 4.2.3.

4.1.4 Tabular overview

Following the distinction between a range of viewing or exposure conditions, the classes and corresponding safe or potentially hazardous exposure conditions can be organized as shown in table 4.1. The expression 'low risk' is intended to characterize a level of risk that in practice usually should be negligible, i.e. a safe exposure situation, however, for special conditions (depending also on the wavelength, etc) such as somewhat prolonged viewing or for the case of skin exposure, tight focusing, might produce an injury.

Table 4.1. Representation of the meaning of the classes following IEC 60825-1 according to exposure conditions. Potentially hazardous exposure conditions (within the NOHD) are marked by an exclamation mark.

	Long-term eye exposure		Short-term (accidental) eye exposure			
	Optical viewing instruments	Naked eye	Optical viewing instruments	Naked eye	Diffuse reflections	Skin exposure
Class 1	Safe	Safe	Safe	Safe	Safe	Safe
Class 1M	!	Safe	!	Safe	Safe	Safe
Class 2	!	!	Safe	Safe	Safe	Safe
Class 2M	!	!	!	Safe	Safe	Safe
Class 3R	!	!	Low risk	Low risk	Safe	Safe
Class 3B	!	!	!	!	Low risk	Low risk
Class 4	!	!	!	!	!	!

4.1.5 Manufacturing requirements

Depending on the laser safety class that was determined by the manufacturer following the specifications in IEC 60825-1, the manufacturer has to implement a number of safety related engineering features, such as for Class 3B and Class 4 laser products a remote interlock connector, an emission warning, a beam stop and a key switch. Labelling on the product is required for all classes except Class 1 and Class 1M. There are also requirements regarding the content of the user information (the manual).

The requirements regarding engineering features and labelling on the product as specified in IEC 60825-1 apply to all kinds of laser products, be it laser shows, medical lasers, toys, high-power laser welders, laser pointers or traffic speed control laser 'guns', and they also apply to LEDs. However, products that emit power levels that are always below the AEL for Class 1 including single fault conditions and service conditions, are exempt from the standard. This exemption is, for instance, intended for low power visible LEDs that are used as indicator lights on all kinds of consumer products, or as alphanumerical display in clocks, or in infrared remote controls.

Additional to the general manufacturing requirements for all kinds of laser and LED products laid down in IEC 60825-1 (this standard is therefore also referred to as a 'horizontal' standard), additional manufacturing requirements may be specified in product type specific standards, such as when a laser is used in a machine, in a medical product, for telecommunication. These are discussed in some more detail in section 4.7.

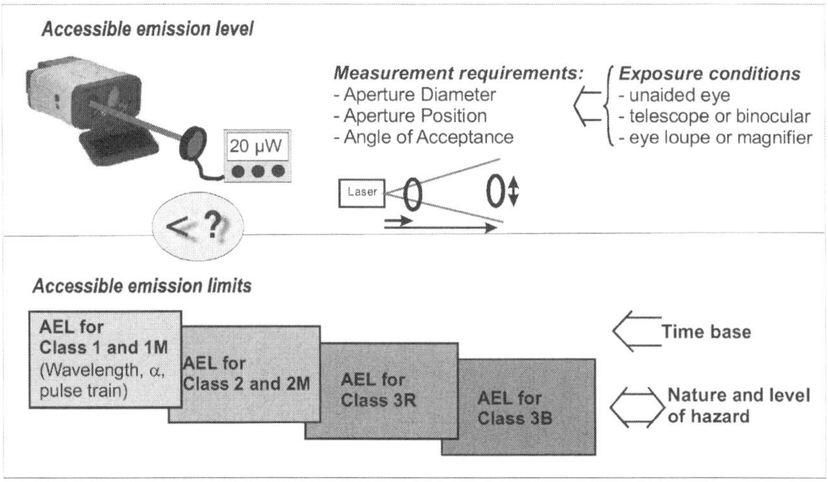

Figure 4.2. For the classification of a laser or LED product, the power (or energy per pulse) that passes through the measurement aperture has to be compared to the set of AEL values for the various classes.

It is recalled at this point that it is important to distinguish between manufacturers requirements of a laser product regarding the classification design, labelling and information for the user in the manual according to IEC 60825-1 (and possibly also product type specific standards) on the one hand and issues related to the use of a product on the other. The manufacturing requirements are internationally standardized and are binding and have to be applied for all laser products irrespective of their use. A detailed discussion of (national) legal requirements regarding the use of a certain type of laser product or a certain laser product class is not possible within the scope of this book but some general issues are further discussed in section 7.1.3.

4.2 Classification scheme

With each safety class, product emission limits are associated that are derived following the specific meaning of the respective class. The set of maximum allowed power and energy values for each of the classes and the accessible emission limits (AELs) are usually defined in units of watts and joules. For the classification procedure, the radiant power or energy has to be assessed (usually measured but it may also be calculated) at a specified distance and with a specified measurement aperture diameter and measurement angle of acceptance, and the values are subsequently compared to the different AELs, as is schematically shown in figure 4.2.

For a given wavelength, Class 1 has the lowest emission limits, Class 3B has the highest. As Class 4 is the class with the highest associated hazard with unlimited power, there are no Class 4 AELs. When the emission level determined with the aperture is below the AEL for Class 1, the product is assigned Class 1. When it is a visible laser and above the AEL for Class 1, but below the AEL for Class 2, it is assigned Class 2, and so on. When the accessible power levels are above the limits for Class 3B, then the laser is assigned to be Class 4.

Classification is based on power emitted from the product, or rather, power (or energy) that is accessible through an aperture that represents a specific viewing situation. The MPE values are given as irradiance and radiant exposure and irradiance and radiant exposure levels to be compared to the MPE values need to be averaged over the area of the limiting aperture, as discussed in chapter 3. Following the concept of Class 1 (safe for exposure of the eye), the power level that is accessible from Class 1 lasers needs to be limited to a value low enough so that it is impossible for an exposure to exceed the MPE for the eye. This is achieved by setting the AEL of Class 1 equal to the MPE of the eye multiplied with the area of the limiting aperture. Just as discussed for Class 1, each set of AEL values for the different classes represents a certain type or level of hazard, corresponding to the hazard associated with the respective class. The derivation of the set of AEL values for the classes is discussed in more detail in section 4.2.1.

As an additional parameter in the classification scheme, the time base characterizes the (typical or worst-case) exposure duration that is associated to the class and for which the hazard exists. The time base most importantly comes into play for the distinction between Class 1 and Class 2: both classes have the same associated general category of hazard, namely 'no hazard to the eye', however, this category of hazard applies to different exposure durations: for Class 1, there is no hazard to the eyes for long-term exposure, while for Class 2, there is no hazard to the eye only for short-term exposures. Consequently, as the difference between Class 1 and Class 2 lies not in the general type of hazard but only in the associated exposure duration, both sets of AEL values are directly derived from the MPEs for the eye. The difference is in the time base, that is 0.25 s for Class 2 and 100 s or in some cases longer for Class 1. The values for the time base specified for the different classes is discussed in section 4.2.2. In terms of nomenclature, the temporal dependence of the MPEs refers to exposure duration, while the temporal dependence of the AELs refers to emission duration. Consequently, as defined in IEC 60825-1, the time base is defined as the emission duration that is to be considered for the classification.

In order to correctly characterize the hazard associated with highly divergent and large diameter beams it is important to keep the impact of the diameter of the measurement aperture and the measurement distance in mind: it is not the total power or energy that is compared to the AEL but the power or energy that passes through a measurement aperture with a specified diameter at a certain position with respect to the product. When the beam diameter at the position of the aperture is larger than the aperture diameter, a power level that is correspondingly lower than the total beam power is compared to the AEL. Consequently, the total power output of a laser or LED product that is highly divergent or that emits a beam having a large diameter can be substantially above the allowed AEL for the class. All too often, this is overlooked and then it is puzzling how, for instance, an LED array that emits a total power of several watts can be classified as Class 1. The specified diameter of the aperture and the measurement distance are representative of the exposure or viewing conditions that are considered in the classification scheme: exposure of the naked eye, exposure while using eye loupes and while using telescopes. Each of those exposure conditions are represented by specific aperture diameters and measurement distances, so that the power level assessed as passing through the aperture is indicative of the power that is 'accessible' with or during the respective viewing condition. This accessible emission (power or energy) level can subsequently be compared to different AEL values, as schematically indicated in figure 4.2. The discussion also shows that the AEL (accessible emission limit) values specified for each class, should not be simply seen as maximum allowed output power, but that they quite appropriately refer to a limit for the accessible emission (accessible for different worst-case exposure conditions). For large sources, not only is the aperture diameter and distance relevant and has an impact on the measured power or energy level, but also the prescribed angle of acceptance. For a source that is larger than the angle of acceptance, only part of the emitted radiation should be collected, and the measured power or energy level will again be lower than the total output power. The prescribed maximum angle of acceptance is particularly important for large area diffuse sources or for arrays as is also discussed in section 3.12.5.6.

In summary, the classification scheme can be considered to consist of three 'components':

- the set of AEL values for each class, that are defined to limit a certain nature and level of hazard associated with a given class;
- the time base, that associates a certain typical or worst case exposure duration to the class;
- the specified aperture diameter and measurement distance, that is representative for a certain viewing or exposure situation (naked eye, eye loupe, telescope). For extended sources, the maximum angle of acceptance also plays a role.

The importance of the different measurement set-ups for the three viewing conditions becomes most prominent for the classification of a product as Class 1M or as Class 2M: the set of AEL values for Class 1M is actually the same as for Class 1 (in the standard, there is only one table which is titled 'AEL values for Class 1 and Class 1M'), and the AELs for Class 2M are the same as for Class 2. For the 'non-M' and 'M' classes, different measurement set-ups are defined: following the meaning of the classes, for classification as Class 1 or Class 2, the power or energy level accessible with the two measurement conditions representing the telescopes and eye loupes need to be below the respective AEL.

This corresponds to the meaning: 'no hazard for the eye even with optical viewing instruments'. For classification as Class 1M or Class 2M, the accessible power levels determined as described above are above the AEL value, while the power level accessible with the measurement condition representing the naked eye needs to be below the AEL. The AEL is the same in both cases, but different power levels, representative of different exposure conditions, are compared to this AEL. The measurement requirements as relevant for a general understanding of the classes are further discussed in section 4.2.3 and in greater detail with all the intricacies as important for actual classification of all kinds of laser or LED products in section 4.3.

There are a number of assumptions inherent in the classification scheme in terms of time base and aperture diameter and placement, some of them could be considered 'worst-case' assumptions. These may lead to laser power or energy restrictions that may be seen as over-restrictive for a specific product. Examples are a time base of 100 s or in some cases even 30 000 s for Class 1 and Class 1M, and measurement criteria that are derived from close up viewing of the product, including with high power eye loupes and large binoculars. In addition, the classification has to be performed under the principle of maximizing the output of the laser product which is accessible during use of the product. Also, single fault conditions have to be accounted for, i.e. the classification has to hold for any reasonable foreseeable single fault that may increase the level of radiation. Here, 'single fault' means that the situation of two fault conditions occurring at the same time does not have to be accounted for, and 'reasonably foreseeable' means that the probability of the fault occurring should not be unrealistically small.

4.2.1 Derivation of the AEL values

Following the meaning of Class 1, Class 1M, Class 2 and Class 2M, namely that the exposure of the eye is safe, the respective AELs are directly related to the MPEs. The only difference between the AEL values on the one hand and the MPEs for the eye are that the AELs are given in terms of power or energy through an aperture, and the MPEs are given in terms of irradiance or radiant exposure as averaged over a limiting aperture (section 3.6.1). The multiplication of the MPE values with the area of this limiting aperture is how the AELs are derived. As there is only one set of AELs that applies to both Class 1 and Class 1M and to both Class 2 and Class 2M, in this section, for simplicity, we sometimes refer to Class 1 and Class 2 only. The differentiation between Class 1 and Class 1M (and Class 2 and Class 2M) is further discussed in section 4.2.3.

Expressed mathematically, the AEL values for Class 1 and Class 1M are obtained by

$$\mathrm{AEL}_{\mathrm{Class\ 1\ and\ Class 1M}} = \mathrm{MPE}_{\mathrm{eye}} \times \mathrm{Area}_{\mathrm{limiting\ aperture}}.$$

Correspondingly, the appearance and organization of the AEL table for Class 1 in terms of wavelength, time dependence and dependence on angular subtense of apparent source is identical to the MPE table for the eye. The only difference is an additional dependence of the AEL values on emission duration in the wavelength range 1400–4000 nm in comparison to the MPE table for the eye where, for exposure durations between 0.35 and 10 s the diameter of the limiting aperture depends on the exposure duration.

The background of the recalculation of the ocular MPEs that are given in units of irradiance or radiant exposure to AEL values that are specified as power and energy is that, once they are defined as power or energy values, the potential use of optical viewing instruments such as binoculars can more flexibly be accounted for by adjusting the diameter of the measurement aperture. For instance, for the wavelength range 400–1400 nm, an aperture with a diameter of 7 mm would characterize the power entering the naked eye while an aperture with a diameter of 50 mm would characterize the power accessible with a 7 × 50 telescope or binocular. The Class 1 AELs for wavelengths smaller than 302.5 nm and larger than 4000 nm are not defined as power or energy values but rather, the MPEs are directly used, as optical viewing instruments are considered non-transmissive for wavelengths below 302.5 nm and above 4000 nm and the product's emission is directly compared to the MPEs for the eye.

Following the direct derivation of the AEL values from the MPEs for the eye, the comparison of the irradiance or radiant exposure averaged over the limiting apertures with the MPE is equivalent to a comparison of the power or energy accessible through the same aperture size with the AEL. The only requirements for these two approaches to be identical is that the time base defined for classification is used as emission duration for the MPE assessment and that the location and diameter of the aperture is the same. In other words, the question 'is the power (or energy) measured through the limiting aperture less than the AEL for Class 1?' is mathematically equivalent to 'is the irradiance (or radiant exposure) averaged over the limiting aperture less than the MPE for the eye?', since the MPE for the eye and AEL for Class 1 is directly related via the area of the limiting aperture.

While the AELs for Class 1 are defined for durations of up to 30 000 s (about 8 h) and for the full wavelength range considered optical radiation of 180 nm to 1 mm, Class 2 is limited to visible wavelengths and to a time base typically associated with unintentional exposure to bright visible light, namely 0.25 s. However, the calculation of the AEL for Class 2 is the same as for visible Class 1 lasers, namely the multiplication of the MPE for the eye for visible wavelengths with the area of the 7 mm limiting aperture applicable for that wavelength range (see table 4.2). The resulting AEL for Class 2 for continuous wave lasers (for small sources) is 1 mW (for extended sources, the AEL is increased by the factor C_6 as discussed in the MPE section). For pulsed or scanned emissions, i.e. for pulse durations less than 0.25 s, the AEL values for Class 2 are equal to the AELs for Class 1. In other words, the only distinction between the AELs for Class 1 and the AELs for Class 2 is that the AELs for Class 2 are defined in the visible region only and are defined up to an emission duration of 0.25 s only.

Therefore, for instance, if a visible laser product were designed to emit no longer than 0.25 s within 100 s and if the radiant emission as assessed with the required measurement set-up would stay below the Class 2 AEL, it would actually be a Class 1 product. In other words, it is impossible to have a Class 2 laser that emits for less than 0.25 s. It is only when a product emits for longer than 0.25 s within 100 s, or continuously, that due to the longer time base of Class 1, and the decrease of the allowed power levels with longer emission durations, the allowed power for Class 1 falls below the allowed power for Class 2.

The basis for Class 3R is the safety factor that is inherent in the MPEs, so that exposure to a level somewhat above the MPE is still considered relatively low risk in terms of permanent damage to the eye or skin. The AEL of Class 3R in the visible wavelength region is five times the AEL of Class 2 and 2M, i.e. 5 mW, and outside the visible region equals five times the AEL of Class 1 and 1M. Correspondingly, exposure from Class 3R laser products can be up to a factor of 5 above the MPE for the eye for 0.25 s exposure duration for a visible laser beam, and up to a factor of five above the MPE for the eye for 100 s exposure duration for wavelengths between 700 and 1400 nm (infrared radiation in the retinal hazard region). Since the MPEs of the eye and the skin are identical outside of the wavelength range of 400 nm to 1400 nm, exposures from Class 3R lasers in the UV and far-IR wavelength range are not only up to five times above the MPE for the eye but also above the skin (with the applicable time base as exposure duration, i.e. 100 s for the infrared and 30 000 s in the UV). Class 3R was not defined for wavelengths smaller than 302.5 nm, as the safety factor for the eye was considered not large enough for the wavelengths around 270 nm.

There is no simple direct relationship between the AEL values for Class 3B with the MPEs for the eye or the skin, however, the AELs for Class 3B are set so that viewing of diffuse reflections is usually below the MPE for the eye and skin exposures usually do not result in serious skin injuries. The AELs for Class 3B for times above 0.25 s is set quite simply at 0.5 W for all wavelengths above 315 nm, is 1.5 mW for wavelengths less than 302.5 nm and changes exponentially between the values valid at 302.5 nm and at 315 nm, i.e. from 1.5–500 mW. Since Class 3B is not related to the safety of intrabeam exposure where the source is imaged onto the retina, but is related to diffuse reflections and to skin exposure, the AEL values for Class 3B do not depend on the angular subtense of the apparent source α and therefore also do not contain the factor C_6.

4.2.2 Time base

It may be inferred from the general description of Class 1, Class 1M, Class 2, Class 2M and Class 3R that these classes have some *typical* or *maximum* associated safe exposure duration: Class 1 and 1M are safe for 'long-term' intentional viewing, and Class 2, 2M and Class 3R are safe only for 'short-term' or 'accidental' exposure. The time base is defined in IEC 60825-1 as the emission duration to be considered for classification. For the case of laser products that emit

Table 4.2. Overview table for the relationship of the AEL values to MPE values ($A_{\text{limit ap.}}$ refers to the area of the limiting aperture as defined in table 3.1 for averaging of the irradiance or radiant exposure in an MPE analysis).

AEL	Wavelength restrictions	Time base restrictions	Relationship to MPE
Class 1 and Class 1M	None	None	$\left.\right\}$ AEL $= \text{MPE}_{\text{eye}} \times A_{\text{limit ap.}}$
Class 2 and Class 2M	400–700 nm	0.25 s	
Class 3R	302.5–400 nm and >700 nm	None	$\left.\right\}$ AEL $= 5 \times \text{MPE}_{\text{eye}} \times A_{\text{limit ap.}}$
	400–700 nm	0.25 s	
Class 3B			— Exposure from diffuse reflections—less than MPE_{eye} — Only minor skin injuries to be expected

pulses or where the radiation is scanned, every possible emission duration within the time base must be considered for classification. For instance, for products that emit single bursts of pulses with some required pause between bursts, such as distance or speed meters, for classification it is required to 'fire' as often within the time base, as the design of the product allows. Since the AEL values for the Classes 1, 1M, 2 and 2M are equivalent to the MPEs for the eye, a certain time base therefore also means that the emission of the product determined at the worst-case position stays below the MPE for an exposure duration equal to the time base. Therefore, in a way, the time base can also be interpreted as maximum 'assumed' or 'typical' exposure duration for a certain class.

In contrast to the exposure duration assumed for an MPE analysis, for the classification, the time base is a fixed value and is rather on the worst case side for Class 1 and Class 1M. For Class 1 and 1M for wavelengths greater than 400 nm the time base is either 100 s or 30 000 s. 30 000 s (about 8 h) is used when the laser or LED product is designed for long term intentional viewing, otherwise the time base is 100 s. Examples where intentional long term viewing is inherent in the design of the product would be LED display panels or retinal display projection systems. However, when the time dependence of the AEL values for Class 1 and Class 1M is examined, it is revealed that the distinction between the two time base values in terms of viewing is only relevant for sources that emit in the blue wavelength range where photochemical damage is dominant, as for thermal damage, the AEL values do not decrease for emission durations beyond 100 s but is expressed as a constant power level. The AELs for photochemical retinal damage are actually also expressed as constant power level for emission durations beyond 100 s, however, the measurement angle of acceptance increases for exposure durations up to 10 000 s, so that the measured power as determined with that varying angle of acceptance increases for durations beyond 100 s (provided that the source is larger than the angle of acceptance). For wavelengths less than 400 nm, i.e. for the UV wavelength range, where exposures are additive over the whole day, the time base for Class 1 and Class 1M is always 30 000 s.

The time base for Class 2 and Class 2M laser products is 0.25 s, which is a representative duration for the case of an unanticipated, i.e. non-intentional exposure (non-intentional at least by the exposed person) to visible radiation where it can be assumed that due to the bright visual stimulus, aversion responses such as the blink reflex will limit the exposure duration. Emission from Class 2 and Class 2M products outside the visible wavelength range needs to be below the Class 1 and Class 1M AEL, respectively. The current edition of the standard IEC 60825-1 does not expressively discuss the choice of time base values for the evaluation of the combined emission; we would like to argue that in the case that the non-visible part of the emission is additive to the visible component (i.e. for wavelengths between 700 and 1400 nm), a time base value of 0.25 s should be used to analyse the combined exposure (see the case study in section 4.8.5 for a more detailed discussion).

Table 4.3. Time base values as defined in IEC 60825-1:2001.

	$\leq$ 400 nm	400–700 nm	$\geq$ 700 nm
Class 1 and Class 1M	30 000 s	100 s 30 000 s if intentionally long-term viewing inherent in design	
Class 2 and Class 2M	as Class 1	0.25 s	IR alone time base as Class 1; for additive: 0.25 s (see text)
Class 3R	30 000 s	0.25 s	100 s
Class 3B	30 000 s	100 s	

The time base for Class 3R is 0.25 s in the visible, 30 000 s for wavelengths in the UV and 100 s for wavelengths above 700 nm.

According to IEC 60825-1, the time base for Class 3B is 100 s for wavelengths above 400 nm and 30 000 s for UV wavelengths, however, this distinction is not relevant as the AEL is a constant power level for emission durations from 0.25 s to 30 ks for all wavelengths. Therefore, the interpretation of the time base as 'assumed typical' or 'maximum' exposure duration, or as duration for which an exposure is safe (whatever 'safe' means for the class under consideration—for Class 3B it would be 'safe or low risk for the skin' and 'safe or low risk from diffuse reflections') is not relevant for Class 3B. For Class 4, no time base is applicable as all products that exceed the AEL of Class 3B are Class 4.

The time base values as defined in IEC 60825-1 are summarized in table 4.3.

4.2.3 Measurement requirements

The classification scheme is based on a comparison of the power or energy that passes through an aperture with the AEL values, that are given as power or energy values (see also figure 4.3). The circular measurement aperture to be used for classification is also referred to in IEC 60825-1 as the aperture stop. It is important to note the different concept of the aperture stop defined for classification (assessment of the accessible power or energy for comparison with the AEL) and the limiting aperture that is defined for the averaged assessment of the irradiance or radiant exposure for an MPE analysis. For some products, the angle of acceptance of the measurement instrument might play a role in the determination of the emission level that is compared to the AEL: when the extent of the source is larger than the specified measurement angle of acceptance, only that fraction of the total power emitted by the part of the source that is within the measurement angle of acceptance (also referred to as the field-of-view—FOV) is considered.

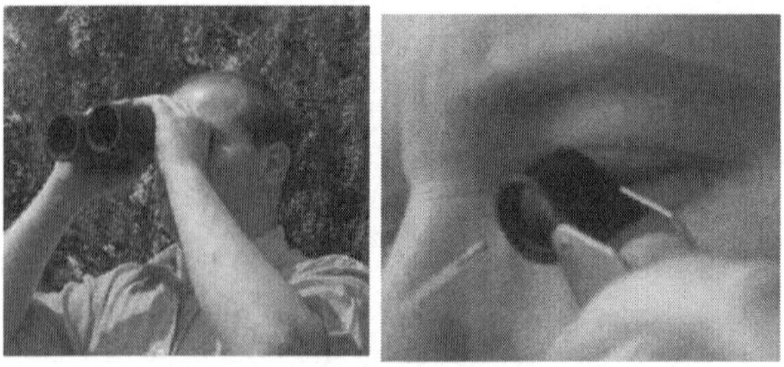

Figure 4.3. Optical instruments can increase the hazard of laser radiation when compared to the unaided eye: binoculars and telescopes by collecting radiation from collimated beams with large diameter on the one hand and eye loupes by allowing a smaller viewing distance on the other.

The set of rules regarding aperture stop diameter, measurement aperture placement and angle of acceptance are collectively often referred to as *measurement requirements*. As the level of determined emission may depend strongly on the placement and the diameter of the aperture stop (for instance for a highly divergent source) and for extended sources also on the angle of acceptance, and it is this value that is compared to the AEL value to assess the laser safety class, the measurement requirements play an important role in the correct classification of all but the simplest laser sources. While the details of the measurement requirements are important only for the classification procedure itself, an understanding of the basic principles helps to understand the potential hazard associated with the various classes for specific laser products and applications.

As noted in previous sections, the international classification scheme following IEC 60825-1 accounts for exposure of the naked, unaided eye as well as for exposure while using optical viewing instruments. When the eye is exposed to laser radiation behind optical viewing instruments, for certain beam geometries, the hazard level can be increased as compared to exposure to the same beam with the naked eye.

It is important to distinguish between two categories of optical viewing instruments which lead to an increase of the exposure of the eye (and therefore of the hazard) in a different way: optical viewing instruments with large collecting optics such as binoculars or telescopes (figure 4.3, left), which reduce the beam diameter of a collimated beam, and eye loupes, which allow a source (for instance LEDs or fibres) that emit a diverging beam to be viewed from a close distance (figure 4.3, right). The measurement requirements are particularly important to distinguish between the two conditions for optically aided viewing and exposure with the naked eye, and in that respect to also distinguish between Class 1 and 1M (and Class 2 and 2M).

4.2.3.1 Unaided eye

For exposure of the unaided (naked) eye, the most critical wavelength region is the retinal hazard region from 400–1400 nm where the radiation passing through the pupil is focused to a small spot on the retina. In order to assess the exposure in that wavelength range, i.e. the power or energy that enters the eye, an aperture with a diameter of 7 mm is used. This aperture diameter is derived from the maximum, dark adapted, diameter of the human pupil and therefore represents a worst-case value. In terms of the (worst-case) viewing distance for the naked eye, a distance of 10 cm is used in IEC 60825-1. The distance of 10 cm is representative of the closest distance that can still be accommodated, i.e. where the source is still imaged as a sharp image on the retina. In practice, this closest viewing distance depends on the refractive power of the human eye and it changes with age: 10 cm is a representative figure for younger people or for short-sighted people with a refractive power that is higher than the average adult. For normal-sighted adults, the typical shortest viewing distance is rather 15–20 cm and for older people it increases even further. For conventional sources and for many laser sources, the closest accommodation distance can be considered as the worst-case distance: when viewing of a source at distances closer than the minimum accommodation distance is attempted, the refractive power of the eye is not sufficient to produce a sharp image so that the radiation at the retina is distributed over a larger area, resulting in a less hazardous viewing situation than for a focused image at a viewing distance of 10 cm.

Consequently, for radiation in the retinal hazard area, the measurement requirement for the exposure condition 'naked eye'[1] is defined in IEC 60825-1 as an aperture stop diameter of 7 mm and an aperture distance to the location of the apparent source of 10 cm (for a discussion on the apparent source see section 3.12.1). When the criterion for classification of a product into a certain class is 'is the emission safe for the naked eye?' then the above measurement requirement is to be applied, i.e. the emission of the product needs to be assessed with a 7 mm aperture at a distance of 10 cm from the apparent source and the power or energy that was determined is to be compared to the AEL for Class 1 or Class 2. As explained in section 4.2.1, such a measurement requirement is identical to an ocular MPE assessment at the same distance and with the MPE calculated for the respective time base value. This not only applies to the retinal hazard region where the limiting aperture defined to average the irradiance is compared to the ocular MPE has a diameter of 7 mm, but it also applies to other wavelength ranges as long as the aperture stop diameter defined for classification is the same as that for the limiting aperture. Therefore, for the 'unaided eye' condition, for wavelengths outside the retinal hazard area, the aperture stop

[1] In the current version of IEC 60825-1, this condition does not have a dedicated name but is specified as applicable for the measurement of power or energy for Class 1M and Class 2M and for the measurement of irradiance or radiant exposure. It might be referred to as 'condition 3' in future editions of the standard.

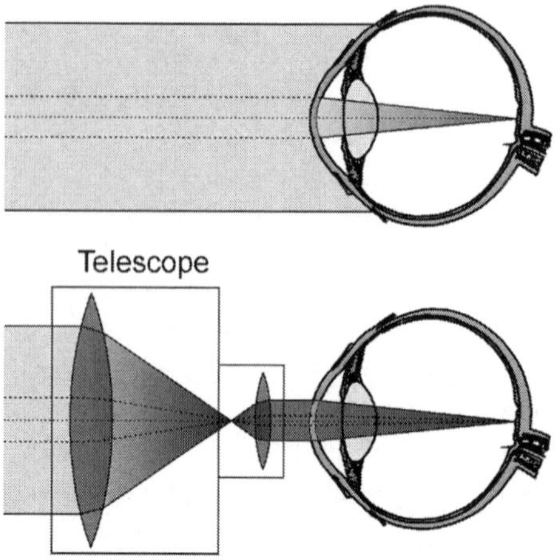

Figure 4.4. A telescope or binocular increases the hazard of a well collimated beam with large diameter, as more energy is collected and would be incident on the eye when compared to exposure of the naked eye. The optical set-up is referred to as a Gallilean telescope which produces inverted images. The principle of binoculars is the same only that a turning prism produces an upright image and also shortens the overall design.

diameter is identical to the diameter of the limiting aperture, table 3.1, defined for the averaged irradiance value (the effective irradiance) is compared to the MPE for the eye. Regarding the measurement distance for the unaided eye for wavelengths outside the retinal hazard area, these are specified as 10 cm for wavelengths between 302.5 and 4000 nm, and as 0 cm distance outside that wavelength range. The background of the wavelengths 302.5–4000 nm is that, in this range, optical viewing instruments are assumed to be fully transparent while outside of this range, they block optical radiation and therefore cannot increase the hazard to the eye.

This measurement condition is important for Class 1M and Class 2M classification, all other classes are based on measurement conditions that characterize the potential use of optical viewing instruments.

4.2.3.2 Telescopes and binoculars

When the beam diameter of a collimated beam is larger than 7 mm, then the pupil of the naked eye with a diameter of 7 mm only intercepts part of the total power in the beam. However, when the large diameter beam is viewed with a binocular with input optics of several centimetres, a correspondingly larger part of the total beam power is collected and incident on the eye, as is schematically shown in figure 4.4.

As binoculars or telescopes are only used at a distance and usually cannot focus to distances much shorter than 2 m, viewing with a telescope or binocular is relevant only for well-collimated large diameter beams. For a divergent beam, the exposure behind a binocular at a distance of 2 m and further will be far less hazardous than an exposure of the naked eye. A telescope or binocular only increases the hazard of large diameter collimated beams: if the beam diameter is smaller than the pupil diameter of the eye, the hazard is not increased by a telescope or binocular, as the total beam power enters the eye, with or without optical viewing instruments.

In IEC 60825-1 a binocular with 5 cm input diameter is assumed, which for an exit pupil diameter of 7 mm gives an optical magnification of 7. The radiative power that can be collected with such a binocular is about 50 times higher than compared to the naked eye, provided the beam diameter is 5 cm or larger.

The measurement condition to represent viewing with binoculars or small telescopes is defined in IEC 60825-1 as a 50 mm diameter aperture stop at a distance of 2 m from the product, when the wavelength is in the range 400–1400 nm (for other wavelength ranges see the detailed discussion in section 4.3.4). This measurement criterion is referred to as *condition 1* in IEC 60825-1—a condition that could also be called the 'telescope condition'.

4.2.3.3 Eye loupes

An eye loupe (a magnifying lens used for close-up viewing) has the function of magnifying the image and of increasing the refractive power of the eye so that a given source can be imaged at closer distance than would be the case for an unaided eye. For a divergent laser radiation such as emitted from the end of an optical fibre or as emitted from LEDs, the fraction of total power of the emitted beam that passes through a given aperture (for instance a 7 mm aperture representing the pupil) is much larger for closer distances than at the minimum accommodation distance of the eye, as shown in figure 4.5.

The higher the refractive power of the eye loupe is, the closer an object can be viewed and the larger it is in the sense of angular extent in respect to what it would be at the usual viewing distance. The measurement condition to represent viewing with eye loupes is defined in IEC 60825-1 as a 7 mm diameter aperture stop at a distance of 14 mm from the apparent source (for a small source—different distances apply to extended sources); this measurement criterion is referred to as *condition 2* and could also be called the 'eye loupe condition'. The distance of 14 mm is a worst case viewing distance that is really only possible with special very high magnification eye loupes, such as shown in figure 4.5. The viewing distance with loupe of 14 mm corresponds to a magnification of 18. This magnification value is obtained by relating 14 mm to the viewing distance for unaided viewing of 25 cm that is usually used to characterize the magnification of loupes. The choice of eye-loupe viewing distance results in the same opening

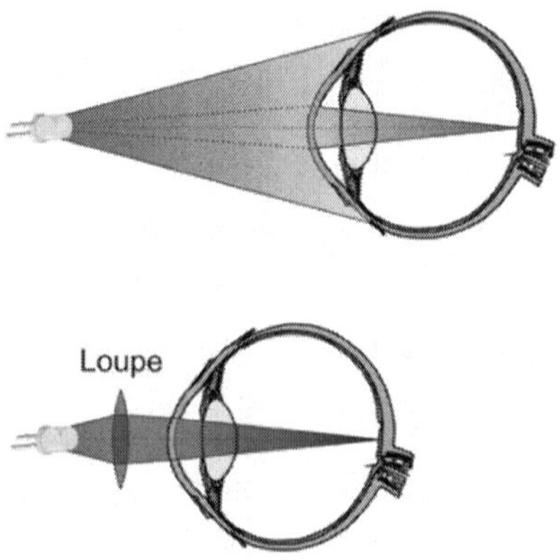

Figure 4.5. A loupe can increase the hazard of highly divergent sources by allowing the source to be imaged at a closer distance (lower drawing) than would be the case for the unaided eye (upper drawing), where the closest distance of accommodation is about 10–15 cm. Due to the shorter distance, more energy is incident on the eye than would otherwise be the case.

angle of the 7 mm aperture at 14 mm as the 5 cm aperture at 10 cm distance $(7/14 = 5/10)$.

For radiation in the retinal hazard region of 400–1400 nm, eye loupes can only increase the hazard of highly divergent sources, as for collimated, i.e. parallel beams, the power measured through the aperture stop depends very little on the position of the aperture in the beam.

4.2.3.4 Use of measurement conditions for classification

Since Class 1 and Class 2 have the meaning 'safe even when optical viewing instruments are used' for these classes the power determined with both optical viewing instruments needs to be below the respective AEL. Condition 1 and condition 2 also apply for the classification of a product as Class 3R and Class 3B. For these classes, the emission of the product has to be assessed with aperture diameters and placements according to both condition 1 and condition 2, and both determined power or energy values need to be below the respective AEL value in order to be assigned to that class. In practice, it will be obvious in most cases which one of the two conditions will result in the higher power value and which one will give the less critical value: condition 1, the telescope condition, will be critical for beams with small divergence and with diameters larger than 7 mm, while condition 2, the eye loupe condition, will be critical for highly diverging sources such as LEDs or line lasers. With 'critical' we mean that if the power or energy per pulse measured with the corresponding condition is below a certain AEL, it will also be below that AEL with the other, less critical condition, which

Table 4.4. Classification concept following different measurement conditions.

$$\left\{\begin{array}{c}\text{Condition 1} < \text{AEL}\\ \textit{and}\\ \text{Condition 2} < \text{AEL}\end{array}\right\} \longrightarrow \begin{array}{c}\text{Class 1}\\ \text{Class 2}\\ \text{Class 3R}\\ \text{Class 3B}\end{array}$$

$$\left\{\begin{array}{c}\text{Condition 1} > \text{AEL}\\ \textit{or}\\ \text{Condition 2} > \text{AEL}\end{array}\right\} \text{but `naked eye condition'} < \text{AEL} \longrightarrow \begin{array}{c}\text{Class 1M}\\ \text{Class 2M}\end{array}$$

in practice need not be evaluated. For collimated beams with a diameter of less than 7 mm, the full beam passes through both optical viewing instruments and these conditions will then yield the same power or energy value as the naked eye condition, namely the total beam power. This reflects that for such a beam, optical viewing instruments do not increase the hazard as compared to the naked eye.

As the AEL values are the same for Class 1 and Class 1M on the one hand and for Class 2 and Class 2M on the other, the differentiation between Class 1 and Class 1M (and Class 2 and Class 2M) is actually based on different measurement requirements. As described above, when the power or energy determined with both condition 1 and condition 2 is below the corresponding AEL for Class 1 or Class 2, the product is assigned to be Class 1 or Class 2, respectively. When the power or energy determined with one of the 'optical viewing instrument' measurement conditions is above the AEL, then this means that an exposure with the respective type of optical viewing instrument (eye loupe or binocular) at the position given by the measurement requirements is above the MPE for the eye for the respective time base. In this case (i.e. the potential hazard with optical viewing instruments), it is still possible that the exposure of the naked eye is safe, i.e. the level of power or energy determined with a 7 mm aperture stop at 10 cm from the apparent source could be below the AEL and then the product would be assigned to be Class 1M or Class 2M. This concept is summarized in table 4.4 and graphically shown in figure 4.6. To limit the power that is accessible through optical viewing instruments, in order for a product to be classified as Class 1M or Class 2M, the power measured with the 'optical viewing instruments' condition needs to be below the AEL of Class 3B.

It is important to note that only one of the two conditions results in a power level that is above the AEL: a large diameter collimated beam would fail condition 1 while a highly divergent beam would fail condition 2 (it might be that in very special cases a product fails both condition 1 and condition 2 and could still pass the naked eye condition, i.e. be classified as Class 1M or Class 2M, but it is certainly not a requirement for Class 1M or 2M to fail both optical viewing instruments conditions).

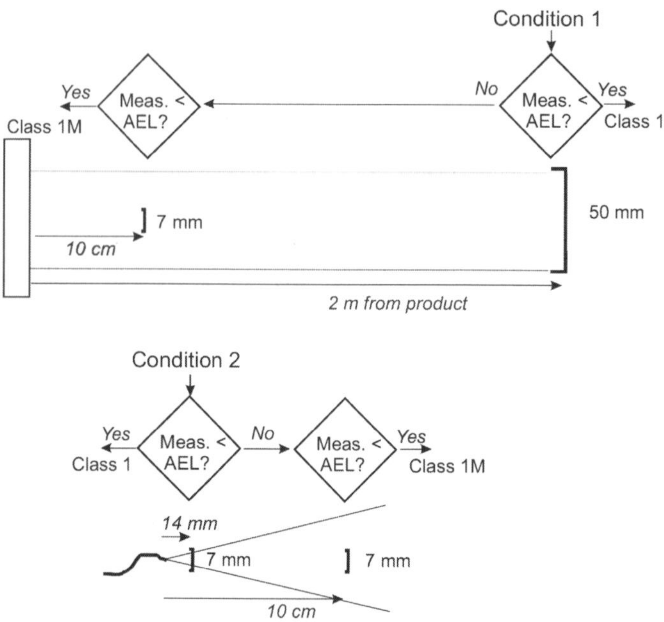

Figure 4.6. Measurement requirements for Class 1 and Class 1M (the same concept also applies to Class 2 and Class 2M): while there is only one AEL table that applies to both Class 1 and Class 1M, the distinction is made by different aperture diameters. The distances are measured from the position of the apparent source unless otherwise indicated. The naked eye condition (7 mm diameter aperture at 10 cm from the apparent source) is the same for both parts of the figure. The values given in the graphic for aperture diameter and distance strictly only apply in the wavelength range 400–1400 nm.

4.2.3.5 Ranking of product classes

The fact that for Class 1M and Class 2M products the radiation levels that are accessible with optical instruments correspond to emission levels associated with Class 3B laser products points to the difficulty of a ranking of the classes in terms of 'level of risk'. It depends on the viewing condition, if Class 1M or Class 2M is more hazardous than Class 3R. However, for the process of classification, a ranking of the classes was specified in IEC 60825-1: Class 1, Class 1M, Class 2, Class 2M, Class 3R, Class 3B, Class 4.

For the process of classifying a certain laser product, the emission as assessed with the different measurement conditions is compared to the AEL values starting with the 'lowest' classes according to the above ranking. Consequently, a product that satisfies the 'naked eye condition' when compared to the AEL of Class 1 and Class 1M but produces Class 3B-levels of accessible emission for one of the 'optical viewing instruments conditions' is classified as

Class 1M. When the classification process was performed by considering only the emission levels accessible with the 'optical viewing instrument' conditions and the measurement conditions for the 'lower' Class 1M were not considered, then the product is classified as Class 3B.

4.2.4 Classification scheme summary

In summary, the classification scheme is somewhat involved but has a logical basis when the three components: AEL values and their relationship to the MPEs, time base and measurement requirements are considered. Following this understanding of the classification scheme, an overview of the classes is once again presented in a tabular form that is shown in table 4.5.

It is important to note that it is not the total power that is compared to the AEL of the different classes but the accessible emission level which is determined with a specific aperture diameter at a specific location. Therefore, for classification, the value of the AEL is just as important as the prescribed measurement condition, and these two really should be seen as linked parameters. The importance and linkage of the measurement condition with classification becomes most evident for Class 1M and Class 2M, where the AEL is actually the same for Class 1 and Class 2, respectively, but the differentiation comes from the different measurement conditions. It is even the case that wording is used which relies on this implicit reference to the appropriate measurement conditions, for instance in the expression (as sometimes used in IEC 60825-1) 'radiation in excess of the AEL for Class 1' or 'radiation in excess of the AEL for Class 1M'. This is the short form to mean 'the accessible emission level as determined with condition 1 and condition 2 is below the AEL for Class 1 and Class 1M' in the first case and 'the accessible emission level as determined with the naked eye condition is below the AEL for Class 1 and Class 1M'.

4.2.5 Embedded laser products

A laser that emits power levels that in itself are hazardous, can still be assigned a lower ('safe') class, such as Class 1, when the laser is fully enclosed, i.e. laser radiation is prevented from leaving the housing of the laser product. Such a laser product is called embedded laser product. Obviously, in reference to the discussion of the previous sections, zero accessible emission is also below the AEL for Class 1.

The embedded laser itself can be quite powerful, as long as it is ensured that the housing prevents laser radiation from being accessible during use even for the case that a fault occurs. If access panels are provided to be removed by the user during use or maintenance, then these access panels need to be linked to a failsafe or redundant interlock system which prevents access to laser radiation, i.e. terminates the laser radiation as soon as the panel is opened. There are further requirements for such an embedded laser product that have to be fulfilled in order

Table 4.5. Tabular overview of meaning as associated to classes in the sense of the level of hazard (or rather, the level of safety, as it is not included what kind of exposure would be hazardous). While the meaning of the classes can be read from left to right through the three columns, they are split up into columns to show the logic of the three components of the classification.

	'Type and level of hazard' → AELs	'Typical or maximum exposure duration' → time base	'Exposure or viewing condition' → measurement condition
Class 1	Safe for eye	for long-term viewing	also with optical viewing instruments
Class 1M			for the naked eye only
Class 2		for short exposures only	also with optical viewing instruments
Class 2M			for the naked eye only
Class 3R	Safe for eye (with safety factor reduced)	for short exposures only	also with optical viewing instruments
Class 3B	∼ Diffuse reflections safe ∼ Low risk for skin injury	for long-term viewing	also with optical viewing instruments

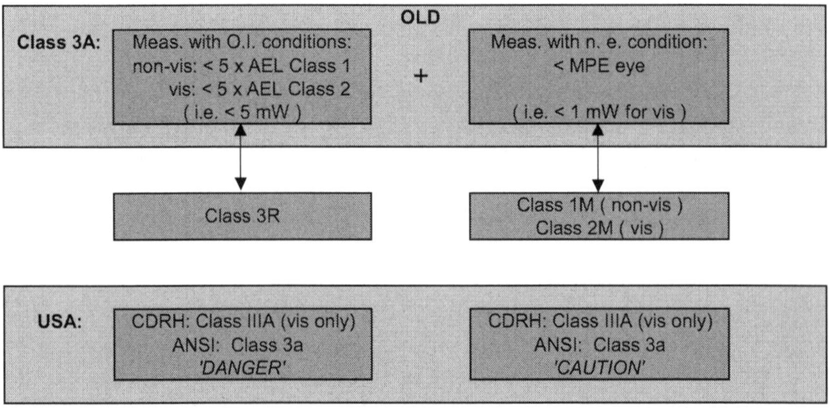

Figure 4.7. Concept of the 'old' Class 3A and basic relationships to the new Class 3R and Class 1M and Class 2M, as well as to current US Class IIIA.

to be assigned Class 1 according to IEC 60825-1 and these will be discussed in section 4.4.1. It is also noted that the embedded laser product may be a Class 2 or Class 3R device, for instance if an alignment laser is used together with the actual higher power embedded laser, and if this alignment laser is not switched off when the access panel is opened.

4.2.6 Old Class 3A and USA Class IIIa

Since products which are classified according to the pre-2001 edition of IEC 6025-1 will be around for some time (there is no need or requirement to reclassify the products that are in use), we also discuss the 'old' Class 3A. The meaning of Class 3A was the same as what is now Class 1M and Class 2M, i.e. that exposure of the naked eye is safe but a potential eye hazard exists when exposure occurs with optical instruments. For the definition of Class 3A, the maximum power level that is accessible with optical instruments was, however, limited to only 5 mW, which is lifted to 0.5 W for Class 1M and Class 2M. While the AELs for Class 3A referred to the MPE values and corresponding measurement apertures and distances (i.e. 100 mm from the apparent source), Class 1M and Class 2M refer to the respective AEL values which, however, when measured with the naked eye measurement condition 3 are identical to the MPE evaluation. The concept is shown in figure 4.7.

Since the 'new' Class 3R is defined to be evaluated with the optical instruments conditions 1 and 2 and the accessible emission limit is defined as five times Class 1 in the non-visible and five time Class 2 in the visible wavelength range, the old Class 3A was really a combination of the new Class 3R (regarding the maximum accessible level—accessible with optical instruments—

being limited to five time Class 1 and Class 2) and of the new Class 1M and Class 2M (accessible emission for the naked eye below the MPE).

A common source for misclassification for products that were sold outside the US was that the old IEC Class 3A was different to the current Class IIIA as it is defined by CDRH (which is equivalent to the ANSI Class 3a—ANSI uses Arabic numerals while CDRH uses Roman numerals for the classes). While CDRH also specifies for their Class IIIA that the power accessible with optical instruments is less than 5 mW (CDRH defines Class IIIA for the visible only, while ANSI defines 3A for visible and non-visible), they made the type of logo on the warning label dependent on whether the MPE is exceeded or not: if the MPE is exceeded the product would get a 'Danger' label, if the MPE is not exceeded, the product would only get a 'Caution' label. Thus, in effect, the meaning of the 'Danger' Class IIIA is equivalent to (visible) Class 3R and the 'Caution' Class IIIA is equivalent to Class 2M. The equivalence between the classes, however, is not complete, as the measurement distance specified in CDRH (and ANSI) are different to the measurement distances of IEC.

4.2.7 Overview table

If all the different aspects of the classification scheme are treated separately, there is a total of eleven product classes with different types and levels of hazards associated with them, as shown in table 4.6.

It is important to note that the overview table only gives very rough and simplified information on the level of hazard that is actually presented by a specific laser product that is used in a specific application. While Class 1 laser products are considered safe for all kinds of conditions of reasonably foreseeable use, all the other classes represent some kind of hazard. Obviously, the level of hazard that is assigned to a specific laser class does not realize itself for each laser product in the same way and also does not behave like a step function as is implied by the sharp borders between the classes as represented by the AEL values. The potential hazards associated with Class 1M, Class 2, Class 2M and Class 3R manifest themselves only in special exposure situations and these classes therefore could be grouped as 'safe except in special cases'—where the special cases are different for the different classes, and relate to exposure duration for Class 2 and Class 3R and to exposure with optical viewing instruments for Class 1M and Class 2M. However, for Class 3B and Class 4, that are generally understood to represent the 'hazardous' laser classes, it is important to understand that the nature, level and geometrical extent of the hazard varies dramatically depending on wavelength, power level, beam geometry and also on engineering controls such as partial enclosures that would limit the accessibility of the hazard. These more applied aspects and the understanding of the hazards associated with the classes for different products and applications are discussed in section 7.1.1.

Table 4.6. Overview of the most important aspects of the laser classes following IEC 60825-1.

	Type of lasers	Associated potential eye or skin hazard	Typical allowed power for cw lasers*	Measurement requirements*	NOHD or ENOHD**
Class 1	Laser is completely enclosed. Radiation is not accessible during use (embedded laser product)	Generally safe during use. Hazards according to power of enclosed laser when interlocks overridden or in service situations.	Power of enclosed laser theoretically not limited	Enclosure needs to withstand laser exposure	no NOHD no ENOHD (during use)
Class 1	Very low emission level	Generally safe because emitted power level is safe even for long-term intrabeam viewing. Intrabeam viewing with optical viewing instruments such as magnifying glasses or telescopes is also safe.	40 μW for blue and 400 μW for red.	7 mm diameter 14 mm and 50 mm diameter 2 m	no NOHD no ENOHD
Class 1M	Collimated and large beam diameter	Safe for long-term intrabeam viewing with the naked eye but potentially hazardous when viewed with binoculars.	Same as for Class 1 but different measurement requirements	7 mm diameter 10 cm	no NOHD has ENOHD
Class 1M	Highly divergent (e.g. line lasers, LEDs, diffuse sources, bare laser diodes)	Safe for long-term intrabeam viewing with the naked eye but potentially hazardous when viewed close up with magnifying lenses (eye loupes) or if optics are used to collimate the beam.	Same as for Class 1 but different measurement requirements	7 mm diameter 10 cm	no NOHD no ENOHD

Table 4.6. Overview of the most important aspects of the laser classes following IEC 60825-1.

	Type of lasers	Associated potential eye or skin hazard	Typical allowed power for cw lasers*	Measurement requirements*	NOHD or ENOHD**
Class 2	Low emission level. Visible wavelengths only.	Safe for short-time (accidental) direct exposure with naked eye and optical viewing instruments. Short time exposure can be assumed for bright light. Prolonged staring might lead to eye injuries, especially for blue laser wavelengths.	1 mW	7 mm diameter 14 mm and 50 mm diameter 2 m	no NOHD no ENOHD
Class 2M	Collimated and large beam diameter	Safe for short-time direct exposure with the naked eye, potentially hazardous when viewed with binoculars or if optics are used to collimate the beam.	Same as for Class 2 but different measurement requirements	7 mm diameter 10 cm	no NOHD has ENOHD
Class 2M	Low visible emission. Highly divergent (e.g. line lasers, LEDs, diffuse sources, bare laser diodes)	Safe for short-time direct exposure with the naked eye, potentially hazardous when viewed close up with magnifying lenses (eye loupes) or if optics are used to collimate the beam.	Same as for Class 2 but different measurement requirements	7 mm diameter 10 cm	no NOHD no ENOHD
Class 3R visible	Low visible emission. Typically alignment lasers.	Accidental exposure usually not hazardous but eye injury possible for intentional intrabeam viewing.	5 mW	7 mm diameter 14 mm and 50 mm diameter 2 m	has NOHD has ENOHD

Table 4.6. Overview of the most important aspects of the laser classes following IEC 60825-1.

	Type of lasers	Associated potential eye or skin hazard	Typical allowed power for cw lasers*	Measurement requirements*	NOHD or ENOHD**
Class 3R non-visible	Low non-visible emission.	Accidental exposure usually not hazardous but eye injury possible for intentional intrabeam viewing.	3.3 mW for 810 nm, 40 μW for 350 nm, 12 nW for 302.5 nm	7 mm diameter 14 mm and 50 mm diameter @ 2 m	has NOHD§ has ENOHD
Class 3B	Medium power lasers	Exposure (including short-time accidental exposure) of the eye to the direct beam may cause serious eye injuries. No or very limited skin hazard and viewing of diffuse reflections is normally safe.	500 mW	7 mm diameter 14 mm and 50 mm diameter 2 m	has NOHD§ has ENOHD
Class 4	High-power lasers	Exposure (including short-time accidental exposure) of the eye to the direct beam and close viewing of diffuse reflections (within diffuse reflections NOHD) may lead to serious eye injuries. May also cause serious skin injury (within skin hazard distance). Presents a fire hazard.	not limited		has NOHD# has ENOHD

* Power or energy measured with specified apertures needs to be below AEL. AEL and measurement requirement is given for small (point) sources ($\alpha < 1.5$ mrad).
** Concept of NOHD and extended NOHD for telescopic viewing (ENOHD) is discussed in sections 5.4 and 5.6, respectively.
§ For unrealistically long exposure durations there might also be a skin hazard zone for Class 3R non-visible and for Class 3B. Class 3B might also have a hazard zone associated with diffuse reflection, however, it is not very large and also only exists for a long exposure duration.
# Class 4 lasers also have a hazard zone and distance that applies to exposure of the skin and for viewing of diffuse reflections.

4.3 Manufacturer's classification procedure

Caveat: This section is intended to provide information that is helpful for the application of the international standard IEC 60825-1 but is not a complete list of all requirements and should not be regarded as a replacement of the standard.

4.3.1 Introduction

While the international standard IEC 60825-1 and equivalent national editions do not have the status of law, standards are generally considered as representing the 'technical state of the art specification'. However, in cases of litigation, it would probably be difficult to justify why a product did not comply with the corresponding international standard. The international standard IEC 60825-1 is adopted worldwide by many national standards committees, and depending on specific national laws can be fully legally binding. In Europe, EN 60825-1 is a harmonized standard listed for the fulfilment of the requirements of the machine directive, the low-voltage directive and the medical product directive, and of corresponding national laws (either directly or indirectly via normative references in standards that are listed, such as IEC 60601 for the electrical safety of medical products). In the US, a national laser product requirement document is in place that has legal status and that following a current revision will adopt the IEC classification scheme with some minor differences that are discussed in section 4.5.

In this section, we provide information relevant to the classification of laser and LED products according to the current edition of IEC 60825-1. While it is the responsibility of the manufacturer of the complete product to classify the laser product, if the user changes the product in a way that changes the class, then the user has the responsibility for reclassification of the product. This is obviously especially relevant when the product becomes more hazardous, such as when an access panel is removed or an interlock protection is disabled.

4.3.2 General issues

IEC 60825-1 is applicable to both laser and LED products. While the title only refers to laser safety, in the scope of the standard, LED products are defined as equivalent to laser products, and LEDs are included whenever the word 'laser' is used. Within the body of the standard only laser products are referred to even though all requirements apply to LED products as well.

A laser product or LED product is exempt from all further requirements (including labelling and information for the user) of the laser safety standard, if it does not contain an embedded laser or LED (i.e. a laser with higher power inside the housing where the classification is based on making the laser radiation inaccessible) and if the accessible emission limits do not exceed the AEL of Class 1 (i.e. measured with the optical instruments conditions) under all conditions of operation, maintenance, service and failure (fault conditions). An example for this exemption are low-power LEDs such as used as indicators on electronic equipment or as displays in clocks. In the strict sense, this exemption means that such an exempt product should not be referred to as 'Class 1 laser (or LED) product', but as an exempt laser (or LED) product. In practice however, such products are more generally referred to as 'Class 1' laser products.

Products that are supplied to manufacturers (OEM products), who then integrate the laser into a final product, do not have to comply with the standard, as it is the final product that must comply.

Where the laser is incorporated into a larger system, where the larger system forms the actual product, then this larger system needs to be classified. Depending on the system design, the class of the larger system can either be 'higher' (for instance if a divergent beam is collimated) or 'lower' (for instance because of transmission losses in exit windows, or due to increased distances between the exit aperture and the apparent source) than the class of the actual incorporated laser. Also, if more than one laser is incorporated in the actual product, the classification is performed for the whole product and the product is assigned a single laser class. The manufacturer should, however, provide relevant safety information not only for the 'main' (most hazardous) laser but also for the other lasers. For instance, almost all surgical laser products have, in addition to the main 'working' laser, a visible alignment laser which is activated in the standby mode before the working laser is turned on. The alignment laser often has an output power of up to 5 mW, and as such also represents a limited level of risk (equivalent of Class 3R), which should not be overlooked even though the whole laser product is classified as Class 4, based on the emission of the main high-power beam.

4.3.2.1 Human access

The classification of laser and LED products is based on the level of radiation that is considered accessible by humans. In this sense, 'accessible' not only means that the radiation is measured through specified aperture diameters and positions, but the term also relates to what is defined in the standard as accessible by humans in relation to enclosures being intended to prevent access. Human access is defined in the current edition of the standard IEC 60825-1 as follows. When radiation is intended to emerge from the enclosure, then it is obvious that this radiation is accessible. For the enclosure of laser radiation (basically by the housing) for Class 2, 2M or 3R, human accessibility is defined on the basis of a straight probe with a diameter of 12 mm and a length of up to 80 mm. When the probe can be inserted through an opening in the housing and when it can intercept laser radiation, then this radiation is considered to be accessible and must be taken into account when determining the product class, by comparing it to the AEL. When the level of radiation inside the housing is at a level equivalent to Class 3B or Class 4, then human access is defined on the basis of direct reflection of radiation through any opening in its protective housing from a single mirror that is placed inside the housing. Here, direct reflection from a single mirror means that it is not necessary to consider multiple reflections 'around' baffles or corners inside the housing. This definition of human access is particularly important for the classification of an embedded laser product where the radiation of the embedded laser (the laser inside the housing) is fully enclosed so that the classification is based on the non-accessibility of the radiation.

4.3.2.2 Operation, maintenance and service

For classification it is important to note the difference between the three operational modes of a laser product: operation, maintenance and service. Operation refers to the intended use of the laser, that is to the performance of the product over the full range of parameters, as specified by the manufacturer, but does not include maintenance or service. Maintenance refers to adjustments and other routine procedures intended to be performed by the *user* and described in the user manual, which are necessary to ensure normal operation of the laser. Service, in contrast, refers to more specialized procedures that are intended to be performed by a qualified service engineer, and are described in the service manual. Thus, basically, the manufacturer can decide what is defined as user maintenance and what is defined as service. In practice an important difference is the level of training: to carry out operation and maintenance generally requires a lower level of product-related training than is necessary to perform service procedures. However, it is of course possible that the person who is the user could be trained by the manufacturer to undertake servicing, and such training would need to include relevant safety issues as well.

Currently, the class of a laser product is determined on the basis of operation only, and does not relate to user maintenance (or to service). Thus, it is possible that for a fully enclosed laser materials processing machine that fulfils all the requirements of the standard necessary for classification as Class 1, the user maintenance procedure includes overriding an interlock on an access panel with subsequent direct access to the high-power beam. Since the general understanding of Class 1 is that it presents such a low level of hazard that no safety related training nor any other procedural controls are necessary, the authors (and many other safety experts) strongly feel that classification ought to include user maintenance as well, so that no task that needs to be carried out by the (untrained) user involves access to radiation above the AEL of the relevant class. It also follows logically from the general understanding of the meaning of Class 1 'safe for the user, no controls necessary', that, since maintenance is by definition intended to be carried out be the user, maintenance should also be safe for a Class 1 laser product, and therefore classification should include maintenance. That is, the product ought to be classified according to the radiation accessible under operation *and* maintenance. Currently this is not the case; the only requirement regarding maintenance is to provide adequate instruction for proper maintenance. Thus, based on the current specification in IEC 60825-1, it is possible to classify an embedded laser where user maintenance involves working with a high-power laser beam as Class 1. However, it is interesting to note that if such a product is a machine, it would not comply with the requirements specified in ISO 11553 [1] where for user maintenance, a maximum exposure level equivalent to exposure to a Class 1M laser is required, irrespective of the class of the laser processing machine. Also, general product safety legislation in many countries requires safety for the user and specifically includes maintenance procedures for special consideration. In practice, therefore, many lasers that are classified as Class 1 are designed such that user maintenance does not involve access to a hazardous level of radiation. This issue is currently under consideration by the IEC laser safety committee.

4.3.2.3 Primary doctrine: maximize emission

The uppermost principle for the classification of a laser product is that classification should be carried out under those conditions that result in the maximum possible (i.e. most hazardous) level of accessible emission. While the measurement as such need not be performed at distances closer than those specified in the standard, all other possible parameters that might influence the accessible emission need to be considered, including (but not only) the following.

- Characterize the emission also for start-up (for instance, some lasers might produce spikes with very high peak powers when switched on—these spikes might not be produced for each switch-on procedure).
- The emission needs to be below the AEL over the complete range of specified environmental conditions, for instance, if a laser diode or LED product is specified for temperatures down to $-40\,^{\circ}\mathrm{C}$, the emission at these low temperatures could be higher than at room temperature.
- Mechanical impacts that can be reasonably foreseeable for the specified use of the product (for instance, a handheld product where the classification is based on a diffusor plate would need to be analysed in terms of the rigidity and quality of the fixture of the diffusor plate).
- The user controllable adjustments need to be set to maximize emission. This rule does not apply to adjustments that can only be accessed during service, for instance adjusting a potentiometer inside the product which is accessible only by removing panels with tools, or by adjusting a parameter in the control software that is protected by a password which is given by the manufacturer only to service engineers. However, this rule includes adjustments that the user can perform during user maintenance procedures and that could increase the accessible emission in the operation mode.
- Following removal of panels or parts of the housing that are not protected by interlocks and where no tools are necessary for the removal, when this removal is part of operation (see also discussion in previous section on maintenance and service).
- Defeat of interlocks if this is part of the operation of the product.

- The classification also includes fault conditions, as discussed in a separate heading later.

4.3.2.4 Misuse of concept and terminology

There are a number of assumptions inherent in the classification principle, which, depending on the typical use of the product, could be considered 'worst case' assumptions. These assumptions may lead to laser power or energy restrictions that may be seen as over-restrictive for a specific product. Examples are an assumed exposure duration (referred to as time base) of 100 s for Class 1 and Class 1M, measurement criteria that are derived from close-up viewing of the product, including use of high-power magnifiers and large binoculars, and the requirements regarding embedded lasers as based on a specific definition of human access.

The classification rules need to be followed, even if for a specific product, for the specified operating conditions, some of them might seem over-restrictive. Classification is based on prescriptive requirements defined in the standard and these requirements cannot be 'adopted' for specific cases—'to be considered safe for the user' is not the requirement for classification as Class 1, but there are a number of specific requirements defined in the standard which need to be followed for a laser product to be classified as Class 1.

The authors have encountered some misuse, or attempted misuse of the classification, such as the following examples:

- the product is 'Class 1 for distances greater than 20 cm' (if classified according to the standard the product was Class 3B but had a rather large divergence, so that the NOHD was only 20 cm);
- a laser processing machine with protective shielding on the sides but no top classified as Class 1;
- a laser scanner where the classification was based on the scanning action but which did not have an appropriate scanning safeguard (i.e. emission would not automatically terminate in a short time in case that the mirror stops due to mechanical or electronic failure).

All these examples are based on the assumption that, when used as intended by the person carrying out the classification, the laser will be safe, but this is not the definition of Class 1. Obviously, a Class 1 laser should always be safe when used as intended, but this is not the same as saying that any laser that is safe when used as intended is Class 1. Classification is rule based, and the specified rules have to be followed, even if they do not always seem to be relevant to the intended use of a particular laser system.

A further confusion is sometimes encountered when a class is assigned to an accessible laser beam at some arbitrary location. (This is the situation in the first example given above.) Classification applies to a laser product, not to a propagating laser beam. Strictly, therefore, it is not correct to say, for example, that a given laser beam is Class 3B, rather, that it is produced by a Class 3B laser.

This may seem pedantic, but such misuse of terminology can result in serious misunderstandings as to the actual class of a laser, or of the hazard arising from exposure to the beam at a different location.

4.3.3 Single fault condition

The classification following IEC 60825-1 requires that classification tests include emissions that can arise during reasonably foreseeable single fault conditions. Possible relevant single fault conditions may be the fault of the driving circuit that controls the power output of the laser, where the failure of an electronic component such as an operating amplifier might lead to increased power output or the fault of a scanner or some mechanical fault such as a displacement of a diffuser plate or a filter.

The requirement in the standard means that unless the failure of the component is so unlikely that it is not considered reasonably foreseeable, the accessible emission level for comparison with the class AELs is the one that corresponds to the condition of failure.

Exemption

Single fault conditions where the increased emission only exists for such a short period of time that it is not reasonably foreseeable that exposure would occur during that period need not be accounted for. Surface emitting LEDs are given in the standard as an example of this exemption, which is based on the assumption that surface emitting LEDs that are, during the fault condition, driven with current levels that are much higher than the usual operating currents usually self terminate by burning out. Another example, in the view of the authors, is the shut down time for a scanning safeguard. When the classification of the scanner is based on the scanning action, then a scanning safeguard is required. Such a scanning safeguard monitors the condition of the scanner (for instance the speed of the scanning mirror) and automatically terminates laser emission if a fault occurs (i.e. if the mirror speed falls below a specified level). However, such a scanning safeguard might be difficult to realize technically with a shut down time that is fast enough to prevent emission above the AEL. Depending on the design of the laser product, the required shut down time could be in the order of microseconds, while a technically realistic value is more of the order of several milliseconds. However, when the product is not intended for audience scanning or for scanning of the retina (such as in a laser scanning ophthalmoscope), it is in most cases possible to show that the probability of an exposure within the shutdown time is small enough to be considered 'not reasonably foreseeable'. For the calculation of the corresponding probability of exposure during the fault (before the scanning safeguard terminates the emission), the expected probability that somebody is at the position of the scanning beam and looks towards the laser can be multiplied by the mean frequency of the occurrence of the fault. It follows that for the exemption to apply and to be justifiable, the probability of exposure needs to be small when the frequency of the fault itself is not small enough to justify being considered as 'not reasonably foreseeable', so that the resulting combined probability for an exposure during the fault is considered not reasonably foreseeable.

Single fault

'Single fault' means that the situation of two fault conditions occurring at the same time does not have to be accounted for. While the requirement in the standard refers to single fault conditions only, it could be that faults which do not lead to hazardous situations as single faults, do so in combination. If both of these faults are quite probable, then the fault is also reasonably foreseeable. It should be considered prudent engineering to consider two faults for classification when they are both quite likely to occur and when the hazard is increased when they occur together. This double fault would not be accounted for by the requirement of the standard when they, as single individual faults, do not increase the hazard. While the consideration of such a case of a reasonably foreseeable double fault is not required by IEC 60825-1, it should be dealt with on the basis of good engineering practice and would probably be considered necessary under more general (national) product safety legislation.

What is not reasonably foreseeable?

'Reasonably foreseeable' means that the probability of the fault occurring is not unrealistically small. There is no general value for a probability (or more accurately, a frequency, which is the probability per unit of time) to be considered a 'boundary' between reasonably foreseeable and not reasonably foreseeable. It depends on the type of product, the duration of the fault before the laser emission is terminated and the modes of usage. In the field of failure analysis, which is a well-established field of engineering [2], a common figure to characterize the frequency of failure of a specific component or of a system as a whole is the mean time between failure (MTBF), which is often given in units of a million hours. For instance, the MTBF of a resistor could be 50 million operating hours, or of a certain transistor could be 5 million operating hours. A MTBF of 5 million operating hours means that the average (mean) frequency for a fault equals 2×10^{-7} per operating hour, or an average of one fault per 570 years of continuous operation. This appears to be a relatively low probability, but if the product is used to scan the retina for medical diagnosis, the probability of ocular exposure during the fault equals unity, and in the view of the authors such a MTBF would not be considered as acceptable. Also in terms of total number of expected ocular injuries, what is referred to as global risk (in contrast to the individual risk to an individual patient having their retina scanned), would be relatively high: a simplistic estimate could be performed based on the assumption of eight operating hours per day, so that the fault of the laser product would occur on average once per 1710 years. When we simplistically assume an equal distribution of failures, then a number of 2000 lasers that are in use would mean that there would be more than one incident of eye damage due to the fault per year. However, as another example, if the laser is intended to be activated only when pointed at objects (such as for some types of bar code scanners) or is an automated system in a production line where the laser points downwards, the probability of exposure to the laser beam is correspondingly small and it might be possible to show that the above figure for the frequency of the failure is acceptably low.

It follows that the actual value of the MTBF that can be said to be 'acceptable' in the sense of being so high that the fault can be considered not reasonably foreseeable depends on the way in which the product is intended to be used, with appropriate consideration of misuse, and the frequency that somebody is exposed to the beam.

Summary

The fault analysis that needs to be performed should firstly identify all possible conditions that can lead to an increase of the emission level (which is not limited to an increase of the actual power but also an increase of pulse frequency, pulse length, or a decrease of the angular subtense of the apparent source, for instance when a diffusor plate breaks or is displaced due to mechanical impact). As a next step, the frequency for each identified fault needs to be characterized (in many cases, the component manufacturer can provide relevant MTBF numbers). The expected (i.e. the mean) frequency of the fault condition needs to be judged in combination with the expected frequency and duration of exposure regarding the acceptability to be regarded as 'not reasonably foreseeable'. If the frequency is *not* low enough, then additional measures need to be taken to decrease the probability for the failure or the probability of an exposure. (The probability of exposure can be reduced by including a safeguard that detects fault conditions and automatically terminates laser radiation when a fault occurs.) Otherwise the classification has to be based on the higher emission level that can exist during the fault.

Depending on the laser product, the fault analysis procedure often requires substantially more resources than the actual classification of the product in terms of assessing the accessible emission level. However, the fault analysis and the corresponding documentation is not only a requirement of the laser standard but is a general principle of product safety and should be performed with corresponding level of comprehensiveness and depth so that it holds up to third party (legal) scrutiny.

4.3.4 Measurement requirements

We have already pointed out in previous sections of this chapter that for classification, it might not be the total emitted power that is compared to the

AEL, but rather the power or energy level that is measured at specified distances with specified aperture diameters and specified values for the angle of acceptance. The power level assessed following the specified measurement requirements can be significantly lower than the totally emitted power level, and is often referred to as the *accessible emission (level)* to distinguish it from the total power level. Therefore, with 'accessible emission' we mean the power or energy value that is determined following the appropriate measurement requirements and that is the value that has to be compared to the AEL for the determination of the class of the product.

As some of the measurement distances are specified relative to the location of the apparent source, it might also be necessary to determine the location of the apparent source. Also, those AEL values that are derived from the retinal thermal MPE depend on the value of α, so that for a complete analysis it is not only important to properly measure the accessible emission level but also to properly determine the angular subtense of the apparent source α (unless a minimal source is assumed for simplicity). (See section 3.12.1 for a detailed discussion of the apparent source). This level of complexity, however, is not necessary for well collimated visible beams with diameters of less than 7 mm, as for such lasers it is the total power that is compared to the AEL, and the actual measurement requirements do not play a role. For such sources, the location of the apparent source is also not relevant and the angular subtense of the source is minimal (i.e. a small source where $C_6 = 1$). In the other extreme, an array of infrared LEDs can emit total power levels of several watts and still be classified as Class 1, based the correct choice of measurement requirements and value of α.

In the following we present the measurement requirements for small sources, which, however, also provide a simplified worst case alternative to the analysis for extended sources: when the measurement requirements as specified for small sources are applied to extended sources, they result in higher emission limits. These principle measurement requirements are adopted for extended sources as discussed in section 4.3.5.

The measurement distance and diameter of the aperture stop depend on the wavelength of the radiation under test and are summarized in table 4.7.

Table 4.7 (which is adopted from table 10 in the current version of IEC 60825-1) is organized into three measurement conditions, condition 1 for the 'telescope' condition, condition 2 for the 'eye loupe' condition, and the third condition, which in the current version of the standard is not yet specifically referred as condition 3 but in this book is referred to as such, and is understood as the 'naked eye' condition. In the following, we give some background information and explanation related to the measurement criteria.

Wavelengths < 302.5 and > 4000 nm

For wavelengths below 302.5 nm and above 4000 nm, optical instruments are not considered to transmit any relevant amounts of optical radiation so that for

Table 4.7. Measurement distance and diameter for small sources as currently specified in IEC 60825-1 (the third condition currently has no specific name but might be in future editions referred to as condition 3). The abbreviation 'a.s.' refers to the location of the apparent source.

	For values expressed in power (W) or energy (J)				For irradiance (W m^{-2}) or radiant exposure (J m^{-2}) and for classifying as Class 1M or Class 2M	
	Condition 1 (telescope)		Condition 2 (eye loupe)		Condition 3 (naked eye)	
Wavelength	Aperture stop (mm)	Distance (mm)	Aperture stop (mm)	Distance (from a.s.) (mm)	Limiting aperture (mm)	Distance (from a.s.) (mm)
<302.5 nm	—	—	7	14	1	0
≥302.5–400 nm	25	2000	7	14	1	100
≥400–1400 nm	50	2000	7	14	7	100
≥1400–4000 nm	25	2000	7	14	1 for $t \leq 0.35$ s $1.5t^{3/8}$ for $0.35\ \text{s} < t < 10$ s 3.5 for $t \geq 10$ s (t in s)	100 100
≥4000–10^5 nm	—	—	7	14	1 for $t \leq 0.35$ s $1.5t^{3/8}$ for $0.35\ \text{s} < t < 10$ s 3.5 for $t \geq 10$ s (t in s)	0
≥10^5–10^6 nm	—	—	7	14	11	0

these wavelength ranges, optical instruments are not considered to increase the hazard. Except for Class 3B, the AEL values for these wavelength ranges are identical to the MPE values and are given in units of W m^{-2} and J m^{-2} so that measurements are performed using condition 3. The measurement distance and aperture diameter given under condition 2 are therefore only relevant for a test against the AEL of Class 3B (i.e. to determine if a product is Class 3B or Class 4). Since the Class 3B AELs are not directly related to MPE values, the specified measurement aperture and distance is rather arbitrary, and is not based on biophysical criteria.

Condition 1

For wavelengths outside the retinal hazard region, the value of 25 mm for the diameter of the aperture for condition 1 is not derived from an assumption of exposure to a 25 mm binocular, but from an assumption of a magnification of 7 (such as from a 50 × 7 binocular). The aperture diameter of 25 mm is obtained by multiplying the diameter of the limiting aperture of 3.5 mm specified as condition 3 by the magnification of 7, as the radiation which is contained within 25 mm at the position of the binocular will be incident on the eye within the diameter of the limiting aperture of 3.5 mm.

Condition 2

The aperture diameter of 7 mm is derived from the typical size of high-power eye loupes. For the retinal hazard region, the increase of the hazard for eye loupes is due to collimation of a diverging beam which can be imaged onto a small spot on the retina. When a collimated beam is observed with an eye loupe, the hazard is actually decreased, as for a highly magnifying loupe, the eye is close to the focal plane and a new beam waist is formed close to the focal plane, so that although the beam is focused onto the cornea, the retinal image size is increased. For wavelengths outside the retinal hazard area, the latter case can, of course, increase the hazard to the cornea compared with the naked eye.

Condition 3

The aperture diameters given for condition 3 are simply the diameters of the limiting aperture that are specified for averaging irradiance and radiant exposure for an MPE analysis and thus refer to exposure of the naked eye. The naked eye condition is defined as the test for a product to be Class 1M or Class 2M. As discussed in section 4.2.3.4, when the accessible emission as determined according to condition 1 or according to condition 2 exceeds the AEL of Class 1 or Class 2, the product can be classified as Class 1M or Class 2M when the accessible emission as determined according to condition 3, the naked eye condition, is below the AEL value.

Angle of acceptance

The angle of acceptance (FOV) does not play a role for small sources.

Equivalent arrangements

In the current version of IEC 60825-1, a note allows the use of equivalent arrangements of apertures and distances. As an example, as an alternative to the 7 mm aperture placed at 14 from the apparent source, the placement of a 50 mm aperture at a distance of 10 cm from the apparent source is given. It should be noted, that this alternative is only equivalent for small sources and leads to different accessible emission levels when the source is an extended one.

Measurement position

It is important to note that the measurement distance specified for condition 1 is defined relative to the exit aperture of the product (more exactly, to the closest point of human access), while the distance for the other two conditions is defined relative to the location of the apparent source.

For beams with a large divergence, the measurement distance has a significant influence on the accessible emission, and therefore, for measurements according to condition 2 and 3, the location of the apparent source has a significant influence on the accessible emission. For laser beams under the small source condition (i.e. when C_6 is set to unity), the location of the beam waist can be considered the location of the apparent source. When the beam diverges from the exit aperture, the beam waist is some distance behind the exit aperture, and might even be virtual, i.e. inside the laser cavity or, when the beam is enlarged and collimated by optics, behind the laser cavity. Consequently the location of the apparent source can be determined by analysing the propagating beam, as in the far-field (outside of the Raleigh range, see figure 3.26) the beam seems to diverge from a single point which is at the location of the beam waist. When the location of the apparent source is further behind the exit aperture than the measurement distance (i.e. the apparent source is recessed from the exit aperture), then the measurement should be carried out at the exit aperture, i.e. at the closest point of human access. For example, when the location of the apparent source is determined to be 60 mm behind the exit aperture, then the measurement following condition 2 would be carried out at the exit aperture, while the measurement following condition 3 would be carried out at a distance of 40 mm from the exit aperture.

When an external beam waist exists (i.e. the beam as it exits the product is converging) then the external beam waist is to be considered as the location of the apparent source.

For low divergence beams with a divergence of a few milliradian, the apparent source will be located well inside the laser product or might even be located behind the laser, so that the default measurement position for both condition 2 and condition 3 is at the exit aperture. However, if the beam diameter at the exit aperture is less than the measurement aperture, it is not necessary to actually perform the measurement right at the exit aperture, but the same result is obtained some distance away, which might be a more convenient distance for the practical measurement.

For condition 1, for the special case when the beam converges from the exit aperture to an external beam waist that is further than 2 m from the exit aperture, and the beam diameter at a distance of 2 m is larger than the specified aperture stop for condition 1, the manufacturer should consider if the appropriate measurement position could be at the external beam waist. This special case of a converging beam with a large diameter is not specifically treated in the standard, and the manufacturer could argue that a rigid interpretation of the standard still requires a 2 m measurement distance. However, taking a more reasonable interpretation, the general principle of maximizing the accessible emission levels should apply. If the manufacturer chooses to classify the converging large diameter beam at the distance of 2 m, then it would certainly be necessary for the information regarding the potentially higher hazard than indicated by the class for exposure some distance away from the product to be given in the user manual. Future editions of the standard may specifically require that converging beams be measured at the beam waist.

For condition 1, it is not necessary to measure closer than 2 m, as exposure with a binocular at closer distances is not deemed to be a realistic risk (and few binoculars can focus that close in any case). An aspect of this issue is also that the divergence of a laser beam for which condition 1 is the more critical condition cannot be great, so that for the unlikely case that exposure does occur at closer distances, the hazard would not be greatly increased as compared to a distance of 2 m.

For embedded laser products, i.e. for products where a higher power laser is fully enclosed so that in operation, no laser radiation is emitted from the product and the classification is based on the inaccessibility of the radiation, the measurement of the accessible emission needs to be based on the definition of human access as treated in section 4.3.2.1.

Scanned laser radiation

For measurement of the accessible emission of scanned laser radiation, the current edition of the standard requires measurement with condition 3 only. Exposure to scanned laser radiation with a telescope or a binocular at 2 m distance from the laser scanner, or exposure at close distance with an eye loop, was not considered to be a reasonably foreseeable risk. (It has to be said, however, that this assumption is based on the more common application of scanned laser beams, notably for bar code reading, but there may well be other applications of scanning laser beams in which the possible use of magnified viewing cannot be discounted so readily.) Currently, therefore, scanned laser products are classified on the basis of exposure of the naked eye (condition 3) only, including classification as Class 1 or Class 2, if the corresponding limits are satisfied, even if one of the two other conditions (the 'optical instrument' conditions) would result in accessible emission levels above the respective AEL. As is further discussed in the case study 4.8.4, the scanning pattern that might be produced on the retina (and that can appear to the eye like a line) cannot be directly used to calculate the angular subtense α.

4.3.5 Measurement requirements for extended sources

The measurement requirements that have been discussed in the previous section are adopted for classification of extended sources in the following.

For extended sources, the appropriate determination of the accessible emission can be quite involved. Not only the location but also the angular subtense of the apparent source needs to be determined. For condition 2, the measurement distance depends on the angular subtense α. For extended sources, including multiple sources and arrays, the angle of acceptance of the measurement plays a role as an angle of acceptance that is smaller as the source reduces the accessible emission. Issues related to the angle of acceptance, the determination of the angular subtense α, and evaluation of pulsed emission, multi-wavelength or multi-element sources are discussed in chapter 3, and as they also directly relate to classification when the measurement distance is observed as specified for classification, we do not repeat the discussion here. In the following we discuss the application of the measurement requirements for classification of extended sources as specified in the current version of the standard IEC 60825-1.

4.3.5.1 Condition 1—telescope condition

For condition 1, the measurement distance and aperture diameter does not depend on the angular subtense of the apparent source, but the magnification of the angular subtense of the apparent source can be used to increase the value of the angular subtense that is then used for the determination of the thermal limit. For such a treatment of the source following condition 1, it is important that the (unmagnified) angular subtense of the source α is determined at the same distance as the accessible emission is determined, which is at a distance of 2 m from the exit aperture.

A telescope or binocular with an input optics diameter of 50 mm and an exit pupil diameter of 7 mm has a magnification of 7 times. However, this magnification only plays a role for laser safety or the classification of laser products when the magnification of the image on the retina leads to angular subtenses greater than the minimal angular subtense of 1.5 mrad. For instance, a star (other than our Sun) viewed with the naked eye produces a minimal image at the retina, and when viewed with a telescope (even one with high magnification) the image is still minimal. (When observing stars, the primary reason for using a telescope is not to magnify the stars themselves but to collect more light so that stars can be seen that would be too faint to be seen with the naked eye.) A typical well-collimated, small diameter laser beam also produces a minimal image on the retina and also cannot be magnified with a binocular or telescope beyond the minimal angular subtense of 1.5 mrad. The current version of the standard refers to a magnification of 7, which is the minimum magnification for a binocular or telescope with a diameter of 50 mm. The principle is that the unmagnified angular subtense α, as determined at the appropriate position, can be multiplied by the magnification M, and the magnified value $\alpha' = M \times \alpha$ can be subsequently used in the evaluation of the retinal thermal limit (including the determination of T_2 which is also relevant for the pulse criteria). For this procedure, it is important that the unmagnified angular subtense α is *not* limited to 1.5 mrad, i.e. α is the actual determined angular subtense, which may be less than 1.5 mrad. It is the value of α' which is set to a minimum value of 1.5 mrad in case the magnification does not increase the angular subtense to a value larger than 1.5 mrad. Similarly, it is the value of α' that assumes the maximum value of 100 mrad, so that for a magnification of 7, the corresponding maximum unmagnified value equals 14.3 mrad.

It is very important to note, as is also stated in the standard, that the magnification may be applied to increase the angular subtense only if it can be shown that the retinal spot is also increased by this factor. While beam propagation modelling showed that the retinal spot for a 50 $\times$ 7 binocular is magnified by at least a factor of 7 in respect to the exposure of the unaided eye [3], it can be easily understood that the magnification of 7 is *NOT* appropriate when the beam diameter at the evaluation distance is less than 50 mm, as a smaller binocular (with a smaller magnification) would also intercept the full beam power.

Based on the principles of beam propagation, the following procedure can be used to determine an appropriate magnification factor M. For the case of a diverging beam, where the beam waist is located so that the measurement position of 2 m is in the far-field of the beam (i.e. outside of the Raleigh range), the appropriate magnification M of the retinal image can be derived by dividing the beam diameter by the diameter of 7 mm. The conservative criterion for the determination of the beam diameter for this treatment is to use the 1/e beam diameter definition (i.e. the appropriate determination for a non-Gaussian beam profile would be to determine the second moment beam diameter and to correct this value by division with $\sqrt{2}$). Since some distance away from the Raleigh range, the diameter of the beam waist determines the angular subtense of the apparent source, simple triangular geometry can show that this treatment is equivalent to determining the location where the beam diameter has increased to 50 mm, to determine α there (which is smaller than when determined at 2 m) and to increase this value of α by a magnification of 7. However, if the measurement position is within about three times the Raleigh range, it is important not to use the beam waist diameter to determine the angular subtense α, but to determine the accurate value by measurement or, for a Gaussian beam, by using the beam propagation model.

For converging beams (i.e. when the beam waist is some distance away from the product) and for the case that the measurement position (2 m from the exit aperture) lies in the near field (inside of the Raleigh range), then the above procedure would lead to an inappropriately large angular subtense α. In this case, if the magnification of the retinal image is to be used for increasing the AEL, it is necessary to determine the accessible emission and the unmagnified angular subtense α at the most hazardous exposure position which is located some distance (further than 2 m) away from the laser product. For converging beams, the evaluation position needs to be beyond the external beam waist. At the most hazardous position for a binocular with 50 mm diameter input optics, the angular subtense α can be determined and this value can be increased by a magnification of 7.

For sources such as diffuse reflections and LED arrays that can be treated using geometrical optics (in contrast to laser beams, where beam propagation modelling must be used), it is obvious that for a 50×7 binocular the retinal image will also be magnified by a factor of 7. For such sources, however, it is important to account for the magnification in terms of the angle of acceptance. As for such sources the eye loupe condition (condition 2) will usually be the more critical one, it was overlooked in the current version of the standard to specify a reduced angle of acceptance for condition 1. The angle of acceptance for condition 1 should be the angle of acceptance as specified for condition 3 (see below), divided by the magnification. For instance, for classification based on the thermal limit, the angle of acceptance for condition 1 would be 100 mrad/7 = 14.3 mrad. Biophysically, this reduction of the maximum angle of acceptance reflects that for the naked eye, 100 mrad relates to a retinal diameter of 1.7 mm, which is the (approximate) diameter where the temperature in the centre of the spot does not depend on the actual diameter of irradiation (when the irradiance is the same). Since a telescope magnifies the retinal image, the 1.7 mm retinal diameter that subtends 100 mrad in the eye corresponds to a source angle that is correspondingly decreased by the magnification factor. The necessity for this specification can also be seen when evaluating a LED array at a distance of 2 m that has a diameter of 20 cm: for the evaluation of the naked eye, with an angle of acceptance of 100 mrad, radiation from the full source would be measured through the 7 mm aperture. For such a source, based on the conservation of radiance, the telescope cannot increase the hazard. With the 50 mm aperture and an (erroneously large) angle of acceptance of 100 mrad, the measured power value would be 50 times that of the naked eye, but the angular subtense α cannot be increased beyond the value for the naked eye, as it is already 100 mrad for the naked eye. The correct angle of acceptance of 14.3 mrad, however, results in the same accessible emission as determined with the 7 mm aperture, as the power contribution from only one fiftieth of the source is measured.

The same argument applies for classification against the photochemical retinal limits, where the appropriate angle of acceptance equals γ_{ph}/M (see section 3.6.1 for the definition of γ_{ph}). For sources which are smaller than the angle of acceptance when unmagnified, it is interesting to note that telescopes can actually increase the hazard. Based on the conservation of radiance, one would think that the photochemical hazard, which depends on the retinal irradiance only, is not increased by telescopes. However, it is not the actual retinal irradiance that is relevant for the hazard level, it is the irradiance averaged over an area that subtends the angle γ_{ph}. Consequently, the radiance also has to be averaged over this angle of acceptance and the law of conservation of radiance does not apply to this averaged radiance value. When the source is smaller than the angle of acceptance, a telescope increases the averaged retinal irradiance, as the power that enters the eye is increased by a factor of M^2 but the retinal spot in comparison to the averaging area on the retina as characterized by γ_{ph} is not increased by the same factor (if the magnified image is smaller than γ_{ph} then the effective averaged spot size on the retina is not increased at all and the hazard level is increased by the telescope by a factor of M^2).

4.3.5.2 Condition 2—eye loupe condition

For condition 2, the measurement distance of 14 mm is specified in IEC 60825-1 for small sources only ($\alpha \leq 1.5$ mrad), while for extended sources, the measurement distance is increased up to a maximum of 10 cm for sources larger than 100 mrad, which is the distance of the naked eye condition. The measurement distance can be calculated as a function of the angular subtense α that is determined at a distance of 10 cm from the apparent source. Two different formulae are given for the measurement distance for classification against thermal and against photochemical retinal limits. For the photochemical limits, the formula also depends on the angle of acceptance γ_{ph} which itself depends on the time base. The formulae that describe the dependence of the measurement position on the source size roughly account for the influence of the magnification of the source by the eye loupe.

4.3.5.3 Condition 3—naked eye condition

The naked eye condition is specified as a measurement distance of 100 mm from the apparent source, and the angular subtense of the apparent source is also to be determined at that distance. The distance of 100 mm is based on the assumption of a near point of accommodation of 100 mm (10 cm).

The angle of acceptance for measurement according to condition 3 is equal to the angle of acceptance specified in the MPE section (see also condition 2).

This condition may be changed in future editions of the standard to account for low-diameter, medium divergence beams (less than 100 mrad) where the most hazardous position is some distance away from the beam waist, as described for an MPE analysis in section 3.12.5.9.

4.3.6 Equivalence to MPE evaluation

In the previous section we have discussed the measurement conditions that are to be used for determination of the accessible emission level. The accessible emission level is subsequently compared to the AEL values to determine the class of the product, as already discussed in section 4.2.3.4. Since the AEL for Class 1 and 1M, for Class 2 and 2M and for Class 3R are directly related to the MPE for the eye, the assessment of the accessible emission level and comparison to the AEL values is equivalent in many aspects to the assessment of the effective exposure level and comparison to the MPE for the eye. The difference between a classification and an MPE evaluation is that for classification, specific measurement distances are prescribed, specific time bases are defined and the diameters of the measurement apertures not only reflect the naked eye but also the potential increase of the hazard due to optical instruments. However, the time dependence, wavelength dependence, and the dependence on the angular subtense of the apparent source of the AELs for Class 1, 1M, 2, 2M and 3R is equivalent to those of the ocular MPEs. Only the AEL values for Class 3B, i.e. the test whether a product is Class 3B or Class 4, are not related to ocular MPEs, but these are very much simpler than the other sets of AEL values, so that the classification procedure is more straightforward. Due to this basic equivalence, we refer to chapter 3 and do not repeat the discussion of the appropriate procedures here. Specifically, the following procedures are directly adopted for classification as described for the MPE analysis: for Class 1 and Class 1M, both the photochemical and thermal retinal limits are defined in the visible wavelength range, so that two potentially different emission levels (different because of different acceptance angles) are compared to the two AEL values. For the evaluation of non-uniform sources and arrays, the procedure for classification is equivalent to the procedure described in section 3.12.5.6 and for evaluation against the thermal hazard, it is important that smaller parts of the source and groupings of subsources are considered by using a smaller angle of acceptance and a smaller value of α. Also the classification of pulsed sources is performed with the same criteria as described in sections 3.11.1, 3.12.8 and 3.13.1 where the time base for the respective class replaces the maximum anticipated exposure duration. For non-uniform pulse trains it is important that the evaluation is performed not only for the time base (equivalent to the maximum anticipated exposure duration for an MPE analysis), but also for shorter emission durations (for instance to determine the average radiant exposure and the number of pulses N). The classification of multi-wavelength products should be performed following the procedure described in section 3.14. See also the case study 4.8.5 for discussion of a classification of a product that emits a visible and a near-IR wavelength beam.

4.4 Requirements for the manufacturer

The requirements as defined in IEC 60825-1 that relate to the manufacturer of the laser products can be organized into the following groups:

General hardware: depending on the class, this relates for instance to interlocks, key switches, emission indicators and, of particular importance for embedded laser products, the enclosure.

Labels: depending on the class, the manufacturer is required to affix a number of labels on the product.

Information for the user: the standard requires that the information for the user, i.e. basically the user manual, features specific safety relevant information.

The requirements as specified in IEC 60825-1 apply to all laser products, irrespective of the type and use (which is often referred to as a horizontal standardization). For some specific types of laser products, such as medical lasers or lasers and LEDs used as sources in fibre optics communication, additional requirements are specified in product type specific standards. The above three groups of requirements will be discussed in the following sections, followed with sections that relate to US specific manufacturer requirements, as well as to additional requirements for special product types, such as medical lasers.

Protection against hazards other than those originating from exposure of the skin or the eye from optical radiation is not specifically covered by IEC 60825-1, but it is a requirement in the standard that other relevant product safety standards need to be fulfilled, depending on the type of product. If no specific product safety standard exists that applies to the product, then the provisions as specified in IEC 61010-1, the safety standard for laboratory equipment, shall apply.

In the following sections we present an overview of the requirements and discuss appropriate interpretations, however, not all detailed requirements are reproduced so that this section has to be seen as supplementary information to IEC 60825-1, not a replacement.

4.4.1 General hardware

The standard IEC 60825-1 requires the laser product manufacturer to implement a number of built-in safety features. In the following, we give an overview of the requirements in the same order as they are listed in the standard, and provide some explanatory comments.

Protective housing

Each laser product shall have a protective housing which prevents access to radiation which is not intended to be emitted by the product. In other words, the product has to be designed with one or more specified exit apertures, but otherwise radiation should not be accessible. In this sense 'accessible' is based on the definition of 'human access' as discussed below.

Panels and other removable or opening parts of the housing that provide access to the beam when they are removed or opened, have to be interlocked when they can be opened without the use of tools. If tools (for instance a screwdriver or an allen key) are needed to open the housing, then an interlock is not needed. Panels that provide access to laser radiation also need to be labelled (even for a Class 4 system). However, a simple way around this labelling requirement of the housing is to place a second inner panel inside the actual protective housing so that when the housing is opened, the inner panel prevents access to the beam. It is not required to label this second inner panel. However, as a warning for the service engineer it is good practice to do so.

The design of the protective housing is particularly important if the product is to be classified as an embedded laser product (usually as Class 1), where the embedded laser without enclosure would be allocated to a higher class than that of the enclosed laser product. Specifications for the design of guards which make up the protective housing of a product where the power of the embedded laser is sufficient to potentially destroy the guard when it is incident on the protective housing can be found in IEC 60825-4 and corresponding national standards (see also section 4.6.2 for a discussion of IEC 60825-4). It should be noted that the enclosure has to withstand laser radiation even in the case of reasonable single fault conditions, which include reflections from work pieces, software errors and programming errors for robots. For an embedded laser product, the definition of human access depends on the level of radiation that can become accessible, as discussed above in section 4.3.2.1.

Access panels and safety interlocks

An access panel is part of the protective housing which provides access to the laser radiation inside the housing. When this access panel is intended to be removed or opened for operation or maintenance by the user *and* if removal of the panel gives access to laser radiation levels as listed below, then the access panel has to be safety interlocked. The table in the standard requires an interlock in the following way:

- if the laser product is to be classified as Class 1, 1M, 2, or 2M, then panels require an interlock if the accessible radiation is of hazard level of 3R, 3B or 4;
- if the laser product is to be classified as Class 3R, 3B or 4, then it requires a panel interlock if the accessible radiation is of hazard level of Class 3B or 4.

The panel and the interlock has to be designed in a way that prevents access to radiation through the opening in excess of Class 1M or 2M depending on the wavelength (meaning that the measurement requirements are those for the naked eye and the measured accessible emission level is compared to the AEL of Class 1 and Class 1M for non-visible radiation and Class 2 and Class 2M for visible radiation). In effect, the interlock has to terminate the laser radiation when the panel (the lid, the door) is opened. If the interlock is not actuated immediately and the laser radiation is terminated with some delay or if the interlock has some slackness so that there would be an opening (a slit) before the laser emission is terminated, the easiest engineering solution is to form either the housing or the panel with an apron or bracket so that radiation cannot be directly emitted through the gap.

Once an interlock has been activated and laser operation is terminated (such as by the opening of an access panel while the laser is operating), simple resetting of the interlock (by merely closing the access panel again) should not itself restart the laser unless the laser is in a safe condition. A deliberate, separate, reset action by the user should be required.

Either a safety interlock has to be failsafe, i.e. to be of such a quality that a failure of the interlock results in a safe condition (i.e. the laser radiation is terminated), or the interlock has to be redundant, i.e. at least two interlocks must be used. (We should assume, at least, that this is the intention of the standard, as it defines what a fail safe interlock is, and although it does not actually require that such an interlock is used this is covered by the general requirement to consider single fault conditions.)

When an override mechanism is provided for maintenance and service, which renders the interlock inoperative so that the laser emission is not terminated when the panel is opened, the following requirements are specified in the standard:

- the manufacturer shall provide adequate instruction about safe methods of working;
- the system has to be designed so that when the override is in operation, the access panel cannot be returned to its closed (normal) position. This is in practice often fulfilled by using an override key or jack that needs to be inserted in a socket so that the key protrudes from the housing and prevents the panel from being closed;
- a label (see below) shall be clearly associated with a readily overridable interlock;
- use of the override has to result in a distinct audible or visible warning whenever the laser is energized or (in the case of pulsed lasers) whenever the capacitors are not fully discharged. Visible warnings have to be still visible when protective eyewear is used—for instance, a red warning light for a red laser is not appropriate, as the corresponding laser eye protection will make the red warning light invisible.

It is general good engineering practice to reduce the accessible laser radiation as much as possible when the laser product is put into the override mode. For instance, the power of the laser can be automatically reduced to a level which is just sufficient to perform the task which makes it necessary to override the interlock.

The authors strongly recommend to manufacturers that they handle the provision of override keys with a degree of caution appropriate to the level of accessible laser radiation that is present when the interlock is overridden. Good product engineering practice and most product safety legislation requires that both operation and user maintenance is safe. When an override key that allows access to radiation with hazard levels equivalent to Class 3R, 3B or Class 4 is provided, the manufacturer should ensure appropriate training before the key is handed over. Based on general product safety regulation and especially for laser processing machines, the system should be designed so that user maintenance procedures do not require access to the beam after overriding the interlock.

Remote interlock connector

Each Class 3B and Class 4 laser product has to feature a remote interlock connector. This is usually a socket at the back of the laser which when the laser is delivered has a jack plugged in where the terminals of the connector are internally short-circuited. This connector can be connected to a door interlock or to an interlock to some guarding or housing so that when the guarding or the door is opened, the laser radiation is automatically terminated. According to the standard, when the interlock is opened, accessible radiation shall not exceed Class 1M or Class 2M depending on the wavelength, however, in practice the laser radiation will be terminated completely (either by switching off power to the laser itself or by operating a shutter).

While it is not required by the standard, it is nevertheless recommended as good engineering practice that, after the circuit has been broken through the operation of a remote interlock that is connected to the remote interlock connector, restarting the laser should require the use of a separate, deliberate reset action unless the laser is in a safe condition when the interlock is closed. Where this is not the case then, for example, while laser emission may be terminated when an interlocked door into a laser room is inadvertently opened, laser emission will restart as soon as the door is closed, clearly not a very sensible safety arrangement if the unauthorized person who opened the door is now inside the room!

Key control

Each Class 3B and Class 4 laser product has to feature a key-operated master control, which is often in the form of a key switch. Instead of a real switch, passwords and magnetic cards are an alternative. The key control is a good example for the importance of user controls to make the product actually safe (classification and compliance with manufacturer's requirements by themselves do not): it is the manufacturer who has to provide a key control for each Class 3B and Class 4 laser product, but it is up to the user to responsibly use this engineering safety feature by removing the key from the key switch and by storing the key in a way such that untrained and unauthorized personnel cannot operate the laser. Too often, one finds laser keys left in place when a laser is not being used; the key control then serves merely as a convenient on-off switch and not as a safety feature as intended by the safety standard.

Laser radiation emission warning

With emission warning, the standard means an audible or visible warning when the laser system is switched on or if the capacitor bank of a pulsed laser is not fully discharged. Thus, contrary to the title of the paragraph in the standard ('laser radiation emission warning') the warning has to be activated independently of the actual emission, it merely refers to the laser system being switched on. (For medical lasers, there has to be a warning that actually indicates emission of laser radiation, as well as a warning that indicates the 'ready' status of the laser product.) Visible warning has to be visible through protective eyewear (see above on warning for override status of an interlock). Class 3R laser products only need an emission warning when the emission is outside of the visible wavelength range. Class 1, 1M, 2 and 2M laser products do not need an emission warning. It is interesting to note that the standard requires a fail safe or redundant emission warning device. While a LED can be practically considered fail safe due to the long lifetime, a simple light bulb or audible warning cannot really be considered fail safe, as this would mean that in failure mode, the warning would have to continue or laser operation be prevented. Consequently, such warning devices would need to be redundant or to incorporate failure detection to prevent laser operation.

Beam stop or attenuator

Each Class 3B or Class 4 laser product has to feature some means of interrupting the emission of laser radiation without actually switching off the laser or the energy to the laser. Usually this requirement is fulfilled by providing a shutter or beam stop that can be either operated manually or automatically, for instance when the laser is first in the stand-by mode and the actual laser emission is activated by an additional command, foot switch or button, and upon release of the activator, the laser product is again in the stand-by mode. For very high-power Class 4 lasers, however, more sophisticated water-cooled beam stops must be employed which are capable of withstanding the laser beam without damage.

Controls

The controls of the laser product have to be located so that operating the controls does not make exposure to levels of radiation equivalent to Class 3R, Class 3B or Class 4 necessary. Basically, this requirement means that the controls, possibly including some display which is only visible from one direction, should not be located next to the laser aperture.

Viewing optics

Any viewing optic or window incorporated in the laser product shall be designed so that human access above levels of radiation equivalent to Class 1M is not possible. This applies to all laser classes, i.e. even a Class 4 system, when it features some (partial) housing, and this housing features some viewing window. It also applies for instance to microscopes that are built into the system (for instance for micromachining purposes). The way of achieving the requirement is either via permanently installed attenuating filters which reduce laser radiation to sufficiently low levels, or by a variable shutter or attenuator that is incorporated in the viewing optics (for instance in a microscope) so that the shutter closes the viewing optics when the laser is on. Requirements regarding resistance against potential laser radiation incident (including in failure modes) are specified in IEC 60825-4 (see also section 4.6.2).

Scanning safeguard

Laser products that are classified based on some scanning motion of the emitted radiation (which results in a pulse pattern as the accessible emission) have to feature a scanning safeguard which terminates or reduces laser emission so that a scan failure or a reduction of the scan speed does not lead to accessible emission levels above the associated class. This requirement should be seen in conjunction with the general requirement that classification includes the single fault condition as discussed in section 4.3.3.

Alignment aids

The manufacturer has to provide safe means of performing alignment of beam path components (lenses, mirrors) when this is part of routine user maintenance. For instance, mirrors and lenses could be adjusted by screws which are located so that the beam remains within a tube or protective guard. Alternatively, the laser power can be reduced appropriately or a lower power visible laser beam can be provided as an alignment aid.

Walk-in access

When the laser product is large enough so that the enclosure is basically a cabin, and a 'door' is provided which provides 'walk-in' access, then the inside of the laser product has to have the following features. Where the laser inside the housing emits radiation so that levels which would be classified as Class 3B or Class 4 could become accessible, then there has to be a facility provided so that the person inside the enclosure can prevent laser emission from occurring. This can be by means of a switch or a mechanical beam stop, for example, but there should also be a means of terminating unwanted emission, should it occur, such as by use of an emergency stop button. The second requirement that is necessary is a warning device that has to be placed inside the housing so that it provides adequate warning (for instance acoustic or visual) of emission of radiation. This requirement is not only necessary when the accessible radiation inside the product is equivalent to Class 3B or to Class 4 level of hazard but also for non-visible Class 3R levels.

It is important to note that these requirements for a walk-in access do not relate to the classification of the product, i.e. they are required independently of whether the product is actually classified as Class 3B or Class 4 or whether it is classified as an embedded laser product of Class 1. For instance, a laser product could be classified as Class 3B or Class 4 even if it has an enclosure (big enough to provide walk-in access) because one or more requirements necessary for classification as an embedded Class 1 are not fulfilled. The above requirements should not be misinterpreted as being sufficient in order to classify the product as an embedded Class 1. Classification is based on human access and when it is possible to go inside the housing and close the door, and there are no means provided to automatically prevent emission when a person is inside, then such a product cannot be an embedded Class 1 laser. If such a product is to be classified as embedded Class 1 (or it could be Class 2 or Class 3R, when an alignment laser is used inside the product which is not affected by the automatic prevention of emission of the higher power laser), then appropriate automatic detection has to be installed to reliably detect if a person is inside. This can be realized for instance with pressure sensitive floor mats. The design needs to be fail-safe or redundant. For classification, it is not appropriate to replace such an automatic system of preventing access to the radiation by procedural controls, such as a lock system where each person with permitted access has a separate key and a separate lock is placed upon entry such that the door cannot be closed. (Obviously, the door needs to be interlocked, and if an override mechanism is provided, then this also needs to fulfil the requirements as discussed above.) While such a key system can be sufficient to satisfy legal requirements for machine and work place safety, it is not sufficient for classification of the laser product as Class 1. Laser classification is based on the design of the product independent of user procedures to prevent human access above the respective AEL, and the product has to be classified accordingly. This discussion also relates to the issue that a laser processing machine can be Class 4 but satisfies the general requirement for a 'safe' machine, which is further discussed in section 4.6.1.

If a product with walk-in access is classified as an embedded laser, overriding the interlock and overriding the automatic prevention of emission when somebody is inside is not allowed to be a procedure which is specified as a condition of user operation of the product. However, following current classification rules, the override might be specified for maintenance procedures to be carried out by the user, although the authors do not recommend this and it might also violate other standards, such as the machine safety standard or product safety legislation, as discussed in section 4.3.2.2. Overriding safety mechanisms so that hazardous levels of radiation can be accessed should only be necessary for service procedures and the override key should be handed over by the manufacturer only to appropriately trained personnel that are authorized by the manufacturer.

4.4.2 Labels

A number of labels are specified in IEC 60825-1 to be carried by the laser or LED product. The wording of the labels needs to be in the appropriate language of the country where the product is sold. Although this is not specifically stated in the standard, however, it is a usual requirement of general product safety legislation, and also follows from the translation of the laser safety standard when it is adopted as national standard so that the label would have to comply to the national edition of the standard that is valid in the country where it is sold. In the following, we give examples of warning labels in the English language. Because of the language issue and because it is difficult to describe the level of hazard that is represented by the various classes in a short statement, alternative, simplified forms of labelling that may require more detailed safety information in the user manual are under consideration.

As throughout the standard, for a LED product, wording containing ‘laser’ also applies to LEDs, so that also in the warning labels, for an LED product, ‘laser’ is replaced by ‘LED’.

The background of the labels needs to be yellow and the border and the wording black, with some exceptions as stated below. It is obvious (and it is specified in the standard) that the labelling should be permanently fixed, legible and properly positioned, but it is surprising what can be encountered:

- classification and explanatory labels positioned at the back of the laser product (which is usually placed against a wall);
- labelling that was poorly fixed so that it came off following standard cleaning (particularly in medical environments);
- the writing and black border was completely removed by standard cleaning (i.e. the label was a bare yellow rectangle).

If the product is so small that labelling is impractical, the required labels can also be provided along with the user information or it should be on the package of the product.

A product that is exempt from the laser safety standard as discussed in section 4.3.2 (when it is not embedded and the accessible emission level does not exceed the AEL for Class 1 including under fault conditions and service) then the product is also exempt from any labelling.

4.4.2.1 Laser hazard warning label

Each LED or laser product, except for Class 1 and Class 1M, has to bear the laser hazard warning label in the form of a triangle. The design of this label is standardized in IEC 60825-1, and is reproduced in figure 4.8, but it might be adjusted in relative size.

Figure 4.8. The laser hazard warning label as defined in IEC 60825-1 (it needs to be on a yellow background).

4.4.2.2 Explanatory label

An explanatory label with a basic shape as prescribed in IEC 60825-1 (again black on a yellow background) needs to contain the following warnings and information. For Class 1 and Class 1M, instead of labels on the product, the corresponding statements may be included in the information for the user (i.e. in the user manual). For a Class 1 laser or LED product, when the manufacturer chooses to affix the label on the product, then it does not need to be black on a yellow background.

Classification and corresponding warning wording for each class

Specific wording is defined for each class which attempts to describe 'what should not be done', for instance for a Class 1M laser product that emits in the visible wavelength range:

LASER RADIATION
DO NOT VIEW DIRECTLY WITH OPTICAL INSTRUMENTS
CLASS 1M LASER PRODUCT

As already mentioned, for a LED product, the wording would be LED RADIATION instead of LASER RADIATION. When the emitted laser radiation is in the non-visible part of the spectrum, the wording would be INVISIBLE LASER RADIATION instead of LASER RADIATION. If the emitted radiation is both in the visible and in the non-visible, and the level of accessible emission in the non-visible is above the AEL for Class 1 (i.e. measured with the 'optical instruments conditions') then the wording has to be VISIBLE AND INVISIBLE LASER RADIATION. These regulations also apply to the aperture label and the label for access panels as discussed below.

Radiation output properties

The explanatory label needs to contain information on the maximum radiative output of the product, as well as the wavelength. When the output is pulsed, the maximum energy per pulse and the pulse duration needs to be specified. The information given here usually does not suffice to calculate the NOHD or the correct optical density of the protective eye-wear, but is helpful as basic information. This information is not necessary if it is a Class 1 laser.

Classification standard and date of publication

The number of the laser safety standard to which the classification and the engineering requirements comply, as well as the date of publication of this standard is to be specified.

An example for an explanatory label for a surgical CO_2 laser with a visible aiming beam is:

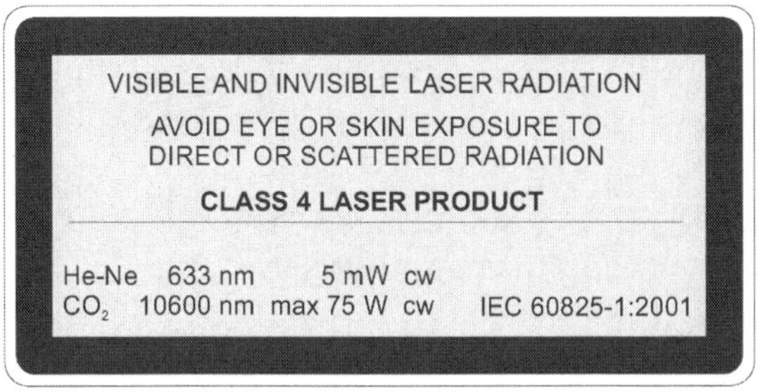

There are no specifications for the presentation of the information, for instance regarding units, or whether the actual type of laser is to be included in the information. The label shown should also not be understood as a suggestion for a properly designed label, it was actually found on a surgical laser (however, the year of publication of the standard was missing, this was added for the presentation). It is also quite common to find that the class warning and the radiation output properties are on two separate explanatory labels. In that case in the view of the authors it is more appropriate to include the information on the classification standard together with the class warning.

4.4.2.3 Aperture label

For Class 3R, 3B and Class 4 laser and LED products, an aperture label is to be affixed close to each aperture through which laser or LED radiation in the excess of the AEL for Class 1 or Class 2 is emitted. The manufacturer has the choice between two wordings, one is

LASER APERTURE

and the other is

AVOID EXPOSURE—LASER RADIATION IS
EMITTED FROM THIS APERTURE

On products with bare fibres, the label is often affixed next to the fibre connector on the laser product housing and the wording is often adopted to 'Laser aperture at the end of the fibre'. There are special regulations for medical laser products as discussed in section 4.2.7.

4.4.2.4 Access panel label

Each access panel, which includes the connecting parts of the housing or enclosure, which when removed or otherwise displaced permits access to laser radiation in the excess of the AEL for Class 1 shall have a label affixed that starts with 'CAUTION' and then states the type (the class) of radiation that is accessible when the panel is removed (irrespective of the actual class of the laser product), and then also contains the wording that is the usual warning wording for the respective class. An example is:

CAUTION—CLASS 2 LASER RADIATION WHEN OPEN
DO NOT STARE INTO THE BEAM

(In the opinion of the authors this is a really a misuse of the concept of classification, since you strictly cannot have 'Class 2 laser radiation' but only a 'Class 2 laser product'. It would be preferable, as was the case with the earlier edition of the standard, to omit the class from this label but to keep the corresponding warning, i.e. 'Caution—Laser radiation when open. Do not stare into the beam'.)

Interlocked access panel label

When the access panel as described in the previous paragraph is interlocked, and the interlock can be readily overridden, then labels have to be placed in close proximity to the opening that is created when the panel is removed. The label(s) have to be visible prior to and during the interlock override. The wording of the label is the same as for non-interlocked access panels as described in the previous paragraph, but an additional line is added:

CAUTION—CLASS 2 LASER RADIATION WHEN OPEN
AND INTERLOCKS DEFEATED
DO NOT STARE INTO THE BEAM

4.4.3 Informational requirements

The standard IEC 60825-1 specifies a detailed list of safety related information that needs to be contained in the user information, i.e. in the user manual. The required information includes detailed parameters regarding the laser source, reproductions of the labels, a description of the location of the labels on the product and specifications regarding the safe use and maintenance of the product. For the case of a Class 1M or Class 2M laser product, the standard also requires a warning regarding what type of optical instrument can lead to hazardous exposure levels. It is sometimes overlooked by manufacturers that compliance with the standard not only means to classify the product, label the product and incorporate safety features such as key switches and interlocks, but also the provision of adequate safety information, as detailed in the safety standard.

Detailed safety relevant information also has to be contained in the service manual.

4.5 US requirements

CDRH is the governmental agency within the FDA in the USA that regulates commercial distribution of laser equipment. The system is based on a code that enforces both classification and engineering features on the product as well as the registration of the product by the manufacturer with the agency. As such, the CDRH code applies only to the manufacturer. Additionally, in the US, the American National Standards Institute, ANSI, develops and publishes laser safety standards that apply to the laser user. The laser safety standard series is called ANSI Z136 and the base document is ANSI Z136.1. This document not only contains guidelines for the user but also contains a classification scheme that is somewhat different to the current CDRH scheme and to IEC 60825-1. However, at the time of writing this book, the responsible ANSI committee started to work on the adoption of the IEC classification scheme, also in the light of the statement by the CDRH that it will adopt the IEC classification scheme in their current revision of the CDRH standard. The inclusion of a product classification scheme into an ANSI user document where there is a legal requirement to follow the CDRH standard (with a different classification scheme) has the background in users building their own laser products for in-house use or users changing the laser so that the classification also changes. However, for the manufacturer, for selling the laser in the USA, the CDRH standard is obligatory.

4.5.1 Registering laser products in the US

Laser products that are sold or offered for sale in the USA must comply with the safety requirements of the US Federal Laser Product Performance Standard 21 CFR 1040.10 and 1040.11. They must also carry certification and identification labels in accordance with 21 CFR 1010.2 and 1010.3, and be reported to the

Center for Devices and Radiological Health (CDRH), a division of the US Food & Drug Administration, in accordance with 21 CFR 1002.10. All these requirements are defined in Title 21 of the Code of Federal Regulations (CFR), Subchapter J—Radiological Health. The main exceptions to these requirements are lasers that are sold to other manufacturers as components for incorporation into a product which will itself be subject to these requirements, or laser products intended only for export (in which case they must satisfy the requirements of the importing country).

Under the provisions of the Federal Laser Product Performance Standard, which are mandatory, manufacturers must design and manufacture their products to be in compliance with the standard, test their products to ensure compliance, certify that each product complies with the standard, maintain records of tests, sales and all issues relating to the radiation safety of their products and submit product reports to CDRH.

Reports have to be submitted to CDRH for each new product, or for each modification to an existing product, demonstrating that the product complies with the Federal standard. These reports must be prepared in a format specified by CDRH, and be accompanied by any additional documentation, such as drawings, descriptions and test data, that is needed to support the claim of compliance being made by the manufacturer. Laser product reports are in effect detailed answers to a specific list of questions appearing on the report pro forma, which is available from CDRH. These questions are arranged under ten separate sections, and cover the following topics. The answers given must demonstrate how the product complies with the relevant requirements of the Federal performance standard.

Part 1: Manufacturer and report identification (including, where appropriate, details of the importing agent).

Part 2: Product and model identification (i.e. the product name and model number, including details of any incorporated laser product).

Part 3: Compliance with the labelling requirements.

Part 4: Compliance with the informational requirements (i.e. user instructions).

Part 5: Description of the product (including details of the beam path and the power or energy levels, and the procedures for operating, maintaining and servicing the product).

Part 6: Levels of accessible laser radiation and classification of the laser product (i.e. the class that is claimed and the justification for it).

Part 7: Compliance with the performance requirements (e.g. protective housing, safety interlocks, remote interlock connector, key control, emission indicator, the specification of any necessary protective eyewear that must be worn while operating the product, beam attenuator, location of controls, viewing optics, scanning safeguard, manual reset, plus special requirements for medical, surveying, levelling, alignment and demonstration laser products).

Part 8: Quality control tests and testing procedures (during production).

Part 9: Life and endurance testing (demonstrating that the product will remain in compliance throughout its useful life).

Part 10: Instrumentation and calibration (describing the equipment used for carrying out measurements and justifying the conclusions reached).

The emphasis of the reporting process is to provide the necessary documentary evidence to support the claim of compliance with the CDRH requirements, and to provide convincing arguments that every product shipped, not just a prototype that has been tested, will conform to the Federal standard. After receipt of the report, CDRH issues an accession number for the reported product. It should be noted, however, that the reporting process relies on a claim of compliance by the manufacturer. While CDRH may intervene and request further documentation, ask for certain changes to be made to the product, or even require its withdrawal, CDRH does not 'certify' a product as approved, and may raise concerns over its safety compliance at any time.

Historically, the US Federal standard has differed in a number of respects from the requirements of the international laser safety standard (IEC 60825-1). The system of classification, though similar, uses a different nomenclature. The CDRH laser product classes are Class I, Class IIa, Class II, Class IIIa, Class IIIb and Class IV. There are differences in some class limits, compared with IEC, and CDRH also adopts a different measurement geometry for classification.

CDRH uses a very conservative time base for assessing Class I (10 000 s in the retinal hazard region), but has an additional class, Class IIa, covering only visible emission, for which a time base of 1000 s is applied. Class IIIa (up to five times the 1 mW limit for Class II) is also restricted to visible emission. The visible band is defined in the Federal standard as the wavelength range 400–710 nm, not 700 nm as used in the IEC standard.

The effect of these differences is that while many lasers—the majority of those in Class I (when based on complete enclosure), Class II, Class IIIb and Class IV—have the same equivalent classification under the IEC system, others do not. There are also significant differences in the warning labels, and in some aspects of the performance requirements. These include the requirement that any laser incorporated into the product that is removable and capable of operation without modification while removed shall itself be subject to the Federal standard, and also the provision of a manual reset for all Class IV laser products to enable resumption of laser emission after interruption by a remote interlock or by the temporary loss of electrical power.

The differences between CDRH and IEC requirements, particularly those of classification and labelling, are not as widely recognized as they ought to be, and it is unfortunately common practice for laser products that have been classified and labelled solely to US requirements to be sold in Europe and elsewhere outside of North America without reassessment, modification or relabelling.

4.5.2 Changes to CDRH requirements

The issue of dual classification and dual labelling for laser products in an international market has been a problem for laser manufacturers for some time. Recognizing this, CDRH announced its intention to modify its requirements to bring them more into line with those of IEC. In January 2001 CDRH published interim guidance [4] indicating that it will not object to laser products being sold in the United States that conform to certain requirements of IEC 60825-1 in lieu of those specified in 21 CFR 1040, including the revised IEC system of classification and labelling. Specifically, it stated that it will accept the IEC equivalent of the following CDRH requirements.

- Definitions
- Classification
- AELs
- Tests for determination of compliance
- Protective housing
- Safety interlocks
- Remote interlock connector
- Key control
- Laser radiation emission indicator
- Beam attenuator
- Location of controls
- Viewing optics
- Scanning safeguard
- Labelling requirements
- User information
- Medical laser products

Other existing requirements of the Federal Laser Product Performance Standard are, however, retained, even where they differ from those of IEC 60825-1. This is because they are beyond the scope of the IEC standard, are sufficiently different from the equivalent IEC requirements, or are included in the IEC standard only as recommendations in the user's guide section rather than in the normative section. Those requirements where the provisions of the Federal standard will continue to apply are as follows.

- Certification (the requirement that laser products must be reported to CDRH and carry a certification label indicating compliance)
- Identification (a label on the product indicating the name and address of the manufacturer and the place and date of manufacture)
- Variances (the procedure by which CDRH may grant variances from the provisions of the Federal standard where this is considered by CDRH to be justifiable)
- Applicability (the type of products that fall within the scope of the Federal standard—essentially all laser products sold, other than components such as

laser diodes sold without power supplies, which must nevertheless still be registered and listed with CDRH)

- Removable laser systems (the requirement that any incorporated laser, if capable when removed of being operated without modification, shall itself comply with the Federal standard)
- Manual reset mechanism (required where termination of laser emission occurs through an unexpected loss of electrical power or through the operation of a protective interlock)
- Purchasing and servicing information (detailed informational requirements as defined in the Federal standard)
- Modification of a certified product (any previously certified product that undergoes modification by a laser-product manufacturer must be reported to CDRH and re-certified)
- Surveying, levelling and alignment laser products (these can be any Class up to IIIa, which the authors interpret as covering IEC Classes 1, 2 or 3R)
- Demonstration laser products (same as for surveying levelling and alignment products)

It is to be hoped that this trend towards the complete harmonization of international requirements in laser product safety will continue, and that the remaining differences between the IEC and CDRH standards can, in the not too distant future, be eliminated.

4.6 Enclosure and classification

For high-power lasers as used in laser materials processing, appropriate guarding is the primary means to ensure safety regarding optical radiation. When the guarding prevents any human access to laser radiation during use, it is possible that the product is classified as embedded laser products of Class 1. We summarize issues related to such a classification and to frequent misunderstandings in the following section. This section is followed by a more detailed discussion of IEC 60825-4 which specifies requirements for guards. These requirements are not only relevant if the guard is part of the enclosure of an embedded laser product of for instance Class 1. However, even if the laser processing machine is classified as Class 4, based on machine safety standards and in some countries legislation, guarding is necessary to reduce the hazard to an acceptable level. Also in this case, the user will expect the guard to comply to corresponding requirements and provide adequate protection from exposure to laser radiation.

4.6.1 Embedded Class 1 laser products—nice but not necessary!

It is possible to classify laser materials processing machines as Class 1. However, a number of requirements that are specified in IEC 60825-1 need to be fulfilled.

These requirements are all discussed in the previous section, here we only give a summary and try to place the meaning of non-compliance, i.e. of classification as Class 4, into perspective. For the following discussion it is important to distinguish the classification of a laser product according to IEC 60825-1 on the one hand and the 'classification' as 'safe' installation or machine following good engineering practice and a risk analysis on the other.

Classification according to IEC 60825-1 is clearly based on human access (which is stringently defined) and on any reasonably foreseeable single fault condition. To prevent human access, which for a laser processing machine also includes possible reflections from inside the housing through any opening in the housing, and also includes single fault conditions, the enclosure needs to be complete. For the case of walk-in access, for classification as Class 1, the design needs to prevent radiation from becoming accessible inside the laser product, as discussed above. When a laser product is classified as Class 1, then in terms of user control measures this means, that the product is so safe that the user basically does not need to know that it actually is a laser product, i.e. no user control measures, and no safety related training is necessary (at this stage of the argument, we assume that the classification includes user maintenance, which is currently not the case).

We would like to clarify here, that it certainly is a meaning of an appropriately classified Class 1 laser product, that it is safe for use (and in future editions of the standard it is expected that it will also mean safe for maintenance), but in terms of acceptable levels of safety for a specific application and product, it is often not necessary, impracticable and too expensive to make the machine a Class 1 laser product in order to fulfil the basic requirements of machine safety. However, the reverse logical is not appropriate, i.e. it is certainly not justified to classify a laser product as Class 1 merely based on the result of a general risk analysis (which is nonetheless important to perform, and generally more important than the actual classification according to IEC 60825-1).

The general goal with respect to materials processing laser installations is to enclose the installation by protective shielding and guards to prevent exposure to the laser radiation. This is also supported by the European machinery directive which requires that no hazardous radiation is emitted from the machine. However, this does not mean that a materials processing laser *has* to be a Class 1 product in order to fulfil the requirement of the machinery directive. In order for a laser product to be classified as Class 1, all of the requirements as specified in IEC 60825-1 have to be fulfilled, however it might not be necessary for a laser installation to fulfil all of these requirements and still be considered as safe for the general use of the machine in the sense of the machinery directive. For instance, the enclosure of the laser might not be complete, i.e. the guards around a materials processing installation could be limited to vertical guards around the installation, but there might be no 'roof'. While in terms of safety on the factory floor this might be fully sufficient, the product cannot be classified as Class 1. Also regarding walk-in access, Class 1 would require prevention of access by design, not only by organizational procedures on the part of the user. An additional aspect is the consideration of single fault conditions and the design of the standard. While the standard for guards specifies classes for guards that are deemed appropriate based on the inspection interval, we would like to argue that this kind of classification for guards with respect to inspection intervals cannot be used as time base for classification. The guards of a laser processing machine that is classified as Class 1 need to withstand any reasonably foreseeable exposure (including single fault conditions) independent of the inspection interval. That is, the guard has to withstand corresponding levels of laser radiation until the laser emission is *automatically* terminated, or it has to withstand the radiation indefinitely. Again this follows from the principle that classification is based on design of the product, not on behaviour of the user. For the manufacturer it might be difficult to accept that a well-designed machine that has a high level of safety standard is 'penalized' by not being classified as Class 1. However, when user requirements are treated reasonably, Class 4 usually 'only' means that the safety of the installation needs to be evaluated in a risk analysis (usually by the laser safety officer) and appropriate user controls are adopted following this risk analysis. The result of the risk analysis might well be that no further control measures are necessary except usually some level of training of the 'man at the machine' regarding do's and don'ts (for instance when walk-in is possible and emission is not automatically prevented). The problem here is not the classification procedure as currently specified in IEC 60825-1, but that Class 4 is considered as generally extremely hazardous and as automatically requiring everybody to wear eye protection, etc. In that sense, it should be rather seen as normal that a laser processing machine is rather Class 4 than Class 1, but still be designed safe for the intended use. What is important is that for the specific machine, operation and maintenance can be performed with an acceptable level of safety, as specified for instance in ISO 11553, irrespective of the class. If the laser machine can be designed to Class 1, then this is 'nice', but it is in many cases not necessary.

4.6.2 Requirements for laser guards IEC 60825-4

Requirements for laser guards that enclose the process zone of an industrial laser processing machine are specified in IEC 60825-4 (Safety of laser products, Part 4: Laser guards). These requirements apply both to permanent enclosures and to those used temporarily (e.g. for servicing), and cover all component parts of the guard, including transparent viewing windows, laser curtains and walls, where these form part of the protective enclosure.

Two important parameters are defined in the standard for laser guards. The first is the foreseeable exposure limit (FEL), which is the maximum level of laser exposure at the inner surface (i.e. laser side) of the guard, that may occur (or can be anticipated) under normal and reasonably foreseeable fault conditions. The term 'limit' is somewhat misleading, as it is not really a limit but the maximum foreseeable exposure that may occur at the inside of the guard for a specific laser set-up and application. The term 'maximum foreseeable exposure' would be more appropriate.

The second parameter defined in the standard is the protective exposure limit (PEL), which is the maximum level of exposure that the guard can withstand at its inner surface in order that the accessible level of radiation at the outer surface does not exceed the AEL for Class 1 (where appropriate aperture diameters are implied). A maximum period of exposure also needs to be specified, since the PEL will typically be lower for a longer period of exposure. The assessment of a guard's PEL must take into account not only its resistance to laser-induced damage but also the optical transmission of the guard at the laser wavelength.

To be suitable, a guard must have a PEL (determined by testing of the guard material) that is higher than the FEL (assessed through an evaluation of the laser process). The PEL is a property of the guard, and for a given guard material and thickness depends on the wavelength, exposure duration and also on the beam diameter, since for the same level of irradiance at the guard, a larger beam diameter will lead to earlier burn through than a smaller beam where radial cooling lowers the temperature. The FEL is the exposure that may occur at the inner surface of the guard for a given laser and process. A guard may be characterized in general terms as a proprietary laser guard, where the guard manufacturer specifies the PEL together with the exposure duration and laser wavelength (using a standardized beam diameter as specified in IEC 60825-4), or a guard can be tested for integration into a specific machine, where it should be tested at the level of the FEL. Where it is being supplied as a proprietary laser guard for use in any appropriate machine, then it must be tested at the level of PEL that is specified for the guard. A proprietary guard must be labelled to indicate the full specification of the PEL, and adequate user information supplied with it. (Different PELs may need to be specified to cover several exposure conditions, e.g. for different wavelengths and for pulsed and cw lasers.) The PEL has to be set at a value of no more than 0.7 times the tested exposure. Where a guard has different components (such as incorporating a viewing window), then the different parts of the guard can have different values of PEL. Testing may be carried out using samples of the guard material of at least 50 mm in size.

The exposure duration for which the PEL needs to be larger than the FEL can depend on the kind of application and the degree of surveillance carried out (i.e. in order to identify the existence of an errant beam that is incident on the inner surface of a guard). Guards are classified according to the maintenance inspection interval. These classifications are as follows.

Class T1 30 000 s inspection interval (e.g. for automated machine usage).
Class T2 100 s inspection interval (e.g. for short-cycle operation with intermittent inspection).
Class T3 10 s inspection interval (for processing under continuous surveillance).

Following the evaluation of a specific process, taking into account possible single fault conditions (for example a programming error where the laser is pointed at the guard and not at the processing materials), the required PEL might be very high, especially for Class T1. For very high levels of irradiance or radiant exposure it could be difficult, if not impossible, to fabricate a suitable guard that could withstand exposure to an errant beam for a sufficient length of time. In order to deal with such a high FEL, it is possible to utilize active guards. These incorporate sensors to identify the presence of a laser beam on their inner surface and so generate a signal to terminate laser emission. One example of an active guard is the use of pressurized double-skin assemblies in which an errant beam burns through the first layer, causing a sudden drop in pressure which is detected by a pressure sensor. Infrared sensors are also sometimes employed. These scan the inner surface of the enclosure and identify hot spots caused from heating by the incident beam. All active guards must, of course, prevent break-out by the laser beam for at least the time required for the system to sense that a fault has occurred and to shut down laser emission. This time interval is known as the laser termination time. For active guards, it is, however, a requirement that any reasonably foreseeable fault within the guard system must not result in the loss of its safety function.

4.7 Application specific requirements

International standards that contain product type-specific manufacturing requirements are for instance for medical laser products IEC 60601-2-22 [5], for fibre optic telecommunication devices in IEC 60825-2 [6] or for laser machines ISO 11553 [1]. Examples of additional engineering requirements for medical lasers are displays of the emitted power or energy and an emergency switch, or for optical telecommunication installations the use of automatic shut-down procedures should their be a cable break.. These specific standards are briefly reviewed in the following section. Additional product type specific standards are in development, such as for hand-guided laser processing. It is important to note that any product type specific standards usually only specify *additional* requirements to IEC 60825-1, rather than replace the requirements that are specified in IEC 60825-1 (but they can in some cases specify equivalent safety features, such as the ‘stand by’ status for medical lasers in place of the shutter that is required in IEC 60825-1). Not only product standards require engineering features, it is possible that certain safety relevant engineering features are also required by national legislation, such as the directive for hazardous work equipment requires a panic button (mushroom type switch) for hazardous work materials. While not specifically listed in the directive, most Class 3B and Class 4 laser products would count as hazardous work equipment, thereby for instance making panic buttons necessary, even if they are not required in IEC 60825-1.

4.7.1 Laser processing machines (ISO 11553 and EN 12626)

Laser processing machines, i.e. high-power laser equipment used in industry for performing materials-processing tasks such as cutting, drilling, welding and surface treatment, are subject to the provisions of ISO 11553 (Safety of machinery—Laser processing machines—Safety requirements) in addition to the requirements of IEC 60825-1 and IEC 60825-4. In Europe this standard is currently published as EN 12626, with the addition of some referenced standards that are only applicable in Europe. EN 12626 is used to demonstrate compliance of a laser processing machine with the requirements of the European Machinery Directive.

ISO 11553 and EN 12626 cover functionally complete systems that incorporate lasers having sufficient power or energy to melt, evaporate, or cause a phase transition in the workpiece. (Medical lasers, and those used in photolithography, stereolithography, holography or data storage are specifically excluded.)

The laser machinery standard, which is intended for manufacturers or suppliers of equipment covered by the standard, provides a check-list for the identification of potential hazards, the undertaking of a risk assessment, and the implementation of appropriate protective measures. It lists the inherent hazards that can be generated by a laser processing machine. These include:

- mechanical hazards
- electrical hazards
- noise hazards
- thermal hazards
- vibration hazards
- radiation hazards due to laser emission, ionizing radiation, collateral and secondary radiation (emitted by the work piece due to beam effects)
- hazardous substances and materials

These ancillary hazards are discussed in more detail in chapter 6.

The standard also lists those hazards that may be created by external influences. These can arise from such causes as:

- temperature
- humidity
- vapours, dust or gases
- electromagnetic and radio-frequency interference
- source voltage interruption or fluctuation
- hardware and software incompatibility

Potential hazards have to be evaluated, a risk assessment carried out covering all phases of the machine's life, and appropriate corrective measures incorporated in the machine 'by design and manufacture'. Specific requirements apply regarding accessible levels of laser emission, and during processing work people should not be exposed to levels of laser radiation exceeding the AEL for Class 1. The manufacturer should also indicate the class of accessible radiation during servicing, and recommend appropriate precautions.

This standard is unfortunately one of several laser safety documents that mistakenly specify exposure limits and levels of accessible radiation in terms of the AEL. AELs (for laser emission between 302.5 and 4000 nm) are defined in units of power or energy whereas exposure levels are expressed in terms of irradiance or radiant exposure (see chapter 4). Furthermore, the laser class properly applies only to a laser product, since it depends on measurements made at specific locations with respect to the laser source and the emission aperture (as discussed in section 4.2); it should not, therefore, be applied to levels of laser radiation at arbitrary positions in space and without specifying the diameter of the measurement aperture. The intended meaning in ISO 11553 is that during processing the maximum level of exposure to accessible radiation should not exceed the MPE (using the time-base applicable for classification), taking into account the possible use of optical viewing instruments. There is a further requirement that during teaching, programming and program verification (i.e. during user maintenance) the level of exposure should not exceed the MPE without the use of optical viewing instruments. These requirements of the machine safety standard should therefore not be misunderstood to mean that the product needs to be classified as Class 1—it might well be classified as Class 4 when it does not meet one or more of the stringent requirements defined for classification of an embedded laser product.

Under this standard there must be a means of isolating the laser beam to prevent it entering the beam delivery system, and all guards and enclosures that are intended to be opened or removed and which could give access to internal laser radiation should be interlocked. The laser machine must be fitted with an emergency stop control to terminate laser emission, stop all moving parts of the machine, switch off the laser power supply and discharge all stored energy.

Where a human presence in the process zone is necessary to carry out adjustments or checks while the laser is on, then the person in the process zone must have complete control of laser operation by means of a hold-to-run device (which, when released, terminates laser emission).

The manufacturer is required to inform the user of the materials intended to be processed by the machine (since the processing of other materials might introduce additional hazards to those accounted for by the manufacturer), to provide information on processing hazards arising from the specified materials and to incorporate a suitable means for collecting the fume that is generated by the laser process. The safe removal and disposal of the fume, in accordance with appropriate national requirements, is the responsibility of the user.

Conformance with the requirements of the standard has to be verified by the manufacturer by visual inspection and functional tests. Adequate safety information must be provided with the equipment, and suitable safety-related training made available by the manufacturer to the user. The standard also makes clear that the person or organization installing the laser processing machine is responsible for the safety compliance of the whole machine, including subassemblies.

4.7.2 Medical laser products, IEC 60601-2-22

In addition to the need for classification, labelling and the various class-related engineering features for laser products defined in IEC 60825-1, Class 3B and 4 lasers[2] used for medical and surgical applications have to satisfy additional criteria. These requirements reflect the special conditions under which such laser equipment is used, and also the necessity for preventing harm from occurring to patients, whether through the unintended operation of the laser or by the use of an inappropriate level of emission.

The additional requirements are given in IEC 60601-2-22: 'Medical electrical equipment, Part 2. Particular requirements for safety—Specification for diagnostic and therapeutic laser equipment'. These additional requirements are discussed below.

For medical lasers, the aperture label as required by IEC 60825-1 can be replaced by the triangular laser hazard warning sign, placed in close proximity to the laser aperture. When it is impractical for the label to be placed directly at the laser aperture, for instance for bare fibre lasers, or for hand pieces and other applicators, the label may be on the main housing together with wording indicating that the laser aperture is at the end of the fibre or applicator, or alternatively to the wording, a pictogram showing a fibre, as specified in IEC 60601-2-22.

The user instructions included with the laser equipment have to give an indication of the maximum NOHD (nominal ocular hazard distance) applicable to each separate beam-delivery accessory. A description of each beam-delivery system should also be included. Details of the beam divergence, pulse duration and maximum output should be given, to include any anticipated increase in these values over time. (In the case of beam divergence, however, it would actually be a *decrease* in its value that would increase the level of hazard.)

An important issue with therapeutic (i.e. surgical) lasers is the level of the laser emission and the accuracy with which this is known. Such lasers therefore require a means of setting the output level (to within 20%) and of calibrating the measurement system. Procedures for doing this together with a recommended schedule for regular calibration should be given in the user instructions.

Where key control of laser operation is required (i.e. for Class 3B and Class 4 laser products) a note has to be included in the manual warning that laser equipment, when not in use, should be protected against unauthorized operation

[2] The current version of IEC 60601-2-22 refers to the 'old' classification scheme, where Class 3R is part of Class 3B. As such, the current version of IEC 60601-2-22 is applicable for Class 3R lasers as well. However, in a current draft amendment (IEC 76/266/CD), the standard is updated to take account of the new classification scheme and specifically excludes Class 3R lasers.

by removal of the key. The specification of any required eye protection has to be given, and guidance included, where appropriate, for removing fume. There also needs to be a warning statement that fume may contain viable tissue particulates.

Guidance on the handling of optical fibres must also be included in the manual, where appropriate, together with a warning that failure to follow the recommendations may result in damage to the fibre and/or harm to the patient. Users need to be advised that the aiming beam can provide a useful indication of the integrity of the delivery system before use of the main beam. A warning has to be included against fire and explosion risks that could arise through the use of inflammable anaesthetics, inappropriate use of oxygen, cleaning and disinfecting solutions, and by ignition of endogenous gases.

There are a number of specific engineering features required for medical lasers by IEC 60601-2-22, in addition to those specified by IEC 60825-1. These include the need for a laser-ready indicator (i.e. showing that the laser is in a state ready to fire) as well as a laser-on indicator (showing that the laser is actually emitting). The 'laser ready indicator' has to be visible (not audible) and following the current version of IEC 60601-2-22 there has to be delay of at least 2 s between switching the laser into the ready mode (and the laser indicator being lit) and the actual emission of laser radiation being possible. The 'laser-on' indicator can be visible or audible, and does not need to be fail-safe or redundant. This indicator is not necessary when the generation of the laser radiation produces a noise of sufficient sound level, such as is often the case for Excimer lasers. These two indicators replace the laser radiation emission warning as specified in IEC 60825-1 and required for all laser products of Class 3B and 4. Future editions of the medical laser standard will probably not require the 2 s delay.

Medical lasers should have a target indicating device, which normally takes the form of a visible aiming beam, which should be limited to powers allowed for Class 2 unless the resulting laser spot is not clearly distinguishable, in which case it may be increased up to 5 mW (i.e. power levels equivalent to Class 3R), but such increase should only be possible following a deliberate and positive action by the laser operator (i.e. it should not be permanently set at this level). In future editions of the medical laser standard, it is expected that the requirement for positive action is limited to ophthalmic lasers, i.e. it is then permissible for other surgical lasers to have the aiming beam permanently set to powers up to 5 mW.

A medical laser product of Class 3B and Class 4 needs to feature an emergency laser stop, independent of all other systems that can stop the laser (i.e. the main key switch of the laser is not sufficient for this requirement) and either has to be marked with 'Laser Stop', or a symbol defined in IEC 60825-1 (a STOP sign with the laser 'star-burst'), or it can be in the design of the standardized general mushroom-type emergency stop. Class 3B lasers that are not used for surgical or ophthalmic procedures, that are in the wavelength range 600–1400 nm, and that fulfil the following requirements, are exempt from the requirement of an emergency laser stop:

- the radiation that is emitted is less than five times the MPE for the skin and has less than 50 mW average power;
- if the average power is more than 50 mW, it is exempt if it does not exceed the MPE for the skin.

Thus the exemption is typical for low level laser therapy systems.

For Class 4 lasers, there should be a manual reset, enabling resumption of laser emission after any interruption that is caused by the tripping of a remote interlock or after an unexpected loss (lasting more than 1 s) of the main electrical supply. (The normal remote interlock connector required by IEC 60825-1 for any laser in Class 3B or Class 4 does not need to have a manual reset, although it can often be advisable to fit one.) However, because of the possible harm to the patient that an unexpected interruption of the laser beam might cause in the middle of a surgical procedure, door interlocks are usually not used in the surgical applications of lasers, but it is nevertheless important to have adequate procedural controls in place as discussed in chapter 8. A remote interlock connector is not required for battery operated medical laser products that are handheld (such as many low level therapy lasers).

As well as having a means, incorporated into the product, of measuring the level of laser emission (which can be through the use of any electrical or optical quantity directly related to the laser output), there must also be a display showing the pre-set level of the laser power or pulse energy (quite obviously, so that the operator can set the desired level of output and know reliably what it will be before the beam is turned on).

Where a foot-switch is used for laser operation, this must be shrouded to prevent inadvertent operation of the laser. Timers, which may be used to enable delivery of laser emission for a pre-set period, must have a back-up system to ensure termination of laser emission in the event of a fault before the pre-set period has been exceeded by more than 20%. The laser must also be fitted with an emergency-stop facility.

In the USA, medical laser products must conform to requirements specified in CFR 1040.20 (as well as those normally applicable to all laser products which are defined in CFR 1040.10), but in addition the specific medical laser procedure must be authorized by the FDA. The requirements of CFR 1040.20 for medical lasers cover the need for monitoring of the laser output of all Class IIIa, Class IIIb or Class IV lasers (to within 20%), except for Class IIIa alignment and positioning beams unless used for irradiation of the eye for ophthalmic purposes. A procedure and schedule for calibration of the laser output must be given, and all apertures (other than those of Class I lasers) must be labelled.

4.7.3 Optical telecommunications

The safety of optical fibre communication systems is not adequately addressed by the use of IEC 60825-1 alone. Such systems do not constitute a single entity that can be identified as a laser product and classified in the conventional way. During normal operation of an optical fibre network the laser radiation is completely enclosed within the fibre core and inside the terminal equipment, but access to laser emission, caused either by the disconnection of a fibre cable or by a cable break, can be gained at large distances from the laser source. Furthermore, the hazard that would arise following such an event can vary widely at different points around the network. This can be due to the splitting of power into several fibres and to the use of fibre amplifiers. To deal with this, IEC 60825-2 (Safety of laser products—Part 2: Safety of optical fibre communication systems) has been developed, and applies to installers and operators of these systems.

IEC 60825-2 uses the concept of hazard levels, which are applicable to all positions in an optical fibre network at which access could be gained to laser emission through a reasonably foreseeable event, such as a component failure, a fibre cable break, a cable disconnection or an operator error. The laser emission that could occur following such an event is assessed in the same way that the laser emission from a laser product is assessed for purposes of classification. That is, the measurement conditions and AELs defined in IEC 60825-1 and which have been described earlier in this chapter are applied to the emission arising from such an event, but instead of assigning a product class to the system as a whole, a hazard level (i.e. hazard level 1, hazard level 1M, hazard level 2, hazard level 2M, hazard level 3R, hazard level 3B or hazard level 4) is assigned to the location at which the particular hazard could occur. Thus, a particular fibre connector could be assigned a hazard level 1M, meaning that if it were disconnected, human access could be gained to a radiation hazard equivalent to that produced by a Class 1M laser.

The standard also defines three types of location in which such events (giving rise to access to emitted laser radiation) might occur. These are as follows.

- *Location with controlled access*—A location where access is controlled and available only to authorized persons who have received adequate training in laser safety and in the servicing of the system involved. Examples include cable ducts, street cabinets, switching centres, test rooms in cable ships and buried or submerged cables.
- *Location with restricted access*—A location where access is restricted and is not open to members of the public. Examples include secured industrial and commercial premises, restricted areas on trains, ships or other vehicles, and overhead cables.
- *Location with unrestricted access*—A location where access is unrestricted. Examples include general office and unsecured industrial premises, domestic premises, public areas on trains, ships, etc and open public areas.

Restrictions are then placed on the hazard level permitted in each type of location. These allowable hazard levels are:

- Hazard level 1, 1M, 2, 2M, 3R or 3B in a controlled location.

- Hazard level 1, 1M, 2, 2M or 3R in a restricted location.
- Hazard level 1, 1M, 2, or 2M in an unrestricted location.

In certain cases (hazard level 3R and 3B in a controlled location; hazard level 3R in a restricted location; and hazard level 1M, 2 and 2M in an unrestricted location) the specified hazard level is only permitted if some form of tool, such as a screwdriver or spanner, is needed to gain access to the hazard. All optical cables must carry appropriate markings to distinguish them form electrical cables. In all cases other than for hazard level 1, detachable optical connectors must be labelled with a sleeve, tag or tape to indicate the existence of a laser hazard (with the laser 'star-burst' symbol) and the applicable hazard level. Where a given hazard level is not permitted (including hazard level 4 in any location) an automatic power-reduction facility (APR) may be used to control the hazard level. This detects a break in continuity and reduces the emitted laser power within a specified time limit (1 s for unrestricted locations and 3 s for restricted or controlled locations).

In the case of multiple wavelengths along a single fibre, or the use of a ribbon fibre cable (which constitutes a multiple source), these should be assessed in the same way as for other laser sources.

Since the hazard level is based on the hazard that could arise following a reasonably foreseeable event (such as a component failure, fibre break or disconnection), an assessment of what is 'reasonably foreseeable' may require a formal assessment of the risk (the probability of the fault or other hazardous event occurring), particularly with regard to the occurrence of failures in the electronic circuitry. An Annex to the standard gives guidance on fault analysis and the assessment of failure modes.

As in all other laser applications, adequate safe working procedures, including training, should be adopted for test and measurement operations that require access to hazardous levels of laser emission.

In the USA special requirements apply to the users of optical fibre communication systems, and these are defined in ANSI Z136.2 (American National Standard for the Safe Use of Optical Fiber Communication Systems Utilizing Laser Diodes and LED Sources). Under this standard an optical-fibre communication system (OFCS) is allocated into one of four service groups (SG1, SG2, SG3a and SG3b), depending on the maximum level of emission that could occur following a fibre break or disconnection. These service groups are based on the ANSI system of laser classification. Unlike the approach adopted in the IEC standard, however, the service group allocated under ANSI Z136.2 applies to the entire fibre system, rather than to the specific location at which the hazard could arise.

4.7.4 Laser light shows

Another field of laser application where the public is potentially exposed is laser light shows used in discotheques, cinemas or in concerts. The legal aspects are somewhat different to household appliances, as for laser light shows the laser is owned and operated by an enterprise of some sort (although laser light shows up to 5 mW, i.e. Class 3R, are already marketed for the private sector, and with lasers becoming cheaper, even higher power lasers (classified correctly as Class 3B but obviously inappropriate for private use) are becoming available). Laser light shows usually fall under national legislation applicable for the safety of such kinds of events and establishments. Some countries have developed national standards to define technical requirements for laser light shows in support of a more general legislation (such as in Germany [7] or Austria [8]). In the UK a guidance document for laser light shows (HS(G)95) is issued by the Health and Safety Executive. Also, an international guideline document was developed by the IEC laser safety committee in the IEC 60825 series, namely IEC/TR 60825-3 [9]. This document is not a standard but has the status of a technical report. Generally, two safety approaches exist for laser light shows and are also followed in IEC 60925-3: either the beam is set up or protected so that it is not accessible by the public, i.e. exposure cannot occur, or, if exposure occurs, then the level of exposure in the area accessible by the public has to be below the MPE for the eye. If the first approach is taken, then there has to be some minimum safety distance between the beam and the public area (the floor, balconies etc) that ensures that the beam (that might have a power of several watts) is not accessible even when people are on each other's shoulders, on tables, or the like, and might hold out mirrors. Typical safety distances are 3 m from the floor for observed shows (i.e. where a light show operator is present who can switch off the show for instance if somebody steps on tables, etc) or 6 m for non-observed shows. In cases where exposure of the public is intended, the laser can still have an output power of several watts (and thus the laser product itself is Class 4), it has to be assured by the scanning action, by special mirrors that spread out the beam or by other effects that the exposure in the public area is below the MPE. For such an MPE analysis, usually an exposure duration of 0.25 s is used, however, it might be prudent for lasers in the blue wavelength range, where exposures may add up over time, to consider longer exposure durations. Some countries such as Sweden do not allow public exposure (even Pink Floyd had to change their show when performing in Sweden) and in the US, a special variance needs to be obtained from CDRH. Additional requirements or regulations may have to be observed for outdoor laser operation, be it for laser light shows or for systems such as lidars, to assure airspace safety. To prevent dazzle or flash blindness that occurs for exposure even below the MPE for the eye, special power restrictions for visible outdoor lasers have been defined in the US, that are particularly low for sensitive flight sections such as take off and landing, i.e. the allowed power levels around airfields are much lower than the MPE for the eye that usually applies to exposure of the public.

4.8 Case studies

In the following, we discuss some specific cases of classification that should help to both show the basic principles of classification as well as to prevent some common misunderstanding and misclassification. The case studies go from the simplest case of a laser pointer where the classification can take a few seconds (the time it takes to measure the output power) to very complicated cases of line laser or arrays with pulsed emission. The following case studies are intended to point out common issues of these types of products and problems rather than being complete and detailed classifications for specific products.

The 'source parameters' of the following case studies imply that uncertainties are accounted for and also that the parameters are maximized in the sense of environmental conditions (for instance, the maximum power is determined for the full temperature range specified for the specific product, i.e. for semiconductor devices typically the lowest temperature). We also do not specifically discuss single fault conditions, i.e. the accessible emission levels as discussed can either apply to the maximum levels accessible during use, if no reasonably foreseeable single fault condition can raise the level above this value or they can apply to maximum levels accessible during fault conditions, when the product is classified based on the accessible emission during faults.

4.8.1 HeNe alignment laser

This case study is an example of a product emitting a well-collimated, small diameter (for instance 1 mm) laser beam. For such a beam, the angular subtense of the apparent source can be assumed to be a small source so that $C_6 = 1$.

The laser output is continuous with a power of 0.9 mW and with a wavelength of 632.8 nm.

Due to the small divergence and small beam diameter, the full beam power is measured through the 7 mm aperture stop, and also condition 1 and condition 2 give the same accessible emission level (as does condition 3).

The AEL for Class 1 and Class 1M equals 0.4 mW (for this wavelength, the photochemical AEL is not relevant). The product's output power value is above this AEL, therefore the product cannot be Class 1 (it could not be Class 1M either, as all measurement conditions give the same accessible emission level).

The AEL for Class 2 and Class 2M equals 1 mW. The emission level is less than that and consequently, the product is assigned as Class 2.

4.8.2 Low-level therapy laser

This case study of a laser that is often used for low-level laser therapy is an example for a highly diverging small source laser such as typical for a bare (i.e. uncollimated) laser diode. The wavelength is 670 nm, the total output power equals 9.0 mW. The angular subtense for such a source is usually determined to

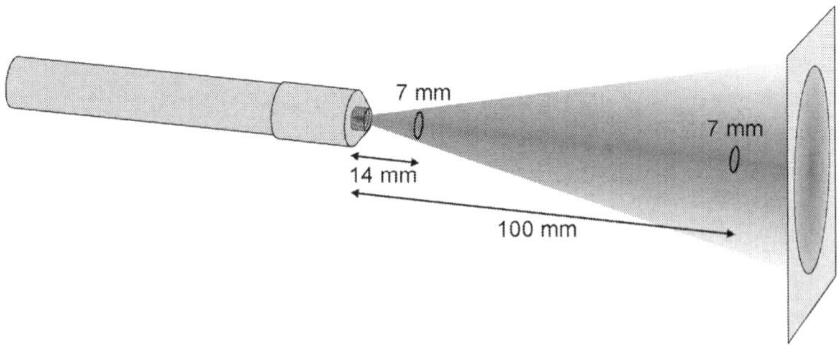

Figure 4.9. Schematic drawing of the beam of a bare laser diode with the two measurement distances of 14 mm for condition 2 (eye loupe condition) and 100 mm for condition 3 (naked eye condition).

be less than 1.5 mrad, i.e. the source can be treated as a point source (which is typical for a bare laser diode, as the apparent source size cannot be larger than the actual laser diode chip unless the beam is shaped to form a line). The location of the apparent source is also the laser diode chip.

For such a divergent source (in our example approximately 500 mrad, but it is not necessary to know or characterize the divergence for the classification), as shown schematically in figure 4.9, it is obvious that condition 2 (the eye-loupe condition) will be more restrictive than condition 1 (the telescope condition).

The power measurement at a distance of 14 mm from the chip (condition 2) gives a power of 7.8 mW, i.e. the beam diameter at a distance of 14 mm is such that not all the total power passes through the 7 mm aperture stop. Since the AEL for Class 3R equals 5 mW, this measurement alone would imply that the product is Class 3B. However, what is left to check is the naked eye condition, i.e. condition 3. At a distance of 100 mm from the chip, the power that is measured to pass through the 7 mm aperture equals 0.34 mW. This value is below the AEL for Class 1 and Class 1M, and consequently, the product without accessories is classified as Class 1M. It might be puzzling at first that the product is classified as Class 1M when the measurement with condition 2 would result in the classification as Class 3B. However, it is exactly the meaning of Class 1M, that the worst case exposure for the naked eye, i.e. at 10 cm, is safe (at least for unintended exposure) but exposure with optical instruments, in this case with highly magnifying loupes, results in hazardous levels (Class 3B equivalent levels) of exposure.

In the previous paragraph we have chosen the wording '*without accessories* the product is classified as Class 1M', since these kind of therapy lasers often come with accessories in the form of different application tips that can be attached to the tip of the laser product. The classification has to consider all accessories that are supplied. Attachments might reduce the accessible emission, but they might also increase it. For instance, an attachment with a collimating lens drastically reduces the divergence, usually without much loss of power, such that the power measured at 100 mm exceeds the AEL for Class 3R of 5 mW, and when such an accessory is available, the product would consequently be classified as Class 3B.

Since it is a medical product, the manufacturer has to specify the NOHD, as required in IEC 60601-2-22 and has to do so for each accessory. The NOHD, in our example, for the bare laser diode without accessories is zero, since the NOHD is stated for the naked eye, and for the accessory with the collimating lens it might be 50 m for a high quality collimating lens (that would also have to compensate for the astigmatism of a bare diode beam), while it might be only 30 cm for a standard quality lens that leaves the beam distorted and not very well collimated.

4.8.3 Line laser

A laser beam in the form of a 'line' is produced when a well collimated circular laser beam is passed through a cylindrical lens as shown in figure 4.10. This produces a beam that fans out with a very high divergence in one direction, and in the other direction is not affected by the cylindrical lens and remains collimated, i.e. the increase of the 'thickness' of the line with distance is minimal (the thickness of the line is usually less than what is shown in the figure). Due to the extremely astigmatic beam we have to consider different positions of accommodation of the eye, which corresponds to different locations of the apparent source. The location of the apparent source has to be understood here in the wider sense of being the location of the 'object' in the optical sense that, for a given exposure position in the beam, produces a certain image on the retina, i.e. the eye images this object by accommodating to the position of the object. For classification, the position of the measurement apertures depend on the location of the apparent source, thus, for different accommodation conditions and corresponding locations of the apparent source, different measurement positions in the beam result.

When the eye is relaxed, i.e. accommodated at infinity, as indicated in figure 4.10(*a*) (where the focal length of the standard eye in air equals 17 mm) then the beam is focused in the direction which has a low divergence, and a line is produced at the retina, as the beam in the other direction, with a higher divergence and an origin much closer, cannot be brought to a small spot on the retina at the same time. For a given position, the length of the line depends directly on the diameter of the aperture, i.e. for the case of looking into the laser beam (with appropriate filters), on the pupil diameter. This can be directly observed when one looks into the laser beam in a dark room. When the lighting is raised, the line that is seen becomes shorter as the pupil constricts. The nearer the eye comes to the cylindrical lens, the wider the line on the retina becomes, and in the extreme, when the origin of the spreading line would be located at the cornea, this condition of 'Maxwellian viewing' would result in practically no optical power of the eye. That a very thin line is produced even when the eye is right up at the cylindrical lens is also indicative that the eye accommodates to infinity rather than into the

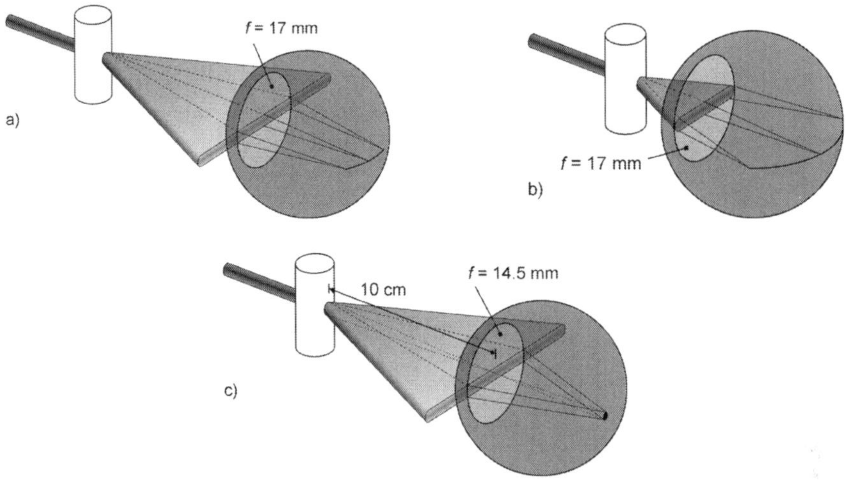

Figure 4.10. Different positions of the apparent source in the sense of the accommodation condition need to be considered for the classification of a line laser. It depends on the power that enters the eye (the measurement aperture) and the angular subtense of the source which one of the conditions is the most critical.

cylindrical lens, as in the second case the spot would blur for distances less than the near point.

So far, we have considered that the eye is relaxed, i.e. accommodates to image infinity. The appropriate analysis of such a source also considers that the eye accommodates to other positions along the beam, and the other critical position is when the eye accommodates to the position of the cylindrical lens which is the origin of the spreading line. In that case, as shown in figure 4.10(*c*)), the eye's optical power is high enough to image the origin of the source and bring the diverging direction to form a spot on the retina. Since the origin of the diverging direction is somewhat distributed inside the cylindrical lens, the typical minimal image which is formed for this condition is a short line that is perpendicular to the direction of the line.

For classification, it is important to consider that the AEL depends, via C_6 and T_2, on the angular subtense of the apparent source while the measurement distance for the determination of the laser power compared to the AEL also depends on the location and angular subtense of the apparent source. An example is discussed in more detail in the following, where the wavelength is assumed to lie in the red.

The following two viewing conditions are distinguished.

(I) Imaging of the beam as originating from some distance behind the cylindrical lens to produce a minimal image width (thickness of the line) on the retina. For this condition, the location of the apparent source is some distance behind the cylindrical lens.

(II) Imaging of the beam as originating from the cylindrical lens to produce a minimal image width on the retina in the direction of the beam plane, producing a line on the retina which is normal to the line (perpendicular to the plane of the fan-shaped beam). For this condition the location of the apparent source is within the cylindrical lens.

To experimentally characterize the apparent source, a number of persons viewed the beam with neutral density filters at different distances from the laser. It was reported by each person that a line in the direction of the beam plane was seen (viewing condition (I)). However, classification has to consider all accommodation positions. An exposure situation where the eye is for some reason focused onto the cylindrical lens, or at least to the distance of the cylindrical lens, might produce an image for some time (i.e. until the eye produces a line with the minimal, sharpest image in the beam plane) in the shape of a line normal to the beam plane.

(I) Imaging of beam to produce a line in the plane of the beam

When the eye accommodates to form a thin line, it is relaxed and accommodates to infinity, and the apparent source lies some distance behind the cylindrical lens. The apparent source in this case takes the form of a line in the plane of the beam. The length of the line at the retina is directly related to the diameter of the pupil of the eye which is assumed to be 7 mm following the laser safety standard IEC 60825-1 (it can be shown that smaller pupil diameters that would lead to smaller values of α and smaller power values entering the eye are less critical than the case of a 7 mm pupil).

Position of the apparent source

The position of the apparent source can be determined by imaging the beam with a lens, to minimize the angular subtense in the 'thin direction' of the line. The position of this smallest image is typically found to lie in the focal plane of the lens that is used to transform the beam, and therefore the corresponding optical object, the apparent source, for the purpose of the classification is assumed to be at infinity behind the laser product. Consequently, the measurement position for classification according to IEC 60825-1 (condition 2) is at the closest point of human access, which is at the outside surface of the cylindrical lens.

Angular subtense of the source at the measurement position

The angular subtense of the apparent source can be determined by imaging the source with a lens onto the camera chip of a beam profiler, or for distances where

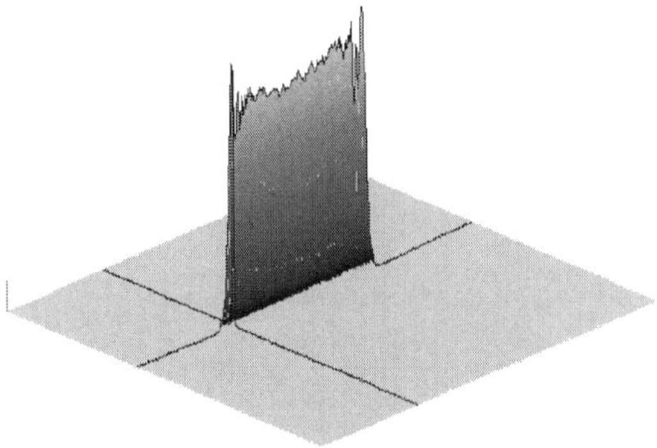

Figure 4.11. Profile of the 'image' of the line laser that corresponds to the condition of accommodation to infinity.

the image is too large for the camera chip, onto a Vernier Caliper. An aperture of 7 mm diameter is placed in front of the lens.

The width of the linear apparent source was not measured and is assumed to be minimal, i.e. 1.5 mrad. The angular subtense of the length of the linear apparent source can be determined at a number of distances to the laser, i.e. to the cylindrical lens. The results of the measurements in most cases confirm (analysed at the testing house of ARC Seibersdorf research) that the angular subtense scales inverse linearly with the inverse of the distance to the cylindrical lens, and for distances closer than a few centimetres, the angular subtense in the long direction is larger than 100 mrad, i.e. larger than α_{max}. An example of such a beam profile is shown in figure 4.11.

With an angular subtense of α_{max} in one axis and $\alpha_{min} = 1.5$ mrad in the other axis, α and C_6 for this viewing condition are determined to be

$$\alpha = 50.8 \text{ mrad} \qquad \text{and} \qquad C_6 = 33.8.$$

AEL and accessible emission levels

The AEL for Class 1 for the above values of α and C_6 is (only the thermal AEL is relevant for the red wavelength range)

$$\text{AEL Class 1} = 10 \text{ mW}.$$

For such a diverging source, condition 2 is more critical than condition 1, i.e. the accessible radiation is determined with an aperture of 7 mm diameter. Since the location of the apparent source for viewing condition (I) is at a distance behind the laser which is greater than the measurement distance specified for condition 2, the measurements need to be performed by placing the detector at the laser exit aperture. If this is not physically possible, the distance dependence of the accessible emission can be characterized to correct for a distance of zero.

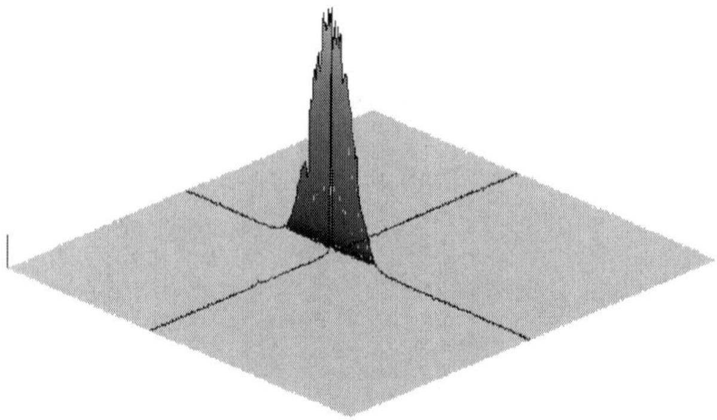

Figure 4.12. Profile of the 'image' of the line laser that corresponds to the condition of accommodation into the cylindrical lens.

For our example, we assume a measured radiant power value of 3.3 mW when the angle of acceptance is not limited for simplicity (for a maximum angle of acceptance of 100 mrad, the measured power value would be smaller, however, it is below the AEL for Class 1 even when the acceptance angle is not limited, which simplifies the measurement).

(II) Imaging of beam to produce a line normal to the plane of the beam

When the eye images the source of the diverging line, the apparent source can be expected to lie within the cylindrical lens. The apparent source in this case is typically a line normal to the plane of the beam and the position of the apparent source is within the cylindrical lens.

Angular subtense of the source at the measurement position

The angular subtense of the apparent source can be determined by an imaging lens and a beam profiling camera; the profile is shown in figure 4.12.

The angular subtense of the length of the linear apparent source is determined at a distance of 10 cm to the location of the apparent source, i.e. basically to the cylindrical lens. The width of the linear apparent source (now perpendicular to the one shown in figure 4.11) is typically minimal, i.e. 1.5 mrad.

The angular subtense of the image in the extended direction depends on the beam profile in the cylindrical lens, and for our example is assumed to be 14 mrad.

With an angular subtense of 14 mrad in one axis and $\alpha_{min} = 1.5$ mrad in the other axis, α and C_6 for this viewing condition is determined to be

$$\alpha = 7.8 \text{ mrad} \qquad C_6 = 5.2.$$

AEL and accessible emission level

The AEL for Class 1 for the above values of α and C_6 is calculated to be

$$\text{AEL Class 1} = 2.0\ \text{mW}.$$

The formula given in IEC 60825-1 for the measurement distance for the thermal limit for condition 2 for this value of α results in a measurement distance from the apparent source of 29 mm. In our example, and at that distance, the measured radiant power value equals 0.74 mW. Therefore, also for the second viewing condition (assuming the location of the apparent source), the accessible emission level is below the corresponding AEL for Class 1, the product is therefore classified as Class 1.

This case study shows that for highly astigmatic beams, the classification procedure needs to consider (at least) two different positions of the location of the apparent source. For such beams, it can be envisaged that the two main directions of the beam have two origins: the direction across the line profile (perpendicular) to the line is collimated and the corresponding origin is far behind the laser product, while in the plane of the line, the beam is highly diverging and the corresponding origin is in the cylindrical lens. The eye can only accommodate to one location and can focus to one direction at a given time so the other direction will be defocused. (For a more complicated product, even more positions of apparent sources than the two extreme cases as discussed here would have to be analysed.) To each accommodation position, a specific location of the apparent source is associated and corresponding measurement requirements result regarding the determination of the angular subtense of the source to determine the AEL and regarding the measurement distance to determine the accessible emission level. By taking the ratio of the AEL to the corresponding accessible emission levels for the two locations of the apparent source, one can compare relative levels of hazard. For the present case study, it turns out that the second case, i.e. the accommodation into the cylindrical lens, is the most critical (when the power of the product is increased, this case will be first to exceed the AEL for Class 1). While the closer measurement distance for the apparent source located at infinity behind laser (I) results in a higher value of the accessible emission level for the case of accommodation into the cylindrical lens (II), at the measurement level distance for (II) the angular subtense of the apparent source is correspondingly smaller.

4.8.4 Scanner

The following discussion relates to laser products where a beam is deflected from a turning mirror or reflecting polygon, which are generally referred to as ‘scanners’. The beam, as it is emitted from the stationary product, thus does not point into one fixed direction but moves in space (a handheld laser that emits a stationary beam cannot be considered as scanned on the basis of the natural tremble!). We concentrate our discussion on the simplest case of a well-collimated (i.e. low-divergence) small diameter, continuous wave beam (such as the one described in the first case study for the stationary case) that is scanned in one direction. However, the main points of the discussion can also be extended to apply to more complicated scanning patterns, pulsed emission and more complicated beams that form extended sources even in the non-scanning condition.

A beam that scans across a measurement aperture (or the pupil of the eye) produces a pulsed pattern of accessible emission (or exposure). When the diameter of the beam at the position of the aperture is smaller than the aperture, the peak power of the pulse is simply the cw power of the laser beam. The pulse duration depends on the speed of the beam as it scans across the aperture (which in turn depends on the distance to the mirror and on the angular velocity of the mirror). The analysis of such a pulsed accessible emission pattern is discussed in section 3.12.8 and is not repeated here. In this discussion we concentrate on the appropriate measurement distance and the angular subtense for classification. Unfortunately, this topic is somewhat 'frustrating', as currently, the only appropriate way of classification often has to be based on some limiting worst case assumptions. We have found that the necessity for these assumptions is often misunderstood, and may have resulted in incorrect classifications of some scanning laser products. The basic problems relate to the *location* of the apparent source on the one hand, since this determines the location of the measurement aperture, and, for the case of the thermal limit, to the *angular subtense* of the apparent source.

Regarding different accommodation positions and therefore the corresponding *locations* of the apparent source, the case for scanners as shown in figure 4.13 is equivalent to line lasers as discussed in the previous case study.

For the viewing condition shown in figures 4.13(*a*) and (*b*) the eye is accommodated to infinity, and since the beam is assumed to be well collimated, this produces a small spot on the retina that scans across the retina. However, regarding the appropriate value of the *size* of the retinal image for calculation of the thermal limit, it is important to consider that it is not a real line that is formed on the retina, but rather a small spot that scans across the retina. Since the visual perception, when one looks into a scanner (particularly when close to the mirror), is very similar to that of a line laser viewed close up, it is tempting to treat the scan line on the retina in the same way as the real line in the previous case study to determine a and C_6, i.e. to use the length of the scan to calculate α. However, this is not correct: it is important to distinguish between the spot size of the focused beam that scans across the retina (and it is this size that should be referred to as α) and the scan pattern, i.e. the length of the scan. While it is obvious that a beam that moves across the retina is less hazardous than when it stays at the same position, it is not possible to quantify 'how much' less hazardous a scanning beam is. That the scan length cannot be used to calculate α, however, is also obvious, as a small spot with a given power that scans across the retina is much more hazardous than

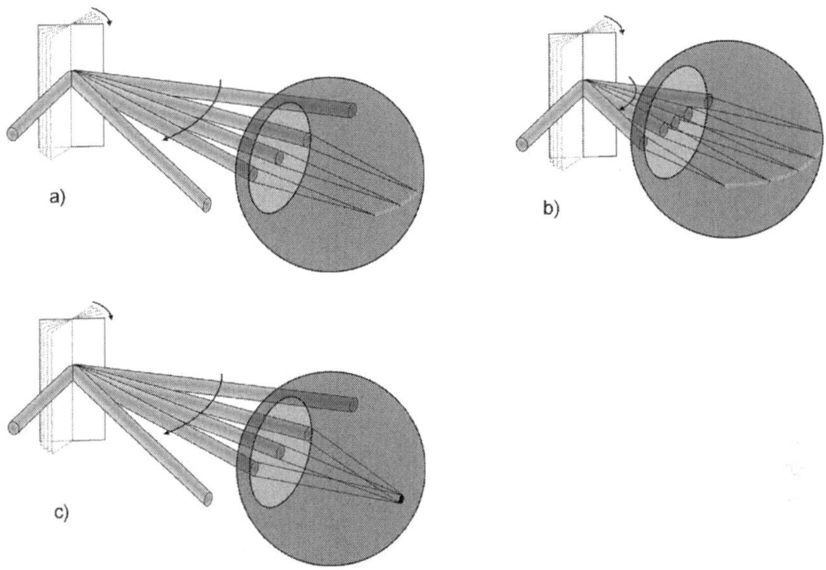

Figure 4.13. For a well-collimated beam that is scanned across the pupil of the eye, if the eye accommodates to infinity, a minimal spot is formed on the retina which is scanned across the retina ((*a*) and (*b*)). If the eye accommodates onto the scanning mirror (*c*), then the beam that enters the eye is imaged onto the same spot on the retina, but the spot is usually larger as in cases (*a*) and (*b*). It is important to note that a small spot that moves across the retina is much more hazardous than when the power of the beam is spread over a line, as shown in the previous case study. Therefore, the scan length cannot be directly used to determine α.

an actual line where the power that enters the eye is spread across the whole line. One can make the difference obvious by considering laser welding or laser cutting where a tightly focused beam is moved over the workpiece, to weld, for instance, a length of 2 m within 1 min. When the same laser beam is fanned out to form a line with a length of 2 m, and this line laser irradiates the workpiece for 1 min, welding would hardly be possible. While in both cases, the energy per irradiated area is the same, the local irradiance for the scanning case is much higher than for the line laser, i.e. for the scanning case, the energy is delivered to the given spot along the scan in a correspondingly shorter time, leading to much higher, but transient, temperatures. When the scan length is erroneously used to calculate α, i.e. when the scan pattern is considered as apparent source size, then this can be compared to assuming that the welding laser produces the same temperature as a fanned out laser. It is only for the evaluation against the photochemical limits that the two cases produce the same effect, as the photochemical interaction depends on the retinal radiant exposure only and not on the time it takes to deliver the corresponding number of photons.

In summary, the thermal hazard of scanned exposures lies somewhere in between the stationary beam on the one side and an actual line laser on the other. The 'position' in terms of level of hazard between the two extremes mainly depends on the scan speed: for very slow scan speeds, the hazard is comparable to a stationary beam, while for scan speeds so fast that the scan duration is less than the thermal confinement time (characterized by T_i), from the viewpoint of heat flow, the whole line is heated at the same time and then the scanning laser has the same effect as if the exposure came from a line laser (with the same power entering the eye for the same duration). Only in this case is it possible to use the scan length to determine α and C_6. It might be possible some time in the future that an appropriate method becomes available to characterize the lower hazard that is represented by a scanned retinal exposure in comparison to the same beam being stationary, but this information unfortunately is currently not available. Therefore, in terms of the value for the angular subtense of the apparent source, currently it is necessary (unless the scan is faster than T_i) to assume that the beam is not moving across the retina (but the pulse duration and energy per pulse is still given by the scanning motion across the pupil).

The above discussion relates to the issue of characterizing the angular subtense of the source and therefore the appropriate thermal AEL as a function of scanning pattern. For classification, additional to the angular subtense of the source we have the added level of the impact of the *location* of the apparent source on the measurement distance and therefore on the determination of the accessible emission level that is compared to the AEL. The issue is very similar to the discussion of the line laser of the previous case study. A complete classification has to consider different accommodation positions with correspondingly different locations of the apparent source. Figures 4.13(*a*) and (*b*) show the case for the eye being relaxed, i.e. the accommodation is at infinity. The corresponding location of the apparent source for this case is at infinity, which can be determined experimentally by using an imaging lens (but an indication is also that one can move right up to the mirror and still image a sharp scanning line, which would not be possible if the 'origin' were the mirror, as this could not be accommodated at close distance). For classification of scanners, IEC 60825-1 specifies that only the naked eye condition needs to be considered, i.e. 7 mm at a distance of 100 m from the apparent source. However, for this viewing condition, the apparent source is located at infinity, and thus, classification has to be performed at the closest point of access, i.e. if the mirror is accessible, basically at the mirror. The closer one gets to the mirror, the longer the pulses become and for a constant beam power, the larger the energy per pulse becomes. This increase of accessible emission level is to some degree compensated by an increase in the scan length, as indicated in figure 4.13(*b*). However, the problem is that for a scanning laser, in most cases it is not possible to account for the decrease of the hazard due to the scanning pattern that forms on the retina for this kind of viewing condition, and C_6 has to be characterized for the non-scanning case. Consequently, for this case study with a well collimated beam, $C_6 = 1$. This clearly overestimates the hazards, but unless the scan speed is very fast, it is not justified to use the scan length to determine α, as was already discussed in the previous paragraph. In practice, however, it is rare that the mirror is positioned such that it is accessible, as usually it is somewhere inside a housing and the exit aperture (covered with some transparent plate), as part of the housing, is some distance away. The classification needs to be performed at the closest point of human access and the further the mirror is recessed, the lower the accessible emission level becomes.

It can in some cases be shown that the worst case, closest distance for classification regarding the thermal retinal limit and the assumption that the eye is focused at infinity is at about 7 cm from the mirror, which certainly reduces the accessible emission limit with respect to measurement at the closest point of human access if the mirror is accessible (but C_6 still needs to be set to unity in most cases).

So far we have considered that the eye accommodates to infinity, but, by analogy with the line laser, a classification also needs to consider other positions of accommodation and corresponding locations of the apparent source. There is a special point along the scanning beam about which the scanned beam pivots. This is typically the scanning mirror, but if the product features some projecting optics behind the mirror, it may also be the exit pupil of the projection optics. If the eye accommodates to that position, the beam that is scanned over the cornea and pupil is imaged to one spot/location on the retina, i.e. no scanning pattern on the retina is produced, as is shown in figure 4.13(*c*). This viewing condition where the location of the apparent source is the mirror is equivalent to figure 4.10(*c*) for the line laser. The corresponding classification procedure is to position the 7 mm aperture at a distance of 100 mm from the mirror. The angular subtense can be characterized by positioning a lens and a CCD array such that the mirror is at the object position when the CCD array is at the conjugate image position. Thus, the extent of the angular subtense for this condition depends on the beam diameter at the mirror and for a well collimated beam with a beam diameter at the mirror of, say, 1 mm is typically larger than when the eye is assumed to image at infinity, i.e. than when the CCD array is positioned in the focal plane of the imaging lens, to characterize the angular subtense α for the case shown in figure 4.13(*a*) (where α applies to the beam as such, i.e. for the non-scanning case). When (*a*) and (*c*) is assumed to be at the same distance from the mirror, than the pulse pattern that is formed by the scanning motion across the aperture and pulse energies are the same in both cases. The difference is in the location at which the eye accommodates, i.e. for which imaging position the apparent source is characterized. When the eye focuses at a position other than the pivoting point (the mirror), then the beam will scan across the retina. When the eye accommodates to infinity, the retinal spot will have its minimal dimension. It depends mainly on the scan speed, if the smaller spot that scans across the retina is more hazardous than the larger spot that remains on the same position on the retina for the case that the eye accommodates to image the pivoting point. At this stage, without further biophysical and theoretical studies, it is not possible to appropriately characterize the level of hazard for the scanning case.

4.8.5 Near-IR and visible beam

This case study relates to a product that emits both a beam in the near-infrared with a wavelength of 1064 nm, and additionally in the same direction, for instance for purposes of alignment or pointing aid, a 532 nm 'laser pointer'. Both beams are well collimated, i.e. have low values of divergence.

The goal that is pursued by the manufacturer is that the product is classified as Class 2 or Class 2M when the infrared beam has a beam diameter larger than 7 mm. It is required for a Class 2 laser product that radiation emitted outside the visible wavelength range is less than the AEL of Class 1, and Class 2M lasers shall not emit above Class 1M outside the visible wavelength range.

The two wavelengths are additive, therefore to be below the AEL, the inequality (3.21) needs to be satisfied, which can be rewritten for this example as

$$Q_{1064}/\mathrm{AEL}_{1064} + Q_{532}/\mathrm{AEL}_{532} < 1$$

where Q is the energy (or power) measured according to condition 1 for a Class 2 product and measured with a 7 mm aperture at 10 cm distance from the apparent source for Class 2M. However, since this requirement reflects the additivity of the two wavelengths, i.e. the case that they are both together and at the same time are incident on the retina, and the time base for Class 2 is 0.25 s, it can be argued that also the time base for the non-visible part to determine the AEL_{1064} for the above inequality may be 0.25 s. That is, instead of calculating the AEL_{1064} for the usual time base for non-visible wavelengths of 100 s, it can be justified that for the evaluation of the combined emission and exposure, both the visible and the non-visible AEL are evaluated for a 0.25 s time base. It can also be the case that the near-infrared beam is pulsed, and the 532 nm beam is cw, and in this case the above inequality also applies, as long as the appropriate accessible emission levels are divided by the corresponding AEL values. For instance, if the accessible emission level of the near-IR beam is 50% of the AEL for Class 1 evaluation for a time base of 0.25 s, i.e. $Q_{1064}/\mathrm{AEL}_{1064} = 0.7$, then there is 0.3 of the inequality 'left' for the visible beam, i.e. with an AEL of 1 mW for the visible cw beam, the allowed power for the visible beam would be 0.3 mW for the product to be Class 2. When the near-IR beam exceeds the AEL when measured with a 50 mm aperture, but is below the AEL for measurement with a 7 mm aperture, then the product would be classified as Class 2M.

However, it is important to note that the infrared laser beam alone would have to satisfy the classification as Class 1 (or Class 1M if the whole product is Class 2M) with the usual time base of 100 s. This is not only a requirement for Class 2 and Class 2M that the emission outside the visible wavelength range is less than the AEL for Class 1 and Class 1M, it is also easily understandable on a practical level. For instance, it is conceivable that exposure occurs to each of the wavelengths alone, for instance because the geometrical and temporal coincidence might not be complete or because red laser glasses are worn which improve the visibility of (other product's) Class 2 red laser beams which are regularly used in the construction business (i.e. these are not eye protection goggles but goggles that filter all but the red light and often come with red Class 2 alignment lasers to improve visibility in bright environments). Such a laser goggle would block the green laser pointer but transmit the 1064 nm wavelength.

For the case that the power and beam diameter of the infrared laser results in a classification as a Class 2M product, because the infrared laser on its own would be Class 1M, then also the visible part can be measured with a 7 mm diameter aperture at 100 mm from the apparent source to determine the energy to be compared to the respective AEL for Class 2M. Since the case study assumes well-collimated beams, the angular subtense of the apparent source, even with the assumption of binocular viewing and some magnification (depending on the beam diameter) is assumed to be minimal. Also the location of the apparent source typically lies some distance behind the laser product so that measurement is to be performed at the closest point of human access for condition 3.

References

[1] ISO 11553 1996 *Safety of Machinery—Laser Processing Machines—Safety Requirements* (ISO: Geneva) basically identical to EN 12626

[2] IEC 60812 1985 *Analysis Techniques for System Reliability—Procedure for Failure Mode and Effects Analysis (FMEA)* (Geneva: IEC)

IEC 601025 1990 *Fault Tree Analysis (FTA)* (Geneva: IEC)

IEC 60300-3-9 *Dependability Management—Part 3: Application Guide—Section 9: Risk Analysis to Technological Systems* (Geneva: IEC)

McCormick N J 1981 *Reliability and Risk Analysis* (Washington, DC: Academic)

[3] Grabner U, Vees G and Schulmeister K 2003 *Beam Propagation Hazard Calculations for Telescopic Viewing of Laser Beams (Proc. ILSC)* (Orlando: LIA) pp 116–25

[4] CDRH 2001 *Laser Products—Conformance with IEC 60825-1, Am. 2 and IEC 60601-2-22; Final Guidance for Industry and FDA (Laser Notice No 50)* (Rockville: CDRH)

[5] IEC 60601-2-22 *Safety of Laser Products—Part 1: Equipment Classification, Requirements and User'S Guide* (Geneva: IEC)

[6] IEC 60825-2 *Safety of Laser Products—Part 2: Safety of Optical Fibre Communication Systems* (Geneva: IEC)

[7] DIN 56912 1999 *Showlaser und Showlaseranlagen—Sicherheitsanforderungen und Prüfung* (Berlin: DIN)

[8] ÖNORM S 1104 1999 *Laser Strahlenschutztechnische Anforderungen bei der Erzeugung von Lichteffekten mittels Laserstrahlung vor Publikum oder bei der Vorführung von Laser-Einrichtungen* (Wien: ÖNORM)

[9] IEC 60825-3 TR *Safety of Laser Products—Part 1: Guidance for Laser Displays and Shows* (Geneva: IEC)

Chapter 5

Beam propagation and exposure assessment

5.1 Measurement versus calculation

The interaction of optical radiation with the tissues of the body, as described in chapter 3, can be complex. This often leads to a requirement for a very precise specification of the radiation properties in order that the significance of a given level of exposure can be ascertained. Emission characterization and exposure assessment are therefore fundamental issues of concern in laser safety.

The quantitative assessment of laser radiation hazards is based on defined measurement criteria. Laser safety standards (e.g. IEC 60825-1) are essentially measurement standards insofar as product classification and exposure assessment are concerned. As discussed in earlier chapters, emission and exposure limits are defined on the basis of the level of radiation that is present at a given location—the measurement distance—and which is contained within a defined circular area at that location. This area is the specified measurement aperture in the case of product classification or the limiting aperture in the case of exposure assessment. In the case of extended (including multiple) sources, the emission and exposure limits apply to the level of radiation contained within the relevant aperture that is received from within a defined field-of-view or acceptance angle.

Although the determination of levels of laser radiation for purposes of hazard assessment may require the employment of actual radiometric measurements, using the techniques discussed in chapter 2, it can sometimes be more appropriate to base these assessments on calculation in circumstances where the essential emission characteristics of the laser (e.g. its wavelength, emission power, beam size and beam divergence) are already reliably known.

Radiometric measurements can require the use of sophisticated instrumentation which needs to be properly calibrated. Many laser users do not have such equipment readily available. Furthermore, since radiation levels that are close to a safety limit can be quite low (because the safety limits themselves are often low) making accurate measurement under ambient lighting conditions is very difficult. In addition, the need for precise alignment of the detection system with the laser beam can introduce significant errors, since only small displacements of the measuring equipment may produce considerable variations in the measurements being made.

This chapter is, however, primarily concerned with the evaluation of individual laser beams, and also with laser emission from optical fibres. The assessment of LEDs is usually more satisfactorily carried out by direct measurements of the emitted radiation. This is because LEDs are not simple point sources, and the specification of their emission given by suppliers is often not sufficient to enable an adequate assessment to be carried out.

Nevertheless, an LED can be excluded from further consideration if it can be shown that its total emission does not exceed the Class 1 AEL. It can also be excluded if the radiance of the LED (in units of watts per square metre per steradian) does not exceed the value of the MPE (in watts per square metre) divided by Ω_{max}. (Ω_{max} is the solid angle subtended by a source having an angular subtense of α_{max}, i.e. 100 mrad, and is equal to 7.85×10^{-3} sr.) The value of MPE/Ω_{max} represents the most conservative radiance limit for any size of source. The source radiance of an LED can be determined by dividing the radiant intensity of the source (in watts per steradian), if known, by the emission area, which can itself be measured using a CCD camera. If either the power or radiance limits are exceeded, it does not necessarily follow that the LED is unsafe, only that further and more detailed evaluation is required.

In the case of lasers, calculation rather than actual measurement of the levels of laser emission or laser exposure can be the preferred option whenever the laser source itself has already been well characterized and the effect of any external components used to transform (that is, to focus, reflect or transmit) the beam can be well defined. This is often the case for laser users, who can utilize the emission parameters of the laser that are provided by the manufacturer in order to evaluate possible exposure levels. This may be necessary for specifying eye protection or for calculating the hazard distance. Calculations can sometimes also be appropriately used for classification whenever a laser of known specification is incorporated into a larger system to form a complete laser product (since classification applies to the whole system, at the highest level of integration of the laser employed).

Chapters 3 and 4 have discussed how levels of laser radiation are used for purposes of either exposure assessment or product classification. When assessing the safety of a given level of laser exposure, the value of irradiance (W m^{-2}) or radiant exposure (J m^{-2}), when averaged over the area of the relevant limiting aperture, is compared with the applicable MPE (maximum permissible exposure). The exposure may be considered safe if its value, based on the defined criteria, is below that of the MPE. The value of irradiance E (or of the radiant exposure H), when averaged over the area of the limiting aperture, is equal to the power P (or the energy Q) contained within the area of the limiting aperture, divided by the area of the limiting aperture. It can sometimes be more convenient to consider exposure quantities in this way rather than to directly determine the actual value of E or H. An exposure that is below the MPE can only be considered safe, however, if optical viewing instruments, which can increase the level of exposure that the eye receives, are not used. The effect of such optical viewing aids is discussed later in this chapter. A laser's class, on the other hand, is established on the basis of the maximum level of radiation (power Por energy Q) that can pass through the relevant measurement apertures at the specified positions.

Nearly all quantitative safety assessments, therefore, require that the amount of radiant power or energy that passes through a given circular aperture is determined. This can be done by calculation if the characteristics of the emitted laser radiation are sufficiently well defined. This will normally mean that the following emission data are known.

- Laser wavelength (λ, in nm).
- If cw: total emitted power (P in W).
- If pulsed (assuming constant pulse rate):
  - pulse energy (Q in J)
  - pulse duration (t_{pulse} in s)
  - pulse repetition rate (f in Hz).
- Beam diameter at some defined location (d in m).
- Beam divergence (θ in radians or degrees).
- The profile of the beam (the variation in irradiance or radiant exposure across the beam, in a plane perpendicular to the beam direction). With many lasers the profile can be close to a Gaussian distribution, as discussed in section 5.3.

A word of caution is needed here, however. There is an inherent conflict between a safety specification and an operational specification. In undertaking any sort of safety assessment we need to know the maximum level of radiation that the laser can emit; the most hazardous conditions that can exist. This may correspond to the upper limit in the case of some parameters (e.g. emission power, pulse repetition rate) but the lower limit for other quantities (e.g. beam diameter and divergence). Operationally, however, in order to achieve the intended purpose for which we are using the laser, we may wish to know the minimum output that can be guaranteed. Whether the minimum or maximum level of emission is considered to be the worst case may therefore depend on whether we are evaluating the laser from its safety or performance perspective. Clearly, then, in using any data concerning the laser as a basis for a safety assessment, we must ensure that the values we use represent the most hazardous conditions that are realistically possible.

When performing calculations as part of a safety assessment, it is also important to ensure that the quantities inserted into formulae are specified in appropriate units. Huge errors can result from the use of inappropriate units, for example by using distances in millimetres instead of metres, or angles in milliradians rather than radians. Great care must therefore be exercised in order to ensure that all the units used in a given equation are mutually consistent.

In the equations that follow in this chapter, the correct solution will always be found by using the base units of distances in metres and angles in radians. However, these are not always the units in which the various parameters are commonly expressed. For example, laser beam diameters are more often given in millimetres, and beam divergence angles in milliradians. With care, such units can often be used in equations without conversion, where they balance each other at the top and bottom of an equation. For example, at a distance of one metre, an angle of one radian subtends an arc of length one metre. (This is actually the definition of a radian.). But equally, at a distance of one metre, an angle of one milliradian (one thousandth of a radian) subtends an arc of length one millimetre (one thousandth of a metre). The two factors of 1/1000 cancel out, and the ratio of a given arc length to a given angle is the same, whether the length is expressed in metres and the angle in radians, or the length is in millimetres and the angle is in milliradians. (When the angle involved is small—up to several milliradians—the length of arc approximates, of course, to a straight line.) In general, therefore, in equations relating distances along the beam axis, distances (such as the beam diameter) perpendicular to the beam axis, and beam angles, these can be expressed, respectively, in units of metres, metres and radians, or alternatively in units of metres, millimetres and milliradians. These latter units (metres along the beam, millimetres across the beam, and milliradians for angles) will often be found useful in many of the commonly used expressions that characterize laser beam propagation.

Quite often, for collimated emission, we are dealing with laser beams whose diameter is well below that of the relevant aperture. In such cases the power or energy that would pass through the aperture is simply the total power or energy of the beam, and the assessment of classification or the level of laser exposure is then much more straightforward. Considerable effort can often be saved if such simplifications are recognized, rather than that a pre-set assessment routine be always followed. Sometimes, in safety assessments, detailed calculations are laboriously pursued only to reach conclusions that would have been obvious if the relevant conditions had been considered more carefully. (Note, however, that in the case of a beam that is smaller than the limiting aperture, the level of exposure—E or H—to be compared with the MPE is the power or energy of the beam divided by the area of the limiting aperture and not by the area of the beam.)

Frequently, however, we are dealing with beams whose diameter at the aperture location exceeds that of the aperture itself, and we then need to determine carefully just how much of the beam (what proportion of its total power or energy) would actually pass through an aperture that is located centrally on the axis of the beam. Later sections of this chapter will describe how this can be done.

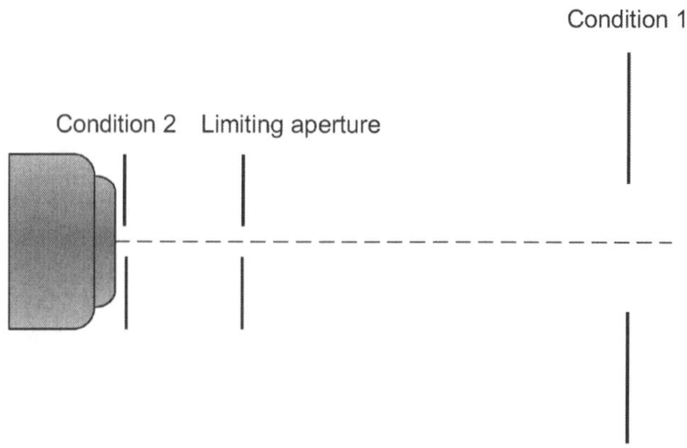

Figure 5.1. The measurement conditions for laser classification. Condition 1 is a 50 mm aperture (for wavelengths between 400 and 1400 nm) or a 25 mm aperture (at all other wavelengths between 302.5 and 4000 nm) located 2 m from the closest point of human access. Condition 2 is a 7 mm aperture located 14 mm from the apparent source (for apparent sources subtending an angle of less than 1.5 mrad, which covers the majority of lasers). The limiting aperture, which is used in the assessment of Class 1M and Class 2M (and for other classes at wavelengths outside the range 302.5 –4000 nm), varies from 1–7 mm (as specified in IEC 60825-1), and is located 100 mm from the apparent source (at all wavelengths between 302.5 and 4000 nm, otherwise the measurement distance is zero).

5.2 Classification apertures

The measurement apertures used for classification in the wavelength range 302.5–4000 nm and that are applicable to point laser sources (i.e. lasers having an apparent source-size of less than α_{min}, which covers the majority of lasers) are shown in figure 5.1. For a laser to be in Class 1, Class 2, Class 3R or Class 3B, its accessible emission must be below the relevant AEL under both of these measurement conditions. The other measurement aperture (the limiting-aperture or 'naked-eye' condition) is used in the assessment of Classes 1M and 2M. A laser that exceeds the AEL for Class 1 or Class 2 (but not the AEL for Class 3B) under either the telescope condition (condition 1) or the eye-loupe condition (condition 2) will be Class 1M or 2M if it does not exceed the Class 1 or Class 2 AEL under the naked-eye condition.

For many beam configurations, however, an examination of the measurement conditions required for classification will indicate that some (or indeed all) of the apertures are irrelevant, and can therefore be ignored.

Figure 5.2 shows a collimated beam that is smaller than each of the three measurement apertures. Because it is a collimated beam, the apparent source is to the left of the laser and so both the eye-loupe condition aperture and the limiting

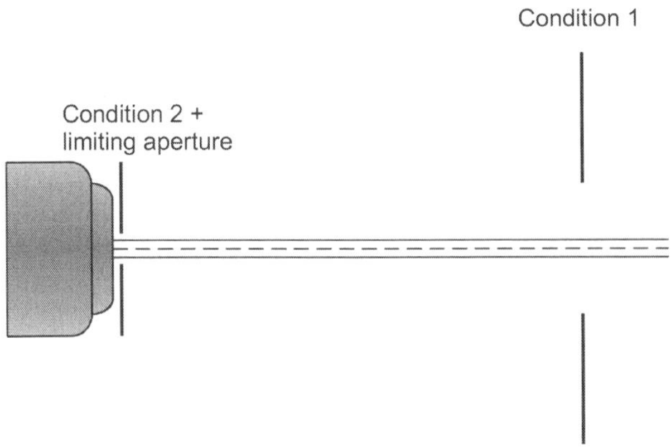

Figure 5.2. Classification measurements for a small-diameter collimated beam. In this case the apparent source is well to the left of the diagram, behind the emission aperture, and so both the condition 2 aperture and the limiting aperture (for naked-eye viewing) are positioned adjacent to the laser in order to be as close as possible to the required measurement position with respect to the apparent source. However, since in this illustration the beam is smaller than any of the apertures, they cannot obstruct any of the emitted radiation and so classification is based on the total emission of the laser. It also follows that any radiation passing through the limiting aperture must also pass through the apertures of both condition 1 and condition 2, and so a laser producing such a beam cannot be Class 1M or Class 2M.

aperture have to be placed adjacent to the laser output. This is the closest we can get to the required measurement position of 14 mm (for the eye-loupe condition aperture) or 100 mm (for the limiting aperture at wavelengths between 302.5 and 4000 nm) from the apparent source. Since the power (or energy) measured under the telescope condition or the eye-loupe condition will be equal to the total emission power (or energy) of the laser, it is the total emission that determines the laser's classification. Furthermore, if the beam is, as shown, also smaller than the relevant limiting aperture, it cannot exceed the AEL of Classes 1 or 2 under conditions 1 or 2 without also exceeding these limits under the naked-eye condition. It cannot, therefore, be Class 1M or 2M, whatever the level of output. For example, a collimated beam within the wavelength range 400–1400 nm that is less than 7 mm in diameter is classified simply on the basis of its total emission, and cannot be Class 1M or Class 2M.

Figure 5.3 shows a larger-diameter beam that overfills the aperture for the eye-loupe condition. Clearly, in this case, the power or energy passing through the aperture of the telescope condition will be greater, and so this condition will be the determining factor for Classes 1, 2, 3R and 3B. For the laser to be Class 1M or 2M, it must exceed the AEL for Class 1 or Class 2 (but not for Class 3B)

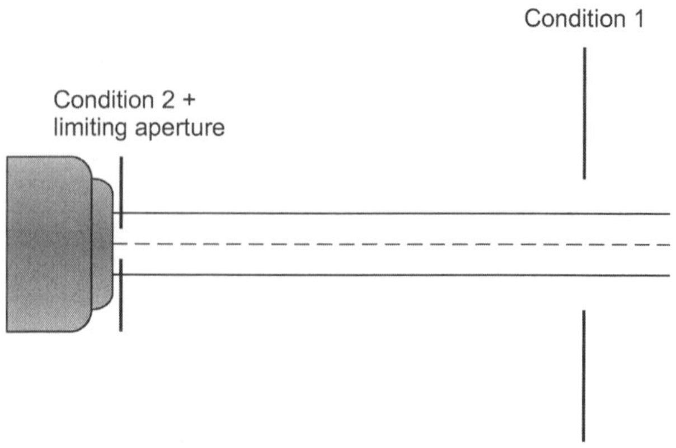

Figure 5.3. Classification measurements for a large-diameter collimated beam. Since the beam is collimated, condition 2 is irrelevant, as all radiation passing through the condition 2 aperture must also pass through the condition 1 aperture. It is the power or energy passing through the condition 1 aperture (or the limiting aperture in the case of classes 1M and 2M) that will determine the laser's classification.

under the telescope condition, but not exceed the Class 1 or Class 2 AEL under the limiting-aperture condition. For this beam configuration, therefore, the eye-loupe condition can be ignored.

In the case of divergent beams that effectively originate from a single point (e.g. bare laser diodes or the emission from singlemode optical fibres), it can be helpful to consider the collection angles of the various measurement apertures. These are shown in figure 5.4. With point-sources having highly divergent emission, the position of the apparent source is located at the actual emitting surface, and so the measurement apertures can be positioned at the required minimum distances from the source. Figure 5.4 illustrates that for such divergent emission the relevant AEL must not be exceeded under the eye-loupe condition for Classes 1, 2, 3R and 3B. This is because a higher power or energy would be measured under the eye-loupe condition and therefore the telescope condition can be ignored. For Class 1M or 2M, the determining factor (for a laser that fails Class 1 or Class 2 but not Class 3B) is, as before, that the AEL for Class 1 or Class 2 must not be exceeded under the limiting-aperture condition.

Not all laser beams, however, (and few LEDs) fall neatly into one of these three categories (that is, a narrow collimated beam as shown in figure 5.2, a large-diameter collimated beam as shown in figure 5.3, or a widely-divergent beam originating from a single point as shown in figure 5.4). In other situations a more careful analysis, applying all the relevant measurement conditions, must be carried out using direct measurement, calculations, or a combination of both.

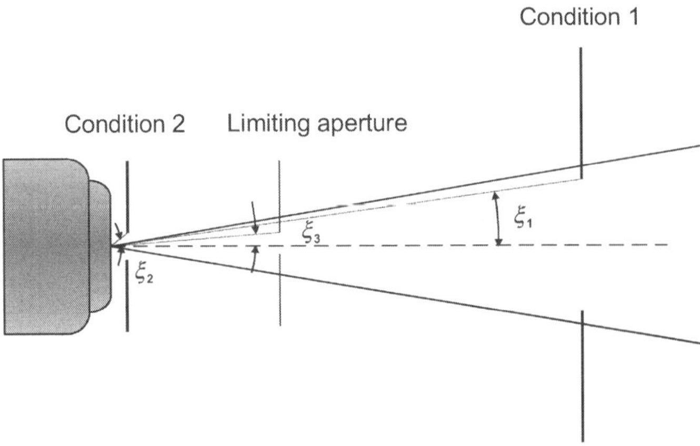

Figure 5.4. Classification measurements for a point source having highly divergent emission (e.g. a laser diode). It can be instructive in the case of divergent sources to consider the collection angles of the various measurement apertures. Condition 1 is irrelevant, since its collection angle ζ_1 is larger than the collection angle ζ_2 for condition 2. Any radiation that passes through the condition 1 aperture must also therefore pass through the condition 2 aperture, and so condition 1 cannot be a limiting condition.

5.3 Beam profiles

5.3.1 Gaussian beams

Most lasers have a beam profile (the distribution of irradiance or radiant exposure across the beam, perpendicular to the beam-propagation direction) that is far from uniform. This means, of course, that assessments of the proportion of the total radiation in the beam that passes through a given aperture cannot be determined simply by comparing the area of the beam with the area of the aperture.

Many lasers generate a beam having a cross-sectional distribution that is close to Gaussian, meaning that its profile can be described mathematically by a Gaussian-type equation. Such a profile has a maximum 'localized' irradiance in the centre of the beam, tailing off to zero at the edges, as shown in figure 5.5. (Although irradiance has the units of W m^{-2}, we can nevertheless define this value at any single point on the curve, and in any given position across a beam, in just the same way that we can express the speed of an accelerating vehicle in units of km h^{-1} at a particular instant in time.) For simplicity, the beam distribution shown in figure 5.5 has been normalized so that the irradiance in the centre of the beam has the value 1.0.

With such an irradiance distribution, the definition of beam diameter is inevitably arbitrary, since it is difficult to locate the precise edges of the beam. In fact, for a true Gaussian distribution, the curve never reaches the horizontal axis

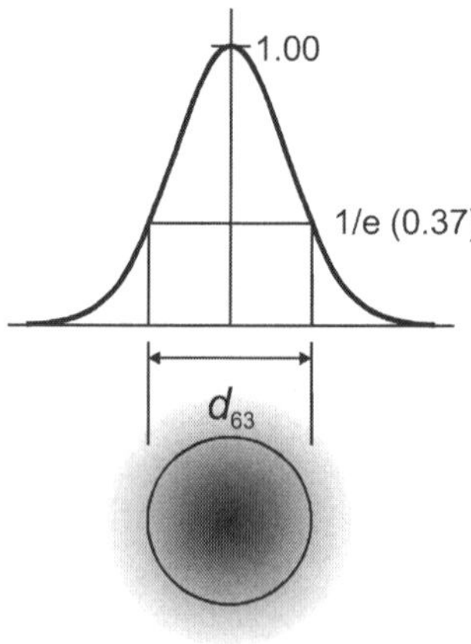

Figure 5.5. The cross-sectional irradiance profile of a Gaussian beam, having a 1/e diameter of d_{63}.

at all, merely getting ever closer to it as it continues to infinity on each side. But this is only a theoretical consideration: for practical purposes the Gaussian curve describes the behaviour of many real laser beams very well. The impracticality of clearly defining the edges of the beam, and hence the beam diameter, means that some convention must be adopted in order that an effective diameter can be specified. The commonly used criterion in laser safety assessments is that of the 1/e diameter. This is based on the positions at which the beam irradiance has dropped to 1/e, or 0.37, times the central, on-axis value. (Like pi, 'e' is a natural number. It has the value 2.718... and is the base number of the function that describes an 'exponential' decay, whereby a quantity is continuously halved and halved again.)

If the beam diameter is defined at the 1/e positions, i.e. by a line joining the two points on each side of the curve in figure 5.5 that have a value of 0.37 times that of the peak, then a circular aperture having this diameter and centred on the axis of a Gaussian beam will enclose 63% of the total beam power. For this reason the 1/e diameter is also known as the d_{63} diameter. Although the 1/e criterion may appear to be a rather arbitrary definition of beam diameter, its usefulness for describing the parameters of Gaussian beams can be illustrated as follows. If the total power (P) in the beam is divided by the area (A_{63}) defined by the d_{63} diameter, then the resultant value, which has the units of power per unit area, is equal to the peak (on-axis) value of the beam irradiance (E_0). It thus defines the maximum or (from the safety perspective) the worst-case value.

In circumstances where the beam is very much larger than the aperture of interest (this aperture is the measurement aperture in the case of classification or the limiting aperture in the case of exposure assessment), then the maximum (on-axis) value of the irradiance (E_0) or the radiant exposure (H_0) can be used as the basis for determining the power (P_a) or energy (H_a) passing through the aperture. This is because the irradiance or radiant exposure across the entire aperture is close to the peak value.

In the case of classification we then have

$$P_{\rm a} = E_0 A \qquad \text{or} \qquad Q_{\rm a} = H_0 A$$

where A is the area of the relevant aperture.

(Even if the beam diameter is similar to or smaller than the relevant aperture, the above approach may still be used as a simplification, although it will exaggerate the hazard and so err on the side of safety.)

In the case of exposure assessment (for the large-diameter beam previously assumed), E_0 or H_0 can be compared directly with the MPE to determine whether this level of exposure is safe and, if it is not, by how much it exceeds safe limits.

For a Gaussian beam, the value of the irradiance E_d at any position in the beam cross-section (e.g. across the profile shown in figure 5.5) can be determined from the equation –

$$E_d = E_0 \exp[-(d/d_{63})^2] \tag{5.1}$$

where d is twice the distance from the centre of the beam of the position at which the value of E_d is required. (In equation (5.1) and in similar equations, d and d_{63} can be expressed in any suitable units—as long as they are both in the same units—although millimetres will be commonly used. E_0 and E_d will normally be in units of watts per square metre.)

Frequently, however, laser manufacturers will specify the beam diameter using criteria other than 1/e. These can be converted to the 1/e parameters by re-arranging equation (5.1) to give

$$d_{63} = \frac{d}{\sqrt{-\ln(E_d/E_0)}} \tag{5.2}$$

where E_d is the irradiance at diameter d and E_0 is the value of the on-axis irradiance. Ln is the natural logarithm to base e.

The most common alternative used by laser manufacturers is the $1/\mathrm{e}^2$ condition, which is the diameter at which the irradiance has fallen to 0.135 ($1/\mathrm{e}^2$) times the on-axis value. The area defined by this diameter will enclose 87% ($1 - 1/\mathrm{e}^2$) of the total beam power, and so this diameter is often denoted by d_{87}.

However, the $1/\mathrm{e}^2$ (d_{87}) diameter can be readily converted to the 1/e value by using equation (5.2) and putting $E_d/E_0 = \mathrm{e}^{-2}$, from which $d_{63} = d_{87}/\sqrt{2}$. The case of non-Gaussian beam profiles is discussed in section 5.3.5.

5.3.2 Beam divergence

Any Gaussian beam expands from what is called a beam waist [2, 3]. This waist represents the smallest diameter of the beam, and may be located at the laser

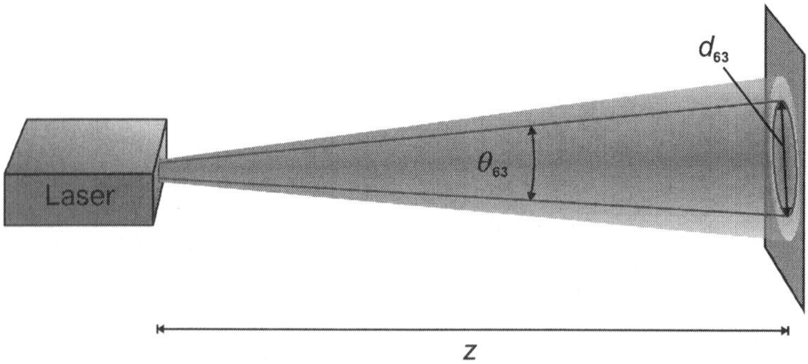

Figure 5.6. The divergence of a Gaussian beam denoted by the angle (θ_{63}) subtended at the beam waist by the d_{63} diameter. (This is in fact the far-field divergence. (See figure 5.8.))

emission aperture, inside the laser resonator or projected outside the laser by means of a lens which brings the emitted beam to a focus. (The waist can even be 'virtual', appearing to be located behind the laser.) Note that the beam waist and the apparent source are not necessarily the same entity. This is discussed in more detail in chapter 3.

In the same way as the diameter of a Gaussian beam can be expressed on the basis of the 1/e condition, so beam divergence, the angular spread of the beam in what is known as the far-field (that is, at distances from the beam waist that are significantly larger than the size of the beam waist), can be defined in a similar way (figure 5.6). The 1/e beam divergence, θ_{63}, is then simply defined as the angle (measured in either degrees or radians) subtended by the beam diameter d_{63} at a distance z from the source. The position $z = 0$ from which the distance z is measured (at viewing positions in the far-field) corresponds to the point from which the beam appears to diverge, and can therefore be considered to be the location of the source (or strictly, the *apparent* source). This location may or may not coincide with the actual emission aperture of the laser. Quite often, it can be inside the laser resonator, but may be anywhere along the beam axis, behind or in front of the laser. In the case of an emitted beam that is focused by a lens, a secondary apparent source will be formed beyond the lens.

At any distance z in the far-field, the d_{63} diameter will be $\theta_{63}z$, where θ_{63} is measured in radians. If θ_{63} is specified in degrees, the d_{63} diameter is $2z\tan(\theta_{63}/2)$. The latter expression simplifies to $z\sin\theta_{63}$ if θ_{63} is reasonably small.

In the case of a laser diode, the divergence of the emitted beam, which usually has an elliptical cross section because of diffraction effects arising from the very small rectangular area of the emission aperture, is often specified at the positions at which the irradiance is exactly half the on-axis value. This is known as the full-width half-maximum value (FWHM) and, for an elliptical beam, will

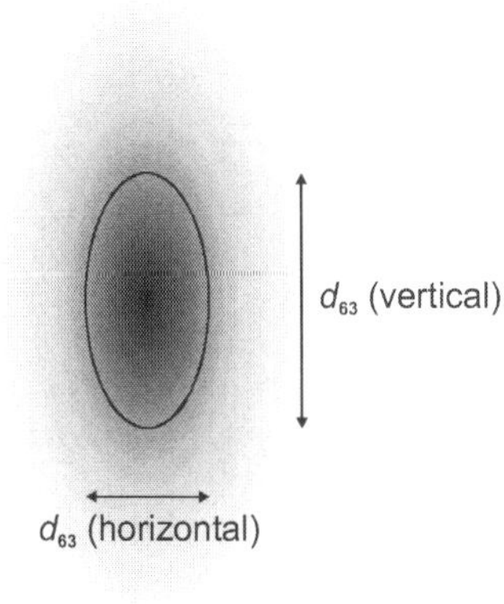

Figure 5.7. Some laser beams, typically those emitted by laser diodes, can have an elliptical cross section, so that the vertical and horizontal diameters (and also the corresponding beam divergence angles) have to be separately specified.

be defined in two orthogonal planes. Using equation (5.2), therefore, and putting $E_d/E_0 = 0.5$, gives a value for θ/θ_{63}, or d/d_{63}, of 0.83. The two orthogonal beam diameters for an elliptical beam are shown in figure 5.7. Such non-circular beams are referred to as astigmatic.

Although exact formulations can be used to perform safety calculations involving elliptical Gaussian beams [1], for most practical purposes it is sufficient to use the approximation of an equivalent circular beam. The effective diameter of this circular beam can be assumed to be

$$d_{63} = \sqrt{(a_{63}b_{63})} \tag{5.3}$$

where a_{63} and b_{63} are the respective orthogonal beam diameters.

The dimensions of the beam emitted from a laser diode, measured in either the vertical or horizontal direction, will increase linearly with increasing distance (i.e. the beam has twice the diameter at twice the distance from the source), as the above equations for d_{63} suggest. However, this is only so because of the very small dimensions of the emitting area compared with the typical measurement distances. (With a laser diode, these distances are likely to be at least several hundred times the dimensions of the emitting area.) Such distances are said to be in the far-field of the laser, in which the equations given earlier for the beam diameter as a function of the beam divergence are a very close approximation. For other laser beams, however, this may not be the case.

The beam diverges away from the waist in a manner described by the following equation (see figure 5.8).

$$d_{63} = \sqrt{d_0^2 + (\theta_{63}z)^2} \tag{5.4}$$

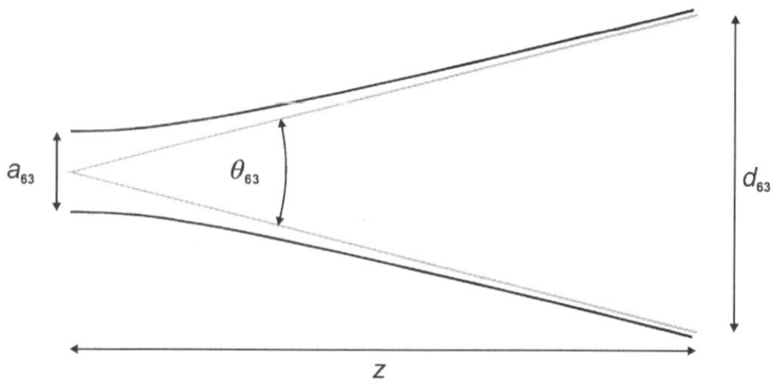

Figure 5.8. As it diverges from the beam waist, the beam has a curved profile. Close to the position of a waist, in the 'near-field' (on the left of the diagram), the beam diameter does not increase linearly with increasing distance, but only gradually assumes its 'far-field' divergence (to the right of the diagram). This can be important in the analysis of well-collimated laser beams. At distances that are large in comparison with the waist diameter, however, far-field conditions apply. This is normally the case at all distances of practical interest for beams having very small waists, such as those produced by laser diodes and the emission from single mode optical fibres.

where d_0 is the diameter of the waist, θ_{63} is the beam divergence andd_{63} is the beam diameter at distance z. The values of d_0, θ_{63} and d_{63} are expressed in terms of the 1/e criterion. In fact, we should define the beam divergence, θ_{63}, more properly here as the *far-field* divergence. In other words it is the value of divergence that the beam will gradually assume at increasing distances from the source.

Initially, in what is termed the *near-field*, the beam expands more slowly than its far-field divergence would indicate. The envelope or profile of the beam in a plane containing the beam propagation axis (as shown in figure 5.8) is curved. These curves, representing the top and bottom of the beam, gradually flatten out as the distance from the beam waist increases, becoming closer and closer to the straight lines representing the far-field divergence. This is not to say that light does not travel in straight lines: the curves merely represent the envelope of the beam (in fact of the d_{63} diameter), and are not intended to designate rays. A similar effect of a curved beam profile can be demonstrated with beams formed from non-laser sources (such as a conventional optical projector, for example).

It can be seen from equation (5.4) that as the distance z from the waist increases, the effect of the d_0^2 term in the equation diminishes. When $\theta_{63}z$ is much larger than d_0 (i.e. outside the Rayleigh range, discussed further in chapter 3) the equation simplifies to the far-field equation

$$d_{63} = \theta_{63}z. \tag{5.5}$$

As mentioned in the discussion of units in section 5.1, while d_{63} in equation (5.5) will be in metres if z is expressed in metres and θ_{63} in radians, d_{63} will be in millimetres if z is in metres but θ_{63} is in milliradians, which is usually a more convenient way of using this equation.

With well-collimated beams, the near-field, within which equation (5.4) applies, can extend for considerable distances from the laser. Use of the simpler equation (5.5) in the near-field would underestimate the beam diameter and therefore *overestimate* the hazard (by assuming that a given laser power is distributed over a smaller area than would be the case). Equation (5.4) will always give the correct solution for the diameter of a Gaussian beam at any distance from the waist (i.e. in the near-field or the far-field). The shorter equation (5.5) is only valid under far-field conditions. Far-field conditions are always created, however, whenever a collimated Gaussian beam is focused using a highly converging lens to produce a very small waist. Since the distance z is always measured from the waist, the actual distance from the laser will in this case be greater than z. Whenever systems external to the laser are used to change the beam geometry, distances used for beam-size calculations must therefore be specified from the correct reference position. This may not necessarily coincide with the position of the laser emission aperture.

Far-field conditions can normally be assumed to apply in the following circumstances.

(a) Emission from a single bare (i.e. unlensed) laser diode.
(b) Divergent emission beyond the point of focus of a collimated beam that is focused by a lens.
(c) Emission from single optical fibres having very-small core diameters (e.g. singlemode fibres).

It is the far-field divergence condition that is illustrated in figure 5.6.

Near-field conditions should be assumed for all well-collimated (low-divergence) beams unless distances are such that the far-field approximation can be shown to be valid.

5.3.3 Fractional power through apertures

As has already been noted, many laser safety calculations, whether for purposes of classification or for exposure assessment, require the determination of the quantity of radiation (in terms of either power or energy) that passes through a given aperture at a given location. In the case of product classification, assessments must be made of the radiation passing through the relevant measuring apertures shown in figure 5.1. For exposure assessment, the power or energy contained within the relevant limiting aperture may need to be determined. This power or energy, divided by the area of the limiting aperture, can then be compared with the MPE to ascertain whether or not the exposure is within safe limits.

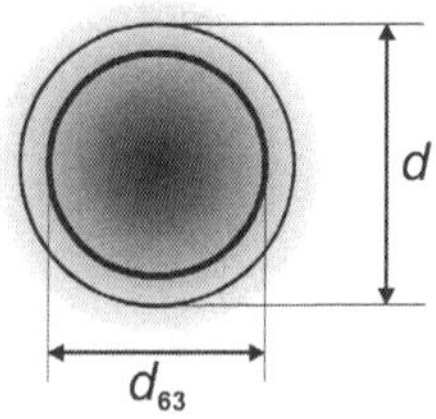

Figure 5.9. The fraction of the total beam power (or beam energy) that is contained within a circular, centrally-positioned aperture of diameter d for a circular Gaussian beam having a 1/e diameter of d_{63} may be calculated using equation (5.6) given in the text. (d can be larger or smaller than d_{63}.)

For such exposure assessments, the limiting aperture should be located at the most hazardous position at which human exposure might reasonably occur. For ocular exposure this is normally taken as no closer than 100 mm from the position of the apparent source. This is because 100 mm is well within the eye's near-point—the minimum distance at which a normal eye can focus—and so accidental exposure at positions closer to the source than this can be considered very unlikely. Furthermore, for radiation within the retinal hazard region, direct viewing of a small laser source (a bare laser diode, for example) from a distance of less than 100 mm can only result in an out-of-focus image of the source being formed on the retina. The incident power or energy at the retina is thus spread over a larger area and is in consequence less hazardous.

Some people can have shorter near points than 100 mm. Such people (called myopes) can have rather closer near-points when their eyes are uncorrected (that is, when not wearing corrective spectacles). Myopes can exploit this when wishing to see fine detail, and often do so without conscious thought, by removing their spectacles so that they can bring their eyes closer to the object of interest and thus form a larger, focused image of the object on their retina. However, taking all factors into account, standards committees have considered the 100 mm distance to be sufficiently conservative.

The proportion of the total power or energy of a Gaussian beam that passes through a circular aperture can be determined from the following equation (see figure 5.9).

$$\eta = 1 - \exp[-(d/d_{63})^2] \tag{5.6}$$

where η is the fraction of the total power or energy passing through an aperture of diameter d. (d can be any value larger or smaller then the 1/e diameter of the beam (d_{63}).) The value used for d_{63} is, of course, the 1/e diameter of the beam at the aperture location.

If P is the total power in the beam, therefore, the power P_a passing through the aperture will be ηP. It can often be useful to use this expression to determine the maximum power (for a given beam configuration) that is allowable for a laser to be in a given product class or for the exposure at a given location to be safe (i.e. no greater than the MPE). In this case the maximum beam power P permitted is P_a/η, where P_a is the applicable power limit and η is the fractional power through the limiting aperture. For product classification, P_a is set equal to the relevant AEL and η is the value applicable to the relevant classification aperture.

To determine the maximum allowable power in order that an exposure to the beam at a given position along the beam be safe, P_a is set equal to the value of the MPE multiplied by the area of the limiting aperture (i.e. the maximum allowable power $P = \mathrm{MPE} \cdot A_L/\eta$, where A_L is the area of the relevant limiting aperture).

5.3.4 Emission from optical fibres

In some laser applications the output of the laser is not utilized directly, but is coupled instead into an optical fibre in order to convey the laser radiation to a distant location. It is in this case the emission from the far-end of the fibre that may need to be evaluated. This can be particularly relevant in optical-fibre communication technology, where emission from optical fibres may need to be assessed for safety purposes. This includes the requirement defined in IEC 60825-2 that the *hazard level* that could become accessible at defined locations is assessed in a similar way to the product class. Evaluation of the emission from optical fibres can also be important in medical applications, where fibre delivery is frequently used.

Optical fibres are formed from two transmissive materials that have differing refractive indices (figure 5.10). The denser medium (higher refractive index) forms the core of the fibre while the less dense medium (lower refractive index) surrounds this to form the cladding.

The refractive index n is a measure of the degree to which light is 'bent' on passing from one material to another, such as from air, which has a refractive index of 1, into water, which has a refractive index of around 1.3. The ratio 1/1.3 is then equal to the ratio of the speed of light in each of the two materials. It is also the inverse of the ratio of the trigonometric sines of the incident and emergent ray angles, measured from a line perpendicular to the interface (the water surface) between the two materials.

The ray angles are always greater in the less dense medium. Where light is incident at the interface from the more dense medium—the water—there is a certain angle of incidence, known as the critical angle, for which the emergent angle is 90 degrees and the refracted light is directed *along* the surface. At sufficiently oblique angles, therefore, where the critical angle is exceeded, light cannot pass through the surface into air but is instead reflected back into the water by what is known as total internal reflection. This effect can be observed by underwater swimmers. Looking vertically upwards from below, the swimmer can see through the water surface. Looking at more oblique angles, however, the surface of the water acts as a mirror, appearing silvery, and it is impossible to see through it. (For a water–air interface, the critical angle is about 50 degrees.)

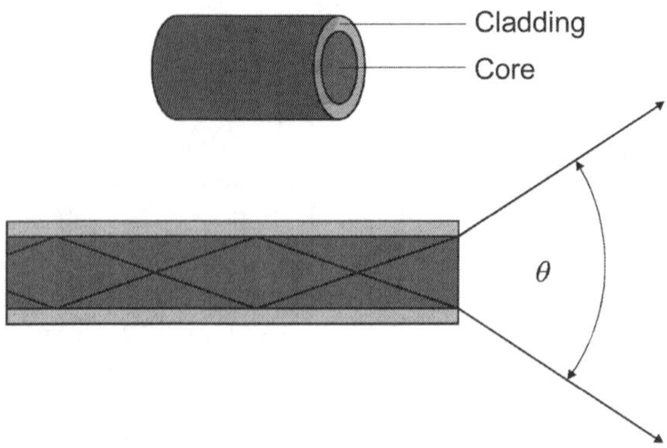

Figure 5.10. The transverse (upper view) and longitudinal (lower view) cross section of a multi-mode optical fibre, comprising a transparent inner core and a transparent outer cladding of lower refractive index than the core. The core diameter is sufficiently large for several 'modes' (ray angles) to be propagated along the fibre. The cone angle of the emission from the end of the fibre is a function of the refractive indices of the core and the cladding.

In the case of optical fibres, the difference in the refractive index between the core and the cladding is much smaller than that between air and water. The critical angle is therefore considerably more oblique and the range of angles over which total internal reflection can occur is much smaller. Nevertheless, light rays within this angular range are 'trapped' within the core, and are therefore carried along the fibre (suffering some loss due to absorption along the way), emerging from the end as a diverging cone of radiation. The range of angles within which this can occur is defined by the numerical aperture of the fibre, or NA.

The NA of an optical fibre is given by

$$NA = \sqrt{n_1^2 - n_2^2} \tag{5.7}$$

where n_1 is the refractive index of the core and n_2 that of the cladding.

The angle θ, which defines both the acceptance and emission angle of the fibre (the range of angles over which light can be coupled into the core and over which it can emerge from the far end), is given by

$$\theta = 2[\arcsin(NA)]. \tag{5.8}$$

Note, however, that this angle is not the same as the range of ray angles actually carried inside the fibre, since refraction effects at each end of the fibre (where light enters or leaves) cause the range of angles in air to be greater than those in the fibre core.

For very short lengths of fibre only certain 'modes', that is, discrete ray angles, can be carried by the fibre, because rays at other intermediate angles are cancelled out due to interference effects (where rays that have travelled over different paths within the fibre arrive at the same point having exactly opposite phases). Over greater lengths, however, 'mode mixing' can occur due to bending of the fibre and the consequent changes to the angles at which individual rays strike the core/cladding interface. This results in complete filling of the available angular cone defined by the fibre's numerical aperture. (Note, however, that with large-diameter fibres the apparent source is not necessarily located at the end of the core, but can be inside the fibre.)

Such a fibre is called a *multi-mode* fibre because of its ability to carry many different modes (ray angles) within the fibre. The emergent beam angle θ which is defined in terms of the numerical aperture and given by equation (5.8), is not, however, the same as the θ_{63} value, which is smaller. θ_{63} is given by

$$\theta_{63} = 2NA/1.73 \tag{5.9}$$

and the 1/e diameter of the beam at any distance z from the fibre can therefore be found from

$$d_{63} = 2zNA/1.73. \tag{5.10}$$

(The factor '2' is left in equations (5.9) and (5.10) to aid comparison with alternative expressions that are sometimes used to define half-angles and beam radii instead of θ_{63} and d_{63}.)

Because different modes within a multi-mode fibre are carried at different speeds (due to the different path-lengths of the modes within the core), the time taken to reach the end of the fibre will vary very slightly. Where very short-duration pulses are being transmitted along the fibre, this effect can give rise to appreciable pulse spreading, limiting the ability of the fibre to carry optical signals over long distances at high modulation rates.

In many optoelectronic applications, therefore, *single mode* fibres are used. These have very small cores of only a few microns in diameter in which only a single mode can be carried (figure 5.11). The geometry of the emerging cone of radiation from the end of the fibre is then governed by diffraction effects (the spreading of light by structures that are on the same scale as the optical wavelength). The 1/e beam diameter at a distance z from the end of the fibre is then given by

$$d_{63} = (2\sqrt{2}z\lambda)/(\pi\omega_0). \tag{5.11}$$

Here the parameter ω_0, the fibre's mode-field diameter, is used rather than the actual diameter of the core to define the diameter of the area within which diffraction occurs. This is because, in single mode fibres, some of the energy is carried in the cladding (i.e. the propagating 'mode' is slightly larger than the core). The emission wavelength is λ, which is expressed in the same units as ω_0.

5.3.5 Non-Gaussian beams

The previous sections of this chapter have discussed the behaviour of Gaussian laser beams. This analysis is applicable to many lasers (especially low-power

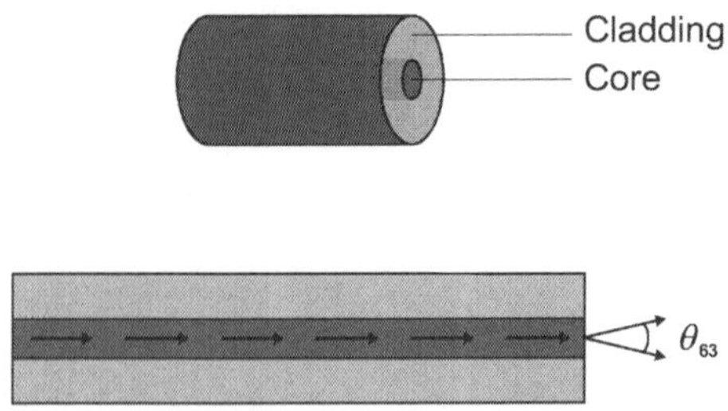

Figure 5.11. Single mode optical fibres have a much smaller core diameter than multi-mode fibres, allowing only a single 'mode' to be propagated along the fibre. The cone angle of the fibre emission (here indicated in terms of the 1/e divergence) is in this case governed by diffraction effects from the end-face of the core.

gas lasers, laser diodes, and optical-fibre emission) for which the output closely follows a Gaussian distribution. One important aspect of Gaussian emission is that a beam that is truly Gaussian will have a Gaussian profile (as in figure 5.5) at all positions along the beam propagation axis, even though the beam diameter will change. (Such a beam is referred to as being in a TEM_{00} mode.) This holds true provided that the beam is not truncated (that is, it does not have its edges 'clipped' by being passed through an aperture of smaller diameter than the beam). Where this does occur, the beam beyond the aperture will no longer be Gaussian (quite obviously, since part of the beam has been 'removed' by the aperture), although the proportion of radiation contained within the truncating aperture will be as described in section 5.3.3.

Many lasers, however, and the majority of LEDs, produce beams that do not have such well-defined properties and cannot be adequately described by a Gaussian function. In fact, beams that appear to have a Gaussian distribution at one position along the beam axis cannot be assumed to be Gaussian unless they also have such a distribution at another arbitrarily-selected position along the beam.

Departure from a Gaussian form often arises because of distortion effects within the laser resonator due to heating. This is especially true of high-energy pulsed solid-state lasers, in which significant thermal distortion can arise within the laser material. In many other cases, high-power lasers are deliberately designed to produce non-Gaussian beams in order that the maximum possible laser power is emitted by the laser resonator. This can enhance the laser's performance without significantly limiting the ability of the emitted beam to be focused down to produce a high level of irradiance or radiant exposure.

Unlike Gaussian beams, the profile of non-Gaussian beams changes as the beam spreads at increasing distances from the source. The analysis of such beams can be complex. A standardized method for dealing with both Gaussian and non-Gaussian beams is based on the second moment diameter [4]. This is a generalization of the $1/e^2$ criterion and can be consistently used to characterize non-Gaussian beams. It treats the irradiance distribution across the beam in the same way as a statistical distribution. A first step in determining the beam diameter based on the second moment method is to identify the centre of the beam profile (similar to the centre of gravity of an object or the mean of a statistical distribution). The second moment diameter is then equal to four times the standard deviation of the irradiance distribution. If the diameter is determined at different positions along the beam, the envelope thus described forms a hyperbola. For Gaussian beams the second moment diameter is identical to the $1/e^2$ diameter.

Any radially-symmetric beam, whether is has a Gaussian irradiance distribution or not, requires three parameters to characterize it. These can be obtained using the second moment method, and are:

(i) the location of the beam waist;
(ii) the waist diameter;
(iii) the far-field divergence.

For asymmetric (i.e. non circular) beams, each of these parameters can be divided into horizontal (x) and vertical (y) components. Additionally, some beams are astigmatic, (fan beams, for example), and have different waist locations in the x–z and y–z planes (z representing the propagation direction).

Other methods, such as those based on $1/e$, $1/e^2$ and FWHM criteria which were discussed earlier in this chapter, do not adequately describe the propagation characteristics of a non-Gaussian beam, nor can they be readily converted into second-moment values. Furthermore, the derivation of such values for non-Gaussian beams can be ambiguous.

To determine the second-moment beam diameter, an accurate profile of the irradiance across the beam has to be obtained. This can be difficult: possible techniques for doing this include using a CCD camera with appropriate image-processing software, the use of a variable aperture, a moving knife-edge or a moving slit, although the use of a CCD camera is preferable. Commercial systems are available that incorporate algorithms for determining beam parameters in accordance with ISO definitions.

A commonly-adopted approach for characterizing non-Gaussian beams is the use of what is known as the beam propagation ratio, which used to be called the 'quality factor' or 'times diffraction limit factor' M^2 [4]. This parameter defines the extent to which a real laser beam departs from the ideal form of a perfect, or 'diffraction limited', Gaussian. (All optical radiation, being a wave motion, is subject to spreading effects due to diffraction. This means that it is impossible to produce a perfectly parallel beam or to focus light down to an infinitely-small spot. The best that can be achieved is 'diffraction-limited' performance, meaning that beam spreading is the minimum that is theoretically possible.)

For a perfect (i.e. diffraction limited) beam M^2 is equal to 1, while for a real (imperfect) beam M^2 has a value greater than 1. The diameter of the spot produced by focusing the beam with an aberration-free lens will then be M^2 times that of a pure Gaussian, and the irradiance produced in the focal spot will $1/M^4$ times that for a Gaussian beam.

A related parameter that is often used is the beam propagation factor K, where $K = 1/M^2$.

The beam propagation ratio is defined by the expression

$$M^2 = (\pi d_{0\sigma}\theta_\sigma)/(4\lambda) \tag{5.12}$$

where $d_{0\sigma}$ is the second moment waist width and θ_σ is the far-field divergence angle determined by the second-moment method [4]. Referring back to section 5.1 and the discussion of units, equation (5.12) is correct when θ_σ is in units of radians and when $d_{0\sigma}$ and the wavelength λ are given in metres. However, expressing the wavelength in metres would be unusual, and so it should be noted that the above equation is also valid if $d_{0\sigma}$ is given in millimetres, θ_σ in milliradians and λ in microns.

As discussed in chapter 3, particular caution needs to be adopted whenever assessing the behaviour of beams whose properties are not sufficiently well characterized, or when using models for beam propagation that are not exact. This is especially so for non-Gaussian beams and emission from LEDs. This can also be the case for multiple, overlapping beams, such as those from laser-diode arrays. (Such arrays can comprise several hundred separate diode emitters which can produce a combined output that is highly incoherent. Diode arrays in which the individual emitters are phase-locked, however, have more controllable and therefore more coherent emission.)

If reasonable approximations cannot be made of the output characteristics of a laser or LED, then direct measurements of the actual beam parameters that are specified in the safety standard for classification or exposure assessment should be carried out, rather than relying on extrapolation or calculation based on a simplified (and possible inaccurate) model of beam propagation.

5.4 Hazard distance

A knowledge of a laser's hazard distance can be very useful in assessing the risk. It indicates the extent of the area within which a hazard might exist. This is especially relevant for highly-divergent beams, where it may turn out that the radiation hazard, regardless of the laser's actual class, is limited to the immediate vicinity of the laser. With other lasers, however, the hazard distance can be considerable, sometimes extending to several kilometres. Whenever lasers are used out-of-doors or in other open-plan environments, a determination of the hazard distance is usually essential.

The distance from the laser over which the radiation hazard of a laser extends is normally based on the hazard to the eye, and is therefore known as the nominal ocular hazard distance (NOHD). The NOHD is defined as the distance from the

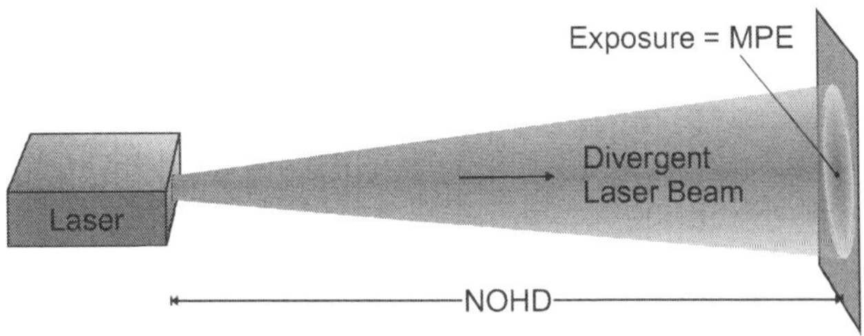

Figure 5.12. The nominal ocular hazard distance (NOHD) is the distance from the laser at which the irradiance or radiant exposure has dropped to the level of the MPE. An exposure of the unaided eye to laser radiation beyond the NOHD cannot therefore exceed the MPE.

laser at which the maximum (usually on-axis) exposure is equal to the ocular MPE (figure 5.12). Within the NOHD, therefore, an exposure to the beam will exceed the MPE and may cause injury; beyond the NOHD it is safe for the unaided eye. Any appropriate time base can be used in determining the MPE and therefore in specifying the NOHD. This can be limited to 0.25 s for emission in the visible band if reliance for protection is placed on natural aversion responses.

An exposure is equal to the MPE if the power contained within the area of the relevant limiting aperture is equal to the value of the Class 1 AEL (provided that the same time-base is used in the specification of both the MPE and the AEL). We can therefore rearrange and combine equations (5.4) and (5.6) to determine the distance z at which the fractional power η contained within the limiting aperture of diameter d_{ap} for a beam of divergence θ_{63} is equal to the level of the Class 1 AEL. The distance z is then equal to the NOHD and so

$$\mathrm{NOHD} = \frac{1}{\theta_{63}}\sqrt{\frac{-d_{\mathrm{ap}}^2}{\ln(1-\mathrm{AEL}_{\mathrm{Class\ 1}}/P)} - d_0^2} \tag{5.13}$$

where P is the total power emitted by the laser, d_0 is the 1/e waist diameter and ln is the logarithm to base e. The NOHD is then the hazard distance measured from the position of the beam waist. The NOHD will of course be in metres if d_{ap} and d_0 are in metres and θ_{63} is in radians, but will also be in metres when d_{ap} and d_0 are expressed in millimetres and θ_{63} is given in milliradians.

The assessment can be simplified wherever the beam diameter (at the NOHD) is considerably larger than the limiting aperture. This will be the case for highly-divergent beams, and also for less divergent high-power beams where the NOHD can be expected to be large. We can then simply determine the distance at which the on-axis value of the beam irradiance is equal to the MPE. This gives the more commonly used expression

$$\mathrm{NOHD} = \frac{2}{\theta_{63}}\sqrt{\frac{P}{\pi\,\mathrm{MPE}}}. \tag{5.14}$$

In this equation, where P is given in watts and the MPE is in watts per square metre, θ_{63} must be defined in radians in order for the NOHD to be calculated in metres. (If θ_{63} is in milliradians then the NOHD will be in kilometres, often appropriate for high-power beams of low divergence, although the mitigating effects of atmospheric transmission can then also be taken into account, as discussed later in this section. Where the MPE is given in joules per square metre, then the energy Q, in joules, must, of course, be used rather than the power P.) Equation (5.14) can also be used instead of equation (5.13) to provide a simpler, conservative (i.e. exaggerated) estimate of the NOHD for any beam that is *not* significantly larger than the limiting aperture.

If θ_{63} is expressed in degrees instead of radians, the equation becomes

$$\mathrm{NOHD} = \frac{1}{\tan(\theta_{63}/2)}\sqrt{\frac{P}{\pi \mathrm{MPE}}}. \qquad (5.15)$$

Examples of NOHD calculations for some of the common beam-configuration situations that are discussed in this chapter are given in table 5.1.

The value of the NOHD that is calculated using equations (5.14) and (5.15) is always measured from the point at which the beam appears to diverge. This point is not necessarily located at the laser's emission aperture, of course. In the case of a beam that is being focused by a lens, for example, the hazard distance beyond the lens that is given by the previous equations will be based on the position of the focal spot. It will be necessary to add to this the distance between the lens and the focal spot in order to determine the actual hazard distance from the lens. (In the case of a lens of focal length f that is being used to focus a well-collimated beam of diameter d_{63}, the angular divergence of the beam beyond the point of focus will be d_{63}/f, see table 5.1.)

Exposure levels that are below the MPE and therefore occur at positions beyond the NOHD can only be considered safe if optical viewing instruments are not used. The effect of such aids on both the exposure level and the hazard distance is discussed in section 5.6.

A knowledge of the NOHD, combined with an understanding of how the laser might move and how the beam could be reflected, can be used to map out the area within which hazardous levels of exposure could exist. This is the nominal ocular hazard area, or NOHA, sometimes referred to simply as the nominal hazard zone (NHZ). This may simply be a circular area centred on the laser and having a diameter equal to the NOHD. If there are physical limits on how the beam can move or could be reflected, however, then the NOHA may be restricted to a smaller area.

In some applications where both eye and skin hazards exist but where eye protection is being worn, it can sometimes be useful to determine the skin-hazard distance, i.e. the distance within which skin injuries could occur. This can be done in the same way as for NOHD and NOHA, but using the skin MPE instead of the ocular MPE. The equivalent 'skin-safe' power to be used in equation (5.13) in

Table 5.1. Examples of NOHD calculations. The NOHD is calculated for a number of common situations, demonstrating how widely the hazard distance can vary for a single laser. The laser is a Nd:YAG (wavelength 1064 nm), and has an output of 1000 W. The beam diameter at the output of the laser (d_{63}) is 5 mm, and the far-field divergence (θ_{63}) is 1.5 mrad. The MPE, derived from IEC 60825-1, is 50 W m^{-2} for an exposure duration of 10 s or more. For more detailed information on these exposure situations, refer to the relevant sections of this chapter.

NOHD	Formula	Schematic
Unmodified laser beam *NOHD* = 3364 m	$NOHD = \frac{2}{\theta_{63}}\sqrt{\frac{P}{\pi\, MPE}}$	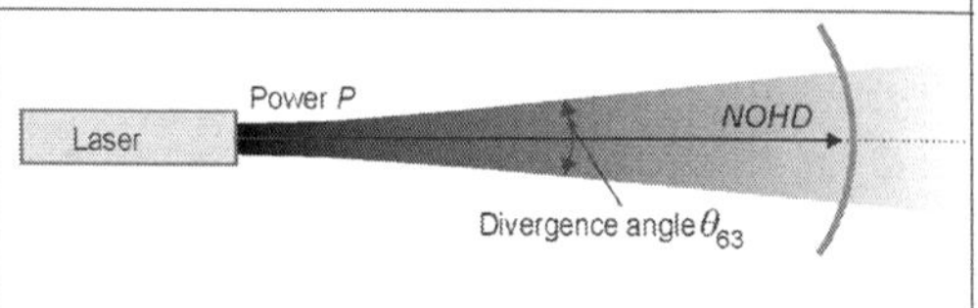
Multimode fibre *NA* = 0.25 *NOHD* = 17.5 m	$NOHD = \frac{1.73}{NA}\sqrt{\frac{P}{\pi\, MPE}}$	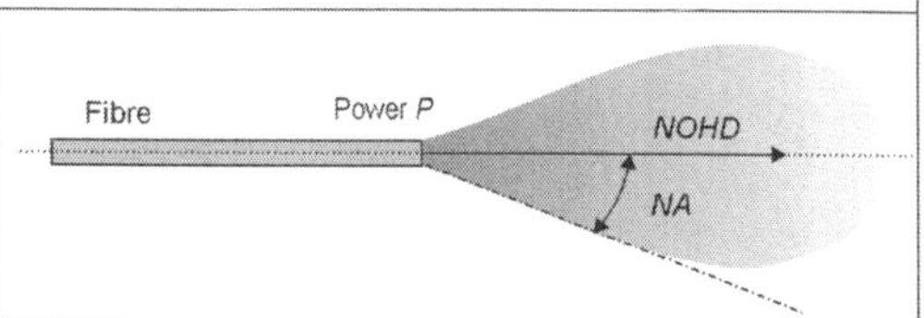
Diffuse reflection *R* = 0.9 *ß* = 0 deg *NOHD* = 2.4 m	$NOHD = \sqrt{\frac{RP\cos\beta}{\pi\, MPE}}$	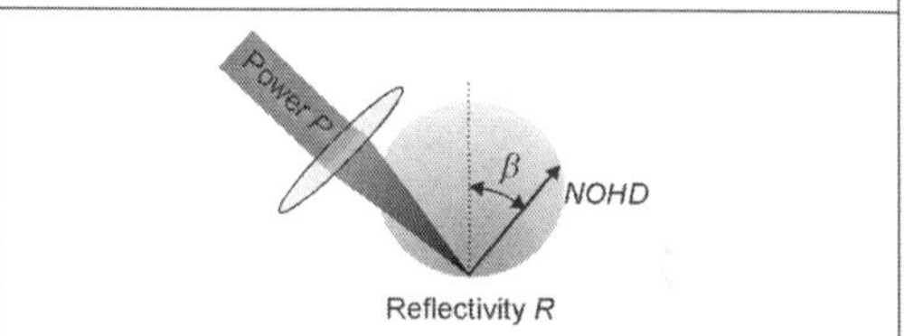
Focused beam *f* = 25 mm *NOHD* = 18 m	$NOHD = \frac{2f}{d_{63}}\sqrt{\frac{P}{\pi\, MPE}}$	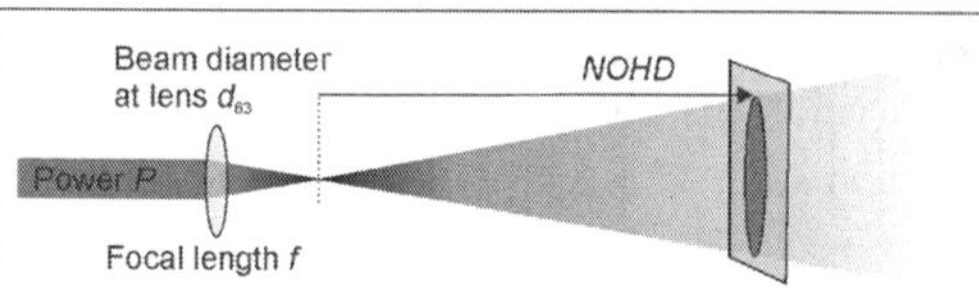

place of the value of the Class 1 AEL will then be equal to the relevant skin MPE multiplied by the area of the applicable limiting aperture.

At wavelengths where the skin and eye MPEs are identical, the skin-hazard distance will, of course, be the same as the NOHD.

Where long beam paths are being used in outdoor applications, account can be taken of losses due to the atmosphere. The principle effects of the atmosphere on a propagating laser beam are:

(a) absorption, caused mainly by the constituent molecules of air;
(b) scattering, due to atmospheric aerosols (minute dust particles and water droplets);

(c) distortion, arising from small-scale temperature fluctuations in the atmosphere called *turbulence*.

While absorption effects are largely constant over time (although highly dependent on the laser wavelength), both atmospheric scattering and turbulence can vary widely.

Levels of atmospheric attenuation due to absorption and scattering are expressed in terms of the extinction or attenuation coefficient ξ, which varies greatly with wavelength and atmospheric conditions. This is the reciprocal of the distance over which the beam would be attenuated to 1/e of its initial power, and is given in units of km^{-1}. The transmittance of the atmosphere τ, measured over a path length of z (expressed in km, since ξ has the units of km^{-1}) is then given by

$$\tau = \mathrm{e}^{-z\xi}. \tag{5.16}$$

Since the on-axis irradiance in the far-field of a laser beam (in the absence of any atmospheric attenuation) is given by

$$E_0 = \frac{4P}{\pi(\theta_{63}z)^2} \tag{5.17}$$

we can combine equations (5.16) and (5.17) to produce a form of what is sometimes called the laser range equation, which takes into account atmospheric transmission losses. The equation is

$$E_0 = \frac{4P\mathrm{e}^{-z\xi}}{\pi(\theta_{63}z)^2}. \tag{5.18}$$

The atmospheric attenuation coefficient ξ at a wavelength of 600 nm varies from 0.06 km^{-1} in exceptionally clear conditions to 0.7 km^{-1} in medium haze and about 10 km^{-1} in moderate fog. Over a pathlength of 5 km, these values correspond to an atmospheric transmittance of 0.74 (i.e. 74%), 0.03 (3%) and almost zero, respectively. These figures refer to direct or 'line of sight' transmission, and are those applicable to laser-beam propagation. Since atmospheric scattering rather than absorption is the principle mechanism of loss within the visible band, however, the transmission of sunlight through the atmosphere under the same prevailing conditions can be much higher than the values given above would suggest. This is due to the effects of multiple scattering, whereby sunlight which is scattered by atmospheric aerosols can still penetrate the atmosphere to ground level. This is readily understood by considering why it is light under daylight cloudy or foggy conditions. There may be no direct line-of-sight to the Sun (i.e. the Sun's disc cannot be seen), yet multiple scattering or diffusion, ensures that a considerable amount of light from the Sun nevertheless reaches the Earth's surface. The power of a laser beam, on the other hand, could be so dissipated under severe scattering conditions as to be undetectable.

In addition to the effects of attenuation caused by absorption and scattering, atmospheric turbulence may also have a significant impact on a propagating laser beam. Turbulence effects can vary greatly, depending on the geometry of the laser emission, the length of the beam-path through the atmosphere, and the prevailing atmospheric conditions. These effects can extend from beam wander (small-scale oscillations of the beam's position) to beam spreading (increased divergence) and beam break-up (creating multiple smaller beams). Rapidly fluctuating 'hot spots' (scintillation) can also occur within the beam, which may create localized levels of irradiance that are greater than the levels that would occur in the absence of turbulence. Since scintillation patterns change constantly, however, this effect can be more serious with a single high-energy laser pulse (which will produce a 'snap shot' of the turbulent conditions at a given moment in time) compared with cw or multiple-pulse emission where the effects of scintillation will be averaged out.

5.5 Beam reflections

Laser radiation hazards can arise not only from direct exposure of the eyes or skin to the laser beam, but also from reflections that may occur at any surface that intercepts the beam. Unintended reflections should generally be regarded as a hazard, as they can redirect a hazardous laser beam in unexpected directions. The following section gives an overview of the optical behaviour of reflecting surfaces, although the evaluation of beam reflections can be given an exaggerated importance in laser safety. Under open-beam conditions where a direct exposure to the beam could be hazardous, protection is better provided by beam enclosures or personal eye protection than by simply trying to avoid the use of hazardous reflecting surfaces. (There are exceptions to this, of course; one example being the use of display lasers in laser light shows. Protection here is often dependent on maintaining minimum clearance distances between the path of the beam and locations at which people could be present. Where direct exposure to the beam could be hazardous it is crucial to ensure that hazardous beam reflections cannot occur.)

A reflection can modify a laser beam in two principle ways. It can reduce the total power (or energy) of the beam and it can also change the distribution of the emitted radiation in space, usually by simply changing the beam divergence.

The reflectance R of a surface defines the proportion of the incident power (or energy), at a specified wavelength, that is reflected from the surface. Thus, where the incident power is P and the reflected power is P_r, the surface reflectance is given by

$$R = P_r/P. \tag{5.19}$$

Reflectance can therefore have any value between 0 and 1. (A surface having a reflectance of 0.5, for example, reflects 50% of the incident radiation.) Since reflectance can be wavelength-dependent, however, it is very important that its value is specified at the laser wavelength.

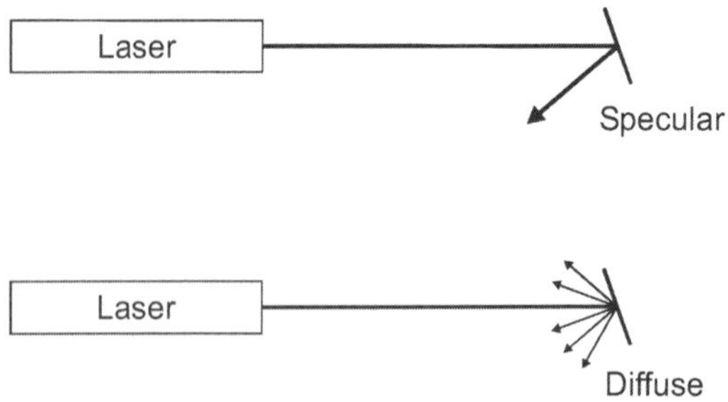

Figure 5.13. Types of beam reflection. Specular, i.e. mirror-like reflection (top) redirects the beam but may not otherwise affect the beam properties. Diffuse reflection, i.e. surface scattering (bottom) redistributes the incident radiation in all directions, forming a secondary apparent and widely-diverging source of radiation (at the point of reflection) and completely destroying the geometrical properties of the original beam. These examples represent two extreme properties of reflecting surfaces; many real surfaces can exhibit a combination of both of these types.

However, in addition to affecting the total amount of radiation in the reflected beam, a reflection can also modify the beam geometry. Two fundamental types of reflection are shown in figure 5.13. *Specular* (mirror-like) reflection simply redirects the beam. Unless the reflecting surface is flat it can also increase or decrease the beam divergence, but the point-source characteristics of the source are unchanged. (Where the surface reflectance at the laser wavelength is close to 1, a specular reflection can be just as hazardous as the direct laser beam, even more so if the beam divergence is reduced by specular reflection at a concave surface.) A *diffuse* reflection, on the other hand, such as that produced by a matt (unpolished) surface, redistributes the reflected radiation in all directions away from the surface, completely destroying the geometrical properties of the original beam.

The reflection characteristics of any surface at wavelengths outside the visible band should not be judged on the basis of its visible appearance. One example of this is the reflection from brushed-metal surfaces. To the eye the surface appears to be a diffuse reflector, but it is a highly specular reflector for the far-infrared beam of a CO_2 laser.

Within the visible band, highly polished steel is an example of a specular reflector while matt white paper is an example of a diffuse reflector. Both may have values of reflectance greater than 0.9 and will both, therefore, reflect most of the light that is incident upon them, but the geometrical characteristics of the reflected light will be very different.

To be able to assess the exposure that could arise at any given distance from a diffuse reflection of a laser beam, the irradiated area of the reflecting surface (that is, the area of the surface covered by the incident beam) has to be regarded

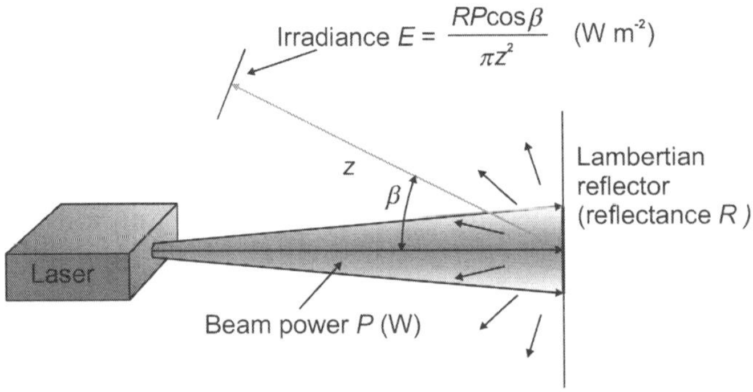

Figure 5.14. Where a reflection is perfectly diffuse (occurring at what is known as a Lambertian reflecting surface), the distribution of the reflected radiation follows a cosine response and the resultant irradiance (and similarly for radiant exposure) at any position along any direction from the reflecting surface can be calculated using the equation given.

as a secondary source of radiation. Unlike the original laser source, this can constitute an *extended* source if viewed from a sufficiently close range, having a finite emission area equal to that of the incident beam, and may be incapable of being brought to a point focus (inside the eye or by any other optical system).

A perfectly diffuse reflecting surface (called a *Lambertian* reflector) re-radiates the reflected radiation into a hemisphere, which has a solid angle of 2π steradians. The reflected irradiance E produced at a distance z from a Lambertian reflector in a direction making an angle β to the normal is given by

$$E = (RP\cos\beta)/(\pi z^2) \tag{5.20}$$

where P is the incident beam power (in W) and R is the surface reflectivity at the laser wavelength (figure 5.14). The irradiance simplifies to $(RP)/(\pi z^2)$ at normal incidence (where β is zero).

Many real surfaces have both specular and diffuse reflecting components. (Those in the visible band that behave in this way can be seen to have a sheen or gloss when viewed at certain angles.) It can therefore be too simplistic to simply categorize any given surface as either one type of reflector or the other. Furthermore, as stated earlier, where direct exposure to a laser beam would be hazardous it is usually inadvisable to rely for protection solely on a belief that only diffuse reflections can occur. Many everyday objects (e.g. tools, such as screwdrivers, and also watches and other jewellery) may be accidentally moved into the beam and can cause hazardous specular reflections.

5.6 Optical viewing instruments

5.6.1 Aided viewing

The use of optical instruments that allow magnified viewing, such as binoculars or even simple magnifying lenses, can increase the level of laser exposure at both the surface of the eye and (in the case of wavelengths within the retinal hazard region) at the retina. This may mean that a laser beam that can be safely viewed by the unaided eye may become unsafe when magnifying optics are used. This is particularly so for laser products in Class 1M and Class 2M, which are intended to indicate that the beam could be unsafe if optical viewing instruments are used. Furthermore, for other laser classes, the hazard distance (the distance from the laser within which the beam is hazardous) can increase through the use of such instruments. As explained in section 5.4, we need to consider not merely the nominal ocular hazard distance (NOHD) within which the MPE can be exceeded, but also the extended nominal ocular hazard distance (ENOHD) within which the use of telescopes or binoculars could be hazardous. Properly prescribed spectacles or contact lenses do not, however, have this effect and need not be considered when undertaking a safety assessment, since their intention is to correct for the eye's focusing errors, and any magnification that they introduce is negligible.

For any optical system to increase the laser hazard, the laser beam must be larger than the relevant limiting aperture, in order that the use of magnifiers can increase the level of radiation contained within the limiting aperture (and thereby increase the irradiance or radiant exposure averaged over the area of the limiting aperture). A laser pointer, for example, has a diameter smaller than 7 mm (the maximum pupil diameter that corresponds to the limiting aperture for wavelengths within the retinal hazard region) and so no additional increase in the hazard is possible. The entire beam of the laser can already pass through the pupil and be brought to a point-focus on the retina, and the use of optical viewing instruments cannot increase the hazard further. With a beam that is larger than 7 mm, however, only part of the beam can normally pass through the pupil, but the use of magnifiers could concentrate more of the beam within the pupil area and thereby increase the hazard.

There are two circumstances in which the viewing of a laser beam through optical instruments can be hazardous, as discussed in chapter 4. For a large-diameter, reasonably well-collimated beam, the use of optical aids such as binoculars or telescopes, often at considerable distances from the laser, can increase the exposure hazard and also the distance from the laser within which an exposure hazard can exist. Alternatively, with a divergent beam, the use of eye-loupes (magnifying lenses), or even some microscopes, can increase the hazard when used to view a laser or LED source at close range. These effects are illustrated in figure 5.15.

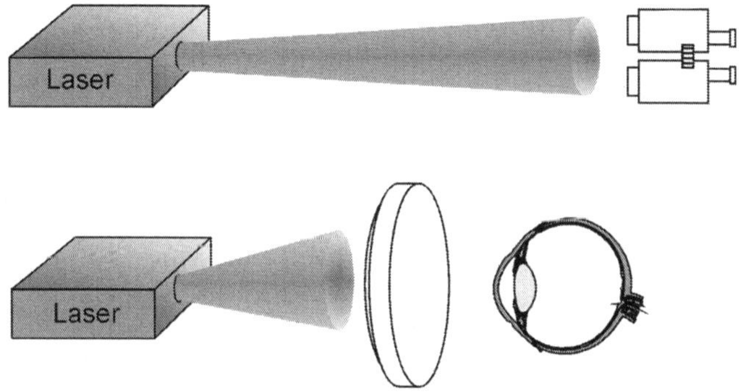

Figure 5.15. The magnified viewing of a laser beam using binoculars at a distance from the laser (top), and using an eye loupe or magnifying lens to view a divergent beam close to the laser (bottom).

5.6.2 Binocular viewing

The use of binoculars (or similar optical systems, such as telescopes, intended for magnified viewing at a distance) can increase the laser radiation hazard at the eye by concentrating more of the beam power (or energy) within the area of the limiting aperture. For laser classification purposes, a measurement aperture of 25 or 50 mm diameter (depending on the wavelength) positioned two metres from the emission aperture is used to simulate the effect of such optical instruments. These measurement apertures are based on the notion of a 7× magnification of the limiting aperture. The two metre distance is intended as a conservative estimate of the closest distance at which such instruments can be focused and used. (Some binoculars can focus closer than this, but they are normally of the compact type having smaller collection apertures, and in any case they are very unlikely to be used at such close range.)

The increase in the level of exposure that could result from such magnified viewing depends on the factor by which the quantity of radiation that is contained within the area of the limiting aperture can be increased. (While the limiting aperture is 7 mm in diameter for radiation within the retinal hazard region, i.e. for wavelengths between 400 and 1400 nm, the ocular exposure can be increased at any laser wavelength that can be transmitted through the binoculars, and not just in the retinal hazard region.)

Binocular-type viewing aids are normally given a specification in the form of $X \times Y$, where X is the linear magnification and Y is the diameter of the outer lens in millimetres. A pair of 8×50 binoculars therefore provides a magnification of 8×, and each of the two outer lenses has a diameter of 50 mm.

The lens diameter can be important for binocular users, since a smaller outer lens, such as those employed in a pair of compact binoculars, can considerably reduce the light throughput of the device under low light level conditions. This is

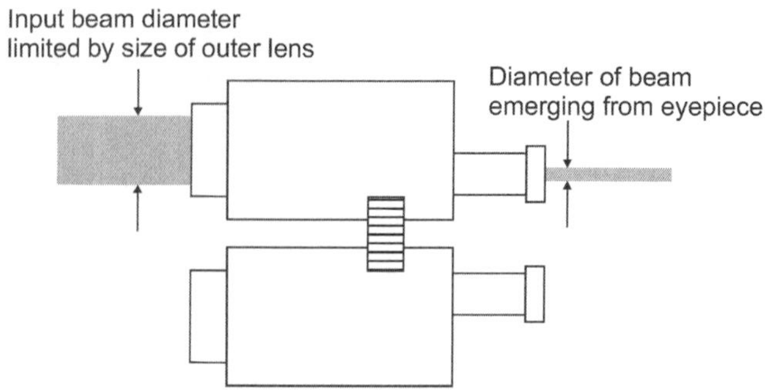

Figure 5.16. Viewing a laser beam with binoculars. The beam that emerges through the eyepiece is reduced in size (in comparison to the diameter of the beam collected by the input optics) by a factor equal to the linear magnification of the binoculars. This can produce an increase in the exposure level at the eye equal to the square of the magnification.

because, with a magnification of 8× as in the example just given, the outer lens needs to have a diameter of at least 8× the pupil diameter in order that there is no reduction in the brightness of the scene being viewed. The 'exit pupil' of the binoculars, that is the diameter of the emerging light-field at the eyepiece, is Y/X, which in this example is 50/8, i.e. 6.25 mm, figure 5.16. A pair of 8 × 20 compact binoculars, however, although having the same magnification as the larger pair, will have an exit pupil of only 2.5 mm. Under low light level conditions the eye's pupils may expand beyond this limit, but will not be able to receive any additional light, and so the scene that is viewed through the binoculars will appear dimmer than when viewed without binoculars.

In order to quantify the increase in exposure from a laser (assumed here to be a point source) that could result from the use of a pair of binoculars, we need to consider the optical effect on the beam caused by the binoculars. A linear magnification of X can increase the exposure (that is, the irradiance or radiant exposure) at the surface of the eye by X^2, but only if the outer-lens diameter Y is sufficient to permit this. In fact, the increase in the exposure will be the *smaller* of either X^2 or $(Y/d)^2$, where d is the diameter of the relevant limiting aperture.

If we consider the 8 × 50 binoculars mentioned previously, the 8x magnification could be taken to imply that the increase in exposure would be equal to the square of the magnification, that is an increase of 64×. (This is because the emergent beam at the eyepiece will have a diameter of 1/8th of the input lens, and hence the radiation collected by the outer lens will have been concentrated down in area by 64 times.) However, the exit-pupil is only 50/8 or 6.25 mm in diameter and so the emergent beam is concentrated within an area that is *smaller* than the limiting aperture (of 7 mm diameter for wavelengths between 400 and 1400 nm). This additional reduction (to below 7 mm) has no further impact on the effective exposure (since the exposure hazard is expressed in terms of the radiation contained *within* the limiting aperture). This means that the increase in exposure is limited to $(50/7)^2$, or 51 times, since this value represents the factor by which the power or energy contained within the limiting aperture has been increased. For the 8 $\times$ 20 compact binoculars, the effective exposure increase is only $(20/7)^2$, or approximately eight times.

If we were concerned, however, with laser emission at, say, 1500 nm, then the limiting aperture (for exposure durations of at least 10 s) is only 3.5 mm. In this case, using 8 $\times$ 50 binoculars, the effective increase in the exposure would be 64 times (i.e. 8^2), since this is smaller than $(50/3.5)^2$ or 204 times, and so the exposure increase is here limited by the magnification and not by the size of the input aperture.

This analysis has assumed, of course, that the input lens of the binoculars is, in the worst case (i.e. when viewing centrally along the beam axis) completely and uniformly filled with incident laser radiation. Where this is not the case, that is where the beam, although larger than the limiting aperture is nevertheless smaller than the input lens of the magnifier or where the incident beam is not reasonably uniform across the input lens, then these effects can be accounted for by applying the analysis for non-uniform beams discussed earlier in this chapter or by direct measurement using an appropriately-sized aperture and by considering the effect of the truncation caused to the beam by the input lens and of the optical magnification.

One factor that has been ignored in this discussion is the optical transmission of the binoculars at the laser wavelength. Such losses will serve to reduce the laser exposure that is actually received by the eye. Good binoculars incorporate high-quality components and have antireflection-coated optics in order to maximize transmission within the visible band. Unless the transmission losses are reliably known, therefore, it is better to ignore these when undertaking safety assessments, since the effect of these losses within the visible or near-infrared is likely to be small. At wavelengths much further into the infrared beyond about three microns, however, and in the ultraviolet below 300 nm, the transmission losses through binoculars can be considerable. The loss incurred, however, will depend very much on the particular optical system employed (including the type of optical glass that is used, the quality of the optical coatings and the number and thickness of the individual optical components). The loss can be estimated with the absorption law (equation (2.20)) when the total glass thickness is specified and the absorption coefficient and reflectance for the wavelength under consideration is known, although in the wavelength range where the glass is heavily absorbing, the loss from reflection can be neglected. The absorption coefficient has been calculated from spectral transmission measurements that were performed for very thin samples of uncoated optical glass of the type Schott BK7 and BaK 4. These types of glass are used very widely in optical instruments including most binoculars and loupes. The absorption coefficient as a function of wavelength is shown in figure 5.17 for the ultraviolet and infrared wavelength range. For wavelengths not shown in the plots, the absorptivity is so low that in practice the

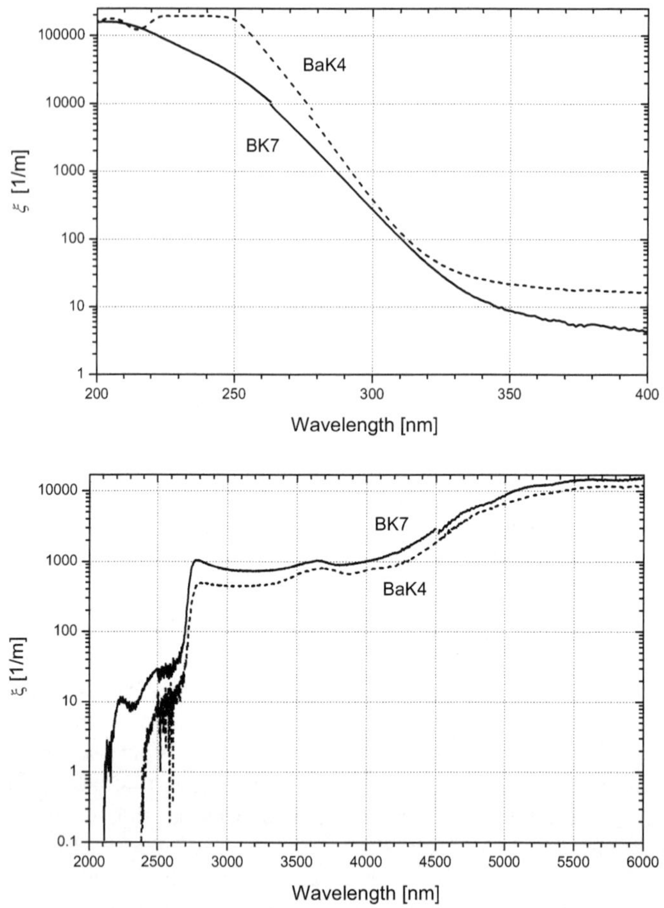

Figure 5.17. The absorption coefficient as function of wavelength for two types of optical glass, for the ultraviolet wavelength range (top) and the infrared wavelength range (bottom).

transmittance can be set to unity. When the thickness of the glass in the instrument is not known, estimates can be used. For instance, typical values of total optical length in glass for binoculars are 5–6 cm for a pocket binocular (where the optical path length in the prism is about 5 cm) and up to 14 cm for a 50 mm high quality binocular (where the optical path length in the prism is typically about 8 cm).

Account should be taken of the possible use of magnifying instruments whenever lasers are being used out-of-doors and where the beam may propagate into uncontrolled (that is, public-access) areas. The effect this can have on the hazard distance (see section 5.4) may also need to be assessed. A reasonable limit to apply is that hand-held binoculars can increase the laser exposure by up to about 50 times. This is sufficient to produce an extended nominal ocular hazard distance (ENOHD) that is up to seven times the hazard distance applicable to the unaided eye (i.e. 7 × NOHD).

In the far-field of a laser beam that overfills the input aperture, the extended hazard distance can be found by adapting equation (5.14), and is given by

$$\text{ENOHD} = \frac{2G}{\theta_{63}}\sqrt{\frac{P}{\pi\,\text{MPE}}} \tag{5.21}$$

where G is the smaller of X or Y/d for binoculars specified by $X \times Y$, and d is the diameter of the relevant limiting aperture.

5.6.3 Close-up viewing

5.6.3.1 Eye loupes

Simple magnifying lenses, and particularly eye-loupes, which are designed for close-up viewing, can increase the exposure received from a divergent beam (such as that emitted from a laser diode or from the end of an optical fibre) when used in close proximity to the point of laser emission. It should be noted that there is no ENOHD for the use of such instruments; quite the reverse, in fact, since the hazard arises from the ability of such lenses to allow sharply-focused viewing at closer distances than would normally be possible with the unaided eye.

An ordinary positive (that is, convex) lens allows the eye to be brought closer to the object being examined and can therefore produce a larger focused image on the retina. For example, a lens having a focal length of 50 mm (or a *power* of 20 dioptres, which is the reciprocal of the focal length in metres), is said to have a magnification of 5×. This is because it allows the eye to view the object from a distance of 50 mm (i.e. as if the pupil were only 50 mm away from the object) rather than from the standardized minimum viewing distance of 250 mm. Since the object is now five times closer it appears to be five times larger. The viewing distance of 250 mm is the nominal assumed position of the *near-point* in front of the eye, although for most younger people it is actually less than this. A more cautious figure of 100 mm is used as the minimum viewing distance in the international laser safety standard. (A 200 mm minimum distance is assumed in US standards.)

Figure 5.18 illustrates the effect of a simple magnifying lens when used to view a divergent point laser source. Without the lens, at a normal viewing distance z_1, the eye's pupil subtends a solid angle of Ω_1 steradians. With the lens, the viewing distance is reduced to z_2, and the collection solid angle has increased to Ω_2 steradians. The resultant increase in exposure at the eye is therefore Ω_2/Ω_1, where Ω_2/Ω_1 is equal to the square of the inverse ratio of the corresponding viewing distances z_1 and z_2, i.e. $(z_1/z_2)^2$. The image is magnified by z_1/z_2. (This is the *linear* or *angular* magnification.) Using a 5× lens to view a divergent laser

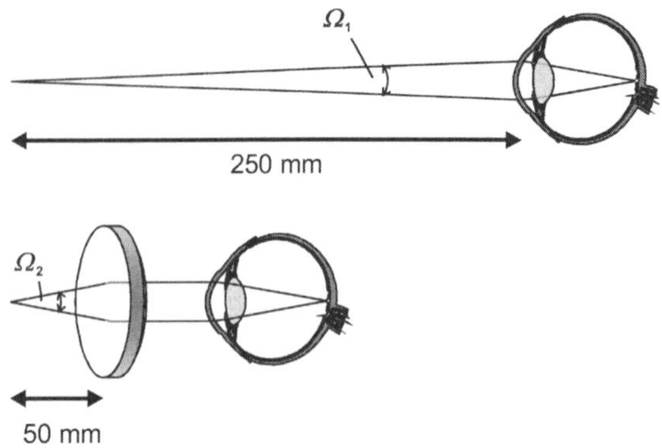

Figure 5.18. Viewing a divergent laser beam with an eye loupe. A magnifying lens having a focal length of 50 mm produces a five-times linear magnification in comparison to the standard minimum viewing distance of 250 mm. This can produce an increase in the exposure level of the eye equal to the square of the magnification, i.e. of 25 times (since the ratio of the solid angles Ω_2/Ω_1 is equal to $250^2/50^2$).

beam will, therefore, increase the eye's exposure by 25 times (when compared with a normal viewing distance of 250 mm). Clearly, however, this assumes that the laser emission is reasonably uniform and that its divergence is sufficient to completely fill the pupil (or, at other wavelengths, the relevant limiting aperture) at the closer position. Where this is not the case, due allowance for these factors can be made. In the same way as for binocular viewing, the effects of transmission losses through the lens may also be allowed for, where these are reliably known for the laser wavelength of interest.

When other optical instruments are used for close-up viewing (e.g. microscopes), the potential increase in the laser exposure will depend on the system's optical design and on its mode of use. In general, however, the exposure increase will be determined on the basis of the increase produced by the instrument in the power or energy at the surface of the eye contained within the area of the relevant limiting aperture. This will depend on the optical magnification of the device as well as on the collection solid-angle of the input aperture, either of which may limit the maximum increase in exposure that is possible in an analogous manner to the effects of magnification and input aperture size in the case of binoculars.

It should be noted that a simple lens can also increase the exposure hazard in other ways, in addition to the particular case of close-up viewing. A lens can be used to decrease the divergence of a laser beam, thus increasing the hazard distance or (especially in the case of Class 1M and 2M lasers) introducing a hazard to the unaided eye where none existed before. Furthermore, a lens can be used to focus the beam of a laser, thereby increasing the level of exposure.

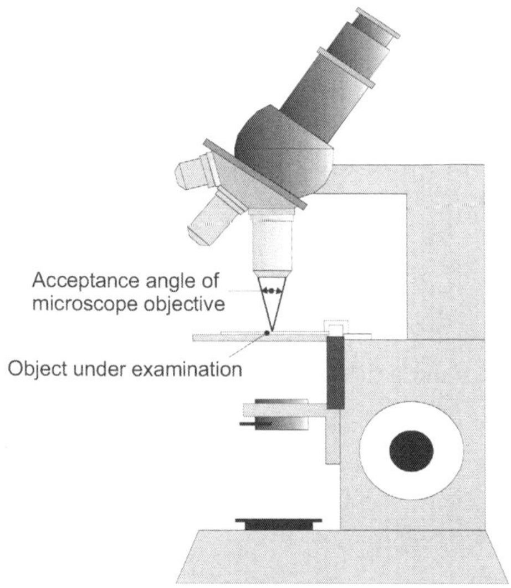

Figure 5.19. The optical arrangement of a simple microscope.

5.6.3.2 Microscopes

Microscopes are regularly used in optical telecommunications to examine the end of optical fibres, and are also employed in some types of laser surgery, although usually as an integral part of the laser equipment. Although protective filters are often employed in both of these applications, it is sometimes necessary to determine the increase in exposure that can occur. In the magnified viewing of optical fibres there is the possibility of viewing a 'live' fibre and thereby receiving a higher exposure than would arise from naked-eye viewing. In the case of surgical applications, laser radiation being directed at the treatment site may be reflected back into the microscope. This can include diffuse reflections from the tissue being treated as well as specular reflections from instruments, etc which are inadvertently introduced into the laser beam.

A typical simple microscope is shown in figure 5.19. This comprises an objective at the input end together with an eyepiece at the output end that is used to view the magnified image produced by the objective. The total magnification is the sum of that due to the objective (which can range from 4× to about 100×) and that due to the eyepiece (up to 20×). At high levels of magnification the microscope objective needs to have a large acceptance angle in order to maximize the light collection efficiency and thereby limit the reduction in brightness (radiance) that inevitably occurs. (This loss in brightness, caused by the inability of the microscope optics to collect light over a sufficiently large acceptance angle, is the reason why ordinary high-power microscopes need special illumination systems.)

The acceptance angle of the objective is usually given in terms of its NA (numerical aperture), as defined by equation (5.8). The NA can vary from 0.1 (corresponding to an acceptance angle of 11.5 degrees) with a low-power objective, to 0.9 (an acceptance angle of over 120 degrees) for a very high-quality 80× objective. (Oil-immersion objectives can have higher levels of magnification with even larger acceptance angles.) For comparison, the emission angle from a single mode fibre is typically no more than 10 degrees. It can be assumed, therefore, when using a microscope to view the end of a singlemode fibre, that the entire emission can be collected by the microscope. In order to be safe, this emission must not exceed the Class 1 AEL, or a filter must be used that will reduce the emitted power to no more than this value.

With multi-mode fibres viewed at low to moderate magnification levels, the fibre NA should be compared to that of the objective and the proportion of the emission that is collected by the objective then calculated. For uniform emission the collected power is given by

$$P_c = (\mathrm{NA_o}/\mathrm{NA_f})^2 P_e \tag{5.22}$$

where P_e is the emitted power from the fibre, $\mathrm{NA_o}$ is the numerical aperture of the objective and $\mathrm{NA_f}$ that of the fibre. A similar assessment can be carried out for other divergent laser sources where the equivalent numerical aperture of the emission is used in place of $\mathrm{NA_f}$.

The assessment of the use of a microscope to view a diffuse laser reflection, as in laser surgery, is more complex. The reflection itself will constitute an extended source (see section 5.6.4) but the microscope will reduce its effective radiance. The extent to which this happens cannot be determined without a detailed knowledge of the optical specification of the microscope objective, although the use of a microscope in such circumstances is more likely to decrease the hazard than to increase it (due both to the increase in the angular subtense of the source and to the decrease in the apparent source radiance).

It is perhaps worth noting that the acceptance angle of the eye-loupe condition used for classification (condition 2) is 28 degrees, corresponding to an NA of 0.24. The use of a microscope with an objective having an NA greater than this (typical of most microscope objectives of 10× or higher power) will therefore exceed the worst-case condition assumed for classification. The possibility arises, therefore, that with a highly divergent laser source (which can include laser diodes) the close-up viewing of the source with a microscope could cause the exposure at the eye to exceed the MPE, even if the source is Class 1. Because of this, changes to the measurement criteria for condition 2 are under consideration that would distinguish between applications in optical telecommunications where microscopes are routinely used and other applications where they are not.

5.6.4 Magnified viewing of extended sources

The foregoing discussion on the use of optical instruments applies to the viewing of small laser sources. For emission from extended sources within the retinal hazard region (that is at wavelengths between 400 and 1400 nm) the potential increase to the hazard that can be produced by the use of magnifying optics may be more limited. Extended sources are those, such as diffuse reflections, multiple sources and LEDs, where the apparent source subtends an angle at the eye greater than α_{min} (1.5 mrad). These sources cannot be focused to a minimal spot-size on the retina and so a relaxation (increase) to the AEL or MPE is permitted in accordance with the rules specified in the safety standard. In such cases the effect of magnification will be to increase further the apparent source-size, so permitting an even greater relaxation in the MPEs, and may negate the actual increase in the level of exposure at the cornea. Note that this can also apply to those sources which, although subtending an angle less than α_{min} can nevertheless exceed α_{min} when magnified.

A more detailed discussion of the issues surrounding extended (i.e. non-point) sources is given in chapter 4.

5.7 Assessment accuracy

Questions are sometimes raised regarding the accuracy with which levels of laser radiation (that is of the power or energy contained within specified apertures which is the basis of much laser-safety assessment) should be determined, regardless of whether this is carried out by physical measurement or by calculation.

The existence of 'safety margins' in the AELs and MPEs should not be used to accommodate uncertainties in measured or calculated quantities. The AELs are firm limits that are used for product classification (and which can have legal implications under European product safety directives) while the MPEs are internationally recommended exposure limits that should not be exceeded. The accuracy required will therefore depend on how close to a given limit a particular quantity actually lies. It is therefore necessary that a particular quantity is confidently below the specified limit, otherwise it must be assumed that it might exceed it (see section 2.8.1). Where the quantity lies very close to a limit, not only must the determination be carried out with sufficient precision, but possible future drifts or fluctuations in the laser's output must also be anticipated and allowed for.

When compiling laser product reports for CDRH (a legal requirement for laser products sold or offered for sale in the United States), issues regarding the calibration and traceability of optical detectors that are used for classification measurements, supported where necessary by measurement error analyses, have to be addressed where these are relevant to the performance and compliance of the product concerned.

In all cases of laser classification or exposure assessment, the accuracy adopted must be sufficient to support any safety claims that are made. This will normally require a well-argued case, which should be documented and based on measurements or calculations (or a combination of both), that a given classification or exposure limit is not exceeded. People making safety claims should have confidence in their conclusions and should be able to defend them against reasoned argument.

References

[1] Li Y and Katz J 1991 Encircled energy of laser-diode beams *Appl. Opt.* **30** 4283–4

[2] Kogelnik H and Li T 1966 Laser beams and resonators *Appl. Opt.* **5** 1550–67

[3] Siegman A E 1986 *Lasers* (California: University Science Books)

[4] ISO 11146 2003 *Lasers and Laser-Related Equipment—-Test Methods for Laser Widths, Divergence Angle and Beam Propagation Factor* (Geneva: ISO)

Chapter 6

Additional laser hazards

6.1 Other hazards of laser operation

Much of this book is concerned with the description, analysis and control of hazards arising from exposure of the eyes or skin to harmful levels of laser radiation. Lasers have the ability to concentrate large quantities of optical radiation over very small areas, resulting in a potential for causing serious personal injury. Nevertheless, other hazards, apart from those of radiation exposure, can be present with many lasers and in certain kinds of laser applications [1]. These associated or additional hazards, which may either be a function of the particular type of laser or dependent on the task for which the laser is being used, are often linked especially with the use of Class 4 laser products. Classification, however, is based solely on the level of accessible laser emission, and takes no account of any additional hazards that may be present. Additional hazards are therefore possible with any class of laser. Even those hazards that arise as a direct consequence of high levels of laser power (often through the interaction of the beam with target materials) can still occur in a Class 1 laser product where it contains an embedded high-power laser.

Most of the additional hazards associated with laser use are not unique to lasers. They can arise with other technologies and equipment. One exception, however, is the generation of air contaminants (fume) in laser processing, which occurs through the vaporization of target material. Air contamination can occur in a form that is different from that produced by other (non-laser) processing technologies, and can involve the production of extremely small particles which can be inhaled deep onto the lungs and remain there for a considerable period.

All additional potential hazards need to be taken into account along with those of radiation exposure as part of a risk-assessment process (discussed in chapter 7) before the selection of any necessary protective controls. In some circumstances these additional hazards can pose a more serious risk to health than the more obvious hazard of laser exposure. Their control should be approached in the same way as with all other kinds of workplace hazard; by first identifying and then, where possible, by eliminating the hazard at source. Where this is not reasonably practicable, then the hazards should be reviewed and an appropriate system of engineering controls, administrative controls and personal protection used to reduce the risk to an acceptable level.

Subsequent chapters of this book discuss the principles of risk assessment, risk reduction (i.e. safety controls) and laser safety management. These issues are mainly considered in the context of laser-radiation exposure. Here we describe some of the additional laser hazards that can exist and also outline the kind of control procedures that may be necessary. However, laser safety is concerned with the management and control of *all* potential laser hazards, and so in carrying out the functions of risk assessment that are discussed later, both the exposure issues covered in earlier chapters and any additional hazards of the type that are reviewed here should be considered together.

This overview includes the more usual categories of additional hazard that may be encountered in the use of laser equipment, but can only give general guidance. It is divided into those additional hazards that are due to the laser beam itself, and non-beam hazards that can arise from other aspects of the laser or its operation. Certain applications or particular laser environments may, of course, give rise to other hazards that are not listed here.

The evaluation and control of additional laser hazards may in some cases require the assistance of health and safety professionals having specialist expertise in the particular hazard of concern.

6.2 Additional beam hazards

6.2.1 Dazzle

It is possible that visible-beam lasers can cause indirect harm even where the exposure level is below the MPE. This is because of their ability to startle or dazzle any unsuspecting person who is exposed to the laser beam, whether deliberately or not. Protection from direct harm from any laser in Class 2 or 2M is normally provided by natural aversion responses. Even so, the exposed individual can be distracted or startled by a sudden, unexpected exposure. This can also occur from exposure to a visible-beam Class 1 laser, and from visible-beam lasers of higher class even beyond the NOHD where exposure might otherwise be considered safe. Such exposure can have serious consequences if the person is performing a safety-critical task (such as driving). It can also cause disturbing after-images in the individual's eyes, and may generate fear if the person believes that their eyes might have been harmed by the exposure.

Even low-power visible-beam lasers should therefore be used with care. Where there is a risk that people unconnected with the laser work and unaware that lasers are in use may be accidentally exposed, even at levels below the MPE, the use of screens, curtains or other suitable enclosures should be considered, or physical limits set on the range of beam movements that are possible.

Visible-beam lasers should never be used to deliberately surprise or alarm other people. The US standard for safe laser use [2] defines criteria for dazzle effects.

6.2.2 Beam-initiated fire and explosion

The high concentration of radiant power within the beam of a Class 4 laser (and also, in certain circumstances, within the focused beam of a laser of lower class), can be sufficient to ignite materials on which the beam impinges. This can include flammable liquids, plastics, wood and fabrics. It can also cause explosions in combustible gases or in high concentrations of airborne dust. These effects can occur more readily (or be made more serious) in the oxygen-rich atmosphere utilized in some laser machining processes, and also in certain surgical applications where inflammable gas mixtures may be present (e.g. anaesthetic or bowel gases).

A limit of 35 mW has been proposed for laser emission in combustible atmospheres for beams (including the emission from optical fibres) having a cross-sectional area less than 7 mm^2, while for beam-areas larger than this the irradiance should not exceed 5 mW mm^{-2} [3].

Special caution should be adopted with all Class 4 lasers, and also with embedded lasers of high power, to ensure that ignition of flammable fabrics and other materials, including gases, cannot occur. Beam stops and enclosures, together with any paint or coatings applied to their surfaces, should not be made of materials that could present a fire hazard. Restrictions may need to be applied to the substances and materials that are allowed to be used in conjunction with the laser.

Where the risk cannot be entirely eliminated, as when using lasers to cut combustible materials such as wood, then suitable fire extinguishers should be kept in readiness and training provided in their use. In addition, waste material should be removed frequently and not allowed to accumulate close to the laser beam.

In operating theatres, drapes that surround the operating site should be kept wet, and a container of sterile water should be located nearby in order to extinguish any small fires that do occur. All instruments, tubing and other equipment used in close proximity to the laser beam should be fabricated from a suitably fire-resistant material.

6.2.3 Other thermal hazards

Even where fire or explosion does not occur, objects (such as beam stops and beam-steering mirrors) in the path of a high-power laser may nevertheless become very hot. This can obviously result in burns if the object is touched, and can also cause it to be damaged or to distort in some way so that the performance of the laser or of associated equipment may be compromised.

Suitable materials should be used in the construction of all items that may come within the path of the laser beam. Adequate heat-sinking or water cooling should be adopted where appropriate on mirror mounts intended for use with high-power laser beams. Guarding or warnings signs should be used to protect people from objects that may become hot.

6.2.4 Fume

Laser-generated air contaminants (LGAC) comprising hazardous particulate and gaseous by-products (hazardous fume) may be released into the atmosphere in high-power laser processing (i.e. cutting, drilling, welding and surface treatment), and also in surgical applications, through the interaction of the laser beam with the target material. This fume is generated by a process of vaporization and recondensation. Very small (submicron) particles can be released which, even if inert, can cause serious respiratory problems. These particles can drift slowly and may remain in suspension in the air for long periods after the laser work has ceased.

Matter may also be generated in the form of fine dust. This can settle on the skin and hair and also on surrounding surfaces, and if allowed to accumulate in sufficient quantity may cause a further hazard when cleaning-up after laser use.

As the laser vaporizes the surface layers of the material, a plasma may be generated (seen as a bright, visible plume) which can attain temperatures as high as 10 000 °C. A complex chain of chemical reactions can occur in this plume, and the fume that is generated can be toxic and carcinogenic. In addition to temperature-related effects, with excimer-laser processing the photon energy can be sufficient to directly change the chemical structure of the material that is released. Regardless of the material being processed, however, the high temperatures produced can create nitrogen monoxide through the oxidization of the nitrogen in air. This can react to form other oxides of nitrogen that are toxic. Furthermore, ultraviolet radiation generated from the plasma can contribute to ozone formation.

The composition of the fume and the quantity of material emitted depend on the processing parameters as well as on the material being treated [4]. Non-optimal processing may produce even more emission than necessary. Typical emission rates are in excess of 20 mg s^{-1} for laser cutting, within the range 1–20 mg s^{-1} for laser welding, and below 1 mg s^{-1} in laser marking and microprocessing applications.

The processing of metals produces mainly metal oxides in the form of aerosols (very small particles that drift in air). These can include lung-polluting substances (e.g. iron and aluminium oxides), toxic materials (e.g. manganese and zinc oxides) and carcinogenic compounds (e.g. chromium, nickel, and beryllium oxides). Organic materials (such as wood, textiles, leather and many synthetic substances) produce largely hydrocarbon emissions, mostly released solely in gaseous form. With all plastic materials a wide range of very hazardous gaseous and aerosol materials, including cyanides and benzine derivatives, can be generated. The processing of polyethylene, polypropylene and polyamide result predominantly in aerosol emission, which can be very viscous and 'sticky', readily adhering to clothing and rapidly accumulating in filter systems. Polystyrole and polymethyl methacrylate (PMMA) produce mainly gaseous emission. Hydrogen chloride, which is highly corrosive, can be produced from polyvinylchloride (PVC). Silica and silicon oxide, which can cause silicosis, are produced as a consequence of glass processing, and hazardous submicron glass particles can also be generated.

In the case of laser surgery, fumes may present a microbiological hazard, and can contain both bacterial and viral particles. The fume can also have an unpleasant smell.

Very low levels of fume can be dealt with by simply ensuring that there is sufficient ventilation in the laser area, although some form of forced extraction system will often be required. The type and quantity of fume produced depend very much on the laser process, and advice may need to be sought, both on the particular nature of the fume and the on the most satisfactory way of removing and disposing of it. Most countries have national regulations governing the air quality of the workplace environment and the way in which air contaminants should be removed.

The source of fume should be enclosed as far as possible. If forced extraction is needed, the collection hood or nozzle should be located close to the laser process zone in order to maximize the collection efficiency. This will also reduce the rate of air-flow that is necessary. Depending on the nature of the fume that is generated, removal of harmful components by filtering may be required before the extracted air is released into the outside atmosphere. The type of filtration system needed will depend on the nature of the fume (e.g. whether it is gaseous or particulate) and on its constituents. Suitable types include fabric filters, electrostatic filters, gas scrubbers and absorption filters. Filters are often banked in series, with the first filters removing the largest particles.

With adequate filtration, recirculation of the extracted air back into the workplace is possible. This can result in significant energy savings in temperature-controlled environments in comparison with systems that simply release extracted air outside. Without a sufficient level of inward airflow, whether through recirculation or by means of normal ventilation, extraction systems can produce a slight drop in air pressure, sometimes resulting in difficulties in opening doors.

Filter systems require periodic maintenance, and airflow monitors can be used to indicate decreasing efficiency as the filters gradually accumulate waste products from the fume. The cleaning and replacement of filters should be carried out in accordance with the manufacturer's instructions and with relevant national regulations covering the disposal of hazardous waste.

Eating or drinking should not be permitted in areas where hazardous fume may be generated.

6.2.5 Additional laser emission

Some lasers employ frequency-shifting techniques to produce emission at a different wavelength from that generated inside the laser resonator itself. This is commonly done, for example, in the case of frequency-doubled neodymium:YAG lasers, where the fundamental wavelength of 1064 nm, in the near infrared, is converted to visible emission at 532 nm. (This is half the wavelength and therefore double the frequency.) As well as producing a green beam at a wavelength of 532 nm, however, the emitted beam may also contain residual emission at the fundamental infrared wavelength. This may not be indicated on the label nor accounted for in the product classification. The available eye protection may only provide protection from the fundamental wavelength, although it is possible to obtain protection that covers both 1064 nm and 532 nm.

Where a laser utilizes frequency-shifting techniques to generate the output beam, it may be necessary to check for the existence of out-of-band emission and to incorporate an optical filter in order to the suppress emission at the unwanted fundamental wavelength, where this is found to be significant. As noted above, the use of dual-band eye protection may also be advisable.

6.3 Non-beam hazards

6.3.1 Electricity

Many (especially high-power) lasers utilize high voltages, and may also incorporate large capacitors that can store significant quantities of electric charge. The energy stored in these capacitors may remain even after the system has been disconnected from the electrical supply. A number of deaths have been caused by electrocution from laser power supplies when lasers have been operated (often during servicing or repair, or in research and development applications) with their protective covers removed.

All electrical components and contacts should be enclosed and all electrical equipment properly earthed in accordance with appropriate national regulations. Equipment should not be operated with its covers removed or with protective interlocks defeated.

Special procedures may be necessary for equipment servicing where this requires access to the inside of the laser equipment or the operation of the laser with any of the enclosures open or removed. Storage capacitors should be fully discharged using a grounding rod prior to any work inside high-power laser equipment. Live working (working in close proximity to live electrical terminals) should only be carried out when absolutely necessary, and should then be subject to established safe-working practices. These can include the use of insulated floor mats and the adoption of single-hand working (to prevent an arm-to-arm circuit through the body). Training in emergency procedures and resuscitation following electric shock should be given. Lone working on electrical equipment should not be permitted.

6.3.2 Non-beam fire and explosion hazards

In addition to the possibility of fire or explosion caused by the action of a high-power laser beam, fire and explosion risks can also arise from other sources of heat, including sparks, ejected hot particles and electrical faults. This risk can be particularly high whenever flammable gases or liquids, including laser-dye solvents, are used in conjunction with the laser work, when combustible waste material is present, or when fume accumulates in filters or exhaust systems. The processing of aluminium or plastic materials can cause fires and explosions in the exhaust ducts and filters.

The mechanical pumps used for circulating the fluorescent dyes in (liquid) dye lasers can ignite the dye solvent due to overheating. This can be caused by electrical faults or by worn bearings, and the pumps should therefore be inspected regularly and maintained in good working order. Because of these risks, dye lasers should not be left running while unattended.

Appropriate design of the working layout should be adopted to minimize fire and explosion risks, by separating potential sources of ignition from combustible materials, insofar as this is possible. An uncluttered working environment and the regular removal of waste are important. Suitable fire-fighting equipment should be readily available, and training given in its use.

6.3.3 Collateral radiation

Certain lasers may emit other forms of electromagnetic radiation in addition to the laser beam. This can include ultraviolet radiation (sufficient to cause 'sunburn') from exposed gas-laser discharge tubes, and radio-frequency emission from RF-excited lasers that are not properly screened.

Hazardous levels of optical radiation (mainly ultraviolet and visible) may also be generated in the plasma produced by high-power beam interactions. Measurements have shown that safety limits for the eyes and skin can be exceeded within a few seconds of exposure to the plasma produced by a multi-kilowatt welding laser [3].

Ionizing radiation (x-rays) may be generated from electrical equipment utilizing voltages in excess of 15 kV.

Appropriate screening and enclosure of all hazardous collateral radiation should be ensured. The type of material needed to form the enclosure will depend on the kind of radiation that is emitted.

6.3.4 Hazardous substances

The material used as the active medium of certain lasers may be hazardous. Laser dyes and some laser gases, especially fluorine and chlorine which is used in excimer lasers, can be highly toxic. In addition, laser dyes can be carcinogenic. Some substances may be irritants to the eyes and skin. Low concentrations of carbon monoxide, sufficient to pose a health risk, are present in the gas-mixture used for carbon dioxide lasers. Even gases that are inert, such as helium and nitrogen, can act as asphyxiants through oxygen displacement if released in sufficient quantity. Solvents that may be used for cleaning and other purposes can be inflammable. They may also be volatile and present a health hazard. Zinc selenide, often used as a lens material with far-infrared lasers, can be hazardous if small particles are ingested following breakage of a lens, or if the lens material is vaporized by the incident laser beam. Many other materials that can be used in conjunction with a laser may also be harmful.

Proper handling, storage and disposal techniques should be used for all hazardous substances in accordance with appropriate national guidelines and regulations. This may require the use of appropriate personal protective equipment (such as gloves and eye protection). Solvents and other chemicals, including laser dyes, should only be used in the minimum quantities necessary, and only in well-ventilated areas. Fume cupboards or glove boxes should be used for preparation where appropriate. When mixing liquids or dissolving powders (e.g. laser dyes), only mechanical pipetting should be used, and care should be taken to prevent any powder becoming dispersed into the air. Once mixed, laser dyes should be stored in unbreakable sealed containers. Eating or drinking should not be permitted in areas where hazardous substances can be present, and hand-washing facilities should be installed and should be used. Wherever appropriate, emergency eye-washing facilities should also be available. Spillages should be cleaned up immediately. Staff should be aware of the emergency action needed in the case of skin or eye exposure to, or ingestion of, any hazardous substances that may be present.

Where compressed toxic, asphyxiant or flammable gases are used, gas cylinders should be housed in an enclosed and properly-constructed cabinet, fitted against a wall and vented to the outside. The gas-line path should be as short as possible, and should be routed and secured so as to minimize the risk of damage occurring to the line. Systems should be leak-tested prior to use. Special fittings may be needed where toxic, corrosive or reactive gases are employed, and purging with an inert gas may be necessary to clear gas lines after use.

With dye lasers, the dye-delivery system should be checked frequently for loose connections and for signs of wear in plastic tubing. It is sensible to locate the entire dye system in a tray to contain any dye leak.

Exposure limits for a wide range of hazardous substances are published by various national agencies and should be observed.

6.3.5 Laser-generated noise

While noise is not often regarded as a hazard associated with the use of lasers, in certain circumstances it can be a safety issue. The repetitive firing of high-energy pulsed lasers, for example, is sometimes accompanied by a loud 'crack' at every pulse, and this may be sufficient to cause hearing damage over extended periods of time. Associated equipment being used in conjunction with the laser can also generate excessive quantities of noise or ultrasonic emission.

Where noise is excessive (i.e. above accepted noise limits) and cannot be eliminated, ear-defenders may need to be worn. Steps should also be taken through the adoption of appropriate procedural controls to limit the duration of individual exposure.

6.3.6 Mechanical hazards

Bulky equipment that has to be moved or which is not properly secured (gas cylinders, for example) can pose a hazard. Trailing cables, gas tubing or water hoses can cause people to trip. Cuts may be caused by sharp edges. Strain or crush injuries can result from the manual handling of large workpieces or other heavy items. Beam delivery arms or flat-bed systems under automatic control may move rapidly without warning.

Appropriate design of the equipment and the enclosure or guarding of moving parts should be used to minimize the risk of mechanical injury. Heavy objects, including gas cylinders, should be properly secured. Cables and tubing should not hang loosely or trail across the floor. Warning signs should also be used, where appropriate, to warn of parts that may move unexpectedly or to indicate the range of movement that is possible. Audible alarms can be used to give warning of any imminent movement of part of a machine. Working arrangements, the design of equipment and the layout of the working area should be such as to minimize exposure to mechanical hazards and the risk of personal injury.

6.3.7 Temperature and humidity

Extremes of temperature or high levels of ambient humidity may affect laser operation and can compromise its safety. Condensation can form on optical components, affecting performance. The ingress of water can also have a serious impact on the laser's electrical system, and may give rise to unexpected failure modes and also the risk of electric shock.

Temperature control and air conditioning may be needed where the effects of ambient temperature or humidity could otherwise jeopardize safe operation of the laser equipment. Tubing that is used to circulate cooling water or other liquids (e.g. laser dyes) should not be routed where leaks could damage equipment or create electrical hazards.

6.3.8 External shock and vibration

System malfunction or the misalignment of optical components, which may give rise to hazardous errant (i.e. misdirected) beams, can be caused through the effects of mechanical shock or vibration.

Equipment should be sufficiently robust to withstand reasonable levels of mechanical disturbance. Anti-vibration mountings and shock absorbers can be used to dampen these effects. In research and development laboratories where optical tables or optical benches are often used, mounting pillars for optical components should be kept as low as possible and all fixtures should be adequately secured. Where long beam-paths are unavoidable, screens and apertures can be used to contain misdirected beams. Open-beam systems where access to the beam is needed can often be contained within an open box-type arrangement (by fitting low walls around the edges of an optical table, for example) to prevent errant beams caused by mechanical disturbance from crossing open areas of floor.

6.3.9 Computer malfunction

If the operation of the laser or any part of its protective system is under computer control, errors in computer programming and the malfunction of computer systems can cause unpredictable and potentially hazardous situations to arise without warning.

The use of computer systems to perform safety-critical functions (such as the control of laser operation, entryways, beam shutters, etc) should be minimized and hard-wired systems of high reliability used instead. Where software-controlled systems are used, they should be thoroughly tested under realistic conditions and employ adequate levels of redundancy to minimize the impact of any system failure.

6.3.10 Ambient noise

High levels of background noise can result in audible alarms being unheeded or spoken instructions being misheard.

Where high levels of background noise exist which cannot be reduced, reliance on verbal instructions or warnings should be minimized, and highly visible warning indicators (e.g. flashing lights) should be substituted in place of acoustic alarms.

6.3.11 Compressed gases

In addition to the hazards already discussed, special attention should be paid to the use of compressed gases. These are commonly used in conjunction with many laser systems. As noted earlier, gas cylinders should be properly secured and not left free standing. Gas regulators should be in good condition and appropriate to the gas being employed. They should not be used to control flow; separate valves should be installed for this purpose. The possibility of, and the consequences arising from, leaking gas should be identified, and minimized in the design and layout of the gas-handling system.

In some cases (e.g. excimer laser gases) the necessity to store hazardous gas cylinders can be avoided by the use of gas generators which produce the gas as it is needed in the quantity required.

References

[1] 1999 *LIA Guide to Non-Beam Hazards Associated with Laser Use* (Florida: Laser Institute of America)

[2] ANSI Z.136.1 2000 *American National Standard for Safe Use of Lasers* (Florida: LIA)

[3] Bothe H, Cammenga H K and Welzel M M 1998 *Proc. 9th Int. Symp. Loss Prevention and Safety Promotion in the Process Industries* pp 860–9

[4] 2000 *Handbook of Industrial Laser Safety* (Seibersdorf)

Chapter 7

Assessment of laser risk

7.1 Workplace evaluation

7.1.1 The laser class

The control of laser hazards to ensure safe use requires a careful assessment of all aspects of laser installation, operation, maintenance and service. This assessment should take into account not just the level of the emitted laser radiation, but also other hazards related to the laser and its associated equipment, the people involved in its use and any impact that the laser's environment might have on safe laser operation.

Previous chapters have discussed the hazards of laser radiation and described how laser products are classified in terms of their accessible emission. This is done by comparing the level of accessible laser output, which should be determined in accordance with the appropriate measurement conditions defined in the safety standard, with the applicable AEL (accessible emission limit). In addition, the level of laser exposure that could arise at the eyes or skin through use or misuse of the laser can be determined and compared with the applicable MPE (maximum permissible exposure). Regardless of the laser's class, the MPE can be applied to determine whether a particular exposure condition is safe, over what distance from the laser a radiation hazard might exist, and what reduction in emitted power or energy is necessary in order to reduce a given exposure to an acceptably safe level.

The class of a laser product indicates, in very broad terms, the harm that could result from exposure of the eyes or skin to its emitted radiation. The type of harm, as implied by the laser class, and the underlying risk arising from exposure to the beam, is discussed below. In this discussion, we consider Class 1M and Class 2M after Class 3R.

Class 1: No harm under intended conditions of use (and, in the case of lasers or LEDs that are incapable of emitting harmful levels of radiation, under any conditions of use).

The Class 1 AELs are based directly on the MPEs, and for wavelengths between 302.m and 4000 nm represent the lowest level of power (or energy) with which it is possible for an exposure to just reach the MPE. At wavelengths below 302.5 nm and above 4000 nm, the Class 1 AEL is set equal to (and specified in the same units as) the MPE. Class 1 lasers permit unlimited viewing of, or exposure to, the emission without the MPE being exceeded. In the visible band, the AEL varies from 0.04 mW at the blue end of the spectrum (where photochemical limits apply) to 0.4 mW in the red, where only thermal effects are significant. The photochemical limits are based on the accumulated energy for periods up to 100 s, whereas the retinal thermal limits have only a shallow dependence on time, with the power varying as $t^{-1/4}$ (or $t^{3/4}$ in terms of energy). Furthermore, the effect of eye movements flattens out this dependency for durations greater than 10 s, and so the thermal limits are held constant for all time periods greater than this.

In the ultraviolet region a time base of 30 000 s (about 8 h) is used for classification, because of the cumulative photochemical effects of ultraviolet exposure. The AEL has a constant energy value up to 100 s for wavelengths between 315 and 400 nm, and up 30 000 s for all other UV wavelengths. This can result in very low values of the AEL; for example, it is only about 0.8 nW for wavelengths less than 302.5, and 8 μW for wavelengths between 315 and 400 nm. However, where exposure occurs for periods less than 8 h, the safety factor can be very large. The maximum exposure that could occur over a period of 10 s from a Class 1 laser having a wavelength less than 302.5 nm would be 3000 times less than the MPE.

Although the skin MPEs in the ultraviolet region are set equal to those for the eye, there is in fact a large safety factor in the case of skin exposure, except for particularly sensitive (non-tanning) individuals.

The measurement conditions used for classification in all classes other than Class 1M and Class 2M assume the possible use of optical viewing instruments. That is to say, the type of ocular hazard associated with a given class is still applicable when such viewing aids are used. In the case of Class 1, therefore, an exposure remains below the MPE even if optical instruments are used to view the beam (actually, to view the laser source from a position within the beam). For many laser beams, the use of viewing instruments will have no effect on the resultant exposure at the eye, simply because the beam is smaller than the limiting or averaging aperture, and therefore the power (or energy) contained within the area of the limiting aperture cannot be increased. For larger area beams, the effective exposure at the eye can be increased through the use of viewing instruments, but for Class 1, Class 2, Class 3R and Class 3B the resultant exposure at the eye would remain within the limits of the class.

It should be noted, however, that the measurement conditions for classification assume a telescopic magnification of up to 7× with a 50 mm objective lens (i.e. 7 × 50), or an eye-loupe magnification of up to 18× (see section 5.6). There may be certain specialized applications where these limits can be exceeded, and in such cases the assumptions inherent in the classification process may not be sufficiently protective. For example, so-called observation binoculars can have a specification of 30 × 80, which would increase the retinal exposure by up to 130 times (well over twice as much as the maximum increase assumed), while large telescopes could increase it further still. On the other hand, an eye loupe as powerful as 18× is unusual (5–10× is more typical), and the only close-up viewing instruments likely to exceed this level of magnification are microscopes.

Class 2: No harm, provided that deliberate staring into the beam is avoided, otherwise damage can be caused to the eyes.

Class 2 is restricted, of course, to lasers emitting in the visible band, and the AEL is based on the Class 1 limit applicable to an emission duration of 0.25 s, i.e. 1 mW. But although the time base for classification is very different to Class 1 (0.25 s instead of 100 s), the $t^{-1/4}$ dependency applicable to the thermal limits means that the actual difference in the allowable power between Class 1 and the 1 mW limit for Class 2 may not be that great. At 400 nm (where the photochemical limits apply) it is a factor of 25 but at 700 nm it is only a factor of 2.5, and so the risk of harm arising from the intentional staring into a red Class 2 laser pointer is very low. Such deliberate fixated viewing, however, requires considerable concentration; it can also be uncomfortable and even painful (rather like viewing the Sun during the day or the undipped headlights of an approaching car at night). In addition, the dazzle and flash-blindness that may be caused by a Class 2 laser (or even by the visible beam from a Class 1 laser) can temporarily impair vision, and can distract someone who is driving or operating machinery, with possible serious consequences.

Class 3R: Harm can be caused to the eyes for other than very short exposure durations, although the risk of injury is generally low given the unlikelihood of prolonged fixated eye exposure.

The AEL for Class 3R is set at fives times the AEL for Class 2 for emission in the visible band, and five times the AEL for Class 1 at all other wavelengths. (Class 3R does not apply for emission wavelengths less than 302.5 nm, although in the opinion of the authors it would have been justifiable for this class to have covered the entire ultraviolet region, because of the huge safety margin in the Class 1 UV AELs for reasonably short periods of accidental exposure.)

Although the time required for an exposure to the beam of a Class 3R laser to exceed the MPE can be quite short (about 0.4 ms in the visible band), the exposure level for longer periods of exposure cannot exceed the MPE by more than five times, if reliance is placed on the aversion response in the case of visible emission. It is because of this that Class 3R lasers are considered to be a low risk. Permanent injury is very unlikely to occur other than through deliberate, fixated staring into the beam of a visible Class 3R laser (by intentionally overriding the aversion response), when the MPE can be exceeded by 12.5 times at 700 nm and by as much as 125 times at 400 nm. An example of such injury occurred when some children were undertaking a scientific project to investigate the constriction of the pupil that was caused by shining a laser into the eye. One of the children suffered a small retinal injury, but was fortunate that his vision returned to normal after a few months. As far as the authors are aware, the only injuries caused by Class 3R lasers (nearly all of which are diode laser pointers) have arisen through deliberately staring into the beam, and not from momentary accidental exposure.

Class 1M: No harm, provided that optical viewing instruments are not used to view the laser source, otherwise damage can be caused to the eyes.

While exposure of the naked eye to the radiation from a Class 1M laser is safe, i.e. below the MPE, the use of optical viewing instruments of the kind represented by the two measurement conditions (a telescope or binoculars in the case of condition 1, or an eye loupe in the case of condition 2) can result in an effective exposure up to 50 times the MPE. For condition 1 this factor is governed by the area of the measurement aperture in comparison to the 7 mm diameter of the pupil; for condition 2 it is represented by the increase in the collection solid angle of a 7 mm pupil at a distance of 14 mm compared with the standard viewing distance of 100 mm. The hazard is only realized, however, if an optical viewing instrument of the appropriate type is used to view the laser source. This will be a telescopic system, such as a pair of binoculars, in the case of a Class 1M laser that exceeds the AEL for Class 1 under condition 1, or an eye loupe or similar close-up magnifier in the case of a Class 1M laser that exceeds the AEL for Class 1 under condition 2. The particular type of viewing instrument that could introduce a hazard should be specified by the laser manufacturer in the information supplied with the product, and may also be indicated on the product label.

It needs to be stressed that there is a fundamental difference between the two types of aided viewing conditions. In the case of telescopes and binoculars, which may pose a risk with reasonably well-collimated beams, the hazard can arise at considerable distances from the laser (i.e. within the ENOHD, see section 5.6.1). People using such instruments may be unconnected with the laser work and unaware of the laser's existence. The use of eye loupes or even microscopes, however, can only be hazardous when viewing divergent laser emission very close to the laser source. The protective measures adopted for Class 1M laser products therefore need to reflect the actual type of hazard together with the likelihood that aided viewing with the relevant kind of instrument might actually occur.

Class 2M: No harm, provided that deliberate staring into the beam is avoided or that optical viewing instruments are not used to view the laser source, otherwise damage can be caused to the eyes.

Class 2M is very similar in concept to Class 1M, except that it is only applicable to visible emission. (Visible-beam lasers can, however, be in Class 1M if they satisfy the conditions for that class.) Class 2M can really be considered as a combination of Class 2 and Class 1M, and much of the discussion above for Class

1M is of relevance. The difference is that without the use of viewing instruments the exposure of the unaided eye is equivalent to that arising from a Class 2 laser, and so protection relies on the natural aversion response to limit the duration of exposure to no more than 0.25 s. It follows that the ENOHD (see section 5.6.1) for a Class 2M laser that exceeds the AEL for Class 2 under condition 1 (the telescope-viewing condition) can be based on either the MPE for 0.25 s (for which protection would be based on the aversion response) or the MPE for 100 s (in which case an exposure would be unconditionally safe).

Class 3B: Immediate harm can be caused to the eyes through direct exposure to the beam, although the viewing of diffuse (scattered) reflections of the beam is normally safe.

The AELs for Class 3B are set at the level at which diffuse reflections could become hazardous. The AELs are, however, considerably more generalized than for other classes, and the Class 3B table has a much simpler structure. At the upper limit of Class 3B (500 mW), diffuse reflections in the visible band may be hazardous if viewed from a distance closer than 130 mm for longer than 10 s. For blue wavelengths (400–450 nm), the level of hazard for diffuse reflections increases if the viewing time is extended, and the safe viewing distance for 100 s is 300 mm. In the infrared region the MPEs are higher and the viewing of diffuse reflections is not hazardous at realistic viewing distances. In the ultraviolet region, at a viewing distance of 130 mm, the exposure duration needs to be longer than 15 min to produce a corneal inflammation (photokeratitis); at 40 cm it takes more than 3 h. Of course, for Class 3B lasers whose emission is considerably less than the Class 3B limit, these exposure durations can be greatly extended.

Class 3B lasers, especially those towards the upper limits of the class, can also create exposure levels in excess of the skin MPE. However, owing to the safety factors inherent in the skin MPEs, and also because of the limited area of exposure, any skin injuries that might arise would be minor. A visible or infrared beam that is focused onto the skin can nevertheless be quite painful, producing an effect similar to a pinprick. In the ultraviolet region, erythema (similar to sunburn) can be caused with Class 3B lasers, and since the onset of this photochemical injury is delayed and the exposure is not accompanied by any sensation (you can't 'feel' the laser beam), it is this aspect of skin exposure to Class 3B lasers, in circumstances where prolonged or repeated exposure might occur, that needs most caution.

Although Class 3B lasers are not usually associated with fire hazards, focused beams or equivalent levels of emission from optical fibres can cause ignition in combustible atmospheres. A maximum emission limit of 35 mW has been proposed for use under such conditions (see section 6.2).

Class 4: Immediate harm can be caused to the eyes (through direct or diffusely-reflected exposure) and also to the skin. In addition, fire can be caused and fume can be generated by interaction of the beam with any material on which the beam impinges, although these associated hazards are not necessarily restricted solely to Class 4 laser products.

Class 4 lasers cover an enormous range, since there is no upper limit to their emission, and consequently there are no AELs for this class. Injuries can extend up to large-scale (and disfiguring) eye damage causing complete blindness and life-threatening skin burns. Lasers at the lower levels of Class 4, however, can have a considerably more limited potential for causing harm, and much depends on the geometry of the emitted beam. The NOHD (hazard distance) of some Class 4 lasers may be only a few centimetres; for others it can be several kilometres. For most applications, other than surgery, protective enclosures will be necessary to guard against accidental exposure during normal use. Special precautions may need to be adopted during maintenance and servicing.

While Class 1 laser products may be considered safe under all reasonably foreseeable conditions of use, all other classes represent some kind of hazard. Clearly, however, the kind of hazard associated with a particular class can vary considerably within the boundaries of that class, and the hazard does not become realized for each laser product in the exactly the same way. The potential hazards associated with Class 1M, Class 2, Class 2M and Class 3R manifest themselves only in special situations, and so these classes could be grouped together as 'safe except in special cases', as discussed above in relation to each class. Class 3B and Class 4 laser products are, however, normally regarded as 'potentially dangerous', although the extent to which they are hazardous and their capability for causing injury will depend on additional factors, including the wavelength and level of their emission, the geometry of the emitted beam and particular features of the product's design.

Classification represents an important aspect of laser safety and it should be regarded as not just a matter of product compliance—an issue for manufacturers rather than users—but as an essential means of conveying basic safety information concerning the laser to those who have to work with it. This means that all laser equipment, regardless of its origins or history, should be correctly classified under the relevant safety standard and properly labelled in accordance with its class. This should really apply not only to purchased laser products, but to all laser systems (that is, equipment incorporating a laser), even if the complete system has been assembled by someone (including possibly the user) other than the original laser supplier. Most importantly, no one who uses or works with a laser should be in any doubt as to what its class is and what the class signifies.

The laser class, however, is by no means the only information that may be needed in order to identify the precautions that might be necessary. There are three main reasons for this.

First, the class contains no reference to the distance from the laser within which a radiation hazard can exist. As an illustration, we would normally regard a Class 4 laser as being considerably more hazardous than a Class 3R laser. However, if the Class 4 laser produced divergent emission and had an NOHD (nominal ocular hazard distance) of only, say, one metre whereas the Class 3R laser produced collimated emission with an NOHD of ten metres, at any position between one and ten metres from the laser, therefore, it would be the Class3R laser that was more hazardous.

Second, there may be hazards associated with the laser other than those arising from direct exposure to the laser beam that are not apparent from the classification. These include ancillary hazards that can arise from the laser process itself as well as from the impact that other activities and external influences could have on the laser's operation (see chapter 6).

Third, the laser class indicates the harm that *could* be caused (by direct exposure to a laser beam), and thus warns the user of the laser's *potential* for causing injury. The class does not quantify the severity of the harm that could be caused; it gives no indication of the circumstances under which a hazardous exposure to the beam might occur; and it doesn't indicate the means by which such an exposure could be avoided.

All of these issues have to be considered as part of a *risk assessment*, a process that is described in more detail later in this chapter. The outcome of the risk assessment, and not simply a knowledge of the laser's class, should form the foundation on which decisions concerning protective measures are taken. For this reason protective measures should not be regarded as prescriptive. They do not follow automatically from the product class; a given class of laser does not always require the same level of hazard control. This can inevitably place a greater responsibility for safety on the laser user (or the user's organization) than a purely prescriptive approach would. However, it should enable the user to avoid unnecessary, excessive or inappropriate controls. Furthermore, under the safety legislation of most countries, we are obliged to adopt precautions against a workplace hazard on the basis of the actual level of *risk*, and not simply on the *potential hazard*.

7.1.2 Does 'safe' mean Class 1? Does Class 1 mean 'safe'?

Although the classification scheme provides a useful guide to the potential radiation hazard that might exist, it can be, and quite often is, misinterpreted. This raises a number of issues related to what might be termed 'class acceptability'. As an example, a Class 4 laser system may meet all the applicable requirements of its class, as specified in the safety standard. (These will include the necessary warning labels, an emission indicator, a beam attenuator and key control.) Does this make it safe? Of course not. The whole point of any laser class other than Class 1 is to indicate that the laser is potentially hazardous. Adequate consideration must also be given as to how it is going to be *used* and how it might be *misused* (whether by accident or intention). At the other extreme, a purchaser, believing that only Class 1 lasers are safe lasers, may demand that a laser system supplied for the purchaser's particular application must be Class 1, which could be an unduly and unnecessarily restrictive requirement.

It is sometimes thought that merely putting an enclosure around the beam of a Class 4 laser will turn into a Class 1 laser. This is not necessarily the case, unless the laser then fully complies with all the requirements specified in the safety standard for a Class 1 laser product. However, even if it remains a Class 4 laser, the enclosure may still serve the function of making it safe for its intended use, and fitting it can be an eminently sensible thing to do.

The foregoing examples demonstrate two common misconceptions. The first is that, for a manufacturer, laser safety is only about correctly classifying the laser and complying with the labelling and other class requirements specified in the standard, leaving the purchaser to worry about making it 'safe' for its intended purpose. The second is that, for a user, given the system of laser classification, it is really only a Class 1 laser that is 'safe'. Both of these perceptions, however, are untrue, and demonstrate a much too simplistic approach to laser safety. All laser systems should, in fact, be 'safe', meaning that they must be capable of being used for their intended purpose, as well as being capable of being adjusted and maintained for that purpose, without significant risk of harm, whatever their class and taking into account all reasonably foreseeable circumstances of use, misuse and failure.

It is possible that a Class 1 laser may have serious hazards other than those of radiation exposure. (This can be especially true of embedded lasers). It may not necessarily, therefore, be unconditionally safe and specific precautions may be needed to prevent harm occurring. Equally, however, certain lasers that are not Class 1 may nevertheless be very unlikely, because of their particular design and intended conditions of use, to give rise to hazardous levels of laser exposure. This can even apply to some Class 3B and Class 4 laser products. What is necessary, for all laser equipment, is that any reasonably foreseeable exposure is below the MPE (taking into account, where appropriate, the use of eye or other protection), and that any other hazards associated with the laser are adequately controlled. A badly designed Class 1 laser may fail to satisfy this fundamental requirement, whereas a well-designed laser of higher class may meet it. The laser product class is therefore an important means of indicating the potential of a laser's accessible radiation for causing harm, but can be insufficient, on its own, for indicating the overall degree of *risk* associated with its use.

7.1.3 Supplier and purchaser responsibilities

A legal responsibility is placed on manufacturers to supply safe products. As mentioned above, this means, in the case of a product incorporating a potentially hazardous laser, that it must be capable of being used (and adjusted and maintained) for its intended purpose in a safe manner. But what is the purpose of a particular laser? If a manufacturer supplies a Class 4 laser with no knowledge of what it is going to be used for, then quite clearly only general warnings of the hazards of a Class 4 laser can be given. It is then the responsibility of the purchaser (which may be the user organization or could be an intermediate system integrator who is incorporating the laser into equipment for resale to a third party) to determine the risks arising from its intended use and the specific control measures that may be needed. If, on the other hand, a laser product is designed to perform a very specific function and is marketed solely for that particular purpose, the supplier must ensure that safety is inherent in the design and that adequate instructions are given for safe use, adjustment and maintenance. In such circumstances it should not be the purchaser's responsibility to investigate the safety issues nor to determine what specific precautions are necessary. This is not to say, of course, that the purchaser has no responsibility for safety. It does mean, however, that appropriate and sufficiently specific guidance, information and instructions concerning the control of hazards should be given by the equipment supplier.

The only exceptions to this are where the laser is to be used for a purpose or under conditions that are different from those intended by the supplier, and where the supplier cannot reasonably be expected to have specific expertise in the user's required application.

Wherever laser equipment is being specially designed and built for a single customer rather than purchased as a standard catalogue product, and particularly if a third party (such as a system integrator) other than the original laser manufacturer or the ultimate purchaser is also involved in any part of the final system design, fabrication, installation or commissioning, it can be very important to reach prior agreement between all parties, and to confirm this in writing, as to where specific safety responsibilities lie and how these are to be coordinated. Too often, in laser system installations where more than one party has been involved, there is a failure to properly address all of the safety issues.

A related issue of some importance concerns the maximum class advisable for certain types of laser use (or, perhaps more relevantly, for certain types of laser user or user environment). Only limited guidance is given in the international laser safety standard concerning the highest class of laser product that is appropriate for a specific application. This states that only Class 1 or Class 2 laser products should be used for demonstration, display or entertainment purposes in unsupervised areas, unless spectators are prevented from exposure to levels exceeding the MPE. There are also other areas of application, however, where lasers of an inappropriate class are sometimes used, although no specific guidance exists. Product manufacturers have a duty here to consider carefully the market for which their products are intended and the way in which they might be used or could foreseeably be misused.

In several European countries the sale of laser pointers is restricted to products in Class 2. In Austria, a maximum power of 1 mW is specified but with no reference to the product class. Consequently Q-switched diode-pumped green laser pointers, which have pulsed emission with relatively high-peak powers, can be legally sold provided their average power does not exceed 5 mW, even though they would be Class 3R. This is problematic because green Q-switched radiation has a lower safety factor than red cw radiation.

There is also concern in safety circles over what is termed 'product migration', whereby potentially hazardous products intended initially for professional use under controlled workplace conditions that are subject to occupational health and safety legislation may be made available, via the internet or through retail outlets to the public. Examples have included display lasers for disco use, laser range-finders intended for yachtsmen, and biostimulation devices (so-called 'soft lasers') which can be as high as Class 3B but which are sold using slogans such as 'should be in every household pharmaceutical kit'. As lasers become more compact and efficient, enabling high-levels of output to be generated from inexpensive units that can be run from a domestic power socket or even from batteries, this problem is likely to grow.

However, even products intended only for occupational use may be unsuitable where it cannot be reasonably expected that the potential users will have a sufficient awareness or understanding of the hazards (regardless of any safety instructions that may be supplied or training that is given), or where the environment of use is such that members of the public and especially children cannot be adequately protected (in spite of any warning signs that may be displayed or prohibitions that may be in force). Sometimes laser products are marketed solely on the basis of their benefits and ease of use, with little effort made to make the purchaser aware of the potential hazards that can exist and of the need to implement appropriate protective measures, including adequate training of all the staff involved in the use of the equipment.

Comparisons are sometimes made with power tools such as electric drills and chain saws, many of which are used by untrained members of the public, but which have the capability for causing serious injury, even death. If these are permissible, why all this concern over laser products? Three answers to this were given in the first chapter of this book. Laser hazards are not 'obvious' in the same sense that the injuries from the misuse of power tools are. (Several people who have suffered a laser injury have afterwards confessed their surprise at discovering the hard way that the laser they were using was so dangerous.) In addition, a person who is accidentally exposed to laser radiation may not be aware of this until a serious injury has been caused. Finally, and perhaps most significantly, lasers have the ability to cause harm at a distance. Someone may be injured without even being aware of the laser's existence, let alone that it had such a capability.

Responsibility for laser safety is therefore shared between manufacturers, suppliers and users. No one involved in designing, manufacturing, specifying, purchasing, installing, commissioning, using or repairing laser equipment should consider laser safety to be solely someone else's concern. When first becoming involved, at any stage, with a particular laser system, it can be wise to keep an open mind as to whether others who are also involved in any aspect of the equipment's safety have in fact addressed their responsibilities adequately. Ask questions and seek confirmation before assuming that other people have fully met their obligations and have properly addressed all of the relevant safety issues!

7.2 Risk assessment

7.2.1 Hazards and risks

Most of the discussion in previous chapters has concentrated on the nature of laser hazards. As already explained, laser hazards include those arising from the direct exposure of any person to the beam, as well as the associated hazards due to other aspects of the laser or to the manner or circumstances in which it is being used.

A *hazard* is any condition with the potential for causing harm. An unenclosed laser beam, for example, or unshielded electrical terminals, may therefore represent a hazard. While the *harm* that could be caused by a hazard is normally taken to mean personal injury (or sometimes even death), it can also cover financial loss, such as through damage to equipment or property, or the loss of working time.

While the identification and characterization of potential hazards is a vitally important aspect of laser safety, the safety precautions that need to be adopted in a particular set of circumstances should not be based solely on the hazards, but on an assessment of the actual *risk*.

Risk is a combination of the *likelihood* of harm occurring and the *severity* of the harm that could be caused. Nearly all everyday tasks and work activities contain some element of risk, simply because there are potential hazards in almost everything we do. Many of these hazards can be so familiar to us that we adopt precautions without much deliberate thought, basing these on our subconscious *perceptions* of the actual level of risk. However, in dealing with hazards that may not be obvious or are not widely understood, and this is certainly the case with many lasers, we need to make sure that we adopt a structured approach to identifying these hazards, analysing the risk, and implementing effective control procedures rather than relying on our own instincts as to what is required.

A not uncommon failing in laser safety is to regard safety control measures as prescriptive in nature, and to consider them individually, in isolation from each other. This, unfortunately, can be implied by the classification system, which can lead people to believe that the necessary controls are dependent solely on the class of the laser, regardless of where, by whom, or how it is being used. The answers to questions such as: 'Should I be wearing eye protection?' or 'Do I need to fit interlocks to the door?' are then considered simply in relation to the laser's class.

This is not to say, of course, that the use of all lasers needs to be preceded by a detailed risk assessment. Lasers that are Class 1, either by virtue of the very low laser power employed or because of their complete enclosure, and also lasers that are Class 2 and which are used responsibly in accordance with the precautions printed on the warning label, would not normally require any formal safety assessment before use. But even with lasers such as these there can be particular circumstances (a likelihood of access by unauthorized people, especially children, is one example) in which additional precautions might be necessary.

In general, therefore, to ensure the safe use of any laser equipment, the following process should be followed.

(a) Identify, and where necessary characterize, all potential hazards (all conditions that could cause harm). This will include laser radiation hazards and any other additional hazards arising from the reasonably foreseeable use, misuse or failure of the equipment.
(b) Eliminate these hazards wherever practicable. This might be achieved, for example, by the use of a lower level of laser power, by total enclosure of the beam and/or of other associated hazards, or even by the use of alternative (non-laser) techniques, if the desired end result can be achieved by less hazardous means.
(c) Where total elimination of any significant (that is, non trivial) hazard is not possible (that is, not *reasonably practicable*), undertake a risk assessment of each remaining hazard by considering the likelihood of harm occurring and the severity of the harm that could be caused in order to determine whether the risk is **acceptable** or **unacceptable** against established criteria.
(d) Implement a system of controls to reduce all unacceptable risks to an acceptable level. (This will include an appropriate combination of administrative framework, engineering controls, safety training, procedural controls and personal protection, which are discussed in chapters 8 and 9.)
(e) Regularly review the continued use and effectiveness of the controls which have been adopted, and modify these whenever necessary.

7.2.2 The risk assessment process

Since risk is a combination of the *likelihood* of harm occurring and the *severity* of the harm that could be caused, risk assessment could be taken to mean a fully-quantitative analysis that assigns numerical values to both of these factors. This would be very difficult in many common laser applications, since both the likelihood (that is, the probability) of the occurrence of any particular hazardous event and the severity of the injury that might be caused can be very difficult to quantify in numerical terms. (What is the probability, for example, that a beam-steering mirror could become misaligned or that somebody might accidentally view the beam through magnifying optics? How severe an injury would be caused by an exposure that was, say, 50 times the MPE?)

An evaluation involving such a quantitative analysis is called a *probabilistic* risk assessment, but is unnecessary in the majority of laser applications. Probabilistic risk assessment can be of relevance, however, in circumstances where, although it is accepted that human exposure to levels of laser radiation above the MPE could conceivably occur, the probability of significant harm actually occurring turns out to be extremely small. One example of such a case is in certain uses of airborne lasers where injury could in principle occur through the use of binoculars. An analysis of all the relevant factors (e.g. the probability of a person being in the direct beam of the laser when it fires, the probability that such a person will have a pair of binoculars, the probability of that the person being within the ENOHD, the probability of using the binoculars at the critical time, the probability of the airborne laser being within the field-of view of the binoculars at the critical time, the probability that an injury will actually result, etc) may show that the level of risk is acceptably small.

For the majority of laser applications, however, we are concerned purely with carrying out a *deterministic* risk assessment. This involves a consideration of all relevant factors of laser use, as well as of foreseeable misuse or system failure, in order to make sure that human exposure to levels of radiation above the MPE is *very unlikely* to occur. We are simply trying to identify the circumstances under which hazardous levels of exposure could arise; we are not trying to formally quantify the probability of this happening or to assign a level of severity to the injury that might result, as we would need to for a probabilistic assessment.

Most accidents happen because things go wrong. This can be through the failure of individuals to work in accordance with appropriate procedures, through the involvement of people who are not properly aware of the hazards or of the means of controlling them or through some fault in the equipment itself. Risk assessment need be no more than the process of identifying the potential causes of things going wrong and thereby giving rise to any of the potential hazards that could exist (as a result of either human behaviour or equipment malfunction), thus enabling appropriate steps to be taken in order to prevent them from occurring. Insofar as laser radiation hazards are concerned, this means simply that all reasonably foreseeable conditions under which a hazardous exposure of the eyes or skin (i.e. any exposure that exceeds the relevant MPE) *could* occur should be investigated, and then the causes that could give rise to this hazardous exposure eliminated or the likelihood of them happening sufficiently reduced. By considering all the relevant and foreseeable conditions of use, we can make an informed judgement of the likelihood of a particular event occurring without needing to assign a numerical probability to it.

There are a number of different techniques for undertaking a deterministic risk assessment. One particular assessment process that can be helpful is to consider laser operation under four separate headings, covering the laser equipment itself, the task or process for which it is being used, the place where this work is being done and the people who are (or who could become) involved in this activity. These four aspects, and some of the questions they raise, are described in more detail below.

- **The equipment**
  - What is the specification of the laser being used?
  - What is the laser's class and wavelength? (This will give an indication of its capability for causing injury.)
  - What is the beam path and how is the beam delivered to its point of use?
  - How far does the hazard extend?
  - How is the laser operated?

- What other hazards are inherent with this particular laser?
- What other equipment is the laser being used in conjunction with?

- **The process**
  - What is the laser (and its associated equipment) being used to do?
  - How is the work being carried out, and how is this being controlled?
  - What additional hazards might this process introduce?
  - Under what conditions can hazards arise and exposure to those hazards occur?
- **The location**
  - Where is the laser being used?
  - What kind of environment is this?
  - Is it well lit, uncluttered, and is the workspace sensibly laid out?
  - What level of access is there into the laser area?
- **The people**
  - Who is involved in this activity?
  - How aware are they of the hazards?
  - How well do they understand the precautions that might be necessary?
  - How likely are they to follow any specified working procedures?
  - Who else might gain access into the laser area?

The process of risk assessment should take all these factors into account. It should identify all the potential hazards that might arise (which should include both those of laser radiation exposure and also of any additional, associated hazards) and the circumstances (that is, the hazardous event) under which each hazard could occur. A judgement can normally then be made as to whether the risk of each identified hazardous event represents an **acceptable** (i.e. very low) or **unacceptable** risk. This assessment needs to be carried out with due care and its conclusions must stand up to scrutiny. They may also need to be defensible in law. Where the risk is judged to be unacceptable, then clearly something needs to be done to control the hazard in order to reduce the risk.

Risk assessment is therefore used to identify the control measures that are necessary, and so risk assessment and control measure selection together represent an iterative process, which may need to be performed several times before adequate controls have been identified.

7.2.3 Risk factors

By *risk factors* we mean those conditions or issues that could have a bearing on the level of risk, and therefore on the protective measures that may be necessary. Risk factors include both the *circumstances* under which the various hazards can occur and the *consequences* of the harm that could result.

7.2.3.1 Circumstances of harm

Not all potential laser hazards are present all of the time. While some may exist throughout the duration of laser operation, others may only arise under particular circumstances. The process of risk analysis should identify not only what harm might occur but also *how* it could occur. Some 'lateral thinking' may be required in order to identify what risk factors might be relevant.

The **equipment** may be poorly designed, difficult to use or complex to operate. Those working with it might therefore be less alert to any safety issues that could arise during its use, or they might not always follow agreed operating procedures if these are particularly involved or time-consuming. There might be potential faults or system failures that could result in the occurrence of additional hazards.

The **process** might be of an experimental or investigative nature, perhaps having an uncertain outcome, making it difficult to determine in advance precise methods of working. Not all the possible consequences of the process might have been anticipated. The way in which the laser will be used might not be entirely predictable. This is particularly the case where the beam is adjustable or the equipment is hand-held. To achieve the results required, there might be a need to modify the process or to rearrange the configuration of the equipment, perhaps introducing unforeseen risks.

The **location** in which the work is being carried out might be cluttered or poorly lit, making accidents more likely to occur. The working area may not be adequately enclosed or screened. Vibration or other environmental factors might disturb the safe operation of the equipment. It might be difficult or indeed impossible to restrict access into a hazardous area.

The **people** using the laser might not be fully aware of all the hazards, or understand how the risks should be controlled. Others might enter the laser area (possibly when no one else is present) having no knowledge of the hazards or of the need to adopt precautions. Some might be particularly curious about the laser or the work being carried out with it, leading them to expose themselves to risks of which they are not aware. Where ordinary members of the public and especially children may gain access to the laser equipment or can enter the hazardous area (which for laser radiation may encompass the entire *extended* hazard zone, taking into account the possible use of binoculars), the risks can be particularly high. Behavioural aspects of safety (human factor issues) are discussed further in chapter 9.

'Time at risk' can also be an important consideration. Short-term work with a hazardous condition (for example, when undertaking a particular task that only needs to be carried out occasionally) may be more acceptable than continuous working with the same level of hazard. Nevertheless, it can often be the occasional task (such as beam alignment or other adjustment) which poses the highest risk, and the fact that it occurs only infrequently should not be an excuse to neglect the need for a proper risk assessment and the adoption of adequate precautions.

7.2.3.2 Consequences of harm

Although the process of risk assessment for laser safety purposes is primarily concerned with an evaluation of the circumstances under which an exposure could foreseeably exceed the MPE or other hazards could occur, the consequences (severity) of the harm that could result from such exposure can also have a bearing on the acceptability or otherwise of the particular risk.

For example, laser injuries (typically burns) to the skin would normally be considered to be less serious than equivalent injuries to the eye. For a given level of laser exposure, however, large-area skin burns (arising from divergent, large diameter, or rapidly scanned beams) would be more serious than smaller burns. Furthermore, very high levels of skin exposure (well above the MPE) can conceivably cause extremely serious bodily injuries, even resulting in death.

Eye injuries are normally most serious when they occur to the retina (that is, from laser emission at wavelengths between 400 and 1400 nm). It is most important to recognize, however, that lasers of all wavelengths can cause serious injury to the eyes, and that there is no ‘eye-safe’ waveband.

Exposure of the eyes to laser radiation at wavelengths greater than 1400 nm, or of the skin at wavelengths greater than 400 nm, can elicit its own ‘aversion response’ through the sensation of heat or even pain that can quickly become apparent. This can mean that prolonged exposure is unlikely to occur, although in the case of exposure levels that are many times the MPE, whether to the eyes or to the skin, serious injury can nevertheless be caused before any avoiding action can be taken, but the taking of avoiding action can involve cognitive issues. Closing the eyes in response to a sudden bright light or removing the hand from something touched which is very hot are both involuntary aversion responses; we do not have to stop and think what we are doing. However, removing ourselves from the path of a laser beam that is causing pain may require some conscious thought; we may not be immediately aware what is causing the pain, or how best to get out of the way. At shorter wavelengths (less than 1400 nm in the case of the eyes or less than 400 nm for the skin) the MPE can be exceeded, even over long periods, without any sensation at all.

7.2.4 Determining the level of risk

In many practical situations, a consideration of all the risk factors that exist, taking into account the hazards that could occur and the circumstances under which they might arise, is sufficient to indicate whether the resultant risk is acceptable or not. Accidents usually happen when there is a lack of knowledge of the hazards that could exist or there has been a failure to anticipate what *might* go wrong. In the majority of laser applications, risk assessment can be based on quite straightforward considerations, provided that sufficient information on the potential of the laser equipment for causing harm is available.

It is occasionally desirable, however, to adopt more formal methods of risk assessment in laser safety. This can be the case with electronic control systems where there is a possibility that a component failure could give rise to an increased hazard. This can occur, for example, in the use of laser diodes, which are often operated at well below their maximum output by limiting the current flowing through the device. A fault arising in the electronic drive circuitry can cause the normal operating output to be exceeded.

A number of different methods of risk analysis can be used, varying from simple analytical approaches to complex mathematical modelling [1–3]. The technique chosen needs to be appropriate for the particular problem being investigated. The accuracy of the conclusions will depend, of course, on the reliability of the data used and of any other information employed.

A list of some of the more common techniques is given below, but interested users may wish to refer to more detailed texts and standards on risk assessment for further information.

Checklists

Checklists can be used as a basic method for itemizing potential hazards or undesirable outcomes that need to be considered.

Event tree analysis

This technique uses inductive reasoning to translate different initiating events into possible outcomes.

Fault modes and effects analysis (FMEA) and fault modes, effect and criticality analysis (FMECA)

FMEA and FMECA are similar hazard identification and frequency analysis techniques that evaluate all the possible fault modes within an item of equipment and determine their effect.

Fault tree analysis

This identifies each undesired event and determines all the ways in which it could occur.

Hazard and operability studies (Hazop)

Hazop studies examine each part of an entire system to determine how deviations from the intended function or performance can lead to undesirable outcomes.

Preliminary analysis

This is a hazard identification and frequency analysis technique that can be used to assess potential hazards at an early design stage.

Reliability block diagram

A graphical frequency analysis technique that uses system models to investigate overall reliability.

Common mode failure analysis (CMFA)

CMFA evaluates the impact of coincidental failure in different parts of a complete system.

Delphi techniques

These techniques are processes for combining expert opinions from different sources into a model for estimating risk.

Hazard indices

These can be used to analyse different system options in order to identify that option having the lowest risk.

Monte Carlo simulation

Monte Carlo simulation uses mathematical modelling to evaluate the effect on performance of system variations and changing operating conditions.

References

[1] IEC 60300-3-9 1995 *Dependability Management—Part 3: Application Guide—Section 3: Risk Analysis of Technological Systems* (Geneva: IEC)

[2] IEC 60812: 1995 *Analysis Techniques for System Reliability—Procedures for Failure Mode and Effects Analysis (FMEA)* (Geneva: IEC)

[3] IEC 61025: 1990 *Fault Tree Analysis (FTA)* (Geneva: IEC)

Chapter 8

Protective measures and safety controls

8.1 Introduction to protective control measures

8.1.1 The use of safety control measures

The most satisfactory way of controlling risk is to eliminate the hazard at its source. This means either avoiding the use of a potentially hazardous method or totally enclosing the hazard. Only where it is not reasonably practicable to do this should alternative means of managing the risk be considered.

In many laser applications it can seem a foregone conclusion that total enclosure of the beam and of any other additional hazards is simply not feasible. Nevertheless, the possibility of avoiding the hazards in this way should always be considered as a first option. Any foreseeable access to hazardous levels of laser radiation (or to any additional laser hazards) therefore needs to be justified, and shown by means of a risk assessment to be acceptable. Many examples unfortunately exist where open-beam laser systems (typically in Class 3B or Class 4) are being used in conjunction with protective eyewear when it would have been quite straightforward, given the kind of process being carried out, for the equipment to have been totally enclosed. It had simply not occurred to the users to do this; their reliance on protective eyewear was simply the way in which they thought that high-power lasers were normally used.

Only where it is not reasonably practicable, therefore, should work requiring potential access to hazardous levels of laser radiation or to other laser hazards be considered. Then a risk assessment should be carried out, as discussed in the previous chapter, and a system of control measures implemented which, taken together, adequately control the risks.

Safety control measures for managing risk should be considered in accordance with the following sequence, with as much emphasis as possible being placed on the first, i.e. on engineering methods of control.

- *Engineering control measures.* These cover those engineering features incorporated in the laser product by the manufacturer together with any additional physical features, including enclosures and interlocks, that may by added by the user (by which is meant the user's organization) to the laser itself or included in the overall laser installation.

- *Administrative control measures.* Administrative control measures cover the overall safety policy of the organization (the 'local rules'), which include the appointment of a Laser Safety Officer, the use of warning signs, and the requirements for safety training (which are all discussed in the next chapter), as well as the designated procedures, often called standard operating procedures (SOPs), that have to be followed in the setting up, adjustment and operation of specific items of laser equipment.
- *Personal protection.* Only where a combination of engineering and administrative control measures cannot alone adequately control the risk should the adoption of personal protection (primarily but not only laser eye protection) be considered.

Control measures should be considered in the order given above as a means for reducing the risk of harm to a safe (i.e. an acceptable) level. However, the adoption of control measures should depend on a risk assessment to demonstrate that, with the proposed control measures in place, the risk would be adequately reduced. It cannot be stressed too strongly that it is not sufficient merely to carry out a risk assessment, to conclude that the risk is unacceptable, then adopt some additional control measures and begin using the laser. Risk assessment is an iterative process, and has not been completed until it is shown that the residual risk following the adoption of a suitable combination of control measures is acceptably low. Furthermore, all control measures should be reviewed together and not considered in isolation; the way in which they impact on each other should be examined. Control measures considered individually may be unable in themselves to adequately reduce the risk but, in combination, may be sufficient. For example, simply making eye protection available is not an adequate means of control. Combined with appropriate training and administrative procedures, however, which specify the circumstances under which the eye protection should be used and limit access into the laser area, the risk of injury may be sufficiently low to be acceptable.

8.1.2 Control measures as a function of the laser class

The class of a laser product gives an indication of its potential for causing harm, based on the level of its accessible laser radiation. Note that the laser class does not take into account any additional hazards that may be present (such as electrical hazards, fire or fume) which may exist in addition to the hazard (if any) arising from exposure to the laser beam, nor does it indicate the distance from the laser over which a radiation hazard might exist.

Laser safety standards have often been interpreted very prescriptively by laser users with regard to the precautions to be adopted when using a laser of given class. However, precautions listed by class are intended only as a guide to what might be necessary, and should not be implemented without first considering their appropriateness to the risks that actually exist and to the particular circumstances of use. This particularly applies to lasers in Classes 3B and 4, which can present very serious hazards. Wherever reasonably practicable, and where human access to the emission is not normally required, they should be totally enclosed to eliminate the hazard at source, obviating the necessity for having safety-trained users and enclosed working areas with interlocked access. On the other hand, even the use of quite low-power lasers may present serious safety concerns where they can be accessed by people (ordinary members of the public and especially children) who might misuse them.

Nevertheless, linking the laser class with a particular set of control measures can be useful as an indication of the kind of precautions that are likely to be necessary, particularly under normal conditions of occupational use and where the hazard class of the laser cannot conveniently be reduced by enclosing the beam (ideally to create a Class 1 system) or by reducing its output. Such a listing is given below, but it is intended only as a guide. A risk assessment may indicate that a laser of given class should be subject to different control measures, which could be either more restrictive or less restrictive than those given here, depending on the particular circumstances.

A risk assessment may be unnecessary for some lower-class lasers where no unusual circumstances exist. In that case the listed control measures may be considered as default protective measures, to be used in all cases unless the conditions of use suggest otherwise. A risk assessment is usually advisable for lasers in classes 1M, 2M and 3R, and for all lasers in classes 3B or 4 (see section 7.1.1).

In this list we have also divided some classes into two, where this seems an appropriate way of describing different kinds of potential hazard that a given class can cover. In most cases the listed control measures relate only to the risk of eye or skin exposure to the emitted radiation, and not to any additional risks that may be present.

- *Class 1 (low inherent power)*
  - No precautions necessary.

  (NB: With some laser diodes it is possible that faults in the electronic drive circuitry could lead to increased levels of emission.)
- *Class 1 (embedded)*
  - Follow the manufacturer's instructions. (A risk assessment may be necessary for the control of any additional hazards, see chapter 6.)
  - Do not operate with any part of the protective covers removed or any interlock overridden.
  - No precautions are necessary during normal use or under routine user maintenance (but note that maintenance or servicing that requires access to the internal laser while it is operating can be hazardous, and appropriate precautions must be adopted).

- *Class 2*
  - Follow the manufacturer's instructions and the warning given on the label.
- *Class 1M and Class 2M (that have failed Class 1 or Class 2 under measurement condition 1)*
  - Appoint a Laser Safety Officer (see next chapter).
  - Undertake a risk assessment.
  - Ensure that users act responsibly, and are adequately aware and appropriately trained.
  - Prevent use of telescopes and binoculars within the ENOHD (see sections 4.2.3.2 and 5.6.2).
  - Display warning signs.
  - Follow the manufacturer's instructions.
- *Class 1M and Class 2M (that have failed Class 1 or Class 2 under measurement condition 2)*
  - Prevent use of optical instruments for close-up viewing, such as eye loupes and microscopes (see 4.2.3.3 and 5.6.3).
  - Undertake a risk assessment.
  - Ensure that users act responsibly, and are adequately aware and appropriately trained.
  - Use in a controlled area (in which the use of hazardous viewing instruments or beam collimation can be prevented) in cases where access to the laser cannot otherwise be restricted.
  - Display warning signs.
  - Follow the manufacturer's instructions.
- *Class 3R*
  - Ensure that users act responsibly, and are adequately aware and appropriately trained.
  - Undertake a risk assessment.
  - Avoid intentional viewing, particularly for visible emission (see section 7.1.1).
  - Do not direct the beam into areas where people unconnected with the laser work may be present.
  - Follow the manufacturer's instructions.
- *Class 3B*
  - Appoint a Laser Safety Officer (see next chapter).
  - Undertake a risk assessment.
  - Ensure that users act responsibly, and are adequately aware and appropriately trained.
  - Establish a system of key security.
  - Use in an interlock-protected enclosed controlled area.
  - Use interlock connector and beam attenuator as necessary.

- Keep open beam paths to a minimum, using guards, screens, etc.
- Use appropriate eye protection.
- Follow the manufacturer's instructions.
- Develop and use adequate procedural control measures.

- *Class 4*
  - Appoint a Laser Safety Officer (see next chapter).
  - Undertake a risk assessment.
  - Ensure that users act responsibly, and are adequately aware and appropriately trained.
  - Establish a system of key security.
  - Use in an interlock-protected enclosed controlled area.
  - Use interlock connector and beam attenuator as necessary.
  - Keep open beam paths to a minimum, using guards, screens, etc.
  - Use appropriate eye protection.
  - Follow the manufacturer's instructions.
  - Develop and use adequate procedural control measures.

It is usually desirable, however, to consider more carefully the risks that are posed by a particular laser under the circumstances in which it is likely to be used, and to establish control measures on this basis.

In this chapter we discuss the necessity for a laser controlled area and then review the use of engineering control measures, administrative control measures and personal protection. Management issues, such as the appointment and role of the Laser Safety Officer, safety training and the development of an overall laser safety policy, are all discussed in chapter 9.

8.2 Laser controlled areas

8.2.1 Types of laser controlled areas

There is sometimes a confusion in terminology between a laser enclosure (discussed further in section 8.3.2.1) and a laser controlled area. An enclosure is, of course, some form of structure intended to surround the hazard and to restrict access to it. A laser controlled area, on the other hand, is the zone around a laser where people may be present and in which protective control measures are required.

A laser controlled area is often taken to mean a dedicated laser room, often having interlocked access and a red warning light by the door, inside of which open-beam laser equipment (often Class 4) is in use (figure 8.1). We would like to define a laser controlled area more generally, to cover all areas in which potentially hazardous lasers are in use and within which some level of safety control is exercised, regardless of the laser's class and whether or not the area is actually enclosed. (This is, in fact, how it is defined in IEC 60825-1.)

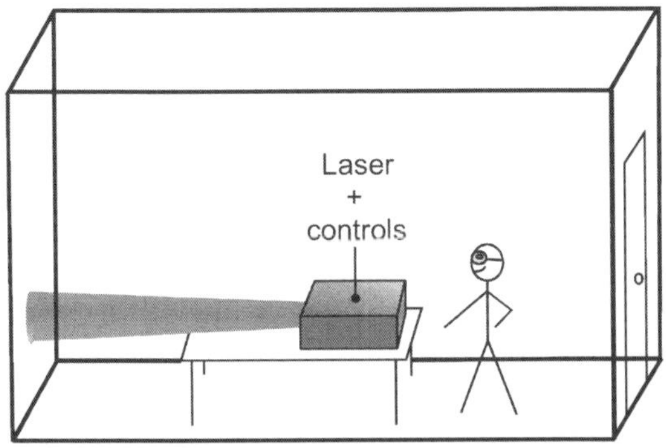

Figure 8.1. A commonly found arrangement for a laser controlled area, consisting of a dedicated room in which laser work is carried out. In this example, although the laser user is shown wearing eye protection (see section 8.6), no other precautions appear to have been made to enclose the beam or to limit the length of its path (see section 8.3.2). The person is also working alone, which is not good practice.

A laser controlled area really needs to be set up around any laser for which specific protective control measures are required. Although the need for control measures should be decided on the basis of a risk assessment, some kind of laser controlled area will normally be necessary for all lasers other than those in Classes 1 or 2. Note that the term 'laser controlled area' is used here and in the international safety standard, although 'controlled laser area' might be more appropriate! At its simplest, it is an area around a laser within which protective control measures are in force.

The intention of a laser controlled area is to establish a zone around the laser equipment within which hazards could arise and over which there is some element of control or restriction. As already explained, a laser controlled area does not necessarily need to be a walled enclosure; an open-plan area may be sufficient, provided that laser hazards do not extend beyond the area and also that an adequate level of control can be exercised within it. Circumstances where open-plan areas would be acceptable include the use of lasers (whether used inside or out-of-doors) for which the hazard distance, taking into account the possible use of optical viewing instruments (typically binoculars or telescopes), does not extend beyond the boundary of the designated area. It is also necessary to ensure, of course, that persons unconnected with the laser activity do not enter the area unless there is an adequate level of supervision or they can be expected to follow the instructions given on warning signs.

As an example, the purpose of having a controlled area for a laser in Class 1M or Class 2M is simply to ensure that the use of optical viewing instruments that could be hazardous is prevented. For Class 1M and Class 2M lasers that exceed the limits for Class 1 or Class 2 under measurement condition 2 (the eye-loupe condition), and which therefore only present a hazard if magnifiers are used or the beam is collimated in the immediate vicinity of the laser, the purpose of the controlled area is merely to prevent such things happening. If all those who work with or have access to such laser equipment understand these issues and follow these restrictions, the purpose of the controlled area is satisfied. The controlled area itself may be no more than the corner of a room or even a particular workbench.

Quite often, of course, a laser controlled area will be physically enclosed, either by the walls of a room or by the use of dedicated enclosures that have been constructed specifically for the purpose. Suitable curtains hung from the ceiling can also be used to screen off the laser from adjacent working areas.

An enclosure around a laser controlled area can serve two functions. It can be designed to keep unauthorized people out and also to contain the hazard (i.e. to keep the laser radiation in). Frequently the same enclosure will serve both functions. Where its sole purpose is to keep unauthorized people outside the hazardous area (in situations where the hazard itself cannot extend beyond the boundary) then waist-height barriers may be sufficient, provided that people can be reasonably expected to respect their purpose and not attempt to climb over or under them!

Where the purpose of the enclosure is to contain the laser radiation that may impinge on its inner surface, then it is necessary to ascertain that the walls, curtaining, windows and entryways are sufficient to adequately perform this function without damage and without any risk to persons outside the area. Windows, in particular, should be covered with a suitable material (which acts as an effective absorber at the laser wavelength) whenever there is some risk of hazardous levels of laser radiation reaching it. In the case of CO_2 lasers, window glass is opaque to their far-infrared emission and so may be sufficient as an enclosure, provided that exposure levels are not sufficient to damage the glass, or the windows aren't opened! The walls, doorways and windows forming the enclosure should be checked for possible gaps. Ceilings and floors may similarly need to be checked were it is reasonably foreseeable that the laser beam could be directed towards them.

Some laser controlled areas (and also the enclosures of some large industrial laser machines) are open at the top, especially where they have been installed within high-bay areas such as in factories. This may be acceptable, but the possibility that (a) laser radiation may be directed or reflected upwards and (b) that someone may be working overhead must first be considered as part of the risk assessment.

It is usually advisable to have wall surfaces in laser working areas that are light coloured and diffusely reflecting. These will help to ensure good ambient levels of illumination while minimizing the hazard arising from any accidentally reflected beam. (As discussed in an earlier chapter, however, simply trying to avoid the occurrence of specular reflections is not by itself an adequate means of control.)

In some circumstances, specially-designed laser-resistant curtains (available from safety equipment suppliers) may be used in place of a rigid, walled enclosure. They should, of course, be sufficient to contain the laser radiation and be able to meet all other criteria that the enclosure is required to satisfy.

A special case of laser area is that of a 'walk-in' enclosure, allowing access into the area when the laser is not working. This is considered further under beam enclosures in section 8.3.2.1.

8.2.2 Controlling access

Whether the doors of an enclosed laser controlled area need to be interlocked in order to terminate laser emission when opened depends on the level of risk, and on how well *procedural* control measures (that is, local rules forbidding unauthorized entry) can be expected to work. It is not advisable, however, to allow those working within potentially hazardous areas to lock themselves inside in order to keep other people out. In an emergency, rapid access may need to be gained into the area.

Methods for interlocking entryways into laser controlled areas are discussed later in this chapter. Before deciding on an interlock system, however, careful thought should be given as to what the interlocking is intended to achieve and how it will need to be used. For example, should tripping of the interlock by the unauthorized opening of the door completely shut down the laser or merely terminate the accessible emission by means of an electrically-operated shutter? Is there a need to override the interlock to allow authorized persons to enter or leave the area while the laser is operating? If so, how is this override to be effected without compromizsing the integrity of the enclosure? Will the laser need to be left operating while unattended?

The answer to these and similar questions will affect the design and operation of the interlock system that is needed.

Where a number of different lasers are in use in the same interlock-controlled area, and especially where lasers may be moved in and out of the area, dedicated laser power sockets (preferably requiring non-standard plugs) can be installed and connected to the interlock system. Not all lasers, however, can be simply shut down without damage by merely cutting of the electrical supply. It such situations it is more satisfactory to adopt a central interlocking system connected to fail-safe beam shutters in front of the laser emission apertures.

8.2.3 Use of warning signs for laser controlled areas

Warning signs (which may be a legal requirement in some countries) should be posted on the outside of laser controlled areas, on or adjacent to the entryways, to indicate the existence of a possible hazard. The sign should include the international laser warning symbol, but it is useful to add some relevant wording as well. The warning symbol is not always recognized by those who may be unfamiliar with laser technology, and so the wording **laser** or **laser radiation** can be usefully included beneath the symbol.

In addition, it can be helpful to include brief instructions on how people are expected to respond to this warning. This could be by such wording as **keep out**; **authorized persons only**; **enter only with permission**; **eye protection must be worn**; or whatever is most appropriate. In particular circumstances (in research laboratories, for example) it can also be helpful to add some additional technical information for the benefit of those who may by curious. This can cover the type of laser in use and/or the kind of process being carried out with it.

Illuminated signs can also be used, indicating whether the area can be safely entered or not. This can be particularly useful when interlocks are employed, as it can help to minimize the inadvertent tripping of the interlock by someone believing the laser to be out of use when it is not.

Too much signage, however, can be counterproductive. Warnings should be clear and unambiguous, and only displayed where they convey useful information. The liberal posting of laser warning symbols around an area may come to mean little more than that laser work is sometimes done somewhere in the vicinity. Where people having no proper awareness of laser hazards routinely enter areas indicated as containing laser hazards, then it is time to change the safety procedures or change the signs; perhaps both!

8.3 Engineering control measures

Engineering control measures should be used to the extent that is necessary for minimizing the risk of harm arising from the use of laser equipment. As has already been emphasized, the hazards arising from laser use should wherever possible be totally contained to avoid any necessity for further protective measures. If, however, it is not practicable to 'engineer out' the hazards entirely, then engineering means should nevertheless still be used as far as is feasible in order to reduce the risk as much as possible.

We will first discuss the engineering design features that are incorporated into the laser equipment in accordance with the laser's class, as specified in the safety standard. Then we will discuss additional engineering control measures that may be fitted by the manufacturer or supplier of the laser equipment, or which are added by the user.

8.3.1 Class-dependent safety features

As discussed in section 4.4, all laser products must satisfy the manufacturing requirements defined in the safety standard. In addition to labelling and user information, these requirements also include the following engineering safety features.

Protective housing

(Required for all lasers except Class 1 laser products. It is also, however, required for Class 1 in the case of certain embedded lasers.)

A laser's protective housing is intended to prevent access to any laser radiation other than the output beam that emerges through the emission aperture. Lasers should not normally be operated, therefore, with any part of the housing removed. Where it is necessary to open or remove the housing (to clean or replace an internal component, for example), the housing should be closed or replaced before turning the laser back on.

Safety interlocks

(Required for embedded laser products that incorporate an internal laser equivalent to Class 3R, Class 3B or Class 4.)

Safety interlocks are fitted in order to prevent access to hazardous levels of internal laser radiation when part of the housing or enclosure of an embedded laser system is intended to be opened by the user, with the laser turned off, to perform a necessary operating or maintenance function. The interlock prevents operation of the laser when the housing is open. Sometimes a key-operated interlock-override facility is included. This is intended for use only during servicing, where there is a necessity for a service engineer to operate the laser with the covers open. Users would not normally expect to have access to the service key. Where servicing by the user is, however, carried out by agreement with the manufacturer, the service key should be kept securely and not with the laser. It should only be available to the person or persons authorized to carry out the servicing. Operation of the laser for its intended function should not be permitted when a service interlock remains overridden or where any part of the enclosure is not in its normal closed position.

The safety standard does in addition, however, permit the use of an interlock override to carry out maintenance activities (that is, routine adjustments and other tasks undertaken regularly by the user in order to maintain the laser in working order). This, in our opinion, is unsatisfactory if it allows the user access to levels of laser radiation exceeding the maximum level permitted by the laser's class. The class of a laser product is intended as an indication to the user of the potential radiation hazard arising during use. If continuing use requires access to higher levels of radiation under the guise of user maintenance, this should in our view be taken into account in the classification.

Laser servicing is discussed in more detail in section 8.8.

Remote interlock connector

(Required for Class 3B and Class 4 laser products.)

This device, normally installed on the laser's power supply, prevents operation of the laser unless a connection is made between the terminals of the connector. This allows the user to connect door interlocks or other safety circuits in order to terminate emission when the circuit is broken (for example, when an interlocked door is opened or an emergency stop button is pressed). This can be especially useful when setting up an interlock-protected laser controlled area. It can be useful, however, when installing an interlock system that is connected to the remote interlock connector on the laser, to incorporate a manual re-set that requires a deliberate action to re-start the laser after operation of the interlock. (See section 8.3.2.2.)

Key control

(Required for Class 3B and Class 4 laser products.)

A key is needed to operate all lasers in classes 3B or 4, and is intended to prevent unauthorized use. This key should be kept secure and specific rules governing its use need to be incorporated into the user organization's procedures. Laser operating keys should not be left in equipment when it is not being used, otherwise the safety function of this feature is clearly negated.

Emission warning device

(Required for Class 3B and Class 4 and also invisible-beam Class 3R laser products.)

This device, usually a warning light but sometimes a warning sound, indicates that the laser is operating or supplied with power and ready to fire. (In the case of a pulsed laser it can mean that the capacitor banks are being charged or are not fully discharged). If the warning device is an indicator lamp, it should be in a prominent position readily seen by all those who are at some risk of hazardous exposure. The warning device can provide a useful indication of the laser's operating state for those not directly in control of laser emission.

Beam stop or attenuator

(Required for Class 3B and Class 4 laser products.)

The beam stop (normally in the form of a mechanical shutter that can be used to prevent the beam emerging from the emission aperture) provides an alternative means of terminating laser emission without the necessity of using the laser power switch. It can be particularly useful for lasers that take a while for their output to stabilize in circumstances where the emission needs to be terminated temporarily. It should also be used as a matter of routine whenever the laser output is not needed for a short time, as this will help to reduce the period during which people are at risk.

8.3.2 Additional engineering control measures

In addition to the engineering features described above and required in accordance with the laser's class, other protective features may be supplied by the

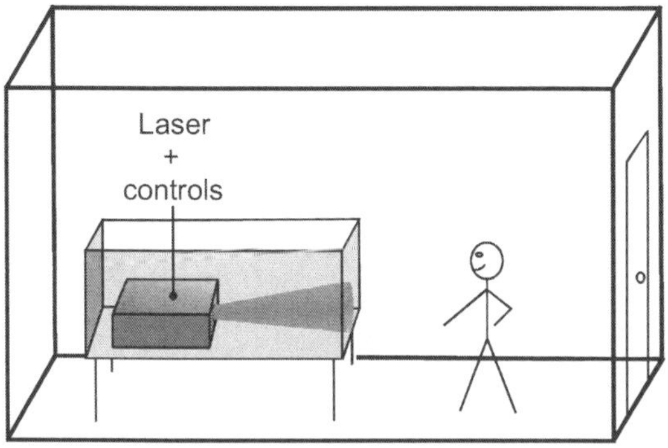

Figure 8.2. A laser controlled area in which enclosures have been added to the laser equipment to reduce the risk of stray beams and accidental exposure. Compare this with figure 8.1, where no such precautions have been taken. If, however, the laser is not completely enclosed as a Class 1 system would be, it is still necessary to restrict access into the area to trained personnel.

manufacturer or can be installed by the user. They may not affect the class of the laser, but can help to provide protection from accidental exposure (see figure 8.2).

8.3.2.1 Beam enclosures

Most of the engineering control measures installed on a laser product by the manufacturer according to the laser's Class do not in themselves make the laser 'safe'; they merely provide aids which the user can incorporate as part of their overall system of risk management. Obviously, a Class 4 laser, even though it may be fully-compliant with the safety standard, still generates a potentially very serious hazard, and much of the responsibility for its safe use falls upon the user.

With high-power beams in particular, the first priority is to implement engineering control measures to eliminate or, where this is not feasible, to reduce access to the hazard. This is done by the use of protective enclosures that must, of course, be sufficiently robust and stable, and capable of containing the hazard. Where a laser is sold to perform a specific function, the fitting of adequate enclosures should be done by the laser supplier. In other circumstances, such as the use of lasers in research applications, this will need to be done by the user. Even where total enclosure is not possible, as much of the beam as is feasible should be protected by beam tubes or by other suitable covers. Open access should be restricted to those parts of the system where this is necessary in order for the particular task or process to be successfully carried out. Open beam paths should be kept short, localizing the hazard within the smallest possible area. Examples of different types of beam enclosure are illustrated in figure 8.3.

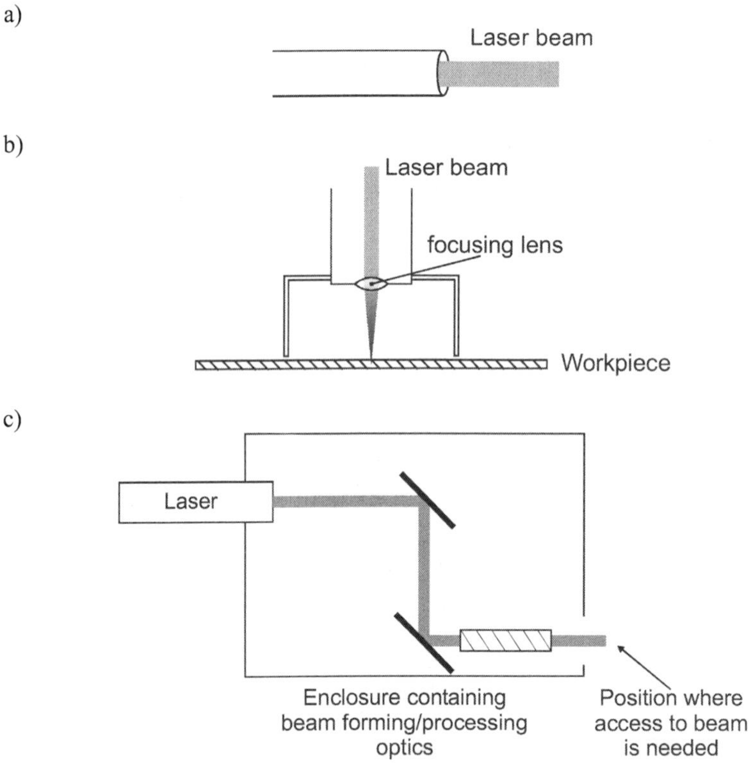

Figure 8.3. Types of beam enclosures: (*a*) beam tube, (*b*) local enclosure surrounding the laser process zone and (*c*) partial enclosure, enclosing those sections of the beam to which access is not normally required.

Metallic enclosures are widely utilized for laser protection, and glass or polycarbonate materials may also be used in wavebands where they are opaque (typically at wavelengths below 300 nm or above 3000 nm). The enclosures should have adequate environmental stability, have sufficient optical density at the laser wavelength, and be capable of withstanding an incident laser beam without damage and without degradation of the enclosure. More guidance on laser enclosures (guards) is given in IEC 60825-4.

It is possible, of course, for beam enclosures to incorporate viewing windows where it is necessary to view the inside of the enclosure. This can be important in some materials processing applications. These windows must, of course, still satisfy the above requirements for providing adequate and sufficient protection.

Some large (usually industrial) laser installations have 'walk in' access, allowing people to enter the inside of the laser enclosure (see figure 8.4). This is different from the normal concept of a laser controlled area, since with a walk-

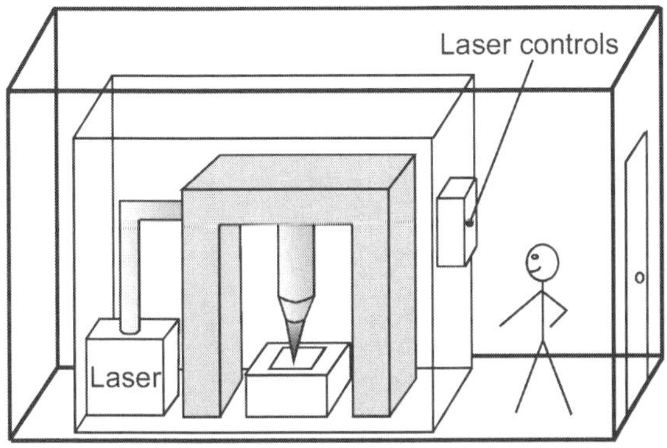

Figure 8.4. A 'walk-in' laser enclosure, often used with large industrial laser machines. Such a system can only be Class 1, however, if the absence of anyone inside the enclosure during normal operation of the laser is assured by engineering means (e.g. by sensors), and is not simply reliant on the adoption of appropriate working practices.

in system the laser is operated from outside the enclosure with the intention that, during operation, no one should be inside. Clearly, interlock-protection of the access into the housing is not sufficient to prevent operation of the laser while someone is inside, since the access door could be closed and the interlock set in the mistaken belief that no one is within the enclosure. Sensors, such as pressure-sensitive floor mats, which are linked to the laser control system and that automatically identify the presence of a person inside the enclosure and so prevent operation of the laser can be fitted. Such a system could then be Class 1, since access to laser emission during normal operation is prevented by engineering means. Where this is not the case, three additional features are necessary. First, some facility must be provided so that the person inside the housing can prevent operation of the laser. This can be accomplished by the use of a unique key or similar device, without which the laser cannot be operated, and which the person takes with them when going inside. Because this is reliant on a procedural control, however, (that the person does actually take the device with them when entering the enclosure) it should be backed up by two additional features. There should be a clear (audible or visible) warning system to indicate to someone inside the housing that the laser is on, and also a means of easily terminating laser emission from inside the enclosure (such as by the use of an emergency-stop button). Although this may be an acceptable solution, the laser cannot be Class 1 if access to laser emission is prevented only by such procedural means.

If laser enclosures are to be left open at the top (sometimes the case with large industrial laser machines) it is necessary to demonstrate by means of a risk assessment that this is acceptable, taking into account the possibility of beam reflections and of the likelihood that cleaning or other maintenance work may sometimes be carried out overhead.

8.3.2.2 Interlock systems

In addition to any interlocks that may be incorporated in the laser equipment itself, users may need to add additional interlocking systems, whether this is to control access into an enclosure that has been fitted around the laser or part of the beam or to control entry into a laser working area. Interlocking should always be adopted whenever a risk assessment has demonstrated that access cannot be adequately controlled by procedural means alone.

Safety interlocks should be of robust and fail-safe design; simple contact switches are not usually sufficiently reliable for safety critical use. The simple override of an interlock (in order to defeat its use merely as a matter of convenience during normal operation) should not be possible. Indeed, the design of an interlock protection system should be such that it is straightforward to use but does not encourage unauthorized defeat. (It is not uncommon to find a protection system that has been permanently disabled by laser operators simply because it was cumbersome to work with. This is undoubtedly as much a function of poor design as it is of a failure of the operators to follow correct procedures.)

Interlock systems used on protective panels or enclosures can be designed to be overridden where there is a genuine and foreseeable need to carry out certain tasks with the covers open or removed while the laser is in operation. Override should, in these cases, be affected by means of a key or by an alternative but suitably secure method. The use of such an override should in all cases be subject to appropriate safeguards in the form of adequate procedural control measures. Defeating or overriding an interlock can often mean that the laser radiation hazard is then greater than is implied by the laser's class or it can give access to additional hazards such as high-voltage electricity.

Where the interlock is used to control access into a laser controlled area, typically an enclosed room, overriding the interlock system is permissible where it is necessary for those who are authorized to work inside the area to enter or leave while the laser is in operation. At its simplest, the override can be activated by means of a push-button switch located on the inside of the area, allowing those inside to leave and authorized users to enter. In cases where at least one person is always inside the room when the laser is operating, the override need only remain on while the switch is depressed. Alternatively, the switch may activate a timer causing the override to remain on for no more than the few seconds necessary for someone to enter or leave. Where an interlock override is fitted, however, the layout of the room must be such that the laser beam cannot pass through the open door. It can also be useful to incorporate a small lobby area just inside the door, with a curtain screening it from the rest of the room. This provides an added protection, and allows the person entering through the door to put on eye protection (if this is required) before moving into the hazard area.

Sometimes it is necessary to allow a door interlock to be capable of being overridden from the outside. This can be accomplished by means of an override key or a coded key-pad. This can be a useful feature where it is desirable to permit authorized persons to enter the area without disturbing those inside. It can also be used where a laser needs to be left running while unattended (although, of course, locking the room is an obvious alternative in this case).

Where there is a circuit interruption to the remote interlock connector that is fitted to all Class 3B and Class 4 laser products (caused, for example, by the tripping of a door interlock connected to it) this may, depending on the design of the laser, either shut down laser power completely or activate an internal shutter in the path of the beam. Where it is preferable, door interlocks may be connected instead to a separate shutter that is external to the laser itself. This should, of course, be of a design that can only fail in the closed position. It must also be capable of withstanding the incident laser beam without damage, such as might be caused by overheating.

When a door or other protective interlock has been activated, recommencement of laser emission should ideally only be possible by the use of a deliberate reset operation (such as by the use of a reset switch), and not by simply closing the door that caused the interlock to operate. Clearly, the purpose of the interlock to protect against unauthorized access could be defeated if it were merely necessary to close the door after entry to restart laser emission. Note, however, that a reset mechanism is not a requirement of the IEC standard for the remote interlock connector fitted to all Class 3B and Class 4 laser systems. It can, therefore, be advisable to install additional control circuitry to provide such a reset function. (A reset is, however, required, in the US.)

Special consideration should also be given to restart procedures in the case of laser equipment that is left running while unattended inside an interlock-protected area. To prevent unauthorized entry *and* restarting of the laser, it may be advisable, depending on the circumstances, to require the use of a restart key (which should not, of course, be left in the laser).

In certain applications, however, and particularly in the medical use of lasers, the adoption of interlocks to protect doors and other entryways may be undesirable. The need for freedom of access for authorized staff and the additional risk to the patient that could be introduced by an unexpected shut-down of the laser, can mean that interlocks introduce more problems than they solve. It is then necessary to base much more reliance on procedural control measures, to make appropriate use of clear and unambiguous warning signs, and to maintain a high-level of staff awareness.

People working inside a laser controlled area should not, of course, be permitted to lock themselves in as a way of preventing unauthorized entry. There may be a need, in an emergency, to gain rapid access. However, as an alternative to the use of a conventional interlock system, and especially in circumstances where an unexpected shut-down of the laser would be very undesirable (and this includes laser surgery), it is possible to use 'power-to-lock' devices, also called door strikes. These allow the door to be opened in the normal manner from the inside at any time by means of a conventional lever-operated latch, but have no lever to withdraw the latch on the outside. The door is kept locked from the outside by means of a moveable plate which is fitted to the doorframe, and which is held against the latch when the system is energized. (The locked state can be indicated by a warning light.) The system can, however, be overridden by the use of an 'emergency entry' button fitted on the outside (which can be coupled to an alarm). Such a locking system prevents casual attempts to open the door from the outside, but does allow any necessary entry without disabling the laser. Although electrically operated, it also ensures that the door is unlocked in the event of a power failure.

With all such protective systems, the manner in which they are both intended and likely to be used should be carefully considered in advance of their implementation. Systems that do not allow their users the flexibility that is needed, particularly with regard to the means of access and override, may lead to the interlock system being disabled, resulting in the complete absence of any engineering control measures. Interlock systems should therefore be fully evaluated prior to their installation, taking into account the function that they are intended to perform, the way in which they are likely to be used, and the reduction in risk which they, in combination with any necessary procedural control measures, will actually achieve.

8.4 Administrative control measures

Administrative (i.e. procedural) control measures are specific instructions or 'local rules', established by the organization concerned, that are intended to be followed by those within the organization who work with laser equipment. As discussed previously, many engineering control measures, where they cannot provide complete protection, need to be augmented by adequate procedural or administrative arrangements. All such control measures should be documented and they should be regularly reviewed and updated whenever necessary.

General administrative control measures defining the overall laser policy of the organization, particularly relevant for organizations having extensive or varied laser use, are discussed in the next chapter as part of safety management. Here, we are primarily concerned with those control measures and restrictions, often called standard operating procedures (SOPs), that are applied to a specific laser system or laser process. It can sometimes be helpful to refer to those rules that are specific to a particular laser or a specific laser process as *procedural* control measures, meaning that they define procedures (things to do and things not to do) that have to be followed whenever a particular laser is used or a specific laser process is being carried out. Other aspects of administrative control cover those matters of general policy that apply to all laser use within an organization. These include items such as training requirements, approved laser working areas, any generally applicable safety control measures, and the criteria for authorization of laser users. An organization's administrative control measures might therefore comprise an overall policy covering all laser use together with a number of separately issued procedures dealing with specific lasers or laser processes. It is a matter of judgement and convenience, depending largely on the size of the organization and the number of different kinds of laser equipment in use, as to how administrative control measures are divided between a single overall policy and separate sets of individual laser specific procedures.

As with all protective control measures, decisions concerning them should be made on the basis of their effectiveness in reducing risk.

8.4.1 The use of product safety features

Several of the engineering features that are required by the safety standard to be fitted to laser products by the manufacturer, and which have been described in section 8.3.1, may need to be used in conjunction with suitable user procedures in order to maximize their effectiveness in reducing risk. They should not be ignored as part of the user-organization's operating procedures merely because they are standard laser features that are the responsibility of the laser supplier. Where appropriate, therefore, where they have a bearing on the resultant risk, such procedures should be included in the documented administrative control measures applicable to those who use or work with the laser equipment.

With reference to section 8.3.1, the user's administrative control measures may need to include restrictions and procedures covering the following aspects.

- The conditions (if any) under which the protective housing can be removed or opened.
- The use of any interlock override facility and the security of any service keys provided.
- The use of the remote interlock connector in conjunction with any user-installed interlocks.
- The security, issue and use of laser-operating keys. The keeping of laser-operating keys permanently in the laser is a common failing amongst laser users. This should only be permitted where there are adequate additional safeguards preventing unauthorized use. (Merely having the laser within an interlock-protected area may be insufficient, if it allows unauthorized entry and use of the laser when it has been left unattended.)
- Awareness of the emission warning device.
- The use of the beam stop or attenuator for the temporary termination of laser emission whenever it is not needed but where the complete shut-down of the laser may be inconvenient.

8.4.2 Other procedural control measures

Procedural control measures define working practices; they should clarify how things are done and (where appropriate) who should do them. They should also specifically prohibit actions that are not necessary but which could be hazardous. Procedural control measures need to be written down, and those affected by them should be aware of them, understand them and follow them

In addition to the use of product safety features discussed in the previous section, further procedural control measures will usually be necessary in order to adequately reduce the risk of injury or other harm. The control measures that are needed will depend very much on the particular aspects of use that were discussed in relation to risk assessment in the previous chapter, namely:

- the type and class of laser equipment and any particular features of its design;
- the way in which the laser beam is delivered to its point of use;
- the application or purpose for which the laser is being used;
- the kind of working environment in which it is located;
- the people who are working with the laser and others who might become exposed to its hazards.

The detail necessary for adequately defining procedural control measures will vary with the kind of laser work being carried out. Where the work is routine and is repeated in much the same way on a regular basis, then the procedures can be very prescriptive, detailing in a logical sequence each specific step that has to be followed. Where, as in a research laboratory, for example, there might need to be more flexibility in the way in which the laser is used, the procedural control measures should nevertheless be sufficient to define the general safety rules that are always to be followed. Individuals who work with potentially hazardous laser equipment, even in research environments, should always be given sufficient training in laser safety *and* be required to work in accordance with well-developed and sufficiently specific safety guidelines.

Procedural control measures should be clear and unambiguous and may need to cover issues such as:

- the names or categories of individuals permitted to operate or work with the laser equipment;
- any prior authorization that is needed;
- any prohibition on working alone;
- any restrictions on where the laser equipment is to be used;
- the procedures that are to be followed with respect to laser operating keys;
- any preliminary checks that need to be made with regard to the working area prior to laser operation (e.g. posting of warning signs; absence of unauthorized personnel);
- any preliminary checks that need to be made to the laser or to associated equipment prior to its operation (e.g. settings or system configuration);

- any personal protection (e.g. protective eyewear) that should be worn (see section 8.5);
- any other specific precautions that should be adopted or procedures that should be carried out prior to use of the laser (e.g. checking that the laser is adequately secured, checking beam alignment, checking that any required beam enclosures are in place, checking that the beam path is safely terminated;
- any actions or procedures that are specifically prohibited;
- any required procedures or sequence of operations covering laser start-up, laser operation and laser shut-down;
- the way in which any adjustments should be made during laser operation;
- any verbal warnings that should be given prior to beam emission (this can be particularly relevant in the case of intermittently-operated lasers);
- the action that should be taken in the event of a suspected accident;
- any procedures that should be followed after completion of the laser work.

Everyone who works in a laser controlled area should be aware that safety can be seriously compromised if the procedures that have been drawn up are not consistently followed. Reliance on procedural control measures should be avoided where it cannot be reliably assumed that the rules will be followed. Instead, engineering solutions must be found. It is a management responsibility (ultimately that of the employer) to ensure the continuing implementation of any necessary control procedures. Clearly, as is discussed in the next chapter, the Laser Safety Officer has an important role here in the regular oversight and monitoring of the organization's safety procedures.

Where particular procedures are regarded, by those who have to work with them, to be unnecessary, inappropriate or unworkable, they should not simply be ignored but instead be reviewed, and where appropriate, updated (or preferably replaced by engineering control measures). Similarly, any changes to the nature of the work being carried out, to its location, or to the equipment being used should require a reassessment of the risks. Working practices and the nature of technology seldom remain constant for long. Safety procedures should keep in step with these changes, and even anticipate them, rather than follow behind when it is belatedly realized that the nature of the risks that they were intended to control have altered.

8.5 Personal protection

8.5.1 Personal protective equipment

Personal protection, the wearing by individuals of what is called *personal protective equipment* or PPE, which can include protective eyewear and other items of protective clothing, may be necessary where engineering control measures alone, or a combination of engineering and administrative control measures, cannot adequately reduce the risk to the eyes and the skin. However, personal protection should never be regarded as an ideal way of controlling risk, and serious consideration should always be given to alternative methods of protecting people from laser hazards.

There are several reasons for avoiding the need to wear personal protection. Personal protection can hinder an individual's ability to see, move or carry out specific tasks, and might therefore introduce other, less obvious, hazards. These restrictions, together with the possible discomfort involved in wearing such protection, may also tempt the wearer to remove it, even if only temporarily. By concentrating on other aspects of the work being carried out, individuals may on occasions simply forget that such protection should be used, or there can be a lack of enforcement, in a belief that it is a matter of individual choice whether the protection is worn or not. Experience suggests, in many working environments, that compliance with requirements for using PPE is not always high. (Consider, for example, the degree of compliance that can be evident on construction and other industrial sites bearing the widely seen instruction that 'hard hats must be worn'. Regardless of the rules, people often make their own decisions on whether to comply or not.)

For these reasons the use of any form of personal protection must be justifiable as an appropriate and necessary method of control. Complete beam enclosure is usually preferable. Where it is necessary to view the laser process or the position of the laser beam, the use of remote viewing systems, such as closed-circuit television or simply a camera connected to a PC should be considered as an alternative. Even though the initial cost may be higher, once installed working arrangements can be simpler and far more flexible.

Nevertheless, there are certain laser activities, and these include medical procedures and some research applications, where the use of PPE, especially in the form of eye protection, is a sensible and necessary precaution. Personal protection needs to be suitable for its intended purpose. Its use must always be subject to specific administrative (procedural) control measures defining the circumstances under which it is to be worn, how it should be checked prior to use and any other necessary requirements covering its procurement, storage, issue and replacement. Training in the proper use of the protection may also be necessary.

8.5.2 Types of protection

Eye protection is the most widely used form of PPE for protecting individuals from laser radiation hazards. It is discussed fully in the next section.

Face shields are sometimes used to provide both eye and face protection. These can be useful to guard against low levels of scattered or diffusely-reflected ultraviolet radiation, but may be insufficient to protect against the direct, intrabeam exposure to a high-power laser beam.

Special clothing, including gloves which can be worn in circumstances where people need to have their hands in close proximity to a laser beam, may be used to provide protection from levels of laser radiation that exceed the exposure limit for the skin. Except in special situations, however, this is a very unsatisfactory method of hazard control. Lasers that present a skin hazard (usually Class 4) are also a potential fire hazard. Any protective clothing therefore needs to be flame- and heat-resistant, in addition to providing sufficient radiation protection to the skin. Protective clothing may restrict the wearer's freedom of movement or their ability to undertake manual tasks requiring some dexterity. Alternative means of providing protection, preferably using engineering control measures or otherwise through a combination of engineering and administrative control measures, should always be considered as a more satisfactory option.

In surgical applications, however, it is often necessary to provide protection of the patient's skin in areas adjacent to the treatment site. This can be done by the careful positioning of wet drapes.

In other applications, the use of closely-woven garments can provide some protection against low levels of ultraviolet radiation, but this is unlikely to be sufficient against direct exposure to an ultraviolet laser beam.

8.6 Eye protection

8.6.1 The use of protective eyewear

Personal eye protection is used in a wide range of laser applications, and is arguably one of the most important aspects of laser hazard control. It is important, however, not because it is necessarily the preferred control method (it often is not), but because of the critical safety function that it is required to serve; that of protecting the eyes from serious harm.

The wide acceptance of laser-protective eyewear, coupled with the knowledge that lasers can seriously injure the eyes, may suggest that personal eye protection is the most suitable (and perhaps in many cases the only) significant means of preventing laser accidents. For many laser users the equation

$$\text{laser} + \text{protective eyewear} = \text{laser safety}$$

seems to summarize all that matters in laser safety.

In fact, as we have seen, the use of any form of personal protection, including eye protection, should only be considered as a last resort; a possible though not ideal alternative when it seems not to be feasible to ensure protection by other, more satisfactory, means.

The use of eye protection under any particular set of circumstances therefore needs to be justified. It should never be considered in isolation from other types of hazard control and never used as the sole method for minimizing risk.

Eye protection can be an appropriate solution whenever individuals are within the NOHA (nominal ocular hazard area) of the laser, taking into account all possible reflections and directions of the beam *and* where it is not reasonably practicable to prevent hazardous exposure by other means. Such protection should only be used, therefore, under circumstances where the following four conditions can all be met.

(i) It is reasonably foreseeable that exposure of the eyes to levels of laser radiation above the MPE could occur.
People should not be required to wear protection that they do not need. Eye protection that is being worn 'just to make sure' usually indicates an inadequate assessment of the actual risk. In any case, the necessary level of protection cannot be determined without a proper evaluation of the potential exposure conditions. It should also be noted that the operation of a Class 3B or Class 4 laser does not necessarily require the use of eye protection, as the risk of eye exposure may have been sufficiently reduced by the use of beam enclosures (as indicated in figure 8.3).

(ii) Alternative means of providing protection are not reasonably practicable.
Wherever feasible, reliance on personal protection should be avoided and proper emphasis placed instead on the use of engineering control measures, combined where necessary with administrative control measures, in order to adequately control the risk.

(iii) There is not a significant coexistent risk of skin injury.
Although it can be sensible to protect the eyes from harm whilst simultaneously accepting the existence of a lower risk of injury to the skin (because the consequences of an eye injury may be more serious), this does not apply at high exposure levels where even skin injuries can become serious (and, on occasion, life-threatening). In such situations alternative methods for hazard control other than the use of eye protection have to be investigated. It is normally only with lasers of moderate output, therefore, that the use of personal eye protection is appropriate.

(iv) The necessary level of eye protection is available.
It is not always possible to obtain eye protection that fully meets the particular requirements. For example, where there is a need to work with a number of different emission wavelengths at the same time, eye protection that is adequately effective at all the required wavelengths may not be available. There can also be other conditions for which the necessary level of protection is unobtainable. In such circumstances alternative ways of ensuring protection must be found. This may require changes to the work being carried out, or even a reappraisal of the lasers being used.

Where the use of personal eye protection is adopted (i.e. where the four conditions above can all be met) it is still necessary that the eye protection be used only in accordance with a predetermined policy. This policy, which should be documented, should cover such issues as the following.

Eyewear selection

Eye protection has to be specified on the basis of established criteria, using a risk assessment to evaluate the exposure conditions for which such protection is required. (This is discussed in more detail in the next section.) The risk assessment should not merely consider the technical issues surrounding the laser exposure and the protection required, but also take into account how the protection is likely to be used, and any other issues that the use of eye protection might raise. It may be appropriate to define who within the organization (such as the Laser Safety Officer) has the responsibility for undertaking this assessment, for approving the eyewear specification, or for authorizing its use.

Eyewear identification

It is very important, where more than one type of laser is being used within an organization, that the eye protection intended for use with each particular laser is clearly identified, and that the required eyewear specification is defined in the written procedures. Eye protectors should have their specification marked upon them by the manufacturer, but this is usually done in an abbreviated form. Some additional marking, indicating unambiguously the specific laser that it is intended to be used with, can often be helpful. Using colour-coded markings on both the eyewear and the laser is one way of achieving this.

Conditions of use

The conditions under which eye protection is required to be used should be clearly defined. (Eye protection might only be needed when certain procedures, beam alignment, for example, are being carried out with the laser.)

Eye protection should never be used for the deliberate viewing of a laser beam (that is, by staring directly along the beam into the laser), but only to guard against accidental exposure.

Wherever possible, everyone who needs to use laser protective eyewear should be issued with their own personal protection, certainly in the case of those who have to wear eye protection frequently and for extended periods of time. Inevitably, there will also be a need for having shared eye protection available, for example, for the use of occasional users or visitors. Infections can, however, be transferred through sharing eyewear, and so regular cleaning using disinfectant wipes should be carried out and this should form part of the written procedures.

Eye protectors need to be inspected regularly for signs of any damage or degradation that might have occurred to the filters themselves or to the frames, headbands and other parts of the protectors. Any defects should be repaired or the eye protection replaced as necessary. Eye protection that has or may have been exposed to the direct beam of a laser should be discarded immediately and replaced, as damage may have occurred which could compromise its protective

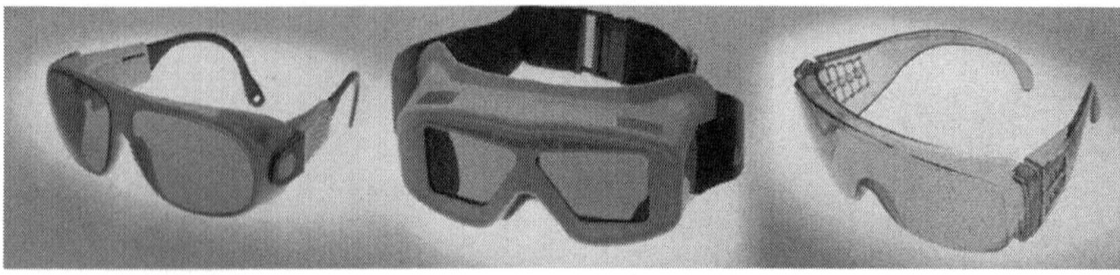

Figure 8.5. Types of laser eye protection. Two styles of spectacle-type eye protection, having arms that rest on the ears and fitted with side protection, are shown on the left and right of the picture. In the centre is an example of goggle-type eye protection, which fits against the face and is held in place by an elasticated band around the head. Goggle-type protection can fit over the wearer's ordinary spectacles.

properties. There is no legal requirement, however, for regular checking or servicing of eye protection by the supplier or by a third party.

Most importantly, the correct eye protection should be worn for the laser being used. Where different lasers requiring different sets of eye protection are in operation, this needs to be checked very carefully, as it is all too easy to unthinkingly use eyewear having the incorrect density or designed for the wrong wavelength.

8.6.2 Specifying eye protection

In the following section we consider the general issues applicable to all types of laser eye protection. The specific requirements that apply in Europe are discussed in section 8.6.3.

8.6.2.1 Basic issues of eye protection

Laser eye protection is available in a variety of styles, in the form of either spectacles (having frames which sit on the ears) or goggles (which are secured by a band around the head); see figure 8.5. As long as sufficient all-round protection is provided (so that exposure of the eyes to laser radiation from around the edges of the protectors is unlikely to occur) then the type that is used can be a matter of personal choice. Indeed, it is often desirable to involve in the selection process those who will have to wear them, as this gives a sense of 'ownership' of the decision, leading to a greater likelihood that the protection will be used. Goggle-types can be more suitable for those who wear ordinary spectacles. It is often possible to obtain laser eye protection that incorporates the wearer's own spectacle correction, although this is usually very expensive. Lightness and comfort are also important issues, especially where eye protection needs to be worn for lengthy periods of time, although where extended use is envisaged, alternative means of control avoiding the need for eye protection should always be considered.

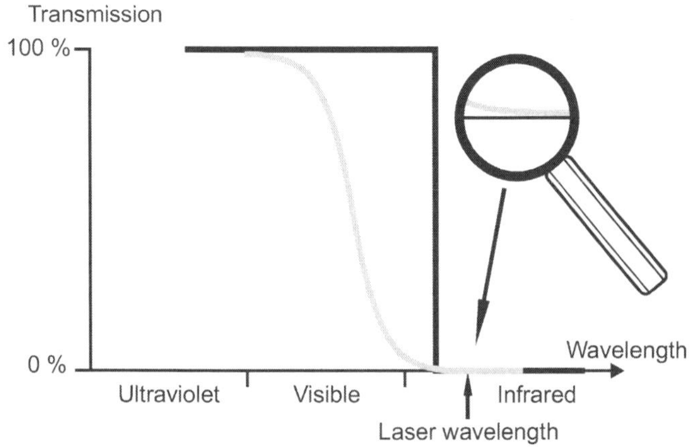

Figure 8.6. Eye-protective filter for a near-infrared laser. The solid line shows the ideal spectral transmission of the filter, having high transmission across the visible region and a sharp cut-off close to the laser wavelength. The feinter line shows the performance of an actual filter, which blocks off some of the visible band but also fails to completely extinguish the infrared laser radiation.

More fundamental than the style of eyewear, however, is the specification of the protection itself. The correct optical characteristics are vital for ensuring adequate protection, yet because of misconceptions about how the protection actually works and what it is supposed to do, inappropriate protective equipment is sometimes used.

We can think of laser eye protection as incorporating optical filters (the 'lenses' of the spectacles or goggles which form the protectors) that 'filter out' the laser radiation. This is, however, very much a simplification. Although at wavelengths well removed from the visible band (in the far-infrared band or well into the ultraviolet region) this is effectively what happens, because of the very high absorption of the optical filters in these wavebands, at other wavelengths it may not be possible to eliminate all of the laser radiation in quite the same way that a kitchen strainer, for example, can remove rice from water.

The factor that can make complete elimination of the incident laser radiation difficult (and it is so obvious that it is frequently overlooked as a consideration) is the need to actually *see* through the eye protection. We are not, therefore, simply separating two different entities, as with rice and water, but trying to reduce the level of optical radiation arising from the laser while at the same time attempting to retain as much as possible of the optical radiation that we need for seeing. These can be conflicting requirements that may make the solution something of a compromise.

Figure 8.6 shows the spectral transmission curve (that is, the variation in the transmission as a function of the wavelength) of a filter intended to provide

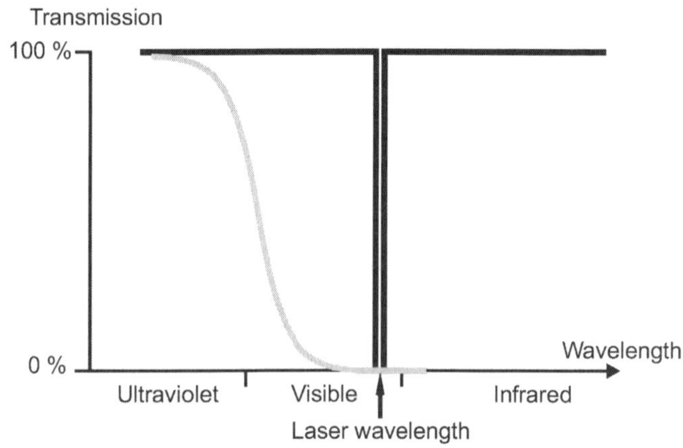

Figure 8.7. Eye-protective filter for a visible laser. The solid line shows the ideal spectral transmission of the filter, having a narrow rejection band centred on the laser wavelength. The fainter line shows the performance of an actual filter which, in order to provide sufficient attenuation at the laser wavelength, has to block off even more of the visible band than the filter shown in figure 8.3.

protection for a laser emission wavelength in the near-infrared region. The curve for an ideal filter is shown in which the transmission at the laser wavelength is zero while the transmission across the visible band is 100%. Such a filter would completely eliminate the incident laser radiation, but would have no effect in the visible band and therefore on our ability to see through the filter. It may prove impossible, however, to manufacture an optical filter having this ideal performance.

The characteristics of an actual filter are also shown figure 8.6. The transmission at the laser wavelength is not zero, and the transmission across the visible band is not 100%. Such a filter does not, therefore, completely eliminate all of the incident laser radiation. At the same time, it has some degrading effect on our ability to see through the filter, by absorbing some of the visible light. This will mean that the filter will appear to the eye to be tinted, and can both reduce the brightness and change the colour of whatever is viewed through it.

Where we need to provide eye protection against a visible laser beam, the options can be even more restrictive. Figure 8.7 shows the performance of an ideal filter for protection against a visible laser beam. This has zero transmission at the laser wavelength while maintaining a high transmission across the remainder of the visible spectrum. Such a filter would remove very little of the visible spectrum. In reality, however, this would be unachievable, and we are likely to have to use a filter whose transmission curve is also shown in figure 8.7. This has a non-zero transmission at the laser wavelength combined with a significant loss across much of the visible band. Such a filter may seriously impede our ability to see and cannot completely eliminate the incident laser radiation.

While most eye-protective filters, like those shown in figures 8.6 and 8.7, work by absorbing the excess beam power, others use reflective coatings which can provide a sharper cut-off between the transmitted and rejected wavebands (see section 8.6.2.3). These can have other disadvantages, however, and are currently not in widespread use.

New types of eye protectors are under development, called optical limiters, which work in an analogous manner to photochromic sunglasses. Unlike sunglasses, however, they can change their optical characteristics very rapidly and over a wide dynamic range in response to a sudden hazardous exposure. They offer the possibility for effective eye protection even at visible wavelengths without compromising normal vision.

Protective filters may also be fitted inside optical instruments (telescopic aiming-sights and microscopes, for example) which are intended for use in conjunction with lasers that could present an eye hazard under optically-aided viewing conditions. Such filters may need to satisfy even more exacting criteria than those used in personal eye protection. Unlike PPE they may be intended for the deliberate (and possibly therefore long-term) viewing of laser emission, rather than merely a short period of accidental exposure. Furthermore, any damage suffered by the filters may be less easily spotted. Such protective filters should, therefore, satisfy stringent requirements and be thoroughly tested and evaluated before use.

8.6.2.2 Optical density

The transmission of any optical material can be expressed in terms of the percentage of the incident light that passes through it to emerge from the opposite side. In fact, because the transmission can vary with wavelength (a piece of blue-coloured glass, for example, will have a much higher transmission in the blue part of the spectrum than in the yellow part), we should denote the wavelength at which the transmission has been measured. In optical technology it is more usual to quote the transmittance of an optical filter in terms of the ratio of the emergent optical power to the incident optical power. A transmittance of 1 therefore corresponds to a percentage-transmission of 100%; a transmittance of 0.1 to 10%, and so on.

In laser eye protection, as in some other branches of optics, we are often concerned with very low values of transmittance, in order that high levels of incident laser radiation are sufficiently attenuated. Rather than expressing a filter's transmittance (at a given wavelength) as 0.001 (or 0.1%), for example, it is then more convenient to use the concept of *optical density*.

Optical density, or OD, is the negative of the logarithm (to base ten) of the transmittance. It is the power of ten by which the optical radiation is reduced by the filter. This is illustrated in table 8.1.

Table 8.1. Optical density of filters.

Transmittance τ	Log τ	OD	Decrease in power
1 (100%)	0	0	$10^0 = 1\times$
0.1 (10%)	−1	1	$10^1 = 10\times$
0.01 (1%)	−2	2	$10^2 = 100\times$
0.001 (0.1%)	−3	3	$10^3 = 1000\times$

A filter having a transmittance of 0.001 (or 0.1%) will decrease the incident power of the laser beam by 1000 times and will have an optical density of 3 (usually expressed as OD 3).

The optical density required to reduce a given level of incident exposure to safe levels is determined on the basis of the factor by which this exposure exceeds the maximum safe exposure (the MPE).

Thus, for an incident exposure expressed in terms of the irradiance E, the transmittance of the filter τ_λ at the laser wavelength, must not exceed the value

$$\tau_\lambda = \text{MPE}/E \tag{8.1}$$

and so the minimum optical density (at the laser wavelength) is given by

$$\text{OD}_\lambda = -\log(\text{MPE}/E). \tag{8.2}$$

Both the MPE and the exposure must, of course, be expressed in the same units (W m^{-2} in the case of irradiance, or J m^{-2} for radiant exposure), and must be determined over the same time base (exposure duration). This time base should be appropriate for the particular circumstances under which an accidental exposure could occur and should represent the maximum reasonably foreseeable duration. For the accidental exposure to the direct beam of a laser, this may only be a few seconds and therefore a period of ten seconds may represent a reasonable upper limit. Where there is a possibility of exposure to lower levels of laser radiation over a longer time period, for example when working with ultraviolet emission where there could be scattering of ultraviolet radiation from components along the beam path, then a much more conservative limit needs to be set, perhaps up to a maximum of 30 000 s (a period of over eight hours). This means that there may well be a difference in the criteria used in the ultraviolet region to protect against accidental exposure to the direct beam of the laser (for which an exposure duration of more than a few seconds is very unlikely), and to protect against lower levels of scattered radiation for which exposure periods may be much longer.

The exposure level (E) should be evaluated in accordance with the procedures defined in the safety standard and which have been discussed in chapters 3 and 5. Proper account, where relevant, should be taken of the procedures for dealing with pulsed lasers and extended sources.

When dealing with laser beams having a diameter smaller than the applicable limiting aperture, it is sometimes more convenient to use the laser power and the Class 1 AEL for determining the optical density of the filter that is needed, rather than using the aperture-averaged exposure level (irradiance or radiant exposure) and the MPE.

In this case the power P of the incident laser beam has to be reduced by the protective filter to no more than the applicable Class 1 AEL (determined, as with the MPE, for the appropriate time base, which is not necessarily the time base specified for classification). The minimum optical density is then given by

$$\mathrm{OD}_\lambda = -\log(\mathrm{AEL}_{\mathrm{Class\ 1}}/P) \tag{8.3}$$

where P is the incident laser power, or equivalently the pulse energy (using the most restrictive of the assessment conditions defined in the standard for pulsed emission).

When working with visible laser beams where it is necessary to be able to see the beam for alignment purposes (a situation that has been responsible for several eye injuries when people have removed their protection in order to see the beam), the beam can be reduced to the equivalent of a Class 2 laser by using the AEL for Class 2 rather than Class 1 in equation (8.3) above. Protection from an accidental exposure will then rely on the aversion response.

The optical density is, however, not the only performance parameter of importance in the selection of protective laser eyewear. Two others are the damage threshold and the visible light transmission.

8.6.2.3 Damage threshold

In most laser protective eyewear, the decrease in the incident exposure is due to absorption in the optical filter. This means that the energy lost from the laser beam is transformed into heat within the filter material. Even with levels of eye protection as low as OD 2, at least 99% of the beam energy is deposited in the filter. (This is not quite true, since a small fraction of the incident energy is reflected from the front surface of the filter.)

Certain protective filters work by reflection instead of absorption, and so deflect the incident laser beam elsewhere rather than absorbing it. (This, of course, may create a further hazard, and so such filters should have convex outer surfaces in order to spread the reflected beam and thereby reduce its intensity.) Reflective filters are usually fabricated by the use of multi-layer dielectric coatings applied to a glass or polycarbonate substrate to create the necessary transmissive and reflective properties. A problem with this type of filter is that its performance is dependent on the angle at which the beam hits the filter, and it can be difficult to fabricate a filter having the necessary protective properties across a wide field-of-view. This may be acceptable for protective filters used within viewing instruments such as microscopes or telescopes, where the field-of-view is well defined and usually quite restricted. A further disadvantage with reflective-type filters is that their protective properties are dependent on a very thin coating applied to the surface. Any damage to this coating, such as scratches or pinholes, can seriously degrade the filter's performance. With absorption-type filters, however, the protective properties are usually spread throughout the entire thickness of the filter material, and any surface defects have no effect on this absorption (although they can, of course, as with ordinary spectacles, effect the wearer's ability to see properly through the filter).

More recent work, however, is leading to improved interference filters having a better angular performance. Since interference filters allow much sharper wavelength cut-offs than absorption types, this is likely to lead to their more widespread adoption in laser safety.

Both types of filter (absorption and reflective) will have a 'damage threshold', an exposure limit above which damage to the filter can occur. In the case of absorption-type filters this is due to the high-levels of laser radiation deposited (absorbed) in the filer material. In reflective-type filters, the damage can be due to the fragile nature of the surface film. If the damage threshold is below the level of exposure for which the filters are intended to provide protection, then their protective properties may be compromised. For this reason, the ability of the filter to withstand the maximum level of exposure for which protection is required must be confirmed.

Optical density and damage threshold are two unrelated properties of a filter. An absorption-type filter of OD 3 and an absorption-type filter of OD 6 will both absorb a similar percentage of the incident laser-beam power (99.9% in the case of the former if we ignore reflection loses, and 99.9999% in the case of the latter, both therefore effectively 100%). It is quite possible that they may have similar damage threshold levels, especially if they both employ the same substrate material (usually glass or polycarbonate) of the same thickness, in which the different levels of absorption are created by different concentrations of dye material within the host substrate. However, the implication of a laser protective filter of OD 3 is that it can provide protection against a laser beam having a power up to 1000 times the MPE; in the case of OD 6 this increases to 1000 000 times the MPE. If the latter filter cannot withstand an exposure that is one million times the MPE, then its higher level of optical density is of no value for laser protection. As an example of this, consider an overhead projector transparency. This is an acetate sheet that is 0.1 mm thick. At the 10.6 μm wavelength of a CO_2 laser its optical density is very high, since it is highly absorbing in the far-infrared. Yet its damage threshold is very low and no one would consider it a very satisfactory form of eye protection!

In order to be able to select appropriate eyewear having the necessary protective properties, we therefore need to know not only the optical density of the filter (measured at the laser wavelength) but also the filter's ability to withstand the maximum exposure for which we require the protection. Laser eye protection should therefore be carefully specified on the basis of both of these criteria. A filter of OD 3, where this optical density is sufficient and for which there is assurance of its ability to withstand the necessary maximum beam power (up to 1000 times the relevant MPE) is more satisfactory, since we know its performance limitations, than a filter of OD 6 for which there is no information of its damage threshold.

To some extent, however, the need for high levels of damage resistance can be overstated. As mentioned earlier, eye protection should not be used where there is also a *significant* risk of injury to the skin. Levels of laser radiation that are likely to damage protective filters may also cause serious burns to the skin, and the need for high values of OD (accompanied by high levels of damage resistance) should always be questioned before acceptance. It can be instructive, when determining the maximum level of foreseeable exposure, to compare this with the MPE for the skin, and to consider whether, if the maximum exposure exceeds the skin MPE by a large margin, the use of eye protection is really a suitable method of hazard control.

8.6.2.4 Visible light transmission

While optical density and damage threshold are important criteria governing the protective properties of laser eye protection, visible light transmission is equally important. Since we need to be able to see through the eye protection, high levels of OD and damage resistance are therefore of no use if they seriously impede our visual ability.

As discussed in chapters 1 and 2 in relation to light measurement, levels of illumination are always quantified in terms of photometric rather than radiometric quantities. These take into account the wavelength dependence of the eye's visual response across the visible spectrum. As we have seen, protective filters, even when designed for laser wavelengths outside the visible band, may nevertheless reduce the overall level of visible light reaching our eyes. They may also change the colour appearance of objects viewed through the protection and reduce our ability to identify coloured lights, by removing more of some parts of the visible spectrum than others.

There are therefore two aspects of visible light transmission that are relevant in the selection of protective eyewear. The first is the overall reduction in the level of visible light, measured photometrically, that is caused by the protective filters. This will normally be expressed by the supplier of the eye protection in terms of the percentage transmission. A visible light transmission of 60% means that, measured photometrically across the entire visible spectrum, the light level viewed through the filters will be 40% less than without them. Eye protection having a visible light transmission below about 30% should be evaluated carefully before adoption, as it may impose an unacceptable loss of visual ability unless used in very well lit environments.

In a similar way, the perceptible colour or tint of many laser eye protectors can create difficulties with colour discernment, which may introduce other hazards in applications where good colour vision is needed. Some medical procedures are an example of this. In addition, many electronic displays and indicator lights employ red LEDs. The use of a red laser requiring red-absorbing eye protection may therefore prevent illuminated LEDs from being seen. The eyewear suppliers may indicate the colour appearance of the filters in their data sheets, but the effect this may have on performing colour critical tasks can often only be assessed by experiment.

In general, the visible light transmission of a filter will be lower and its colour (if any) will be more apparent at higher levels of optical density. For this reason, using higher values of optical density than those required for a particular application may have significant disadvantages, and there is certainly no benefit in using protection of a higher specification (higher optical density) than is needed.

Different designs and types of protective filter, from different manufacturers, can have different levels of visible light transmission and colouration for the same value of optical density and damage resistance. It can, therefore, be worth spending some time carefully evaluating different options before a final purchasing decision is made.

Suppliers of laser eye protection in the United States, under ANSI Z136.1, have to provide further information in the form of transmission curves or tables, in addition to the optical density at the designated laser wavelength(s), covering both the scotopic and photopic (night vision and day vision) characteristics of the filter material. Data on damage thresholds for both the filters and frames should also be given.

In Europe, different requirements apply, and these are discussed in section 8.6.3.

8.6.2.5 Other filter effects

In addition to the possibility of damage due to absorption of the incident laser radiation, two other effects can compromise the performance of laser protective eyewear.

Many filters used for eye protection incorporate organic dyes that are introduced into the substrate material in order to provide the necessary spectral characteristics. These dyes, however, are subject to fading if exposed to excessive levels of light, including sunlight. This can have the effect, even when there has been no noticeable change to the filter appearance, of increasing the filter transmission at the laser wavelength.

There are also nonlinear effects that can arise under conditions of high-energy pulsed (typically Q-switched or ultrashort pulse) laser exposure. When measured at lower levels of laser exposure the filter may exhibit the required absorption properties. Under the influence of a high-energy laser pulse, however, saturated absorption or temporary bleaching can occur, causing the laser pulse to be transmitted through the filter with little attenuation. After transmission of the pulse, the filter returns to its normal absorbing state.

In addition to these optical effects, protective eyewear can, of course, be damaged by other means, and therefore needs to be sufficiently robust, impact resistant and chemically inert.

Femtosecond (10^{-12} s) laser pulses are effectively white, i.e. broadband, and so conventional eye protection is not possible with such lasers.

8.6.3 European standards for laser protective eyewear

Europe has two specific product standards covering laser protective eyewear. These are EN 207 [1] and EN 208 [2]. EN 207 covers the majority of eye protectors, where the requirement is to reduce the incident exposure to no more than the MPE. EN 208 covers the special case of protection against visible radiation where it is necessary to see the beam for alignment purposes. This is achieved by reducing the incident power to the equivalent of a Class 2 laser.

Both these standards define measurement and test criteria that laser eye protection has to satisfy. These include tests for visible light transmission, optical quality, ruggedness and environmental stability, as well as for optical density and laser-induced damage.

All types of eye protection in Europe are covered by the provisions of the Personal Protective Equipment Directive (which is mandatory within the European Union), and must be CE-marked to indicate compliance with what are called the essential safety requirements of the directive. Compliance with these two standards, verified by an independent test house, is the usual route by which conformance with the more general requirements of the PPE Directive is demonstrated. EN 207 and 208 have therefore become the *de facto* legal requirements for all laser protective eyewear sold in Europe.

A key feature of these standards is the link made between the optical density and the damage threshold. For the condition where a given level of exposure exceeds the MPE by a certain factor X, the protective filter should have a transmittance of no more than $1/X$ (i.e. an optical density of at least $\log X$). It should also be capable of surviving an exposure of at least X times the MPE for a specified time (taken as ten seconds or 100 pulses for other than ultraviolet emission). Where a given protector has an actual transmittance of $1/Y$ (i.e. an OD of $\log Y$), and an ability to withstand (for the specified duration) an incident exposure up to Z times the MPE, then the *smaller* of either Y or Z must be at least equal to X for the protection to be suitable. (The symbols X, Y and Z are used here by way of explanation but do not appear in either EN 207 or EN 208.)

As an example, let us assume that, for a given laser, the MPE is 100 W m^{-2}. Let us also assume that an assessment of the exposure conditions has shown that the maximum exposure level that can occur with this laser is 100 000 W m^{-2}. This means that the factor X by which the exposure level exceeds the MPE is 1000. The eye protection must therefore have a maximum transmittance (at the laser wavelength) of 0.001 ($1/X$), which corresponds to a minimum OD of 3 ($\log X$). At the same time, the protective filter must be able to withstand a potential exposure level of 100 000 W m^{-2} (X times the MPE). If a given filter has an OD of 4 (i.e. $Y = 10\,000$) but can only withstand an exposure of 50 000 W m^{-2} (i.e. $Z = 500$), then it will not be suitable, since although it has an adequate optical density, its damage resistance is inadequate. For any actual filter, Y and Z will generally have different values since they are unrelated properties of the filter.

Under EN 207, therefore, eye protection is specified on the basis of the more restrictive of *either* the optical density *or* the damage threshold. Rather than being denoted by its optical density, it is given what is termed a *protective density* or scale number, equal to $\log Y$ *or* $\log Z$, whichever has the lower value. The scale number, which is prefixed by the letter L, is therefore not necessarily the same as the optical density, which may be higher if the filter performance is limited by its ability to withstand the incident laser radiation (which will often be the case). Damage-threshold testing is also required for the frames, and this rather than the performance of the filters can limit the specification given to the protector.

In selecting eye protection to comply with EN 207, the minimum scale number that is required is equal to $\log(E/\text{MPE})$, rounded up to the next whole number, where E is the level of exposure. In this case, however, E is taken as the incident power (or energy) divided by the actual area of the beam, and not averaged over the area of the limiting aperture as required under IEC 60825-1. Although this can result in overprotection, it is necessary in order that the damage-threshold criteria described above are met.

EN 208 operates in a similar way, except that it is based on the Class 2 AEL rather than the MPE. X is then the factor by which the incident laser power exceeds the Class 2 AEL of one milliwatt (for emission durations greater than 0.25 s). As with EN 207, the eye protection is given a scale number on the basis of the more restrictive of either its optical density or its damage threshold. To differentiate it from the scale number used in EN 207, it is prefixed by the letter R instead of L. In selecting alignment eye protection, the scale number that is needed is equal to $\log P$, rounded up to the next whole number, where P is the incident laser power in milliwatts. Selecting a higher value of the scale number than is necessary can make it difficult to see the beam and thus defeat the purpose of using alignment eye protection. For the same reason, alignment protectors intended for high laser powers can be unsatisfactory if their performance is limited by their damage threshold rather than by their optical density, since the optical density will then be higher than required.

In order to rationalize the large number of different filter specifications (laser wavelength, maximum exposure level, maximum exposure time, minimum damage threshold) that might otherwise be required, the European standards, and in particular EN 207, define a simplified set of criteria and establish a prescriptive approach to the selection process. Instead of using the MPE values given in IEC 60825-1, EN 207 uses just nine separate exposure limits, which cover three wavelength bands and three time bases corresponding to different notional types of laser. These laser types are (i) continuous-wave lasers, signified by the letter D (from the German 'Dauerstrich'); (ii) normal pulse including Q-switched lasers, signified respectively by I ('Impuls') and R ('Riesenpuls'); and mode-locked lasers, signified by the letter M. These code-letters (D, I, R and M) should be taken to define particular ranges of exposure duration rather than necessarily describing the type of laser for which protection is provided. (For example, in the waveband 315–1400 nm, 'cw' covers any duration greater than half a millisecond.)

These time ranges and the corresponding values of the MPE are given in table 8.2. The scale number of the protection required under EN 207 is simply the logarithm of the factor, rounded up to the next whole number, by which the actual level of exposure exceeds the appropriate MPE given in table 8.2. (This is equivalent to inserting the MPE from table 8.2 into equation (8.2), then rounding up the answer.)

As an example, consider cw laser emission at a wavelength 1300 nm and a worst-case exposure level of 300 W m^{-2}. (The exposure level should be determined, as required by the standard, on the basis of the *actual* incident irradiance, in practice the peak irradiance at the centre of the beam, rather than the value obtained by averaging over the area of the limiting aperture, although for a large-diameter beam these values may be similar.) From table 8.2, the relevant MPE is 10 W m^{-2}, and so the value of $log(E/\mathrm{MPE})$ is log(300/10), which is equal to 1.48. Rounding this up to the next highest whole number, therefore, indicates that the required scale number of the protection is L2.

Provisions are defined in an annex to the standard for dealing with pulsed emission in a similar way to that used in IEC 60825-1 (by applying the most restrictive of the average power, single-pulse energy, or the single-pulse limit using the $N^{-1/4}$ correction factor).

The adoption by EN 207 of simplified MPEs and standardized exposure conditions inevitably leads to some significant differences in comparison to IEC 60825-1, with the result that the protection levels derived by the use of EN 207 can be over-specified. This is especially so for some pulsed lasers. Furthermore, the link between optical density and damage threshold that is inherent in these standards is sometimes inappropriate. In the ultraviolet region, for example, a time base of 30 000 s is always assumed. This is unrealistic where the requirement is to protect against an accidental exposure to the direct beam, and can lead to levels of protection far greater than those actually needed.

While EN 207 and EN 208 are primarily intended as product standards, they must also be used for the selection of CE-marked eye protection because of the concept of protective density that is adopted and the simplifying assumptions regarding exposure conditions that are made. It might be thought, in circumstances where the use of these standards indicates a level of protection that is considerably higher than is actually required, that the MPEs given in IEC 60825-1 could be used instead as the basis for selecting the level of eye protection that is needed. Because of the connection made in the European standards between optical density and damage threshold, however, the use of CE-marked eye protection for exposure conditions other than those assumed in EN 207 can

Table 8.2. MPE values used for specifying eye protection under EN 207. The nine different values of MPE used for specifying eye protection under the European standard EN 207 are dependent on the waveband of the laser radiation and on the duration of exposure. The code-letter for the laser type is used in the marking of the eye protection, and defines the test conditions under which it is certified. The scale number L of the required protection is equal to $\log(E/\mathrm{MPE})$, rounded up to the next whole number, where E is the level of exposure and the MPE is the value indicated below for the appropriate emission duration. The level of exposure E must be specified in the same units (irradiance or radiant exposure) as that given above for the MPE. E is taken as the incident power (or energy) divided by the area of the beam (which can be based on the d_{63} diameter where appropriate), and not averaged over the area of the limiting aperture as required under IEC 60825-1. This can result in overprotection, but is necessary to ensure that damage-threshold conditions are met.

Waveband	Laser type (designation of test condition)	Emission duration	MPE value used in EN 207
180–315 nm	Mode-locked pulse laser (M)	$< 10^{-9}$ s	3×10^{10} W m^{-2}
	Normal pulse laser (I) or Q-switched laser (R)	10^{-9} to 3×10^{4} s	30 J m^{-2}
	Cw laser (D)	$> 3 \times 10^{4}$ s	10^{-3} W m^{-2}
>315–1400 nm	Mode-locked pulse laser (M)	$< 10^{-9}$ s	1.5×10^{-4} J m^{-2}
	Normal pulse laser (I) or Q-switched laser (R)	10^{-9} to 5×10^{-4} s	5×10^{-3} J m^{-2}
	Cw laser (D)	$> 5 \times 10^{-4}$ s	10 W m^{-2}
>1400 nm–1000 μm	Mode-locked pulse laser (M)	$< 10^{-9}$ s	10^{11} W m^{-2}
	Normal pulse laser (I) or Q-switched laser (R)	10^{-9} to 0.1 s	10^{2} J m^{-2}
	Cw laser (D)	> 0.1 s	10^{3} W m^{-2}

result in inadequate protection with regard to damage resistance, unless additional information (separately defining the optical density and the damage threshold) is given by the supplier to allow the purchaser to undertake their own assessment.

Quite often, European laser users selecting eye protection carry out their analysis using the MPEs given in IEC 60825-1 in order to determine the OD that they need. They then purchase CE-marked protection specified to EN 207 that has the equivalent scale number (i.e. the same numerical value as the OD they require). This, however, is a mistake, for the reason given above, and may mean that the protection does not have a sufficiently high level of damage resistance.

Eye protection conforming to EN 207 or EN 208 must be code-marked by the manufacturer to indicate its specification. For EN 207 these marks are as follows.

- The wavelength or wavelength range for which protection is provided. (This is given in nanometres but the units need not be marked on the eye protector.)
- A letter denoting the test condition (the time base included in table 8.2). This always includes testing for cw emission.
- The scale number, prefixed by L.
- The manufacturer's identification mark.
- An optional mechanical strength symbol signifying additional ruggedness.

The order of the first two of these marks is reversed in products certified under earlier versions of the standards.

For EN 208 the required marks are as follows.

- The maximum laser power.
- The maximum pulse energy.
- The wavelength range for which protection is provided. (This is given in nanometres but the units need not be marked on the eye protector.)
- The scale number, prefixed by R.
- The manufacturer's identification mark.
- An optional mechanical strength symbol signifying additional ruggedness.

An additional mark denoting the test house that has certified the eye protection may also be included, and is required in some countries. The words 'adjustment eye-protectors' (in the national language) are also required in some countries on eye protection conforming to EN 208.

As an example, eye protection denoted by

600–800 D L4 X S

has a scale number 4 under EN 207 for cw emission (indicated by D) at wavelengths between 600 and 800 nm. It has been manufactured by company X and has added ruggedness for harsher environments (indicated by S).

Under EN 208 (applicable to protection against visible-beam lasers where it is necessary to be able to see the beam), eye protection marked

1 W 2.10^{-4} J 514 R3 Y

has a scale number of 3 (indicated by R3) and provides protection (by means of natural aversion responses) for accidental viewing of a 514 nm beam up to a maximum power of 1 W or a maximum pulse energy of 0.2 mJ, and has been manufactured by company Y. It does not meet the added ruggedness requirements (since the symbol S does not appear). Note that the maximum pulse energy is often expressed in the form 2.10-4.

8.7 Working in laser controlled areas

The principle reason for setting up a laser controlled area where it is not reasonably practicable to totally enclose the hazard is to separate the hazard from people who do not need to have access to it, and to isolate it from other, unconnected, work activities. But establishing a laser controlled area is not, by itself, sufficient. It can lead to other risks, in that those who are authorized to work within the controlled area (which is often enclosed) can find themselves working unobserved and undisturbed, and with no need to consider the risk that their activities may impose on other people. Without sufficient planning and discipline, therefore, this can lead to an informal *ad hoc* approach to the work being carried out, possibly resulting in the adoption of unsafe working practices. Even where procedures have been agreed, these can be abandoned if their purpose is not properly understood or where convenience overrides caution.

It comes as no surprise, then, that most serious eye injuries have occurred to those working with laser equipment *inside* laser controlled areas, often in research environments. The area may have been properly set up, adequate warning signs displayed, and even interlocks used to prevent unauthorized entry, yet if those inside do not adopt sufficient precautions in the way in which they carry out their work, accidents can, and do, occur.

Many of these accidents arise from the misuse, or non-use, of adequate eye protection. Wearing protective eyewear for long periods can be uncomfortable and restricting, and the temptation to remove it on occasions can be strong. This is a very good reason for finding alternative means of hazard control if at all feasible. The alignment of a visible laser beam can be difficult, for example, if it cannot easily be seen through the eye protection. Eye injuries have occurred where people have removed their eye protection simply to see the position of the beam more clearly while making an adjustment. In other cases, protective eyewear was not being used at all where it was felt that, because the beam was horizontal and well below head height, such protection was unnecessary. Accidental and unexpected reflections of the beam can easily occur from tools and instruments moved through the beam, from the surfaces of optical components positioned in the beam, and even from watches, rings and other jewellery when hands are placed into a beam (which may not be considered to be a skin hazard) to make adjustments.

It is certainly good practice to keep the beam within well-defined limits (at waist height where this is feasible), avoiding specular (mirror-like) reflections. However, while precautions should always be taken to minimize any uncertainty in the beam position and to prevent stray beams from occurring, an assumed knowledge of the position of the beam should never be used as a reason for not wearing adequate eye (and where necessary, skin) protection. Accidents happen when the unexpected or unplanned occurs. Precautions are therefore needed not merely against the obvious, but also against every unexpected but nevertheless foreseeable event. That is the purpose of a risk assessment.

Beam alignment is one of the most hazardous procedures routinely carried out with lasers, particularly where long beam-paths are involved. But it is hazardous usually because insufficient thought has been given to aligning the beam in a safe manner. Where access to the beam is needed while this adjustment is being made (although remote viewing options should be considered as an alternative), adequate personal protection should, of course, be worn. Nevertheless, an appropriate alignment procedure should be adopted that minimizes the uncertainty of the beam's position and contains it at all times within well-defined limits.

Alignment should be carried out by moving in stages from the laser outwards, using beam-stops to terminate the beam at intermediate positions, and apertures, where necessary, to restrict the beam's angular movement. Figure 8.8 illustrates this alignment process. Alignment should also be carried out using the lowest beam power possible, employing a filter if necessary over the emission aperture to reduce the emitted power. (For very accurate alignment this may need to be of high optical quality with exactly parallel surfaces and positioned in a precision mount perpendicular to the beam.) If the power can be reduced to no more than the equivalent of Class 3R then no eye protection need be worn.

Those who work in laser controlled areas need to have an adequate understanding of the hazards that exist and to be aware of the risks that can arise. They should follow procedures that define what precautions need to be taken and how specific tasks should be performed. Where, in research and development activities, for example, there has to be some level of flexibility and freedom in the way that people are permitted to work, there is nevertheless a responsibility to ensure that everything they do is carefully considered and planned, and is performed within the framework of a structured safety policy.

8.8 Laser servicing

The classification of a laser product is based on the level of the laser emission that is accessible during normal use of the laser for its intended purpose, as discussed in chapter 4. Certain lasers (embedded lasers) can have higher levels of radiation inside the equipment than that used for classification, simply because there is normally no access to the internal radiation. Most often, this occurs when a

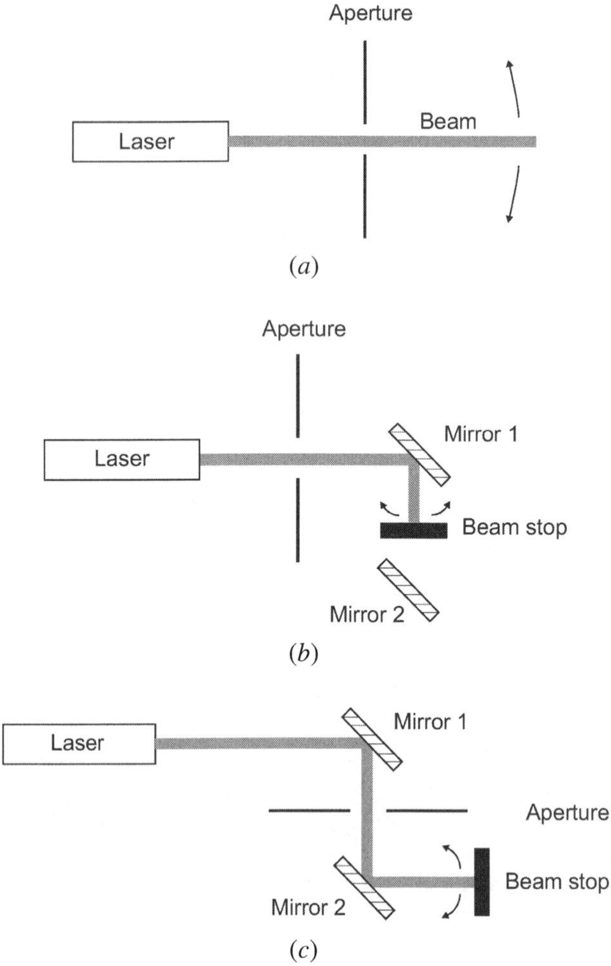

(*a*) (*b*) (*c*)

Figure 8.8. The alignment of a laser beam should be carried out in stages, using apertures and beam stops wherever necessary to restrict the beam (especially with long, complex beam paths), moving gradually away from the laser and at each stage minimizing the uncertainty in the beam's position. (*a*) Initial alignment is carried out by adjusting the direction of the beam so that it passes through an aperture located at the desired position of the beam. A highly misaligned beam cannot therefore pass further than the aperture. (*b*) Adjustment at the first mirror is carried out while blocking the beam from the second mirror. (*c*) This process is repeated at each subsequent mirror, so that at every stage the beam is contained within defined limits.

hazardous laser is totally enclosed, producing a Class 1 laser product. During servicing, however, in order to carry out tests or adjustments, there may be a need to operate the laser with some of the protective covers removed or with other protective features disabled. This can increase the laser's hazard beyond that implied by its class.

If the laser equipment is small enough it should be removed to a restricted area that is suitable for controlling the hazard. Where, as is often the case, this is not possible due to the nature of the laser equipment, then a temporary controlled area may need to be established around the laser while the servicing work is carried out. The servicing work should be subject to a separate risk assessment, and adequate temporary control measures established, including restrictions on access into the area, to ensure that the risk of harm is sufficiently low. A boundary may need to be set up around the laser while servicing work is being undertaken, and screens or other barriers may need to be used in order to prevent hazardous levels of radiation from passing beyond the boundary. A higher level of training and hazard awareness than is necessary for normal operation of the laser will often be required for servicing work.

Organizations maintain responsibility for the safety of any work being carried out on their premises, including that done by visiting service engineers. Agreement on appropriate safety procedures should be reached, and a signed permit system may be advisable for handing the equipment over to the service engineer and for accepting it back fully-restored to normal operation.

References

[1] EN 207 2002 *Personal Eye-Protection—Filters and Eye-Protectors against Laser Radiation (Laser Eye-Protectors)* (Brussels: CEN)

[2] EN 208 2002 *Personal Eye-Protection—Eye-Protectors for Adjustment Work on Lasers and Laser Systems (Laser Adjustment Eye-Protectors)* (Brussels: CEN)

Chapter 9

The management of laser safety

9.1 Health and safety responsibilities

The identification and control of laser hazards should not be carried out in isolation, but undertaken within a management framework as part of the overall health and safety policy of the organization concerned. While there are some differences in the detail of safety legislation from country to country, the emphasis of all safety law is to require manufacturers to supply safe products and for employers to take all reasonable steps to protect their employees from injury or ill health. Specific national requirements may exist for dealing with particular kinds of hazard, but there are also more general obligations placed on employers to ensure a safe place of work through the use of inherently safe work equipment and through the proper control of workplace risks.

Employees, too, share an obligation with their employer to ensure that the work they carry out and the way in which they undertake it does not pose an unacceptable risk to themselves or to other people (including other staff, visitors and the general public). Safety rules should be drawn up and procedures developed in order to minimize occupational risks. Safety should form an integral part of all line management duties and be regarded as a common concern throughout the organization.

Generally, while specialist safety staff may also be employed to provide safety guidance and to perform a safety management function, they do not carry sole responsibility for preventing accidents or other safety-related mishaps. Safety is not merely someone else's concern; it should be everybody's.

When reviewing safety or the success of a safety policy, an organization's safety performance should not be judged solely on the basis of its historical accident record. The particular combination of circumstances necessary for the occurrence of a specific and potentially serious accident may never yet have happened, but this combination could arise next week.

A great deal can be learned concerning an organization's control of workplace risks by considering what might be termed its safety culture; by looking for evidence that accidents are very unlikely to occur. Is safety regarded as sufficiently important and is it driven from the top? Is discussion of safety issues encouraged, is the reporting of apparent deficiencies welcomed, and are these acted upon? Is its safety policy actively promoted and widely understood? Are well-developed and appropriate procedures in place? Are they enforced, and is a failure to comply considered a serious disciplinary matter? Do the procedures keep up with changing circumstances (with new or relocated equipment, changes in staff, etc)? Are internal safety audits routinely carried out? Is sufficient safety training regularly provided? Are people well informed on the safety issues that matter?

It is important that documentation is kept detailing all aspects of a laser safety programme. This should demonstrate that significant hazards have been identified, that the level of risk has been assessed, that all necessary control measures have been implemented and that sufficient training has been provided. Some countries (e.g. Germany and Austria) have very prescriptive requirements covering the documentation that should be maintained.

9.2 The framework policy

Protective measures that are intended to control the risks arising from the use of laser equipment should not be adopted in a piecemeal or *ad hoc* manner, but implemented as part of an overall and systematic safety management programme. It can be helpful to distinguish here between policy and procedures. The former relates to the overall framework within which lasers are used and the latter to the detailed precautions adopted for a specific laser system or laser process.

All organizations should have a written safety policy; those using lasers may need a specific and documented laser safety policy as well. Whether a separate policy covering lasers is necessary will depend largely on the extent of laser use within the organization. If there are just one or two items of laser equipment, used in a routine manner and at most by a very limited number of individuals, then a general documented policy governing laser use may be unnecessary. It can be sufficient to undertake a risk assessment and to implement the necessary controls (which should, of course, still be documented) within the framework of the organization's overall safety policy.

Where laser use is more widespread within an organization, involving several separate lasers or many different individuals, then a written framework policy for controlling laser use is often advisable. This can have the benefit of ensuring a more uniform and consistent approach to laser safety management throughout the organization, and can also avoid the necessity for repeated risk assessments involving similar sets of circumstances.

It is a matter of judgement as to how comprehensive and detailed the framework policy should be; much will depend on the type of laser equipment in use and the range of laser work being carried out. As a guide, it can be useful to include a description of the kind of hazards that might be present, and to define appropriate requirements governing laser equipment, working areas and authorized users.

A typical format might be as follows.

Laser hazards

This could include a brief description of the nature of applicable laser hazards and a short overview of the classification system, in order to provide a background to the policy and a rationale for its existence.

Safety management

Responsibilities should be defined and the role and authority of the Laser Safety Officer explained (see the next section).

Control of laser use

This can be done by restricting laser use:

(a) to laser equipment that has been correctly classified and labelled, and which complies with the applicable safety standard and with any other requirements that are deemed appropriate;
(b) to locations within the organization that have been approved as being suitable for laser use or which satisfy defined criteria;
(c) to people who have been appropriately trained and are deemed to have sufficient competence in the safe operation of the laser equipment;
(d) to laser work for which an adequate and documented risk assessment has been undertaken and for which appropriate controls have been specified (possibly subject to prior approval).

Approval of the equipment, of its location, of the people working with it, and of the control procedures adopted can be given in the form of a 'permit to work' system authorized by the Laser Safety Officer (see the next section) or by another suitable person, as considered appropriate.

Procedures for laser operation

Where there is only a limited variation in the kind of lasers being used and of the type of laser work being carried out, it can be useful to define in the framework document the specific control procedures that are to be adopted. Where the risks associated with use of different lasers within the organization are more varied or where changes in laser work can be anticipated to occur reasonably often, it may be inappropriate to define specific procedures in the main policy document. Instead, only broad guidelines need be included, but there should then be a requirement, as indicated above, to undertake a detailed risk assessment of each separate laser or laser area in order to define the necessary controls and protective measures.

The purpose of the framework policy is to establish and define the way in which laser safety issues are managed. It should be appropriate to the particular conditions and circumstances that exist within the organization concerned, and should be promoted as a document intended to ensure safe laser use. Too often, in regard to safety, assessments are carried out and documents are produced with the sole aim of complying with legislation; to have something to show in defence should an accident occur or an investigation be carried out. This, of course, is an entirely inappropriate attitude. While compliance with the law is, of course, essential, the whole purpose of the safety analysis and documentation that the law requires is to reduce risk and to prevent workplace accidents. Policies and control procedures should therefore be developed with a view to their real effectiveness in achieving this end, and not simply for the purpose of satisfying the letter of the law.

The costs of laser safety (training, management time and safety controls systems) need not be excessive, but should be anticipated and should form an integral part of the budget allocation covering the purchase and operation of laser equipment.

9.3 The role of the laser safety officer

The international laser safety standard requires that a Laser Safety Officer is appointed where lasers in Class 3B or Class 4, or lasers in Class 3R having invisible emission (that is, outside the range 400–700 nm), are in use. A Laser Safety Officer can also be useful in other circumstances, such as to oversee the restriction on the use of magnifying aids with lasers in Class 1M or Class 2M, or to take responsibility for the safety management of the servicing of embedded laser systems. The Laser Safety Officer (LSO) has the responsibility to review the use of lasers within their organization and to designate the controls that should be implemented.

Appointing an LSO is a legal requirement in some countries; in others it is taken as indicative of good practice while the legal emphasis is placed more on the necessity for appropriate control measures.

There is considerable variation in the way in which the role of the LSO is interpreted. Some countries have adopted formal training and accreditation programmes for LSOs, but most have not. In some organizations the LSO is given virtually complete and sole authority to determine the necessary control procedures and is also able to prohibit laser use that is deemed by the LSO to be unsafe. The LSO may be expected to carry out whatever laser analyses and risk assessments are needed, and to draw up the organization's safety procedures. In other cases the role is seen as more advisory, and is sometimes restricted to that of record keeping. This may involve little more than maintaining lists of laser equipment in use, of the safety controls in force and of the names of authorized laser operators.

There are two important issues concerning the role of the LSO that do cause confusion and need to be recognized. The first is that legally, in most countries, responsibility for all matters of workplace safety rests with the employer. The employer may delegate, but nevertheless retains ultimate responsibility, and so the LSO's position has to be considered within this context, that of carrying out responsibilities on behalf of the employer. The level of authority granted to the LSO is a matter for the employer (or for another delegated manager on behalf of the employer) to decide.

The second issue concerns the level of understanding and competence that should be possessed by the Laser Safety Officer in matters such as laser technology, optical physics, radiation measurement, bioeffects and the detailed provisions of safety standards. Some aspects of laser safety, particularly those involving the characterization of laser sources or the quantitative assessment of levels of exposure, can be complex, requiring specialized knowledge and considerable experience in applying it. It is unreasonable to require that every organization using Class 3B, Class 4, or invisible-beam Class 3R lasers should have an LSO who is an 'expert' in all matters of laser safety.

Where laser use is routine, where the risks are well controlled, and where operating conditions are unchanging, there is no need to acquire and to maintain in-house expertise in quantitative exposure evaluation or in the use of control procedures that are not relevant. An example of this kind of situation could be the use, by a small manufacturing company, of a carbon dioxide flat-bed laser cutting machine that is not completely enclosed (and is therefore Class 4) but where the likelihood of accidental beam exposure is very low and sufficient control procedures have been implemented (perhaps on the recommendations of the equipment supplier). It can be sufficient in this case for the LSO to undertake a purely administrative function in order to ensure that the necessary controls remain in place and that all authorized laser operators understand and follow the procedures that are in force. To expect the LSO to understand the measurement conditions for classification, the meaning of angular subtense, or the way to correct the single-pulse MPE for multiple pulsed emission is pointless, and does nothing to aid the cause of safety or to promote the adoption of laser technology.

As with other workplace hazards, organizations should always seek external expert advice whenever safety issues arise that are beyond their own competence to handle. This can be especially relevant when new equipment is first installed. Even where an organization's employee has undergone training in laser safety, an external adviser is likely to have had considerably more practical experience in undertaking risk assessments and in recommending an appropriate system of controls.

In other circumstances, for example in a large research laboratory, in a big industrial concern or in a group of associated hospitals, wherever there is extensive laser use, it would be more appropriate to ensure that the in-house LSO was fully competent in the quantitative aspects of laser safety, as this ability will very probably be required on a regular basis. In order to acquire the necessary level of expertise, the professional development of the LSO is likely to require a combination of training and practical experience over an extended period of time.

The role and responsibilities of the LSO within a particular organization need to be discussed and agreed, and should be written down. The LSO's duties should cover all those aspects necessary to ensure the continuing safe use of laser equipment within the organization concerned.

The 'job specification' for an LSO might cover such issues as:

- whether the LSO has executive responsibility (to authorize and prohibit) or merely acts in an advisory capacity (and if the latter, who actually makes safety-related decisions?);
- to whom the LSO reports and is responsible;
- whether the LSO is expected to develop or merely implement a safety policy;
- what the LSO's responsibilities are for undertaking safety audits and inspections;
- whether the LSO has responsibility for giving or arranging safety training courses;
- what records the LSO is required to keep.

It is, of course, the responsibility of the employer (or another delegated manager) to ensure that the person appointed as the organization's LSO has sufficient ability to undertake the role competently, and that the LSO is given the necessary resources (sufficient time and budget) to satisfactorily fulfil their agreed obligations.

In large organizations it can sometimes be useful to have more than one person involved in an LSO role. There can be assistant or departmental Laser Safety Officers having responsibility for certain defined areas. In this case, of course, there needs to be adequate contact and liaison between all those involved in sharing safety responsibilities, to ensure consistency of approach and the effective management of laser safety across the whole organization. A further benefit of involving other people in a laser safety role is that it provides a means for acquiring skills and capabilities over time in the manner of an 'apprenticeship', ensuring that there is no serious dislocation of safety support should the principal LSO leave or be transferred to other duties.

The appointment of an LSO is, however, only part of the overall approach to managing the safety of laser use; it is a means to an end and not an end in itself. It doesn't provide a licence to abandon caution or to permit unsafe practices to be carried out, simply because an uninformed LSO has allowed it. It does not remove the need for undertaking formal assessments or for adopting a structured approach to managing risk. And, of course, the LSO shouldn't be regarded as a convenient scapegoat for the failings of others or of the management system as a whole!

9.4 Safety training

Training in laser safety is normally required for all those whose actions and behaviour in working with laser equipment could put themselves or other people at risk of harm. This will usually include all those who operate laser systems in Classes 3B or 4, but may also include those working with lasers in a lower class. Those having responsibility for safety, such as Laser Safety Officers and other managers, may also need training to enable them to perform their duties satisfactorily.

Training should be matched as closely as possible to the needs of those being trained. Whether training is given internally or by using outside agencies, training requirements should first be carefully identified.

All those involved in laser use need to have a basic knowledge of:

- the nature of laser radiation (that it is a concentrated form of 'light' or optical radiation, and is not to be confused with ionizing radiation such as x-rays—which is the form in which most people come across and interpret the word 'radiation');
- the hazards (arising from laser radiation and from other aspects of the laser equipment or of the process being carried out with the laser) that can exist (especially those relating to the particular laser being used);
- the meaning of laser product classes, warning signs and descriptive labels;
- how to use (set up, adjust, operate, maintain, shut down, etc) their particular laser equipment safely (what to do and what not to do);
- what to do when a suspected injury occurs.

In the case of a Laser Safety Officer a more in-depth understanding will usually be required than is covered by the basic issues outlined above, although circumstances will vary widely as discussed in section 9.3. It is however likely to be necessary that the LSO should:

- know that optical radiation encompasses visible light and invisible infrared and ultraviolet radiation, and that it differs from ionizing radiation;
- appreciate the basic properties and characteristics of laser emission;
- understand the meaning of laser product classes and the purpose of hazard warning signs and labels;
- know the emission wavelength of the laser equipment used on-site and understand in which waveband (e.g. ultraviolet; visible; near-infrared below 1400 nm; far-infrared above 1400 nm) this wavelength lies;
- know the tissues at risk from exposure to this laser radiation;
- appreciate the possible consequences of exposure to this laser radiation;
- know the distance from the laser equipment or the area surrounding the laser equipment within which hazardous levels of exposure can occur to the eyes or skin under differing conditions of use (including normal operation, adjustment work and, where applicable, both unaided and magnified viewing);

- understand the nature and extent of other hazards (such as electrical, fire and fume) that may arise from the reasonably foreseeable use, misuse or failure of the laser equipment;
- understand the essential requirements of health and safety legislation and the general principles of good safety management, including the need for regular safety monitoring and adequate safety information;
- understand (in circumstances where the LSO is required to specify control procedures) the need to eliminate hazards at source where this is reasonably practicable, and know the priority with which different types of protective control procedures should be considered;
- understand the specific control procedures (for the laser equipment of concern) that are necessary to eliminate the risk of harm occurring or to reduce it to an acceptable level;
- be able to ensure that safe working procedures (covering normal operation, adjustment work, and servicing, as required) have been established, are documented, and are being implemented;
- recognize when their own knowledge is not sufficient and that additional advice or guidance is needed, know where to seek assistance (using internal or external resources) and be able to act on the advice given.

An important decision regarding training concerns the understanding of quantitative aspects of laser safety that may be needed, that is of MPEs and exposure assessment, and also perhaps of AELs and classification measurements.

Quite often, laser operators (and even those appointed as LSOs) having little prior understanding of basic optical principles attend short courses on laser safety which quickly move on to discussions of irradiance, measurement apertures, angular subtense and other detailed aspects of laser safety which may have little relevance to day-to-day laser operation. This can leave such individuals confused, bored and dissatisfied, and having learnt little of the basic safety issues that really do concern them. As in all types of training, it is what is learnt that is important (i.e. the learning outcome), not merely what is taught (the training input).

Equally, however, those, such as some LSOs and what might be called ‘advanced’ laser users, who may need to classify laser products or undertake quantitative risk assessments, can have unreasonable expectations of the outcome of short periods of training. It is difficult to cover all the material necessary for quantitative laser safety within the format of a short (one or two day) course. Furthermore, those having little prior experience in this field usually find that they need to develop their understanding over an extended period of time, by applying what they have learnt during training to the real world of laser safety, before they can reach a reasonable level of competence.

Where training in quantitative aspects of laser safety is given, it may need to include the following topics, as appropriate, and should also provide the opportunity, through tutorials, workshop sessions or by the use of worked examples, for people to develop and reinforce their basic understanding of the subject matter.

- The wave and photon nature of optical radiation.
- The spectral, spatial and temporal properties of laser emission.
- Basic radiometric quantities and the relationship between them (i.e. radiant power and radiant energy, irradiance and radiant exposure, and also possibly radiant intensity, radiance and integrated radiance).
- The principles and techniques of radiometric measurement.
- Laser beam profiles and beam propagation effects.
- Plane angles, solid angles and simple beam geometry.
- The concepts of apparent source, source-size and angular subtense.
- The concepts of MPEs and AELs.
- The use of the MPE and AEL tables, and of the associated data given in the safety standards.
- Measurement conditions for laser classification.
- Limiting apertures for exposure assessment.
- Techniques for dealing with pulsed emission and scanning beams.
- Techniques for dealing with multiple and extended sources.
- Techniques for dealing with multiple emission wavelengths.
- The effect on ocular exposure of magnifying instruments.
- The selection and specification of protective eyewear (and, where appropriate, of other forms of personal protection).

In some circumstances it may also be necessary for the individual to acquire a sufficient level of understanding of the assessment, evaluation and control of additional laser hazards (such as laser-generated fume). Few courses currently cover these aspects in any detail, and so the process of learning will generally be through reading technical papers and following up references for the specific information required. There is also a growing amount of material on these aspects of laser safety on the World Wide Web.

Training in detailed aspects of quantitative risk assessment and hazard controls in laser safety is sometimes sought when what is really needed is professional advice. (This is analogous to training as a car mechanic when you need a new exhaust system for your car.) If it is a one-off situation (a new laser installation, for example, for which protective measures and safety procedures need to be established) it might be more appropriate to seek external guidance from a laser safety specialist than to develop the necessary capability in-house. On the other hand, if there are safety issues that are likely to arise on a more regular basis (because of future laser acquisitions or changing conditions of use) then it would be more sensible to ensure that the necessary expertise is available from within the organization. (Training in car maintenance is sensible if this is something you aim to do regularly.)

Several organizations, academic institutions and private companies in both Europe and the United States run training courses in laser safety on a regular basis. In a few countries these are linked to national accreditation schemes. What is important is that those who work with lasers or have responsibility for safety posses a sufficient level of understanding of laser-safety issues appropriate to their needs. This understanding can be acquired through a combination of formal courses, private reading, computer-aided learning and on-the-job experience.

Training should be repeated as often as necessary as new staff join, and updates provided regularly to reinforce understanding. It can also be useful to hold occasional safety meetings for all of those involved in laser work. This provides the opportunity to refresh people's awareness of safety issues, to raise concerns, to discuss problems and to identify solutions.

Training is very much related to human factor issues, which are discussed in the next section. Involving the individuals who are at risk from laser hazards in the safety-review process can help to give them a feeling of 'ownership' of the controls that are implemented. This is likely to lead to more effective compliance with whatever safety procedures are deemed necessary.

9.5 Human factors

As was noted in chapter 1, lasers can cause harm at a considerable distance from the laser itself, and laser hazards are often not at all obvious. Individuals can therefore be at risk of harm without being aware of it, but safety is more than simple awareness. Human behaviour is complex. Actions are based largely on personal judgement, and so people don't always follow 'rules' particularly well. Human factors, the issues that influence behaviour, play an important part in the way that people deal with risk, and these factors cannot be ignored in the management of laser safety.

The way in which people work is governed to a large extent by their own inherent understanding, perceptions and abilities. These are their *personal attributes* which they bring to their work. An individual's behaviour is also influenced by the nature of the work they have to do and by the tools and other equipment they have to use. These are the *job factors*. In addition, the particular characteristics or 'culture' of the organization in which they work will have a significant bearing on the way they carry out their job. These are the *organizational factors*.

Personal attributes, job factors and organizational issues all combine to influence the way in which people will respond to particular circumstances. People are not machines, and they do not all behave in the same way under the same set of conditions. Merely informing them about workplace hazards and explaining the safety rules may not always result in the desired behaviour.

Much of this is influenced by the way in which people are trained and by how they acquire their understanding and perceptions of laser safety issues. Where there is no formal training, people simply learn on the job and gain experience over time, but they can develop misperceptions of issues that have not been properly explained, or adopt unsafe practices copied from other people. On the other hand, training, where it is provided, may be inappropriate. It may not meet the needs of the person being trained, by concentrating on issues that are not directly relevant. It may be at the wrong intellectual level, or it may simply be badly presented and therefore fail to get across the key issues. Training may also fail to address people's underlying perceptions of risk, by simply explaining what should be done without giving any rationale or justification, leaving people without a proper appreciation of the risks with which they have to deal.

Momentary inattention or a mistaken action, whether through misunderstandings, distraction or fatigue, can result in an accident. People may not recognize potential hazards (arising, perhaps, from a changed configuration of the laser equipment) and may have faulty perceptions of the actual risks involved. They may underestimate the risk of something with which they are very familiar and over which they feel in some control. Equally, risk can be exaggerated, sometimes leading to fear, where the technology is unfamiliar and the issues are poorly understood.

There may be pressure from co-workers or a culture within the organization as a whole to disregard what may be seen as 'unnecessary' or over-protective precautions that slow down work, cost money or impose undue restrictions on the way in which the work is carried out.

All of these aspects of human factors should be considered as an integral part of the risk assessment process. They should be addressed within the overall safety policy and managed through the use of appropriate protective controls. Procedural controls are particularly vulnerable to human-factor failures, and this must be taken into account during the assessment of risk. There should be good reason to believe that any procedural controls that are proposed will be routinely followed. Where this cannot be assured, much more reliance should be placed on engineering controls and on complete containment of the hazard.

Safety is something that needs to be reinforced regularly, at all levels within an organization. While the possibility of equipment malfunction rightly forms part of the risk assessment process, in the majority of accidents human failure plays at least some part, and often it is the most significant part.

Appendix A

Glossary

accessible emission level—the quantity of laser radiation emitted by a laser when measured in accordance with the classification conditions specified in the safety standard and intended to be compared with the AEL.

administrative controls—protective measures based on rules and procedures.

AEL—accessible emission limit, the maximum allowable emission for a given class of laser product, where laser emission is measured in accordance with specifed conditions.

angle of acceptance (also cone angle or field of view)—the limiting angle used in safety assessments to restrict a radiation measurement to that radiation emitted by only part of the source.

angular subtense (of apparent source)—the plane angle (usually specified in milliradian) subtended by the apparent source at a defined distance from the source (symbol α).

ANSI—American National Standards Institute.

aperture (also aperture stop)—a circular hole used to restrict (i.e. limit the cross-sectional area of) a beam of optical radiation, often for measurement purposes.

apparent source—the source from which emitted optical radiation appears to orginate (which may or may not be the actual physical source of radiation). For laser beams it is more appropriate to consider the apparent source as being that which gives rise to the smallest retinal image.

astigmatic beam—a beam having an asymetric (i.e. non-circular) cross-section.

average power—power (emitted or received) averaged over time (for pulsed lasers, both average power and peak power can be of relevance in safety evaluations).

averaging aperture—see limiting aperture.

aversion response—the reflex avoidance action (closing the eyes and turning away) when a person is suddenly exposed to a bright source of light.

beam—see laser beam.

beam profile—the cross-section of a laser beam, in a plane that is perpendicular to the beam propagation direction.

beam stop—a device that intentionally prevents a laser beam from propagating further. With high-power beams, a beam stop may need to be cooled in order to prevent it suffering thermal damage.

beam waist—the position along the beam having the smallest cross-sectional area.

biologically effective exposure—an exposure level assessed in accordance with the measurement conditions (including averaging or weighting criteria) specified in the safety standard and intended to be compared with the MPE. It will often be different to the actual physical level of exposure.

blink reflex—sudden closing of the eyes in response to a bright source of light.

blue-light hazard—the cumulative photochemical effect of short-wavelength visible radiation on the retina.

brightness—see radiance.

CDRH—Center for Devices and Radiological Health (a division of the US Food and Drug Administration responsible for radiation safety).

CIE—Commission Internationale de l'Eclairage (International Commission on Illumination)

class—see laser product class.

collimated beam—a laser beam that has a very low divergence and so is effectively parallel; its diameter therefore increases only slowly with distance.

continuous wave laser—a laser whose output is at a steady continuous level (abreviated cw laser).

control measures—protective precautions adopted by a user organization and intended to ensure the safe use of laser equipment.

cornea—the transparent 'bulge' at the front the eye through which light enters.

cw—continuous wave.

d_{63}—the diameter of the central portion of a Gaussian laser beam that contains 63 per cent of the total beam power or energy. This definition of beam diameter is used because the maximum (on-axis) irradiance or radiant exposure in a Gaussian beam is equal to the total beam power or beam energy divided by the area of the beam defined by the d_{63} diameter.

diffuse reflection—a reflection in which the reflected radiation is scattered in all directions away from the reflecting surface (as with visible light incident onto a sheet of paper). (Compare with specular reflection.)

direct viewing (of a source)—direct (i.e. not diffused or scattered) exposure of the eye to optical radiation, including exposure to a specular reflection. (It is not necessary that the eye be looking directly at the source.)

divergence—the angular spread of a laser beam.

dose—an exposure of the eyes or skin to optical radiation specified in terms of the radiant exposure.

effective exposure—see biologically effective exposure.

electromagnetic radiation—see radiation.

embedded laser—a laser inside an enclosed laser product, such that its full emission is not accessible under normal conditions of operation of the laser product.

embedded laser product—an enclosed laser product that is classified on the basis of the lower level of radiation accessible under normal conditions of operation than the level of the internal (inaccessible) radiation.

emission aperture—the opening in a laser housing through which laser radiation (i.e. the laser beam) emerges.

energy—a quantity of radiation, expressed in units of joules. For constant, uniform emission it is the product of the radiant power and the emission duration.

engineering controls—protective measures based on the use of physical design features, such as enclosures and interlocks.

ENOHD—extended nominal ocular hazard distance, the distance from a laser over which the MPE for the eye may be exceeded, taking into account the possible use of magnifying instruments (telescopes and binoculars).

exitance—the emitted power per unit area.

exposure—optical radiation incident on the surface of the eyes or skin, or the quantity (specified in appropriate units) of such incident radiation.

extended source—an apparent source that has an angular subtense greater than 1.5 milliradians (mrad).

eye loupe—an optical magnifying aid to allow close-up viewing.

fail safe—an attribute of a component or engineering feature in which failure does not increase the hazard of the system of which it forms a part.

far-field—that portion of an emitted laser beam (furthest from the laser) in which the beam diameter increases linearly with distance.

field of view (of a detector)—see angle of acceptance.

fume—the material generated by the vaporizing of target material using the beam of a high-power laser. The fume may contain both gaseous and particulate matter, including (in the case of medical applications) viable biomatter.

Gaussian beam—a laser beam whose cross-sectional distribution of power or energy follows a Gaussian equation; the simplest form of laser beam.

hazard—an entity (such as a laser beam, or the fume generated by a laser process) having the potential for causing harm or injury.

human access—the capability for any part of the human body to encounter a laser hazard. (Specific definitions of human access are given in laser safety standards.)

ICNIRP—International Commission on Non-Ionizing Radiation Protection.

IEC—International Electrotechnical Commission

infrared radiation—invisible optical radiation having wavelengths longer than in the visible band.

integrated radiance—the energy emitted per unit area per unit solid angle (the energy-equivalent quantity to radiance).

intermediate source—an apparent source having an angular subtense between 1.5 and 100 milliradians (mrad).

intrabeam viewing—the direct viewing of a laser beam; the exposure condition arising when a laser beam is incident on the eye.

irradiance—the incident power per unit area.

ISO—International Standardization Organization.

joule—the unit of energy. An energy rate of one joule per second is equivalent to a power of one watt.

laser—a device that generates directional, coherent, optical radiation through a process of stimulated emission. (It is an acronym for light amplification by the stimulated emission of radiation.)

laser beam—the path through space occupied by the radiation emitted by a laser.

laser controlled area—an area within which laser hazards may exist and protective measures are adopted.

laser equipment—a general term used in this book to mean any hardware item, assembly of components, or system that is, or incorporates, a laser.

laser product—any laser device or equipment that is subject to classification under a laser safety standard.

laser product class—the hazard category assigned to a laser product on the basis of the maximum level of its accessible laser emission.

laser radiation—optical radiation emitted by a laser.

Laser Safety Officer—a person given some level of responsibility on behalf of his/her employer for overseeing laser safety.

large source—an apparent source having an angular subtense greater than 100 milliradians (mrad).

LED—light emitting diode, a semiconductor device that emits incoherent light over a moderate range of wavelengths (typically 20 to 50 nm).

lesion—an observable injury (often used with regard to laser injury of the retina).

light—see visible radiation. The term 'light' is also sometimes used more loosely to mean optical radiation generally (that is ultraviolet and infrared radiation in addition to visible light).

limiting aperture (also called averaging aperture)—the aperture within which measurements of irradiance or radiant exposure are averaged.

loupe—see eye loupe.

LSO—Laser Safety Officer.

maintenance— routine adjustments and other basic procedures (other than operation) intended by the manufacturer of a laser product to be carried out by the user of the product. (Compare with operation and service.)

maximum anticipated exposure duration—the maximum period for which a person may be exposed to laser radiation under the foreseeable circumstances of use.

measurement requirements—detailed criteria specifed in laser safety standards for undertaking measurements of laser emission or exposure (mainly related to the measurement distance, aperture diameter and angle of acceptance).

milliradian—a measure of plane angle, used for specifying angular subtense and beam divergence. One radian (1000 milliradian) is equal to the fraction $1/2\pi$ of a circle, and so one milliradian is approximately 0.06 degrees.

MPE—maximum permissible exposure, the highest level of laser exposure at the eyes or skin that is generally considered safe, where the exposure is measured in accordance with specified conditions.

near-field—that portion of an emitted laser beam (closest to the laser) in which the beam diameter does not increase linearly with distance.

NOHD—nominal ocular hazard distance, the distance from a laser over which the MPE for the eye may be exceeded, provided that no magnifying viewing instruments are used.

ocular—pertaining to the eye.

operation—the normal use of a laser product for the purpose intended by the manufacturer. (Compare with maintenance and service.)

optical density—a measure of the attenuation (reduction in transmission) of an optical filter. It is equal to the logarithm of the inverse of the transmittance.

optically aided viewing—the use of magnifying viewing instruments, such as eye loupes, microscopes, binoculars or telescopes.

optical radiation—that portion of the electromagnetic spectrum comprising ultraviolet, visible and infrared radiation, and covering a wavelength range from 180 nm to 1 mm.

optical viewing instruments—optical magnifying devices such as eye loupes, microscopes, binoculars and telescopes.

peak power—the power of a laser pulse. (For pulsed lasers, both average power and peak power can be of relevance in safety evaluations.)

personal protection—items of protection (such as laser safety eyewear) worn by an individual to protect themselves from a specified hazard.

power (radiometric)—a quantity of radiation, expressed in watts. For constant, uniform emission the power is equal to the energy that is emitted divided by the duration of emission. (See also peak power, average power.)

power (refractive)—a measure of the focusing ability of a lens. It is specified in units of dioptre, where one dioptre is equal to the reciprocal of the focal length measured in metres.

protective density—a parameter used in the specification of European laser eye protection that takes into account both the optical density of the protection and its ability to withstand the incident laser radiation.

pulsed laser—a laser whose output is intermittent, usually taken to apply to lasers whose periods of emission (i.e. individual pulses) do not exceed 0.25 seconds. A pulsed laser may emit single pulses on demand, or repetitively at regular intervals, when it may be specified by its pulse repetition rate, in hertz (i.e. the number of pulses per second). Lasers that emit pulses longer than 0.25 seconds are often referred to as being 'pseudo-cw', meaning that they are considered to be more like continuous wave lasers that are switched on and off rapidly.

radiance (also called brightness)—the power emitted by a source per unit area per unit solid angle (or the radiant intensity per unit area). Often used to characterize the emission of extended sources when evaluating retinal hazards, since the retinal irradiance is directly proportional to the source radiance (ignoring variations in the spectral transmission of the eye).

radiant intensity—the power emitted by a source per unit solid angle. Often used to characterize the emission of small ('point') sources.

radiant exposure—the incident energy per unit area.

radiation—a form of energy that can propagate ('radiate') through space, and collectively known as electromagnetic radiation (because it comprises oscillating electric and magnetic fields). It can be characterized by wavelength, and extends from x-rays at very short wavelengths to radio waves at very long wavelengths. Optical radiation is electromagnetic radiation having intermediate wavelengths.

Rayleigh range—the distance from a laser to the position along the laser beam where the beam diameter has increased by the square root of two (i.e. by 1.414).

reasonably foreseeable—that which cannot be disregarded as being very unlikely.

retina—the light-sensitive layer (analogous to the film in a camera) lining the inside of the back of the eye on which an image is formed and in which the process of vision occurs.

retinal hazard region (or retinal hazard area)—the optical waveband from 400 nm to 1400 nm which can be transmitted through the eye to reach the retina.

risk—a combination of the likelihood that a given event or entity will cause harm or injury with the severity of such harm or injury. It is often used in laser safety in the more restricted sense of a deterministic risk, to mean the likelihood that an exposure exceeding the relevant MPE could occur.

risk assessment—an evaluation of the risks associated with a particular set of circumstances, with the aim of determining the appropriate protective measures.

scale number (of eye protection)—the level of protective density provided by an eye protector under European specifications.

second moment—a weighted measure of the diameter and divergence of a laser beam that is used in beam-propagation modelling. It is more generally applicable than other measures used, such as those based on 1/e or $1/e^2$, which only work well for beams having a Gaussian profile.

service—specialized attention to a laser product normally carried out by a service engineer and not intended by the manufacturer of the product to be carried out by the laser user. (Compare with operation and maintenance.)

small source—an apparent source having an angular subtense less than 1.5 milliradians (mrad). It is often referred to as a 'point' source, although a point source is physically unrealizable.

specular reflection—a 'mirror-like' reflection, in which the reflected radiation is highly directional. (Compare with diffuse reflection.)

spontaneous emission—the process by which optical radiation is generated in conventional (i.e. thermal) light sources.

standard—a document issued by a national or international body intended to establish common criteria in the form of definitions, requirements and specifications, etc. A standard may or may not have legal force.

stimulated emission—the process by which optical radiation is generated in a laser.

suprathreshold—an exposure level that is above the threshold for causing injury.

time base—the period of time over which emission of or exposure to laser radiation is assessed.

threshold—an exposure level that represents the boundary between exposure levels causing no injury and those levels that do result in injury.

ultraviolet radiation—invisible optical radiation having wavelengths shorter than in the visible band.

visible radiation—that portion of the optical spectrum, nominally from 380 nm to 780 nm, that is visible to the human eye. In laser safety, because use is made of the concept of an aversion response, the visible band is defined more narrowly as being between 400 nm and 700 nm.

watt—the unit of power. A power of one watt for one second represents an energy of one joule.

Appendix B

Special parameters

Wavelength ranges

180 nm–1 mm	Full wavelength range of optical radiation relevant to laser safety (< 180 nm vacuum ultraviolet, > 1 mm microwaves).
180–400 nm	Ultraviolet radiation (UV); it is further subdivided into UV-A, UV-B and UV-C as shown in table 2.2.
400 nm–700 mm	Visible radiation, as defined in the IEC and ANSI laser safety standards. (In the CDRH standard the visible band is defined as 400 nm to 710 nm.)
700 nm–1 mm	Infrared radiation (IR); it is further subdivided into IR-A, IR-B and IR-C as shown in table 2.2.
400–1400 nm	Retinal hazard region (the wavelength region for which the retina is assumed to be more at risk than the cornea or lens).
302.5–4000 nm	Wavelength region where optical instruments are assumed to be transmissive. (ANSI specifies a different wavelength range.)

Time values

18 μs	Inflection point T_i for ocular MPEs from 400 nm to 1050 nm, derived from the thermal confinement time. (See table 3.11 in section 3.12.4 for T_i for other wavelength ranges.)
0.25 s	Time base for Class 2, Class 2M and for visible Class 3R, also often assumed as the maximum anticipated exposure duration for MPE analysis of visible radiation. Based on aversion reaction to bright light (aversion response).
10 s	T_2 for small sources, the break time for retinal thermal MPEs (and also AELs for Class 1 and Class 1M), beyond which the MPE has a constant irradiance value. The shortest exposure duration for which retinal photochemical limits are considered relevant

For all exposures above 1400 nm for more than 10 s, the MPE has a constant value of 1000 W m^{-2}.

100 s T_2 for large sources (i.e. for $\alpha > 100$ mrad).
The time at which the acceptance angle γ_{ph} for the retinal photochemical hazard starts to increase beyond 11 mrad.

30 000 s Approximately 8 hours (i.e. the working day), the maximum exposure duration for which MPEs are specified.

Angular subtense (source-size) values

1.5 mrad α_{min}, the angular subtense associated with a small source; considered as the minimum value of angular subtense for any source (due to optical and scattering limitations).

11 mrad The acceptance angle γ_{ph} used for assessing the retinal photochemical hazard for exposure durations between 10 s and 100 s.

100 mrad α_{max}, the maximum acceptance angle used for assessing the retinal thermal hazard.

110 mrad The acceptance angle γ_{ph} used for assessing the retinal photochemical hazard for exposure durations beyond 100 000 s.

Correction factors (ANSI and CDRH)

IEC	ANSI	Formula	Spectral region (nm)	Comment	Section
C_1	—	$5.6 \times 10^3 t^{0.25}$	302.5 to 400	MPE for thermally induced UV damage	3.11
T_1	—	$10^{0.8(\lambda-295)} \times 10^{-15}$ s	302.5 to 315	Break time between thermal and photochemical damage in UV	3.11
C_2	C_2	$10^{0.2(\lambda-295)}$	302.5 to 315	Wavelength dependence of photochemical UV ($<$ 302.5 nm and $>$ 315 nm no wavelength dependence	3.11
T_2	T_2	$10 \times 10^{[(\alpha-\alpha\ \mathrm{min})/98.5]}$ s[a]	400 to 1400	Inflection time for retinal thermal limit. Above T_2, MPE is specified as an irradiance value and does not depend on exposure duration	3.12.4.2
C_3	C_B	1.0	400 to 450	Wavelength dependence of photochemical retinal limit	3.12.6.1
		$10^{0.02(\lambda-450)}$	450 to 600	(approximately equivalent to inverse of action spectrum)	
C_4	C_A	$10^{0.002(\lambda-700)}$	700 to 1050	Wavelength correction for thermal retinal limit	
		5	1050 to 1400	(based on decreased absorption of melanin and deeper absorption in choroid compared to visible)	
C_5	C_P	$N^{-1/4}$	400 to 10^6	Decreases single pulse MPE for multiple exposures (N is number of pulses within evaluation duration)	
C_6	C_E	1 for $\alpha \le \alpha_{\mathrm{min}}$ $\alpha/\alpha_{\mathrm{min}}$ for $\alpha_{\mathrm{min}} < \alpha \le \alpha_{\mathrm{max}}$ $\alpha_{\mathrm{max}}/\alpha_{\mathrm{min}} = 66.7$ for $\alpha > \alpha_{\mathrm{max}}$	400 to 1400	Increases retinal thermal limit for extended sources	
C_7	C_C	1	700 to 1150	Wavelength correction of thermal retinal limit	
		$10^{0.018(\lambda-1150)}$	1150 to 1200	(based on increasing absorption in media in front of retina)	
		8	1200 to 1400		

[a] $T_2 = 10$ s for $\alpha < 1.5$ mrad and $T_2 = 100$ s for $\alpha > 100$ mrad

Appendix C

Common misunderstandings

In the following table we summarize some common misunderstandings and mistakes that are often made in undertaking laser safety assessments.

		Discussed in section
Averaging of irradiance	The irradiance (W m^{-2}) or radiant exposure (J m^{-2}) that is used for comparison with the skin and eye MPEs is not simply the total power or energy in the beam divided by the area of the beam (but it is to be averaged over the limiting aperture).	2.2.2 (measurement) and 3.6.1 (biophysics)
α	The angular subtense of the apparent source α is not the divergence of the beam, α is usually much smaller.	3.12.1
α	The angular subtense of the apparent source is not derived from the beam diameter at the exit aperture. (If sufficiently far away from the beam waist, it can be derived from the diameter of the beam waist which might be virtual and inside the laser product or even behind the laser.)	3.12.5.9
α for scanned	For scanned sources, α has to be determined for the un-scanned beam as the scan pattern on the retina cannot be used to calculate α.	4.8.4

α for fibres or waveguides	For fibres or waveguides, it cannot automatically be assumed that the location of the apparent source is the end of the fibre and the source size is the diameter of the fibre. The apparent source might be located somewhat inside the fibre and might be smaller than the fibre diameter.	3.12.1
Location of apparent source	The exit aperture of the laser product is not the location of the apparent source. The location of the apparent source needs to be determined, for example by imaging with a lens. The apparent source for a reasonably well-collimated beam is located some distance inside the laser or more often behind the laser. For a laser beam that has an approximately Gaussian profile, the beam waist can be treated as the apparent source for most exposure distances, both in terms of location and angular subtense.	3.12.1
Most hazardous position	For extended sources, a distance of 10 cm from the apparent source for beams having a divergence of less than 100 mrad is not the worst case position, which is further away from the beam waist where α is correspondingly smaller.	3.12.5.9
Classification apertures	The power or energy that is compared to the AEL to determine the classification of a laser product is measured through specified apertures located at specified distances. The measured power or energy can be much less than the total output of the laser if the beam is larger than the classification apertures.	4.2.3.4
Angle of acceptance	Both the MPE and AEL relate to measurements using defined angles of acceptance, which can reduce the measured values for (measurement), sources larger than the applicable angle of acceptance.	2.4 (measurement), 3.12.5.3 (thermal), 3.12.6.2 (photochemical)

Multiple and non-uniform sources	For evaluation of the thermal hazard presented by non-uniform and multiple sources (arrays), it is necessary to vary the angle of acceptance within the range of 1.5 mrad to 100 mrad and to scan the source with each value for hot-spots. It is not sufficient to simply use the maximum angle of acceptance of 100 mrad. For evaluation of the photochemical hazard, it is not necessary to decrease the angle of acceptance below the specified value, but it is also necessary to scan the source for hot-spots.	3.12.5.6 (thermal), 3.12.6 (photochemical)
Angle of acceptance—imaging	To obtain a well defined angle of acceptance in the general case makes imaging of the apparent source onto the field stop necessary. It is not sufficient to realize this set up with a fixed lens-field stop distance (a fixed image distance), as it is necessary to vary this distance so that the apparent source is imaged, which gives the maximum measurement signal for the detector behind the field stop.	2.4
Angle of acceptance—accessible source	The alternative method to obtain a well-defined angle of acceptance, to place the field stop at the source, is only possible when the physical source is also the apparent source and the source is accessible. This set-up can be used to select single emitters or groups of emitters out of an array.	2.4
Add pulse energy	When MPE or AEL values are specified in terms of energy (J) or radiant exposure (J m^{-2}), all individual exposures within the total exposure duration under consideration need to be added up. For wavelengths in the ultraviolet, this can apply for up to 8 hours.	3.11.1 (ultraviolet) 3.12.8 (retina)
Criteria for pulses	For pulsed sources, it is not sufficient that the average power be less than the MPE or AEL that is given for the exposure duration or time base (such as by assuming that q-switched pulses with an average power no more than 1 mW can be Class 2). Additional criteria are defined with reduced pulse limits depending on the number of pulses within the exposure duration or time base.	3.12.8

Non-uniform pulses	When pulse trains with non-uniform pulses are evaluated or classified, it is important to consider not only the maximum evaluation duration or time base, but also every single pulse and any group of pulses within the pulse train.	3.12.8
α for condition 1	For condition 1 (telescope condition), the angular subtense α has to be determined at the measurement position for the 50 mm aperture (but see next issue for possible magnification of α).	4.3.5.1
Magnification by telescope	For condition 1 (telescope condition), the angular subtense of the apparent source can only be increased by a factor of 7 when it can be shown that the retinal image is magnified by this extent. If the beam diameter at 2 m is less than 50 mm the worst case magnification is less than 7. If the measurement position of 2 m is outside the near field of the beam, the correct magnification factor can be estimated by the ratio of the 1/e beam diameter to 7 mm. Alternatively the angular subtense of the apparent source can be determined at the location in the beam where the beam diameter is equal to 50 mm and this value can then be multiplied by a factor of 7 (which also applies for converging beams). The angular subtense α for condition 1 is to be determined at the measurement position for the 50 mm aperture, i.e. at 2 m.	4.3.5.1
Measurement distance	The measurement distance for condition 2 (eye loupe condition) and for the naked eye (MPE condition) is specified relative to the position of the apparent source, not relative to the product or the exit pupil.	4.2.3
1M and 2M	For classification as Class 1M and Class 2M, either condition 1 (telescopes) or condition 2 (eye loupes) is violated (not both). In table 10 of IEC 60825-1, the values given 'For irradiance or radiant exposure' also apply as the condition (naked eye) for Class 1M and Class 2M where the AEL values are given in terms of power or energy.	4.2.3

1M and 2M	It is the meaning of Class 1M and 2M that exposure with only one of two types of optical instruments is hazardous—consequently, measurements taken with one of the conditions will lead to Class 3R—or Class 3B—equivalent accessible levels of radiation.	4.2.3
NOHD	Many laser beams do not increase in size linearly with distance. Simplified assumptions of uniform beam expansion as the beam diverges from the exit aperture can lead to underestimates of the hazard.	5.4

Appendix D

Some MPE and AEL values

Selected MPE and AEL values from IEC 60825-1 are given in the following table. These values only apply to small sources ($\alpha \leq 1.5$ mrad) and to cw emission. For pulsed emission or for larger sources, different limits apply.

Wavelength	MPE for the eye (W m^{-2})	AEL for Class 1 and 1M	AEL for Class 2 and 2M	AEL for Class 3R	AEL for Class 3B
180–302.5 nm	0.001	0.001 W m^{-2}		n.a.	1.5 mW
315–400 nm	10	7.9 μW		40 μW	0.5 W
400–450 nm	25	0.04 mW	1 mW	5 mW	0.5 W
500–700 nm	25	0.4 mW	1 mW	5 mW	0.5 W
810 nm	17	0.6 mW		3.2 mW	0.5 W
1064 nm	51	2.1 mW		10.4 mW	0.5 W
1380 nm	405	16 mW		78 mW	0.5 W
1400–4000 nm	1000	10 mW		50 mW	0.5 W
4 μm–1 mm	1000	1000 W m^{-2}		5000 W m^{-2}	0.5 W

The MPE limits are calculated for an exposure duration of 30 000 s (about 8 hr) in the ultraviolet region, for 0.25 s in the visible band (relying for protection on the aversion response) and for 10 s for wavelengths above 700 nm. Lower limits will apply in the visible band if protection for longer exposure durations is needed. The MPE limits apply to exposure levels that are averaged over the area of the relevant limiting aperture.

The AEL values are based on a classification time base of 30 000 s in the ultraviolet region and 100 s for wavelengths >400 nm. The AEL limits apply to the accessible laser emission measured at the positions and using the apertures that are specified in the safety standard. The total emission for a given product class may therefore be higher than these limits if the laser beam, at the specified position, is larger than the specified aperture.

Index